1386

BRIAN J. KERNOHAN

FOREST FRAGMENTATION
Wildlife and Management Implications

FOREST FRAGMENTATION

Wildlife and Management Implications

Edited by

James A. Rochelle
Leslie A. Lehmann
Joe Wisniewski

BRILL
LEIDEN · BOSTON · KÖLN
1999

This book is printed on acid-free paper.

Library of Congress Cataloging-in-Publication Data

Forest Fragmentation : wildlife and management implications / edited by James A. Rochelle, Leslie A. Lehmann, Joe Wisniewski.
    p.   cm.
  Includes bibliographical references and index.
  ISBN 9004113886 (cloth : alk. paper)
  1. Fragmented landscapes.   2. Forest management.   3. Wildlife conservation.   I. Rochelle, James Arthur, 1942–   II. Lehmann, Leslie.   III. Wisniewski, Joe.
QH541.15.F73F67   1999
333.73—dc21   99–23815
                                                                                CIP

Die Deutsche Bibliothek - CIP-Einheitsaufnahme

Forest fragmentation : wildlife and management implications / ed. by James A. Rochelle ... – Leiden ; Boston ; Köln : Brill, 1999
  ISBN 90–04–11388–6

ISBN 90 04 11388 6

© Copyright 1999 by Koninklijke Brill NV, Leiden, The Netherlands

All rights reserved. No part of this publication may be reproduced, translated, stored in a retrieval system, or transmitted in any form or by any means, electronic, mechanical, photocopying, recording or otherwise, without prior written permission from the publisher.

Authorization to photocopy items for internal or personal use is granted by Brill provided that the appropriate fees are paid directly to The Copyright Clearance Center, 222 Rosewood Drive, Suite 910 Danvers MA 01923, USA. Fees are subject to change.

PRINTED IN THE NETHERLANDS

Table of Contents

Foreword .... vii - xiii

What Habitat is an Island?
   Fred L. Bunnell .... 1 - 31

Vulnerability of Forested Ecosystems in the Pacific Northwest To
Loss of Area
   J. Michael Scott .... 33 - 41

Fire Effects on Landscape Fragmentation in Interior West Forests
   James K. Agee .... 43 - 60

Past, Present, and Future Landscape Patterns in the Douglas-fir
Region of the Pacific Northwest
   Steven L. Garman, Frederick J. Swanson and Thomas A. Spies .... 61 - 86

Forest Loss and Fragmentation: Which has the Greater Effect on
Persistence of Forest-dwelling Animals?
   Lenore Fahrig .... 87 - 95

Is Landscape Connectivity Necessary and Sufficient for Wildlife
Management?
   Kimberly A. With .... 97 - 115

Edge effects: theory, evidence and implications to management of
western North American forests
   Laurie Kremsater and Fred L. Bunnell .... 117 - 153

The Effects of Forest Fragmentation on Avian Nest Predation
   John M. Marzluff and Marco Restani .... 155 - 169

The Role of Genetics in Understanding Forest Fragmentation
   L. Scott Mills and David A. Tallmon .... 171 - 186

Forest Fragmentation of the Inland West: Issues, Definitions, and
Potential Study Approaches for Forest Birds
   Rex Sallabanks, Patricia J. Heglund, Jonathan B. Haufler,
   Brian A. Gilbert and William Wall .... 187 - 199

Forest-level Effects of Management on Boreal Songbirds: the Calling
Lake Fragmentation Studies
   Fiona K.A. Schmiegelow and Susan J. Hannon .... 201 - 221

Forest Fragmentation Effects on Breeding Bird Communities in the
Oregon Coast Range
   Kevin McGarigal and William C. McComb .... 223 - 246

Using Landscape Design Principles to Promote Biodiversity in a
Managed Forest
   David C. McAllister, Ross W. Holloway and Michael W. Schnee .... 247 - 270

Patch Sizes, Vertebrates, and Effects of Harvest Policy in
Southeastern British Columbia
   Fred L. Bunnell, Ralph W. Wells, John D. Nelson and
   Laurie L. Kremsater .... 271 - 293

Forest Fragmentation: Wildlife and Management Implications
Synthesis of the Conference
   William C. McComb .... 295 - 301

Key Word Index 303

Appendix: Conference Summary Statement

FOREWORD

LET'S KILL A PANCHRESTON
GIVING FRAGMENTATION MEANING

Fred L. Bunnell

If you have puzzled about the meaning of "fragmentation", begin with "panchreston". "Panchreston" is a wonderfully efficient word that means "a proposed explanation intended to address a complex problem by trying to account for all possible contingencies but typically proving to be too broadly conceived and therefore oversimplified to be of any practical use"[1]. Fragmentation is one of several panchrestons that hinder practitioners attempting to translate the concepts of conservation biology into practice. This book is about giving meaning to "fragmentation"; particularly, meaning that is of practical use.

The book derives from a conference. The conference derived from a publication[2] commissioned by The Oregon Forest Resources Institute (OFRI). OFRI had requested a review of the likely impacts of forest practices on terrestrial, forest-dwelling vertebrates in Oregon. In the review, Bunnell et al. (1997) found little evidence that processes of fragmentation and associated concepts, such as negative edge effects, were modifying vertebrate abundance in western forests. Some months later, Lenore Fahrig[3] observed that much of the literature discussing fragmentation failed to distinguish between loss of favorable habitat and spatial distribution of remaining patches of habitat. Confusion was brewing in the words if not the woods. As stewards of natural resources, forest practitioners were being asked to address concepts that were untidily mixed. Moreover, in western forests there seemed little evidence of effects predicted by these concepts. Greater clarity was needed, the conference was developed, and from that, this book.

There were two broad objectives. First, to bring more clarity to the terms describing phenomena at the forest level (phenomena such as fragmentation, connectivity, and edge effects). Second, to collate and present what evidence could be found of the expected outcomes of habitat fragmentation in western forests. For each objective, the challenge was to connect the concepts and data to forest practices. These objectives create the structure of the book. It begins from a broad perspective. Bunnell presents a kind of primer on fragmentation, emphasizing connections with the equilibrium theory of island biogeography, and asks whether there are habitat islands in managed forests. Scott illustrates broad landscape level patterns in the Pacific Northwest, emphasizing habitat loss. Agee provides an overview of how natural disturbances have created spatial patterns within inland forest types, while Garman and coauthors review how spatial patterns have changed in coastal forest types. Fahrig concludes the broad perspective by returning to a question introduced at the outset – Does forest fragmentation or habitat loss have the greater effect on persistence of forest-dwelling wildlife? From the broad overview, the book continues by examining processes determining the consequences of fragmentation. The importance of movement and some of the theoretical consequences of spatial distribution of habitat are discussed by With. Processes around edges, and consequences of these processes, are reviewed by Kremsater and Bunnell and by Marzluff and Restani. The insights genetic studies can offer to studies of fragmentation are summarized by Mills and

Tallmon. Against this background, findings of three broad-scale studies examining consequences of fragmentation are presented: Sallabanks and coauthors for the intermountain northwest, Schmiegelow and Hannon for the boreal mixedwood, and MacGarigal and McComb for the Oregon Coast Range. Two case studies of efforts to apply concepts of fragmentation to forest planning conclude the conference contributions: McAllister and coauthors for the Tillamook in Oregon, and Bunnell and coauthors for southeastern British Columbia. McComb provides a summary of the implications of the combined contributions.

The second objective was amply attained. A wealth of data on some consequences of fragmentation (e.g., patch area, edge effects) are synthesized and general patterns abstracted in the chapters following. A large part of the first objective also was attained. More clarity was brought to the related concepts of fragmentation, habitat loss, connectivity, and edge effects. But the beguiling attractions of any panchreston – conceptual breadth and simplistic description – withstood much of the attempted clarification. There is no universal definition of fragmentation in the chapters following. Though seriously wounded by comments of several authors, the panchreston survived. It is time to kill it. The implications of fragmentation and related concepts to resource management and conservation are far too large to be left "too ... oversimplified to be of any practical use".

Killing panchrestons isn't easy. They exist for a reason. Given the determined efforts of conference organizers to increase clarity, we need to ask ourselves why fuzziness remained. The concept "fragmentation" obviously has broad appeal, perhaps too broad. Moreover, the term is useful. If we are to clarify the concept two undertakings are important. The first is to consider how we are confused about the concept. The second is to connect the concept to land use practices. In this latter undertaking we should recognize that the answer to an oversimplified definition is not a much more complex definition. It is sufficient to collate empirical outcomes, or demonstrated effects of particular actions or groups of actions.

Before examining the ways in which we confuse ourselves, I want to substantiate that there is useful meaning for the term "fragmentation". Even if you cannot find "panchreston" in your dictionary, you will find "fragment". As a noun it means "a part, broken off or detached"; as a verb it means "to break (something) into pieces or fragments". Fragmentation is simply the process of creating those fragments. It is critical that we recognize and remember that fragmenting habitat is not the same as fragmenting a porcelain vase or a mirror. We may be lucky enough to find all the broken pieces of the vase or mirror; the total amount is not reduced. We can never fragment habitat, however, without changing a portion of it to some different form of habitat or land use; some of the original habitat is no longer present. Early writers in conservation biology did not simply recognize this duality, they emphasized it.[4] The use of "fragmentation" in conservation biology means "to reduce to fragments" not simply "to break into fragments". It thus has more similarities to our discussions of now fragmented writings of classical authors than to fragments of broken mirrors. Sometimes we forget this distinction.

As we apply the word "fragmentation" to forests and forest practices we tend to confuse ourselves in four ways. First, we impose distinctions of startling clarity where they do not exist. Second, we measure or scale things relative to us. Third, we act as if nature were far too well behaved and well intentioned to "break (something) into pieces" without our help. Fourth, we encompass all the effects that happen when something is reduced to pieces into a single term, thereby eliminating our ability to distinguish among processes. Rather than discard "fragmentation" as a troubling panchreston we can confront the ways we tend to confuse ourselves. Naturally, the forms of confusion are interrelated.

Even if we talk about fragmentation as being a gradient, we do not analyze it that way. This reflects our desire to use tractable models and our inexperience at analyzing large areas. Because we seek them, we find discrete habitats, seral stages, or structural stages amenable to our simple models of patch dynamics or meta-populations. Often we let various forms of remote sensing or Geographic Information

Systems determine the "fragments" for us. We then apply the ideas generated from analysis of crisply defined patches to the untidy continua of the real world. The world is naturally full of gradients from seral stages to aspects and elevations. Our models, however, do not handle gradients well so we first conceptualize sharply defined patches, then analyze them as if the sharp contrasts were real. When this approach bothers us enough, we tend to seek out less common study systems with more sharply defined boundaries to which the simple models apply more fittingly. That does little to advance the growth of knowledge among the more common, but untidy systems. Our ways of viewing and analyzing fragmentation may work well among isolated woodlots in the countryside of Warwickshire or eastern North America, but connect poorly to the forests of the Pacific Northwest. This form of confusion is related to the way we choose to scale things. No matter how discretely we delineate such patches or fragments, there is little evidence that other vertebrates define patches the way we do (Bunnell, this volume).

We are not alone in assessing or scaling the environment according to our own perceptions. All organisms do. We are more troubled by the phenomenon simply because we are using our perceptions to explain or understand responses of other species. Wiens (1995)[5] provided a thoughtful review of how ecologists have viewed space. He recognized a natural progression from ignoring spatial variation (considering the environment homogeneous) to considering it heterogenous without regard to spatial pattern or heterogenous, recognizing spatial arrangement of patches. It is the patches that cause us grief. We assess them the way we, our satellites and aerial photographs, or greatly simplified models and theories, perceive them. Or we scale them to the convenience of our measurements and sampling. There is no reason to believe that other organisms perceive the patchiness or mosaic similarly. In some instances we know they do not.[6] Foresters are interested in the distribution of discrete stands of different volumes or ages across a forest because that distribution determines their approach to management. Ecologists tend to recognize these or similar stands as distinct, partly as a matter of convenience. It allows each stand to be treated as a homogeneous unit to which the strong heritage of simpler models, such as the equilibrium theory of island biogeography, can apply. The problem is one of scale, particularly in the sense of categorizing or dissecting continuous scalar gradients, such as canopy cover or age. Not only do we tend to make the forest into much more sharply defined units than other organisms perceive, but we make these sharp distinctions at a scale convenient to us. That scale is not necessarily shared with other organisms. Any distinction between a heterogeneous environment with little spatial pattern and that with a defined spatial pattern is organism specific. Group selection may create a somewhat more heterogeneous environment for a wide-ranging carnivore, but discrete habitat patches for a salamander. This scaling problem means that unless we are very thoughtful, it will be difficult to create a consistent definition of fragmentation. Just what is a patch or fragment is remarkably slippery.

It helps us slide into our third form of confusion. Because we intuitively recognize that patches naturally represent a continuum across species, we focus on those pieces of habitat that are clearly a product of human actions. We then find ourselves confused about what to call habitats that naturally occur as scattered pieces (fragments?). We could strive for semantic consistency, but then would run into logical difficulties. Semantically, we could use adjectives and recognize intrinsic fragmentation and human-induced fragmentation. We could apply intrinsic fragmentation to natural conditions and processes that break habitat into pieces without any help from us. A variety of habitat types naturally occur as scattered pieces – wetlands, talus slopes, krumholz in the subalpine. But, if we consider these pieces as fragments of the same sort as the scattered patches of a once continuous forest, we are in conceptual and logical trouble.

Most simply, if these habitats always were naturally patchy, then no new condition or phenomenon has broken them into pieces. They are as they always were. The ponds inhabited by western painted turtles in a forest are clearly separate. But the condition is not new. The ponds never were one big lake that got broken into scattered, little ponds. And neither were the turtles one population, except once too long ago for us to care. That is the point. We should have sufficient respect for natural selection to assume that organisms occurring in naturally patchy environments will respond differently to disturbance than organisms occurring in more continuous environments. Sallabanks and coauthors in this volume note that some organisms are closely adapted to naturally fine-scaled environments of many small patches. They suggest, as did Bunnell et al. (1997)[2], that vertebrates in western forests naturally subject to frequent fires should not be expected to show pronounced responses to habitat fragmentation. Their habitat never was continuous. It is too late to propose that we not apply terms like "fragmentation" or "fragmented" to habitats that naturally occur as scattered pieces. The habit is well entrenched. Moreover, the potential consequences of existing as small, scattered populations still apply, however these patches of habitat evolved. We do need to be acutely mindful that our expectations of responses are different for species living within naturally scattered bits of habitat and species whose naturally continuous habitat has been abruptly broken into pieces. We deny or confuse these expectations when we apply them as if there were no difference.

The fourth broad form of confusion is the entangling of a number of discrete concepts into one word. Assume that a group of species finds older forests much more favorable habitat than younger forest.[7] As the forest was logged, several changes would likely happen at once: 1) total area of older forest habitat would decrease; 2) amounts of edge would initially increase; 3) amount of interior forest habitat would decrease; 4) isolation of patches of older forest would increase; 5) number of patches of older forest would increase; and 6) average patch size of older forest would decrease. We can imagine exceptions. In a forest with much natural edge, the amount of edge could be reduced with time if the regenerating stand were uniform. Similarly, if the forest originally consisted of scattered stands, the number of patches of older stands could decrease with harvesting. We sometimes tend to view these as separate phenomenon, but all six tend to happen at once. For that reason all six are treated in this volume – though separately. But note that numbers 1) and 5) combine to produce 6). These three are completely interdependent; less forest occurring in more patches means the patches are smaller. The others show only partial interdependencies. The problem is that all six are wrapped up in the panchreston "fragmentation". If six separate processes, affecting different groups of species differently, stay entangled in a single term, knowledge will progress slowly if at all. Guidance to practitioners based on knowledge will progress equally slowly.

There is little either researchers or practitioners can do to separate these six consequences of management actions in the real world. They occur together because that is the way the world is built. The danger lies in failing to recognize that some can be partially separated, and that each of these consequences involves somewhat different processes that affect species differently. Management actions should differ depending on which consequences or processes are having the most profound effect.[8] We have to do better if we are to avoid guiding practice from the beguiling comfort of a panchreston.

I tried to simplify our confusion about fragmentation into four broad classes or forms. We can reduce each form of confusion. Two forms of confusion, our tendency to impose unduly sharp contrasts on environments and the scaling issue, are interrelated – both dealing directly with the way we perceive things. Levins (1992)[9] termed the scaling issue "the central problem of ecology". I believe the scaling issue manifests itself in two ways. The first is that there is no single correct scale, so that any distinction of patch sizes can never be entirely correct or suitable for the entire range of species. The second, is that difficulties in selecting a scale have encouraged us to treat environmental heterogeneity as a mosaic of patches. Thus we sharpen the contrast to tease out a mosaic. If Levins is correct, there are only interim solutions. Practitioners need to strive for a range in spatial heterogeneity within the forest; no single spatial pattern will be correct. Researchers need to examine responses of a given species at more than one scale. Given how important movement is to both scaling and fragmentation, it is likely that pattern-process relationships can be extrapolated only over particular scales or domains. That suggests that progress will be faster if we group species by their abilities to move or distances they disperse. Confronting heterogeneity without the convenience of a mosaic of internally homogeneous patches is more difficult, but equally necessary. We view forests as mosaics of discrete stands not just because of our perceptual limitations, but because it is our conceptual heritage. For decades ecologists treated environments as homogeneous, to the point that heterogeneity became an obstacle to scientific progress (Keddy 1991)[10]. As a result we have almost no theoretical base for treating heterogeneity other than as a mosaic of different (internally uniform) patches. It is important that we move beyond that limited conceptual framework, and especially important in forests. In response to concerns about forest-dwelling organisms, many current forest practices now seek to create structural heterogeneity over small areas. This kind of heterogeneity cannot be represented as patches, and places practice well ahead of the theory or knowledge that should guide it.

The third form of confusion I noted is our discussion of "naturally fragmented" environments. When Mount St. Helens erupted it naturally fragmented a more continuous forest. Some forest and other habitat types, however, occur naturally as scattered, ever-changing bits (see Agee, this volume). These are two very different conditions. Our expectations of consequences of the two should also be different. Gathering thinking about these two kinds of phenomena into the same term, both fails to respect natural selection and sustains the unhelpful breadth of the panchreston. It is unlikely that we will gain insight by applying findings from naturally patchy habitats, or species adapted to naturally patchy habitats, to inferences about fragmentation of formerly contiguous habitats.

Our fourth form of confusion – entangling many consequences and processes into one term – does most to make "fragmentation" a panchreston. This confusion suffers from the same conceptual barriers as does scaling, but is equally a product of sloppiness. Most of us know that when we begin harvesting a continuous forest tract, several things happen at once. Still, we tend to lump them all within the single term "fragmentation". Either we don't see how we can tease the different processes apart, or we are comfortable with the panchreston. The former is often real; the latter is unhelpful. It would help the growth of knowledge if we restricted the term fragmentation to the consequences of breaking something into pieces. The specific consequences most closely linked to fragmentation are then area effects and distance effects. Almost always the fragmentation of habitat will be caused by a reduction of the total amount of habitat. Researchers will not be able to provide constructive guidance to practitioners about the spatial distribution of habitat unless reductions in habitat are acknowledged. It is true that the different consequences will remain difficult, though not impossible[11], to separate. If we consistently acknowledge that more than one thing is happening at once and work to assign consequences to processes, we will be more successful at separating related processes and guiding management actions.

I noted earlier that we could clarify the importance of the concept of fragmentation even as we struggled to separate the processes within the concept. This is possible simply by documenting effects of particular actions or combinations of actions. The evidence gathered in the following chapters suggests that the primary detrimental effect of fragmentation is loss of habitat, with the spatial distribution of habitat being secondary (see especially chapters of Bunnell, Fahrig, and McGarigal and McComb, this volume). That has important management implications because it suggests: 1) that delineation of critical habitat elements is important, and 2) that focus on how habitats should be arranged in space is potentially misleading and wasteful. If the spatial arrangement of habitat is less critical than total amount, that permits greater operational flexibility and allows forest managers to exploit the advantages of zoning intensity of forest practice (Bunnell et al., this volume). The chapter of With indicates the theoretical importance of encouraging connectivity within the managed forest matrix, rather than setting aside unharvested corridors. Such an approach can probably be implemented by maintaining some structural elements within the managed forest. Data summarized by Bunnell suggest that for vertebrates this should be easy to do; there is little evidence that separate patches of old growth are currently isolated.

There is ample evidence that patchiness of habitat, which some term "fragmentation", is not universally "bad" (e.g., McGarigal and McComb, Sallabanks and coauthors, this volume). Once we apply the term "fragmentation", we tend to view the outcome as "bad"; perhaps because of the root meaning of something broken. Accumulated reductions in habitat are ultimately harmful, but the broad spatial distribution of habitat is less important than the total amount. As Agee reports in this book, some our actions (e.g., fire suppression) have made formerly discontinuous habitat much more continuous (less "fragmented"). This reduction in patchiness has been almost universally harmful to animals and the forest. The few exceptions are bark beetles, budworm, woodborers and possibly black-backed and three-toed woodpeckers, that flourish in dead and dying forests. Our fearfulness of fragmentation, and lack of conceptual models to address fine-scale heterogeneity, have diverted attention from the important contributions patchiness makes to sustaining species richness. The importance of patchiness is evident not only in inland forests (Agee, Sallabanks and coauthors, this volume), but also in coastal forests (Bunnell, and MacGarigal and McComb, this volume). Current evidence suggests that managers can do much to sustain species richness by increasing fine-scale patchiness, whether this be termed "fragmentation" or not.

A consequence of more, smaller patches is the increased amounts of edge. The potential dangers of edges also appear to have been exaggerated for western forests. Three chapters in this book directly address edge effects. Their net effect appears to be to encourage vertebrate species richness and abundance. Some species may be negatively affected (see Kremsater and Bunnell), and require special management. Just as fire is natural, so are edges. It would argue against natural selection to expect widely spread negative effects in western forests, particularly those forests naturally experiencing frequent fire. By focussing on the processes producing edge effects, the contributions of Marzluff and Restani and Kremsater and Bunnell help to clarify where and why edge effects are important. They provide guidance to practitioners about the kinds of situations where edge effects are most likely to be negative.

Authors of this volume have attempted to separate processes encompassed by fragmentation. There are specific treatments of microclimatic edge effects, predation around edges, patch area, total area, and distance effects. But the real world rarely fits our terms and concepts tidily, and the accommodating breadth of a panchreston is ever enticing. We may have wounded the panchreston, but we haven't killed it; and kill it we must. If not we are doomed to "a proposed explanation intended to address a complex problem by trying to account for all possible contingencies but typically proving to be too broadly conceived and therefore oversimplified to be of any practical use". That aids neither the growth of knowledge nor effective management. It is critical that we make faster progress in clarifying our thoughts, and in documenting specific consequences. The major change in western forests that threatens forest-dwelling vertebrates has been habitat loss associated with forest conversion to agriculture or urbanization (Scott, this volume). Chapters of the book report little evidence that other negative consequences of fragmentation are currently acting. Theoretical treatments, however, suggest potential threshold effects (see chapters of Farhig and With). Those thresholds may be lurking, as yet undetected. We are unlikely to detect them from the comfort of a panchreston.

1. The Random House Dictionary of the English Language. 2nd Edition Unabridged.
2. Bunnell, F.L., L.L. Kremsater, and R.W. Wells. 1997. Likely consequences of forest management on terrestrial, forest-dwelling vertebrates in Oregon. Publication No. M-

7, Centre for Applied Conservation Biology, University of British Columbia, Vancouver, BC. 130 pp.
3. Fahrig, L. 1997. Relative effects of habitat loss and fragmentation on population extinction. J. Wildl. Manage. 61:603-610.
4. Simberloff,,D.S., and L.G. Abele. 1982. American Naturalist 120:41-50; Haila, Y. and I Hanski. 1984. Ann. Zool. Fenn. 21:393-392;. Wilcove, D.S., C.H. McLellan, and A. P. Dobson. (1986). Pp. 237-256 in M.E. Soulé (ed.).Conservation biology: the science of scarcity and diversity, Sinauer Associates, Sunderland, MA.
5. Pp.1-26 in L. Hansson, L. Fahrig, and G. Merrian (eds.). Mosaic landscapes and ecological processes. Chapman & Hall, New York, NY.
6. Bunnell, F.L., and D.J. Huggard. (1999; J. For. Ecol. & Manage. 11(2/3):113-126) provide several of examples of different responses of vertebrates when measured at different scales.
7. The group would contain relatively few vertebrate species (Bunnell this volume), but would include a larger number of moss, liverwort, and lichen species.
8. Davidson (1998) provides simple, clear examples of how choices in management actions could differ (Wildlife Society Bulletin 26:32-37).
9. Ecology 73:1943-1947.
10. Pp. 181-201 in J. Kolasa and S.T.A. Pickett (eds.). Ecological heterogeneity, Springer, New York, NY,
11. McGarigal, K. and W.C. McComb. 1995. Ecological Monographs 65:235-260; Fahrig, this volume.
12. I appreciate the invitation to clarify my thoughts on fragmentation, and my discussions with David J. Huggard. This is Publication Number R-30 of the Centre for Applied Conservation Biology.

What Habitat is an Island?

Fred L. Bunnell

The equilibrium theory of island biogeography suggested mechanisms for the concepts of fragmentation and connectivity. For the theory to be predictive in forests, forest stands must act as isolated islands in a sea of hostile land. There is little evidence that forest-dwelling vertebrates perceive old growth stands as distinctly as we do. Broad taxonomic groups of vertebrates were significantly more abundant in old growth than in young stands in 6 of 27 tests ($p < 0.05$); there were no vertebrate communities consistently associated with old growth. In 321 tests of individual species assumed to be associated with old forests, species were more abundant in old growth than in young stands in 69 of 229 instances (30%), and more abundant than in clearcuts in 61 of 92 instances (66%). Relatively small amounts of residual structure maintain most species assumed to be old-growth associates at levels statistically inseparable from levels in old-growth stands. Nor is there evidence that the younger stands surrounding older forests are hostile to movement. Combined these observations suggest that total amounts of habitat are more important than the distribution of habitat. Key elements for defining total habitat are reported.

Key words: connectivity, habitat, forestry, fragmentation, vertebrates

1. INTRODUCTION

Concepts of habitat islands and fragmentation were stimulated by the equilibrium theory of island biogeography that in turn has roots in Preston's canonical distribution of commonness and rarity. When applied to forest practices these concepts engender controversy. I review the concepts and their progeny, then examine basic notions reflecting origins of the concepts. Specifically, the concepts are based on the idea of discrete habitat patches ("islands") separated or isolated from each other by a "sea" of hostile land. I examine the degree to which vertebrates recognize discrete habitat patches in western forests and the degree to which intervening forest land is a

Centre for Applied Conservation Biology, Forest Sciences Centre, The University of British Columbia, Vancouver, Canada

"hostile sea". From that examination I draw inferences relevant to forest practices. The inferences are drawn for vertebrates. They may be different for cryptogams or arboreal lichens.

2. ISLAND BIOGEOGRAPHY AND ITS PROGENY

Forest practices alter forest structure over a range of scales, but two are relatively discrete – the small stand or treatment unit (1 ha to 100 ha) and the much larger managed forest (100 000 ha or more of different treatments distributed through time and space). There is a large, controversial, and confusing literature on effects that can occur over larger areas. This literature centers around two interrelated concepts: fragmentation and connectivity. Both concepts derive from soundly quantified relations (Darlington, 1957; Huffaker 1958; Preston, 1962a, b; MacArthur and Wilson, 1967). The controversy grows not from source of the concepts (which is credible and helps to explain patterns of extinction), but from the manner in which the concepts have been transferred to forests.

There is nothing tidy about the way our concepts grow – fragmentation and connectivity are inseparable. Roots most closely associated with fragmentation are eloquently described by Quammen (1997). Roots of connectivity extend at least to the classic studies of population fluctuation by Gause (1934) and Huffaker (1958). Much about the concepts grew from observations on relationships between the number of species and area of oceanic islands or samples. Roots are deep. Forster (1778:169) observed "Islands only produce a greater or lesser number of species, as their circumference is more or less extensive". Roots in sampling were summarized as a "rule-of thumb" by Darlington (1943, 1957), and neatly formalized by Preston (1962a, b). Preston termed the relationship "the Arrhenius equation", acknowledging a predecessor. We call it the "species-area curve" but retain Preston's expression: Species are related to Area in the form $S = cA^z$ (c and z are "constants" giving form to the curve but varying among groups of organisms and regions). When Preston offered this formalization there were sufficient data that patterns across a variety of communities could be compared. Samples of biological communities always contained some species that were very rare, many species that were moderately abundant, and a very few species that were very abundant. The pattern was sufficiently consistent that Preston viewed it as a law of nature, with force of canon, and termed it the "canonical distribution" of commonness and rarity (Preston, 1962a). Area, for Preston, was the size of a representative sample.

Preston made an important distinction between "sample" and "isolate". A sample is richer in rare species than an isolate because, "on isolated islands we must have an approximation to internal equilibrium and presumably to a self-contained canonical distribution and, since an island can hold only a limited number of individuals, the number of species will be very small" (Preston, 1962b:411). Nothing slops in from the surrounding area. The major ideas (samples and isolates or islands) were available to be combined into a larger view. MacArthur and Wilson did that, first in "*An equilibrium theory of insular zoogeography*" (1963), then as "*The theory of island biogeography*" in 1967.

MacArthur and Wilson addressed oceanic islands (the isolates of Preston). Data from oceanic islands indicated that new species were continually arriving, old species were continually disappearing (going extinct), but the net effect was no gain or loss in number of species—a natural equilibrium. The big questions were: what created the equilibrium and what patterns did the equilibrium follow? Confronting these questions MacArthur and Wilson (1967) changed approaches to thinking about islands in two major ways: they offered a theory about mechanisms for the equilibrium, and they expanded the implications of insularity and isolation. Both changes had large consequences. MacArthur and Wilson knew that species turnover on islands occurred rapidly, so they sought a mechanism that could operate in the present rather than through age of the islands. Whatever the mechanism, the gains had to balance the losses. They suggested that species were lost through extinction. Species could be gained by two means: first, when a single old species splits into a pair of new species, and second by immigration, when a new species arrives and becomes established. Immigration, they reasoned, is vastly more frequent than is speciation, so the equilibrium that Preston had described must result from local extinction and immigration. They had answered the first question about mechanisms creating the equilibrium. To answer the second question and explain the patterns this equilibrium followed, MacArthur and Wilson (1967) retained Preston's (1962a) species-area curve, but explained patterns among curves in terms of rates of extinction and immigration. From data on oceanic islands they deduced effects of both area and distance.

Area: larger islands contain more species than do small islands. This occurs because small islands experience more extinctions (small populations are more vulnerable to chance events) and receive fewer immigrants (species wandering from the mainland or nearby large islands are not as likely to encounter them—a kind of "target size" effect).

Distance: equivalent-sized islands more remote from the mainland or source population will have fewer species because the extinction rate is the same but the immigration rate is lower (fewer immigrants reach the island).

The second big change in thinking MacArthur and Wilson encouraged was to expand the concept of insularity or isolation. The term "theory" had implications. Unlike a natural law (canon) which others could describe for different species in other places, "theory" invited challenge and expansion. Commenting on isolation they noted: "Insularity is moreover a universal feature of biogeography. Many of the principles graphically displayed in the Galapagos Islands and other remote archipelagos apply in lesser or greater degrees *to all natural habitats*" (MacArthur and Wilson, 1967:3; emphasis added).

By erecting theory and contending that the theory applied to all natural habitats, MacArthur and Wilson invited research and speculation. Expansion occurred first around the notion of habitat islands: Janzen (1968) on host plants as islands for herbivorous insects; Culver (1970) on caves as islands for their resident biota; Vuilleumier (1970) on subalpine bird communities in the Andes; Brown (1971) on mammal communities of remnant forests on mountaintops poking above the Great Basin. Soon the concepts were applied to design of nature reserves (Diamond, 1975; Bunnell, 1978) and to old-growth stands in forested landscapes (Harris, 1984). Thinking also was directed to why small populations (smaller islands contain fewer individuals) tend to go extinct. Shaffer (1981) formalized the concept of minimum viable populations introduced by Main and Yadav (1971); Gilpin and Soulé (1986) and Shaffer (1990) formalized the notions of viable population analysis. Doing that they gathered up some of the notions of connections, sources, and sinks first quantified by Huffaker. When Levins (1970) examined extinction, he introduced the term "metapopulations". A "metapopulation" is a population of smaller populations; that is, a number of populations of a species, occupying some arrangement of insular habitat patches. Each patch is subject to local extinction and recolonization from other patches. There is no "mainland" in a metapopulation. If a patch becomes disconnected (isolated) it suffers the increased risks of any small population.

The explosion of activity associated with the erection of a theory about mechanism has had an unfortunate consequence – a loss of conceptual development. Obviously, we do not test the theory by rediscovering the pattern; the pattern was known for decades before the theory. Studies of woodland patch sizes that examine no mechanism are often documenting simple behavioral responses, not population processes. More recent work, propelled by a sense of urgency, has done much to encourage confusion between behavioral responses and the population mechanisms initially postulated by MacArthur and Wilson. Before considering how this welter of concepts and data apply to forest practices, it is helpful to summarize key elements:

- Fundamental to the concepts is the notion of a discrete kind of habitat removed or isolated from similar habitats. The initial concepts were derived for oceanic islands surrounded by habitat hostile to species on the islands.
- For some concepts (e.g., minimum viable or effective population size) island biogeography is not an issue; it does not matter how the population became small.
- Isolates or true islands show a pronounced area affect.
- Isolates or true islands show a pronounced distance effect[1].

Only the "area effect" is unclear. Thinking about the "area effect" is confused in two ways: the mechanism and its relationship to fragmentation. Consider mechanism. The area effect almost certainly is more than a "target size" effect with larger areas intercepting more immigrants. Larger areas also have more diverse habitats, more potential niches, and will be hospitable to more kinds of species. That is,

---

[1] This is true only if colonizers can disperse across the intervening habitat. Brown noted "Apparently the present rate of immigration of boreal mammals to isolated mountains [in the Great Basin] is effectively zero" (Brown, 1971:477). In such a case recolonization is absent, only extinction is acting, and the species richness can only decline in the near future.

there is a "habitat effect" inextricably imbedded in the "area effect". Similarly, an equivalent area of diverse habitat will host more species than the same area of uniform habitat. Freemark and Merriam (1986) noted that the area effect they found to be consistent with the theory of island biogeography could be completely explained by increased heterogeneity in larger fragments, rather than area itself. The relationship of total habitat area to fragmentation also can be termed a "habitat effect". The problem arises because many studies on "fragmentation" have confused loss of habitat with fragmentation. Most recent authors appear to recognize the fact that fragmentation and habitat loss are nearly inseparable, particularly in forests (e.g., Harris, 1984), but in their discussion lump the consequences under effects of "fragmentation". The distinction is important because fragmentation and habitat loss can have different management implications (Fahrig, 1997). During planning of this conference fragmentation was defined as: "The process of reducing size and connectivity of stands that compose a forest". Three features of that definition merit comment. First, it approximates the common dictionary definition of fragmentation, to break apart into separate (unconnected) pieces. Second, it implicitly assumes that stands comprising a forest are sufficiently different that they are perceived as discrete entities by organisms other than humans. Third, it fails to distinguish between breaking apart and loss of total habitat. It fails to distinguish between the total amount of habitat and the distribution of habitat, although these may have quite different effects. Relationships between total habitat area and fragmentation are illustrated in Figure 1. Fragmentation describes the distribution of habitat (as pieces) not the total area. For most species, the best habitat is concentrated in space, in patches. Figure 1a illustrates habitat loss (the habitat has not been broken apart), but there is a potential "area effect" because the patches are smaller. Figure 1b illustrates habitat loss without an "area effect" because the biggest patches of habitat are still present. There is a potential "distance effect" because the patches are farther apart. There actually is less fragmentation because there are fewer separate pieces of habitat. Most current literature describes the transition to "A" or "B" as increased fragmentation; actually it is increased habitat loss – we have not created more pieces. Figure 1c illustrates both fragmentation (more pieces) and habitat loss. It is difficult to fragment habitat without losing habitat, but it has been done experimentally in grassland or old field systems where mowers can create clever experimental designs. In these instances, where the concepts habitat loss and fragmentation have been separated, we know how some meadow voles respond to fragmentation. They don't; they respond to the total amount of habitat, no matter how it is distributed (e.g., Wolff et al., 1997). I know of only one study that examined the independent effects of habitat area and configuration or distribution in forests. After examining birds in 30 forested landscapes in Oregon, McGarigal and McComb (1995:251) concluded that "with the exception of a few 'edge' species, variation in abundance among landscapes was more strongly related to changes in habitat area; habitat configuration was of secondary importance." Confusion arose when concepts well supported by studies of oceanic islands were applied to forested ecosystems. Studies of "islands" or patches of forest engendered new concepts. Two of these—"forest interior" species and "edge effects"—received particular attention. Both relate to the area or size effect of the island or patch. Forest interior species are assumed to require relatively large areas of contiguous forest. They should not be present in small patches. Negative edge effects are viewed as reducing the size of the "forest island", either by changing conditions typical of the forest interior (e.g., microclimatic effects) or by the extension of mortality processes (such as depredation, parasitism, and competition) inward from the edge. Forest interior species and edge effects are treated by Kremsater and Bunnell, and Marzluff and Restani (this volume). Some patches of a forest (e.g., old-growth stands) appear discrete to humans. Evaluating the degree to which they are isolated from the surrounding forest is more difficult. Unfortunately, none of the predictions of the theory hold if the islands or patches are not isolated. There is no area effect; nor is there a distance effect. Neither do the various models of metapopulations apply. To exist, meta-

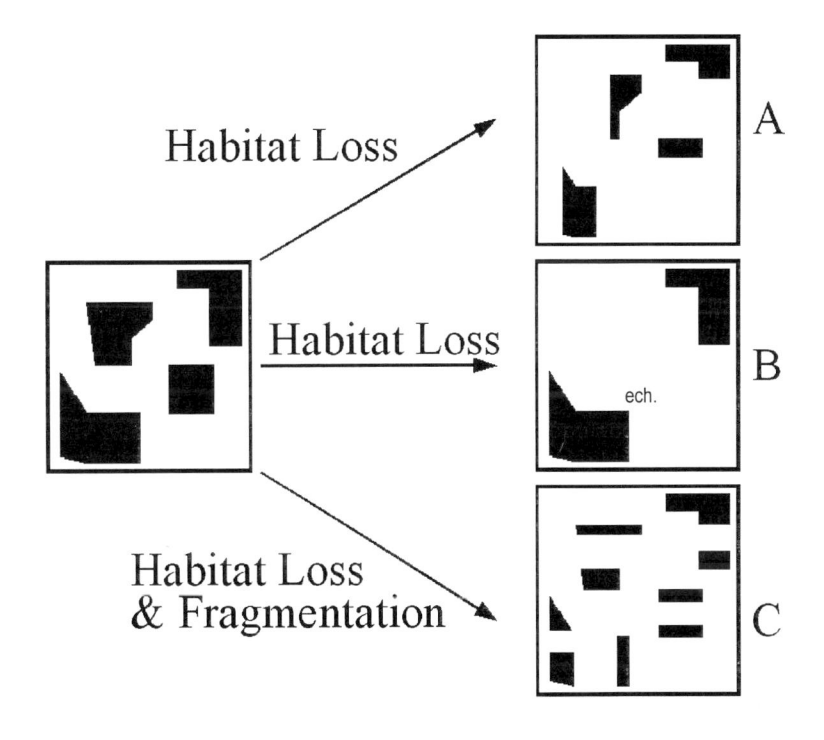

Figure 1. Effects of habitat loss and fragmentation on patch size (area effect) and isolation (distance effect). See text for details. From Fahrig (1997:604), with permission of the author.

populations require a species living in discrete, scattered habitat types among which movement is infrequent. Populations become smaller and more prone to extinction when this movement is interrupted. The converse of isolation is connectedness. Habitats are connected when species can move relatively freely among them. Indirect evidence of connectedness (movement was not assessed directly) is available when species can survive and breed in the intervening habitat (e.g., Tables 1 and 2). I examine the question of movement among habitats, or the degree of isolation of forest fragments, under the heading "Then we need a sea".

3. FIRST, WE NEED AN ISLAND

In the literature on forests this island is usually termed a patch or fragment—a small piece of forest separated from other forest. Apart from obvious differences among organisms' responses, there are two difficulties in defining the island or patch. The first is defining a portion of the forest that is sufficiently different from other portions. The second is evaluating whether this different portion is indeed isolated by surrounding, hostile areas. Several kinds of habitat in a forest might be very different from other parts. They include riparian areas, hardwood stands in conifer forest, and early or late seral stages. I treat the first three briefly, focussing on late seral or structural stages. The initial question is "Do organisms other than humans perceive these patches as discrete entities (islands)?" Connectivity is discussed separately.

3.1 Riparian areas

Riparian areas often are richer in species and more densely inhabited than are upslope areas (Raedeke, 1988; Knopf and Samson, 1994). For a few species (e.g., river otter, bufflehead) close juxtaposition of land and water are essential. Few forest-dwelling species, however, are restricted to riparian areas (e.g., Anthony et al., 1987; McComb et al., 1993a). Bunnell et al. (1998) reviewed studies providing statistical tests of relative abundance of vertebrates in riparian and upslope areas. Contradictory results occur across studies, implying that differences between riparian and upslope areas are not a simple function of proximity to water. It appears that riparian areas are a unique habitat type primarily because of their higher productivity and greater richness of vascular plants. The latter resulting, in part, from frequent disturbance (Naiman et al., 1993, Spackman and Hughes, 1995; Planty-Tabacchi et al., 1996). In western forests increased richness in plants is most evident among shrubs and deciduous trees. In their review Bunnell et al. (1998) found that among bird species strongly associated with riparian areas, 15 of 35 (43%) nested predominantly in hardwood trees or shrubs. The existence of a riparian flora, however, has not led to vertebrate species uniquely adapted to it. Riparian habitat is unlikely to serve as an "island" and actually experiences higher species turnover than do most habitat types (Rice et al., 1983; Knopf, 1986).

3.2 Hardwood stands

In most western forests hardwoods are more common along streams, inland aspen groves are an exception. Hardwoods are important to vertebrates primarily because they are more cavity prone than are conifers and they usually build fewer secondary compounds to defend their leaves. The latter condition means they host many herbivorous insects eaten by birds and bats, and produce a rich litter that benefits amphibians and small insectivorous mammals. As a result several vertebrate species preferentially use hardwood stands (e.g., Ralph et al., 1991; McComb et al., 1993a; Hagar et al., 1995), but few species appear limited to hardwoods. Small inclusions of either hardwoods among conifers or conifers among hardwoods often, but not consistently, serve to increase local species richness (e.g., Willson and Comet, 1996). Like riparian areas, hardwoods contribute to the richness of a region, but serve poorly as "islands".

3.3 Late-successional stands

Vertebrates show affinities or direct needs for various habitat elements comprising a forest (e.g., cavity sites, downed wood). The amounts and sizes of these habitat elements differ within stands of different ages or that have experienced different management regimes. We can recognize three broad groups of vertebrates: generalists, early seral associates, and late seral associates. In forest types of Oregon and coastal British Columbia the proportions of species in each group are about 15-20%, 15-25%, and 20 to 30%, respectively (Hagar et al., 1995; Bunnell et al., 1997,1998). Some species defy any tidy classification by responding differently in different forest types. Generalists need not concern us, and the habitat of early seral associates is readily created. I focus on older seral stages, because we are most concerned with patch types that are most rapidly diminishing and most difficult to create — old growth.

Initial studies of forest fragments were promising. They appeared to be consistent with the theory (reviews of Freemark and Merriam, 1986; Freemark, 1990). These studies were in the northeastern United States or adjacent Canada and the American midwest. Areas of forest, or patches of trees, typically were surrounded by agricultural or urban land. The forest fragment behaved like an island in a sea of land-use practices hostile to forest-dwelling species. Theory was extended to patches of older forest amidst stands of younger forests.

Table 1. Tests of differences in species richness and total abundance across stand age for the four classes of terrestrial, forest-dwelling vertebrates in forests of the Pacific Northwest.

Study	Number Sites	Number Species	Difference in Species Richness[b]	Difference in Abundance[b]	Young	Mature	Old Growth	Shared[a] Species (%)
Amphibians								
Raphael (1984, 1991)	46	12 (5)[c]	NS	M > O > Y	9	11 (1)[d]	10 (4)[d]	83
Bury and Corn (1988a)	30	10 (5)	NS	(O & Y) > M	9	8	9 (1)	90 [0.80]
Welsh and Lind (1988)	42	14 (9)	NS	O > Y	10	12	13 (4)	78 [0.56]
Welsh and Lind (1991)	54 terrestrial	15 (11)	NS	O > Y	12	13 (1)	14 (5)	80 [0.64]
	39 aquatic							
Corn and Bury (1991a)	45	11 (9)	NS	O > Y	10	10	10 (2)	100 [0.78]
Gilbert and Allwine (1991b)	3	9	NS	NS	6	7	8	75 [0.56]
Aubry and Hall (1991)	46	13 (7)	NS	M > Y	7	7 (1)	7	100 [1]
Dupuis et al. (1995)	6	4 [e]	NS	O > M	NA	3	3 (1)	75
Aubry et al. (1988)	45	9 (2)	~	NS	3	4	8	25
Reptiles								
Raphael (1988)	46	14	NS	Y > O	8	11	9	55
Welsh and Lind (1991)	54 terrestrial	16 (1)	NS	NS	8	10	12	42
	39 aquatic							
Bury and Corn (1988a)	18 OR	5	NS	NS	2	1	4	25
	12 WA	3	NS	NS	2	2	3	67

Table 1 continuing

Birds

Raphael (1984)	46	(winter)	50	NS	NS	43	34 (7)[1/]	82	39
Raphael (1984)	46	(spring)	77	NS	NS	66	51 (14)[1/]	90	67
Anthony et al. (1984)[g/]	12	(winter)	19 (7)	NS	~	~	~	~ (1)	~ (1) [0.86]
Anthony et al. (1984)[g/]	12	(spring)	39 (21)	NS	~	~ (1)	~ (3)	~ (2)	[0.71]
Manuwal and Huff (1987)	16	(winter)	15 (13)	NS	O > Y	13 (1)	13 (3)	14	93 [0.69]
Manuwal and Huff (1987)	16	(spring)	46 (45)	NS	NS	~ (2)	~ (6)	~ (7)	[0.65]
Manuwal (1991)	46	(spring)	34	NS	O > Y[h/]	30	32	30	88 [0.78]
Nelson (1988)[i/]	47	(1985)	16 (8)	O > Y	~	11	12 (5)	16	69 [0.38]
Nelson (1988)[i/]	47	(1986)	15 (8)	O > Y	~	9	10 (6)	14	57 [0.25]
Carey et al. (1991)	45	(1985)	49	NS	NS	43	44	49	88 [0.88]
Carey et al. (1991)	45	(1986)	46	(O & Y) > M	(O & Y) > M	38	37	46	83 [0.83]
Lundquist and Mariani (1991)	48	(detections)	8 [1/]	NS	NS	8	8 (1)	8	100 [1]
Lundquist and Mariani (1991)	48	(nests)	9 [1/]	~	~	5	9	5	56 [0.56]

Small Mammals

Raphael (1984)	46		28	NS	M > OG > Y	24	28	21	100
Anthony et al. (1987)	12		16 (5)	NS	(Y & M) > OG	~	~	~	~
Taylor et al. (1988)	47		23 (5)	~	NS	~	~	~ (1)	~
Carey and Johnson (1995)	12		13 (8)	NS	O>Y	13	~	13 (3)	100 [1]
West (1991)	46		20 (11)	NS	NS	~	~	~ (2)	[1]
Corn and Bury (1991b)[a/]	45		20 (13)	NS	NS	11	100 (1)	12	92 [0.46]

Table 1 continuing

a/ Shared species (%) is the percent of species found in the old growth that also are found in young stands. Values in [ ] are Jaccard's coefficient of similarity between young and old growth stands.
b/ Differences in richness tested by analysis of variance of mean number of species per plot across age classes; differences in abundance tested by mean number of all species per plot across age classes (alpha = 0.05 for all tests).
c/ Number of species for which statistical tests of relative abundance were possible.
d/ Number of species more abundant in this age class.
e/ Salamander species only.
f/ Determined by correlation with age of stand.
g/ Original not seen, cited from Hansen et al. (1991).
h/ Primarily due to counts of Vaux's swifts, winter wrens, chestnut-backed chickadees, and western flycatchers.
i/ Five of 18 young and mature stands resulted from clearcutting.
j/ Cavity nesters only.

Extension of the theory assumed that vertebrates responded to older stands very differently than to younger stands in two ways: 1) some vertebrates showed a pronounced preference for older stands; 2) younger stands were hostile and impeded movements (reduced connectivity) between separated, older stands. Here I define patches as areas of old forest surrounded by younger forests, and ask two broad questions about vertebrate relations to those patches:

1) Are there differences in vertebrate community structure across ages of naturally regenerated stands?

Answers should reveal if there are well-defined "old-growth" communities among vertebrates. If such communities exist, then selected management actions could sustain groups of species, and monitoring could track indicator species that represented an entire group.

2) Do individual species appear closely linked to late successional stages?

If there are no clearly defined groups of species, management actions must be targeted to individual species, possibly associated with particular features of late-successional stages.

Community composition is addressed in Table 1 where differences in vertebrate communities among forest age classes of naturally regenerated forests are summarized by taxonomic group. The youngest age class in these comparisons is usually 40 to 75 years old. Differences in species richness were assessed by analysis of variance comparing mean number of species per plot across the three age classes. Where possible I report both the percentage of species shared between young and old growth stands (>200 years) and similarities in community composition of old growth and young stands as indexed by Jaccard's coefficient of similarity (Sneath and Sokal, 1973:131; 1.0 = identical composition). Differences in abundance across age classes for all species in a large group (e.g., all amphibians or all birds) were evaluated by analysis of variance of plot or stand means. Differences in abundance among forest age classes appear primarily among individual species rather than communities of species; there is no apparent "old-growth" community among amphibians, reptiles, birds, or mammals.

Table 2 addresses the second question by summarizing, for individual species, the results of asking two questions: 1) Is abundance greater in old-growth than in natural young stands? and 2) Is abundance greater in natural stands than in recent clearcuts? For the latter question, it usually was possible to compare some measure of abundance in old growth (200 years plus) with abundance in recent clearcuts. When that was impossible, comparisons were made between clearcuts and stands as young as 40 years old. Most younger stands were of fire-origin, and contained residual structures (such as large snags) more common in old growth. Data for individual species were tested by analysis of variance of stands or plots grouped by age class when species were relatively common, and by the G-test (log likelihood ratio; Sokal and Rohlf 1995:686-697) for goodness of fit when they were less common. Data of Table 2 indicate that there are species or small groups of species that are strongly linked to late-successional habitat elements.

3.3.1 Amphibians

Among the 9 amphibian studies reviewed, none showed significant differences in species composition or richness between the "young" and "old growth" age classes (Table 1). The percent of species shared between those classes ranged from 75% to 100% in all but one study (which I could not test). Commonly, but not consistently, old-growth stands contained more amphibians than did young stands (Table 1). Differences in abundance did not appear as differences in community structure because only a few species differed significantly with stand age. Numbers of species tested and showing significantly greater abundance in mature or old growth classes are shown in parentheses in Table 1 ($p < 0.05$). Although amphibians usually were more abundant in older than in younger stands, species rarely disappeared completely from young natural forests (amphibian richness differed little between old and young stands). Moreover, species that occurred less commonly in samples of younger stands were not consistently the same (see Table 2). Differences with elevation and moisture gradients were more pronounced than were differences with stand age (Aubry et al., 1988; Aubry and Hall, 1991; Bury et al.,

Table 2. Summary of differences in abundance for individual species when comparisons were made between old growth and naturally regenerated young stands and between natural stands and clearcuts in forests of the Pacific Northwest.

AMPHIBIANS & REPTILES	Old Growth > Young Stands[a]?								Natural Stands[b] > Clearcuts?				
	1	2	3	4	5	6	7	8	9	10	11	12	13
Aquatic Salamanders													
Northwestern salamander	y		N	N		n		y	Y	Y			
Long-toed salamander	y								Y	Y		r	r
Roughskin newt	y	N	N	N	n	N		n	Y	N			y
Terrestrial Salamanders													
Clouded salamander		Y		Y	N	N	Y	Y		N		R	Y
Slender salamanders[c]		n			N	Y			n			Y	
Ensatina	N	Y	N	N	N	Y		N	Y	N		Y	Y
Western redback salamander	N	n/Y[d]	N	N	y	Y	Y		Y	N	Y		
Frogs and Toads													
Tailed frog		N		Y	N	Y			Y	Y		Y	Y
Western toad						N		R					R
Pacific tree frog		n			N	N		R	n/R			r	R
Red-legged frog	n	N		N		y			n				
Reptiles													
All lizards		R				Y		R	R				R
All snakes		n				y	R	R	R/y				N

1. Aubry et al. 1988; 2. Bury and Corn 1988a; 3. Aubry and Hall 1991; 4. Corn and Bury 1991a; 5. Gilbert and Allwine 1991b; 6. Welsh and Lind 1991; 7. Dupuis et al. 1995; 8. Raphael 1988a; 9. Bury and Corn 1988a; 10. Corn and Bury 1991a; 11. Dupuis et al. 1995; 12. Bury 1983; 13. Raphael 1988a.

BIRDS	Old Growth > Young Stands[a]/?									NaturalStands[b]/ > Clearcuts?			
	1	2	3	4	5	6	7	8	12	9	10	11	12
Cavity-users													
Brown creeper	Y	Y	y	Y	N	N	N	R Y	Y	Y	Y		Y
Chestnut-backed chickadee	Y	Y	y	N	Y	N	N	Y	Y	Y	Y	Y	Y
Northern flicker	Y	y	y	N	N	N	N						
Downy woodpecker	n	y				r	R						
Hairy woodpecker	Y	Y	y	N	N	r	R Y	Y			y		N
Pileated woodpecker	Y	y	n	N	N	N	R		N				
Northern pygmy owl	Y	y				N	y						
Red-breasted sapsucker	Y	Y	y	N	N	r	N				Y		
Red-breasted nuthatch	Y	Y	y	N	N	N	N	Y	R	Y	Y	Y	N
Vaux's swift	Y		Y	Y	Y		n						
Winter wren	y	y	y			N	N	N	Y N	Y	y	Y	Y
Flycatchers													
Hammond's flycatcher	R	R	y				N		Y			Y	Y
Olive-sided flycatcher	Y	y	n			N	R Y						
Western flycatcher	Y	n	y				Y Y						

1. Carey et al. 1991 (1985); 2. Carey et al. 1991 (1986); 3. Manuwal 1991; 4. Lundquist and Mariani 1991 (1985); 5. Lundquist and Mariani 1991 (1984); 6. Raphael 1984 (Winter); 7. Raphael 1984 (Spring); 8. Buckner et al. 1975; 9. Wetmore et al 1985; 10. Bryant 1997; 11. Vega 1993; 12. Hansen et al. 1995.

What Habitat is an Island?

	Old Growth > Young Stands[a]?									Natural Stands[b] > Clearcuts?			
Others													
Common raven	n	y					R	N					
Golden-crowned kinglet	y	y	y		R	y	n	R		Y	Y	Y	N
Gray jay	n	R	y							y	Y		
Ruby-crowned kinglet	y				R		N						
Steller's jay	y	y			R	N	N	N				N	N
Swainson's thrush	y	y			R	Y	N	R			y	N	N
Townsend's warbler	n		R		N	R				Y		Y	
Varied thrush	Y	Y	y		Y	N	Y			Y	Y		

	Old Growth > Young Stands[a]?									Natural Stands[b] > Clearcuts?						
	1	2	3	4	5	6	7	8	9	10	11	12	13	14	1	7	8	9	15	16
MAMMALS																
Insectivores																
Montane (dusky) shrew	Y	Y	N	N	Y	N	Y	r	N	Y	Y	N				Y	Y
Pacific shrew	N	N	N	N	N	N	N	N	N			N			N	Y			n
Pacific water shrew		Y	N	N	N	N	N	N	N						y		Y	Y
Trowbridge's shrew	N	R	N	N	R	N	N	N	N		Y	N			N	N	R	Y	Y
Vagrant shrew		N	N	N	N	N	N	N	N		N	N			R	R	R	Y	Y
Water shrew								r		y								
Shrew mole	y	Y	N	N	N	N	N	N	y		N	N				Y	N	
Bats																
Big brown bat													/Y			
Long-legged myotis													Y/Y[e]			
Silver-haired bat													N/Y			
Myotis A[f]													Y/Y			

	Old Growth > Young Stands[a]?	Natural Stands[b] > Clearcuts?
Myotis B		Y/Y
Rodents		
Deer mouse	Y Y N N Y Y N Y N Y	R R R
Columbian mouse	Y Y Y	
Red-backed vole	N N N N Y Y N R Y N Y	N Y Y Y N Y
Red tree vole	N N Y N Y	Y N
Northern flying squirrel	N N N N	Y
Douglas squirrel	y N r	Y Y
Pacific jumping mouse	N y	R
Western jumping mouse	y Y N	
Carnivores		
Ermine	r N N Y y y	y Y
Fisher	N	
Black bear	N	

1. Corn and Bury 1991b; 2. Gilbert and Allwine 1991a; 3. West 1991; 4. 5. Aubry et al. 1991 (OR Cascade Range); 6. Aubry et al. 1991 (WA Cascade Range); 7. Raphael 1988a; 8. Corr et al. 1988 (H.J. Andrews); 9. Corn et al. 1988 (Wind river); 10. Raphael 1988b; 11. Carey and Johnson 1995; 12. West 1991; 13. Thomas and West 1991; 14. Rosenberg et al. 1994; 15. Hooven and Black 1976; 16. Gashwiller 1970.

[a] Tests: $p < 0.05$, N = no, Y = yes, R = CC > OG; where data did not permit statistical tests, the same relations are depicted in lower case.
[b] Natural stands were old growth when possible, but may be as young as 40 years. [c] *Batrochoseps* spp. [d] Different results from H.J. Andrews and Wind River Experimental forests. [e] Detections in Washington/Oregon. [f] Myotis A includes Little brown and Yuma myotis; myotis B includes California, Keen's, Long-eared, and Western small-footed myotis.

1991a; Corn and Bury, 1991a; Gilbert and Allwine, 1991b; Welsh and Lind, 1991).

Data permitted 38 statistical comparisons of abundance for individual species between old growth and naturally regenerated, young stands. Species were more abundant in old growth in 9 (24%) of the instances (Table 2). Instances in which species were more abundant in younger stands are underestimated by 'R'[2]. Species significantly more abundant in old growth were either terrestrial-breeding salamanders or the tailed frog. Comparisons were possible in 12 cases for frogs and toads; in only 2 instances (tailed frogs) was the species more abundant in old-growth stands. Tests of differences in abundance between natural stands (young through old growth) and recent clearcuts were made 22 times. In 14 instances species were more abundant in older forests. Clouded salamanders, western toads, and Pacific tree frogs were sometimes more abundant in recent clearcuts (Table 2). The group most consistently showing higher numbers in older stands than in clearcuts was again terrestrial-breeding salamanders, in 7 of 11 tests. Clearcutting usually decreased the abundance of terrestrial-breeding salamanders, but amphibians as a group were not strongly dependent on old-growth forests (Tables 1 and 2). When documented, associations with age appear to reflect microhabitat characteristics (see also deMaynadier and Hunter, 1995). Analyses for individual species indicate that relationships with downed wood, shrubs, or deciduous trees were more strongly expressed than were differences with stand age (Aubry et al., 1988; Aubry and Hall, 1991; Corn and Bury, 1991a; Welsh and Lind, 1991; Dupuis et al., 1995).

---

[2] 'R' indicates that the species was statistically more abundant in young stands. Because 'N' indicates a statistically significant "no" answer to the question, in some instances denoted by 'N' the species was slightly more abundant in young stands, but not significantly so (for example, ensatina and western redback salamander in the study of Aubry et al. (1988).

3.3.2 Reptiles

Reptiles prefer warmer and drier environments than do amphibians so we expect different relations with stand age; specifically, reptiles should be more abundant in more open stands. Table 1 reveals few clear patterns due to the paucity of studies and captures within studies. Limited captures lowers the number of shared species reported and makes detection of differences in abundance elusive. Proportions of species shared between unmanaged young and old-growth stands appear lower among reptiles than among amphibians (young stands generally are less rich than are old growth stands, Table 1). There is little apparent difference in relative abundance of the entire reptile community in stands of these ages (but see Table 2). There are ecological reasons, beyond the difficulty of sampling reptiles, that hinder wide application of the observations summarized. Young stands studied by Raphael (1988) and Welsh and Lind (1991) were nearing the rotation age of more productive, coastal sites, and the old-growth stands were drier and more open than would occur in coastal forests. In short, reptiles were found to be more abundant in more open stands, which happen to be old-growth stands in some studies. Comparisons between old growth and clearcuts are more revealing (e.g., Table 2). The majority of tests summarized (6/8) found that both lizards and snakes were more abundant in younger or clearcut stands (R of Table 2). Forest-dwelling reptiles apparently profit from openings large enough to create warmer, drier areas (see also Marcot, 1986; Raphael, 1988; Welsh and Lind, 1991).

3.3.3 Birds

Birds are the richest class of vertebrates. Despite increased potential to detect differences in community composition, birds show no more dramatic differences among forest age classes than do amphibians or reptiles. Of the 12 tests for differences in species richness across forest age classes, significant differences were detected 3 times (Table 1). Nelson (1988) reported only cavity nesters and found significantly more species in old growth than in young stands in both study-years. Carey et al. (1991) found old growth and young stands to contain significantly more species than mature stands in one of two study-years. The percent of species shared between old growth and young stands ranged from 39% to 100% with most values over 67%. In several instances, however, species were present only at very low numbers in young stands. The lack of statistical difference in species richness appears contrary to the greater numbers of species reported from old growth than from young or mature stands in some studies (e.g., Raphael, 1984 in Table 1). That pattern reflects the greater range in structure within old growth stands. When all plots were combined more species were detected, but the mean species per plot did not differ significantly. There were few significant differences in total abundance of birds between age classes: in 2 of 8 comparisons old-growth stands contained more birds than did young or mature stands (Table 1).

I examined 22 frequently surveyed individual species, 21 of which had been reported as closely associated with late-successional stands (Rosenberg and Raphael, 1986; Fenger and Harcombe, 1990; Hansen et al., 1995; McGarigal and McComb, 1995). The northern flicker was included because it is a large primary excavator, providing cavities for other species. Statistical tests were possible in 123 cases. Presumed late-successional species were more abundant in old growth than in young stands in 38 of 92 instances (41%). Primary excavators were more abundant in old growth in 38% of the tests (15/39); secondary cavity users (including brown creeper and winter wren) were more abundant in old growth in 61% of the tests (11/18). Only Vaux's swift (reliant on hollow, open-topped snags) was consistently more abundant in old growth; the varied thrush was more abundant in 4 of 5 tests (Table 2). Results were different when abundance in natural stands was compared to abundance in clearcuts. In 24 of 31 tests (77%) bird species assumed to be closely associated with late-successional stages were more abundant in natural stands than in clearcuts (Table 2). Data of Tables 1 and 2 suggest that there are no well-defined old-growth communities of birds, abundance of birds does not differ significantly with age of unmanaged forests after about 40 years of age, and that few species are restricted to the oldest age class. It is informative that the single species evaluated that was consistently restricted to old growth (Vaux's swift) is not a primary excavator and must rely on large natural cavities. When cavity users are excluded, presumed late-successional species were statistically more abundant in old growth in 12 of 35 tests (34%); 4 of those 12 were tests of the varied thrush.

Marked discrimination with stand age occurs only when natural stands are compared to recent clearcuts. Presumed late-successional species were either significantly depressed or entirely absent from clearcuts (Table 1). Most data analyzed were from stands of fire origin, and authors commonly noted that younger stands contained small numbers of older trees or snags that would not be present under intensive fibre production. Apparently, small amounts of residual structure maintain many species assumed to be late-successional associates at levels statistically inseparable from levels in old growth stands. That observation suggests promise in maintaining late-successional associates in managed stands by retaining suitable levels of required habitat elements. Many late-successional bird species are cavity nesters (26 of 38 statistically significant associations with old growth stands). Provision of cavity sites is important.

Hagar et al. (1995) reported only three bird species apparently restricted to old growth (spotted owl, marbled murrelet, and Cordilleran flycatcher). All were excluded from Table 3, the last for want of data. As spotted owls and marbled murrelets receive more attention, it is becoming clear that these species also are related more to specific forest structures than to stand age itself (e.g., Buchanan et al., 1995). One hypothesis explaining spotted owls' association with older forest assumes that favored prey abundance (e.g., northern flying squirrels) is greater in older forests. Northern flying squirrels were more abundant in old-growth forests in 1 of 6 tests (Table 2). Fewer data are available for the marbled murrelet, but again association appears to be with particular structural features rather than with great age. Of 41 marbled murrelet nests found in Oregon at the end of the 1995 field season, 6 were in 90-year-old western hemlock in which mistletoe had created large nesting platforms (Marshall et al., 1996). That age, however, is older than the desired rotation age on some sites and thus incompatible with intensive fibre production. The needs of the species are unclear because recent surveys in British Columbia have found the species in forests of height Classes 3 and 4, or 20 m to 40 m (Manley 1999).

3.3.4 Mammals

Mammals span a greater difference in body and territory size than do birds, and most tests in Table 1 are restricted to small mammals that are amenable to sampling. Species richness did not differ across stand age class and 92% to 100% of the species were shared between old growth and young stands (Table 1). Total abundance differed in 2 of 6 tests, but in only one instance was the total abundance of small mammals greatest in old growth. Mammals show fewer close affinities than do birds with habitat elements more common in older stands (mammals use of cavities, for example, is generally more opportunistic). Bats are an exception, and in 8 of 9 tests bat activity was higher in older growth forests than in young stands. When foraging, many bat species exploit forest:clearcut edges (see Kremsater and Bunnell this volume).

Across the 16 studies collated, statistical tests were possible in 138 instances (Table 2). Small mammals assumed to be associated with late-successional forests were more abundant in old growth than in young stands in 29% of the tests (29/99; Table 2). They were more abundant in natural stands than in recent clearcuts in 59% of the tests (23/39). Findings are thus similar to those for birds except affinities for old-growth attributes are more weakly expressed. Implications to management are likewise similar: it appears possible to maintain many mammals presumed to be old-growth associates within younger stands. Two groups of mammals omitted or inadequately evaluated in Table 2 often are reported as being strongly associated with old growth — furbearers and forest-dwelling ungulates. Members of neither group appear to require large contiguous patches of old growth (Bunnell et al., 1997, 1998).

4. THEN WE NEED A SEA

Patches act as discrete entities or islands only when they provide favorable habitat and the surrounding habitat is hostile. Without a hostile matrix around the patch there is no island in the sea.

4.1 What is connectivity?

Connectivity exists when organisms can move freely among separate patches of habitat. If organisms cannot move freely the patches and subpopulations they host are disconnected and isolated. As older forests are harvested, remaining patches become smaller and appear increasingly isolated from each other. When isolation is real and patches are disconnected, populations within them should behave as if on oceanic islands and conform to the predictions of the theory of island biogeography (e.g., MacArthur and Wilson, 1967). Provided there are species restricted to older forests, patches of older forest will host more or less distinct subpopulations; that is, they will be a metapopulation *sensu* Levins (1970). The theory describing metapopulations predicts higher extinction rates for these disconnected subpopulations and ultimately lower population persistence (Levins, 1970; Shaffer, 1987, 1990). Stated differently, connectivity among different patches of habitat reduces the likelihood of local extinction and helps to sustain biological diversity. This statement reflects the short-term view. Over the longer term current genetic diversity and species richness developed through isolation of viable populations (Mayr, 1963; Simpson, 1944; Eldredge, 1992) and isolation can lead to rapid evolution and speciation (Grant, 1986; Gould, 1997).

The concept of connectivity generates heated debate. Debate is not about the importance of movement of individuals among separated subpopulations (e.g., Taylor et al., 1993), but about the kinds of habitat that permit movement. Debate is encouraged by two broad issues: biological interconnections and tactical choices. Biological interconnections ensure that it is difficult to distinguish what phenomenon is responsible for an observed decrease in populations. Forestry and other land use practices invariably produce several changes simultaneously. Most simply, as a forest is harvested the remaining older forest becomes concentrated into separated patches. A declining population may be responding to a decrease in total habitat, to the distribution of habitat (it may require larger contiguous areas), or to the lack of movement or connectivity among habitat patches (there is sufficient habitat, but chance declines in discrete patches cannot be recovered through recolonization). We debate which of these three potential reasons is actually responsible. Connectivity is important only in the last case. Connectivity is important when the species reproduces and survives poorly or not at all in the intervening habitat *and* the intervening habitat is a barrier to movement (note that this first condition is identical to that required to invoke the first or second cause for population decline). When intervening habitat is hostile to both survival and movement, and favorable habitat occurs patchily, the importance of connectivity is well established and the impact of connectivity on population viability is both theoretically sound and empirically documented (Fahrig and Merriam, 1985, 1994; Fahrig and Paloheimo, 1988; Taylor et al., 1993 and references therein). The second broad issue of the debate is about how best to facilitate movement (sustain connectivity) among local subpopulations.

4.2 Sustaining connectivity

Given demonstrably harmful effects induced by lack of connectivity, conservation biologists seized means of sustaining connectivity in fragmented landscapes. Two schools of thought emerged: 1) facilitate movement between seemingly isolated patches by creating corridors of habitat between patches; and 2) promote movement among patches by enhancing the intrinsic connectivity of the matrix (i.e., make the area between patches more favorable for movement).

4.2.1 Corridors
The most commonly advocated approach to ensuring connectivity is by using corridors (e.g., Noss and Harris, 1986; Noss, 1987; Harris and Atkins, 1991; Beier and Loe, 1992; Harrison, 1992; Lindenmayer and Nix, 1993). Corridors generally are thought of as linear strips of habitat, differing from surrounding vegetation, that connect two or more similar patches (e.g., Hobbs, 1992). We often assume that corridors must be similar in nature to the habitat they connect. Implementation of corridors (e.g., B.C. Ministry of Forests, 1995; Harris and Atkins, 1991) is propelled by demonstrated negative effects of lack of connectivity when intervening habitat is hostile, the simplicity of real and perceived actions when establishing corridors, and a sense of urgency.

Simberloff et al. (1992) summarized four broad rationales invoked by corridor advocates.

1. Lower extinction rates as these relate to equilibrium theory (island biogeography *sensu* MacArthur and Wilson, 1967). By facilitating movement among patches corridors are seen as increasing immigration (or recolonization) rates. Without connectivity only emigration rates would be acting and the equilibrium number of species in a patch would tend downward (e.g., Brown, 1971). Populations of individual species would not receive immigrants and thus would be susceptible to all influences that encourage higher rates of extinction among small populations.
2. Reduce demographic stochasticity as it relates to metapopulation theory and minimal viable populations. Demographic stochasticity (changes in population structure caused by chance events) is a potential threat to small populations (e.g., Simberloff, 1988; Shaffer, 1987, 1990). Metapopulation theory predicts that vulnerability to these chance events is reduced when otherwise isolated subpopulations are connected and subject to immigration or recolonization. Corridors are assumed to mitigate this threat by facilitating movements between otherwise isolated patches (e.g., Hanski and Gilpin, 1991).
3. Stem inbreeding depression. Small populations contain less genetic variation. When such lack of variety leads to the accumulation of deleterious gene combinations, the population experiences inbreeding depression, and survival is reduced. Some researchers argue that gene flow is required to prevent inbreeding depression from causing extinction and that corridors are

Table 3. Summary of publications providing empirical information on the use of corridors.

Source	Matrix	Study Design	Locations of samples or observations	Species/guild	Findings/conclusions
Wegner and Merriam, 1979	agricultural	observational	within corridors, patches, and matrix	small mammals and birds	demonstrated preferential residency and movements along corridors (fencerows).
Henderson et al., 1985	agricultural	experimental with replication and control	within corridors and patches	eastern chipmunk	demonstrated patch recolonization through movements along corridors (fencerows).
Bennett, 1990	agricultural	observational	within corridors and patches	small mammals	demonstrated residency and movement by dispersers in forested corridors
Merriam and Lanoue, 1990	agricultural	observational	within corridors and matrix	small mammals	demonstrated preferential movement through corridors (fencerows) over matrix (fields)
Bennett et al., 1994	agricultural	observational	within corridors and patches	chipmunk	demonstrated preferential residency and movement through corridors (fencerows) based on corridor habitat quality.
Beier 1995	urban	observational	within corridor	cougar	demonstrated movement through forested corridors by dispersing juveniles
Andreassen et al., 1996	fenced artificial site (matrix was bare ground)	experimental without replication and control	within corridors	root voles	demonstrated varying corridor width effects on movements.
Dmowski and Kozakiewicz, 1990	forest/wetland	observational	within corridors	birds	demonstrated facilitated movement along shrub corridor
Ruefenacht and Knight, 1995	managed aspen forest	experimental with replication and control	within corridors and corridor gaps	deer mouse	survival within forested corridors was unaffected by corridor width and continuity (gaps)
Machtans et al., 1996	managed boreal mixedwood forest	experimental with control	within corridors and matrix	birds	demonstrated corridors (forested buffers) were used more frequently than clearcuts for residency and movement

needed for this gene flow (e.g., Harris, 1984). Others are less convinced (e.g., Slatkin, 1985; Lande and Barrowclough, 1987; Lande, 1988). Again the notion devolves to movement of individuals among otherwise isolated subpopulations.

4. Fulfill an inherent need of individual animals for movement. Home ranges occupied by wide-ranging species can be larger than the size of conservation areas. Corridors may facilitate movements in wide-ranging species among separate patches of favorable habitat; movements that would not occur without corridors. The matrix habitat must be hostile to movement for this effect to be important.

Each rationale has biological plausibility and theoretical appeal[3]. In their review, Simberloff et al. (1992) noted that none of the four rationales was confidently supported by data. In the intervening 5 years, Rosenberg et al. (1997) found little more empirical evidence to support any of the four basic arguments for corridors. Potential disadvantages of corridors have been suggested. Hess (1994) cautioned that corridors may enhance disease transmission. Other potential disadvantages forwarded by Simberloff and Cox (1987) are increased spread of fire and predators, increased animal exposure to humans, and channeling of conservation funds away from more beneficial strategies. Several workers report deleterious edge effects of corridors and the fact that they could become death traps for some species (e.g., Vander Haegen and Degraff, 1996). Generally, assertions about the dangers of corridors have little more empirical support than assertions about their values.

Lack of evidence of the efficacy of corridors as a conservation strategy does not imply lack of evidence of corridor use. Corridors are used, just as other habitat is. To promote connectivity, however, corridors must be used for movement and for movement at rates and distances in excess of what would occur without corridors. Two types of evidence on movement are available: 1) observations of connected and unconnected patches leading to inferences about movement within corridors; and 2) direct observations of movement within corridors (ideally compared to movements within the surrounding matrix).

Indirect evidence was first provided by MacClintock et al. (1977), who are widely cited as documenting the utility of corridors. They found that a small forest patch contained an assemblage of forest bird species similar to a larger patch to which it was connected. The connecting corridor contained several species found in the patches. Species presence in the smaller patch may have been a result of the corridor, but movement within the corridor was not assessed. Identical results would have been attained simply if corridor and patches represented favorable habitat. Other workers have documented higher numbers of birds or mammals in connected versus unconnected patches of woods or unmown fields (Dunning et al.,1995; Haas, 1995; LaPolla and Barrett, 1993). Such studies have documented animal presence in linear patches located between larger patches, and then inferred that the linear patches were acting to promote movement. Movement within corridors, however, was not assessed.

Direct evidence of movement within corridors is sparse and summarized in Table 3. Small mammals in agricultural landscapes preferentially use corridors (fencerows) within a matrix of cultivated fields for movement, dispersal, and breeding (e.g., Wegner and Merriam, 1979). In every instance where sampling of the surrounding matrix was conducted, the matrix also was used and movement was recorded within the matrix (Table 3). These latter observations indicate that movement occurs within corridors, but suggest that corridors are not necessary for movement. There was no hostile sea. Machtans et al. (1996) and Dmowski and Kozakiewicz (1990) provide evidence of preferential movement through corridors by birds and small mammals in forested landscapes, but corridors may not need be continuous to promote movement (Ruefenacht and Knight, 1995). Although two studies provide evidence that corridors facilitate movements in forests, an important ques-

---

[3] Debate is typically about degree of effect. For example, only a few immigrant individuals per generation is sufficient to minimize deleterious inbreeding effects and sustain genetic diversity (Bunnell, 1978; Slatkin, 1985; Lande and Barroclough, 1987).

tion remains unanswered: Do corridors increase connectivity in a biologically meaningful way?

There are few well-designed studies, but see Henderson et al. (1985), LaPolla and Barrett (1993), and Andreassen et al. (1996) who demonstrate effectiveness of corridors when the matrix is agriculture or bare ground. Despite their experimental rigor, these studies provide no evidence that absence of corridors encourages local extinction. While evidence for movement within corridors is accumulating for agricultural and urban landscapes, extrapolating findings and conclusions to managed forest landscapes is questionable (Small and Hunter, 1988, Lindenmayer, 1994). We lack evidence of the efficacy of corridors in managed forests.

4.2.2 Connectivity in the matrix

An alternative approach to mitigating isolation effects is to maintain or enhance connectivity in the matrix. In managed forests that means promoting movements, including dispersal, in transient, successional landscapes (e.g., Tiebout and Anderson, 1996). The approach has less appeal than corridors because it is as difficult to evaluate and lacks the distinction of visible action. Of the 10 studies that assessed movements in corridors, only 3 also measured use of the matrix habitat (Table 3). At least two kinds of actions are possible (e.g., Rosenberg et al., 1997). One is to create larger, high-quality patches so that the degree of connectivity required for population persistence is less (the population within a patch is relatively high). The second is to act to ensure that movement through the matrix is relatively unimpeded. The latter is more daunting primarily because movement is the least well documented of vertebrate attributes. What seems clear is that given the potential dangers documented for corridors, any action that reduces reliance on them is helpful. Although the challenge may appear daunting its scope is undefined. Much of the interpretable work on corridors confirms that some movement through the matrix occurs (e.g., Wegner and Merriam, 1979; Merriam and Lanoue, 1990; Machtans et al., 1996; Table 3). Although corridors were reported to support more movement than the matrix, how much movement through the matrix would have occurred in the absence of corridors is un-known. The problem in managed forests may be smaller than is commonly perceived.

Percolation theory, the study of connectivity in randomly generated structures (Stauffer and Aharony, 1985), provides a theoretical framework for describing and predicting movement through landscapes. As a theoretical framework for movement it provides some consistency with empirical observations (e.g., organisms covering large distances can rely on sparsely distributed resources). Such consistency has encouraged extension of the theory, primarily through "percolating clusters" and "neutral landscape models" (see Keitt et al. 1997 and With this volume). Despite inherent promise percolation models and reality have yet to meet. The assumption of random movements is appropriate when treating dispersing individuals (e.g., Bunnell and Harestad, 1983); the distribution of patches in a managed landscape, however, may be highly ordered (but see O'Neill et al., 1988). Moreover, each species tends to perceive patch scales uniquely (Bunnell and Huggard, 1999). The theory has served to emphasize both species-specific gap-crossing abilities or propensities, and the utility in making the matrix less hostile to movement and survival. In summary, management for connectivity within the matrix appears no more challenging than designing effective corridors, may be fraught with less operational difficulty and provide greater operational flexibility, and could prove much less costly.

5. IMPLICATIONS TO FOREST MANAGEMENT

Lack of connectivity (isolation) and strict affinity to particular patches of habitat are essential if the theory of island biogeography is to apply. For forest-dwelling vertebrates of the Pacific Northwest I derive the following generalizations.

There is little evidence that vertebrates perceive old forest stands as discrete patches. Data of Tables 1 and 2 indicate that very few vertebrates perceive old growth as we do. If habitat is not perceived as discrete, separate patches then the total amount of habitat is more important than the distribution of habitat.

Table 4. Forest-dwelling, terrestrial vertebrates designated "at risk" in Alaska, British Columbia, Washington, and Oregon, and their association with forest elements (taxa may be associated with more than one habitat element).[a]

Habitat element	Alaska[b]	British Columbia	Washington[b]	Oregon	Total Listings[c]
Cavity sites	-	12	15	22	49
Downed Wood	-	1	-	14	15
Shrubs	1	6	3	5	15
Broadleaved trees	-	8	1	4	13
Large live trees	-	2	3	7	12
Adjacent or continuous canopy	-	-	4	3	7
Riparian[d]	1	5	4	16	26
Late-successional[d]	2	3	4	5	14

[a] Sources include Alaska Department of Fish, Game WWW site as of July 1997; British Columbia Ministry of Environment, Lands and Parks (1992, 1996); Rodrick and Milner (1991) for Washington; and Marshall et al. (1996) for Oregon.
[b] Includes candidate species or species of special concern.
[c] Species listed by 2 or more jurisdictions are counted more than once; that serves to indicate the generality of limitation by particular habitat elements.
[d] As noted in the text, these are more clearly habitats themselves, rather than habitat elements.

Vertebrates appear to perceive habitat in terms of habitat elements. Data of Tables 1 and 2 imply that habitat structure is more important than habitat age. If we are to focus on amounts of habitat we need to know critical habitat elements. One way of selecting such elements is to evaluate relationships for species currently designated "at risk" or of "special concern". Few designated species were not associated with one of the seven habitat elements of Table 4. The western gray squirrel in Oregon and Washington requires regular seed and mast production; it is unclear how forest practices are associated with low numbers of lynx, wolf and grizzly bear in Washington. Species designated as "late-successional" in the rationale for listing usually require cavity sites or downed wood. Habitat elements of Table 4 provide a good starting point for assessing amounts of suitable habitat.

Many elements creating associations with late-successional stages can be sustained within stands by forest practices. Large live trees, snags, and downed wood can be provided for during even-aged management by retention of living trees and snags during harvest, or by growing trees rapidly to kill for snag production. Broadleaved elements can be retained in both riparian and upland areas, and their suppression during stand establishment implemented patchily. Generally, the process of retaining these elements is simpler during multi-aged management. A variety of attempts have been made to hasten the production of old-growth attributes in managed stands; most of them focused on the requirements of single species (e.g., Anderson, 1985; Bunnell, 1985; Armleder et al., 1986). When multiple species are considered it remains untested whether a combination of stand treatments is economically more efficient than simply maintaining old growth.

Stand age remains an issue for several reasons. Large live trees and snags are a function of age. Older trees with rough bark may be required by overwintering birds (nuthatches, chickadees, brown creepers, woodpeckers) which glean insects under bark scales. Age or elapsed time is important in generating cavity sites through both period of exposure to wounding events and entry of heart rot, and activities of decay fungi more generally. Elapsed time, or age, also influences habitat features that affect terrestrial vertebrates but are not included in Table 4. Litter or organic matter depth is deeper in older forests (Spies and Franklin, 1991) and deeper layers appear important to several vertebrate species (e.g., Corn et al., 1988; McComb et al., 1993b; Rosenberg et al., 1994; Carey and Johnson, 1995; Thompson, 1996). Many epiphytes (mosses, liverworts, and arboreal lichens) are slow to develop significant biomass. Reasons for slow development of local abundance may include appropriate microclimate, time for rare, long-distance colonization events, slowly shedding bark (slow growth), and deeply fissured bark (Bunnell, 1985). The last three of these are products of age.

Current trends in forest practices will make stand age a poor predictor of vertebrate habitat. The original intent of silvicultural systems was to facilitate regeneration. As the range of values requested from forests has expanded, different systems designed to accommodate more values have been implemented (e.g., "green tree retention", FEMAT, 1993; "variable retention", CSP, 1995). One feature of these approaches is that they create forests of very different structure than do most classic silvicultural systems; in particular, a variety of age classes is present in small areas. One consequence is that, except for classic clearcutting, stand structures will follow variable trajectories. Another consequence is that we can learn little about evolving forestry practices from documenting what has occurred in previously created stands.

No single approach is sufficient. Bunnell (1997) observed that the worst possible approach to maintain vertebrate diversity would be to manage every hectare the same way. "Green-tree retention" (FEMAT, 1993) illustrates the point. Such retention has appeal as a management approach because it allows for commodity extraction while emulating natural disturbance and retaining late-successional attributes. Predation rates on shrub nesters in retention stands, however, were greater than in clearcuts (because retained trees served as perch sites for aerial predators; Vega, 1993). As tempting as it is to adapt a single approach, efforts to maintain the entire array of vertebrate diversity likely will have to include a blend of conventional forestry (such as clearcutting), long rotations and reserves, and creative efforts to establish and maintain late-successional attributes in managed stands.

Connectivity is important but the degree of necessary interchange cannot be known. There is ample evidence from true islands that connectivity is important. For some threats to small populations (e.g., inbreeding depression) the amount of necessary interchange can be estimated and is surprisingly low (e.g., Bunnell, 1978; Slatkin, 1985; Lande and Barroclough, 1987). The major threats to small populations, however, are chance events (e.g., Shaffer, 1987, 1990) which cannot be known.

Data are insufficient to advocate either corridors or matrix management as the better approach to promoting connectivity. Conservation benefits have not been documented for either approach; dangers have been documented for corridors. The danger in relying on the matrix is its presumed hostility, which is not strongly expressed (Tables 1 and 2). Advantages and difficulties of relying on the matrix should be examined (e.g., Rosenberg et al., 1997). Fortunately, there is little evidence that lack of connectivity is a threat in forests of the Pacific Northwest.

Acknowledgments

I thank B.G. Dunsworth, D.J. Huggard, and D.E. Varland for reviews that improved the manuscript. This is Publication Number R-31 of the Centre for Applied Conservation Biology, University of British Columbia.

LITERATURE CITED

Anderson, R. J. 1985. Bald eagles and forest management. *Forestry Chronicle* 61:189–193.

Andreassen, H.P., S. Halle, and R.A. Ims. 1996. Optimal width of movement corridors for root voles: not too narrow and not too wide. *Journal of Applied Ecology* 33:63–70.

Anthony, R. B.(sic), G.A. Green, E.D. Forsman, and S.K. Nelson. 1984. Unpublished report, USDA Forest Service, Olympia, WA. cited from Hansen et al., 1991.

Anthony, R.G., E.D. Forsman, and G.A. Green [and others]. 1987. Small mammal populations in riparian zones of different-aged coniferous forests. *Murrelet* 68:94–102.

Armleder, H.M., R.J. Dawson, and R.N. Thompson. 1986. *Handbook for timber and mule-deer co-ordination on winter ranges in the Cariboo Forest Region.* B.C. Ministry of Forests, Land Management Report No. 13, Victoria, B.C.

Aubry, K.B. and P.A. Hall. 1991. Terrestrial amphibian communities in the southern Washington Cascade Range. *In*: L.F. Ruggiero, K.B. Aubry, A.B. Carey, and M.F. Huff (tech. coords.). *Wildlife and vegetation of unmanaged Douglas-fir forests.* USDA Forest Service Gen. Tech. Rep. PNW-GTR-285, Portland, OR. pp. 327–338.

Aubry, K.B., L.L.C. Jones, and P.A. Hall. 1988. Use of woody debris by plethodontid salamanders in Douglas-fir forests in Washington. *In*: R. Szaro, K.E. Severson, and D.R. Patton (tech. coords.). *Management of amphibians, reptiles, and small mammals in North America.* Proc. of a symp. 1988 July 12-21, Flagstaff, AZ. USDA Forest Service Gen. Tch. Rp. RM-166, Ft. Collins, CO. pp. 32–37.

Aubry, K,B., Crites, M.J., and West, S.D. 1991. Regional patterns of small mammal abundance and community composition in Oregon and Washington. *In*: L.F. Ruggiero, K.B. Aubry, A.B. Carey, and M.H. Huff (tech.coords.). *Wildlife and vegetation of unmanaged Douglas-fir forests.* USDA Forest Service Gen. Tech. Rep. PNW-GTR-285, Portland, OR. pp. 295–303.

B.C. Ministry of Environment, Lands, and Parks. 1992. 1992 red and blue lists for terrestrial vertebrates, B.C. Ministry of Environment, Lands, and Parks, Victoria, B.C.

B.C. Ministry of Environment, Lands, and Parks. 1996. 1996 red and blue lists for terrestrial vertebrates, B.C. Ministry of Environment, Lands, and Parks, Victoria, B.C.

B.C. Ministry of Forests. 1995. *Forest practices code of British Columbia, Biodiversity guidebook.* Ministry of Forests, Victoria, B.C.

Beier, P. 1995. Dispersal of juvenile cougars in fragmented habitat. *Journal of Wildlife Management* 59:228–237.

Beier, P. and S. Loe. 1992. A checklist for evaluating impacts to wildlife movement corridors. *Wildlife Society Bulletin* 20:434–440.

Bennett, A. 1990. Habitat corridors and the conservation of small mammals in a fragmented forest environment. *Landscape Ecology* 4:109–122.

Bennett, A. F., K. Henein, and G. Merriam. 1994. Corridor use and the elements of corridor quality: chipmunks and fencerows in a farmland mosaic. *Biological Conservation* 68:155–165.

Brown, J.H. 1971. Species richness of boreal mammals living on the montane islands of the Great Basin. *American Naturalist* 105:467–478.

Bryant, A.A. 1997. Effect of alternative silvicultural practices on breeding bird communities in montane forests. Report to MacMillan Bloedel Ltd., Nanaimo, B.C.

Buchanan, J.B, L.L. Irwin, and E.L. McCutchen. 1995. Within stand nest site selection by spotted owls in the eastern Washington Cascades. *Journal of Wildlife Management* 59:301-310.

Buckner, C.H., A.J. Erskine, R. Lidstone, B.B. McLeod, M. Ward. 1975. The breeding bird community of coast forest stands on northern Vancouver Island. *The Murrelet* 56:6-11.

Bunnell, F.L. 1978. Basic considerations for study and management programs: constraints of small populations. *In: Threatened deer.* IUCN, Morges, Switzerland. pp. 264–287.

Bunnell, F.L. 1985. Forestry and black-tailed deer: conflicts, crises, or cooperation. *Forestry Chronicle* 61:180–184.

Bunnell, F.L. 1997. Operational criteria for sustainable forestry: focusing on the essence. *Forestry Chronicle* 73:679–684.

Bunnell, F.L., and A.S. Harestad. 1983. Dispersal and dispersion of black-tailed deer—models and observations. *Journal of Mammalology* 64:201–209.

Bunnell, F.L., and D.J. Huggard. 1999. Biodiversity across spatial and temporal scales: problems and opportunities. *Journal of Forest Ecology & Management* 115:113-126..

Bunnell, F.L., L.L. Kremsater, and R.W. Wells. 1997. *Likely consequences of forest management on terrestrial, forest-dwelling vertebrates in Oregon.* Oregon Forest Resources Institute, Portland, OR.

Bunnell, F.L., LL. Kremsater, and M. Boyland. 1998. *An ecological rationale for changing forest practices on MacMillan Bloedel's forest tenure.* MacMillan Bloedel, Vancouver, BC. The Forest Project http://www.mbltd.com/

Bury, R.B. 1983. Differences in amphibian populations in logged and old-growth redwood forest. *Northwest Science* 57:167–168.

Bury, R.B., and P.S. Corn. 1988a. Douglas-fir forests in the Oregon and Washington Cascades: relations of the herpetofauna to stand age and moisture. *In*: R.C. Szaro, K.E. Severson, and D.R. Patton (eds.). *Management of amphibians, reptiles, and small mammals in North America.* USDA Forest Service Gen. Tech. Rep. RM-GTR-166, Fort Collins, CO. pp. 11–22.

Bury, R.B., and P.S. Corn. 1988b. Responses of aquatic and streamside amphibians to timber harvest: a review. K.J. Raedeke (ed.). *In: Proceedings of a symposium*, 11-13 February 1987, University of Washington, Seattle. Institute of Forest Resources, Contribution No. 59,University of Washington, Seattle, WA. pp. 165–181.

Bury, R.B., P.S. Corn, K.B. Aubry, [and others]. 1991a. Regional patterns of terrestrial amphibian communities in Oregon and Washington. *In*: L.F. Ruggiero, K.B. Aubry, A.B. Carey, and M.F. Huff (eds.). *Wildlife and vegetation of unmanaged Douglas-fir forests.* USDA Forest Service Gen. Tech. Rep. PNW-GTR-285, Portland, OR. pp. 341–350.

Bury, R. B., P.S. Corn, and K.B. Aubry. 1991b. Aquatic amphibian communities in Oregon and Washington. *In*: L.F. Ruggiero, K.B. Aubry, A.B. Carey, and M.F. Huff (tech. coords.). *Wildlife and vegetation of unmanaged Douglas-fir forests.* USDA Forest Service Gen. Tech. Rep. PNW-GTR-285, Portland, OR. pp. 353–362.

Carey, A.B., and M.L. Johnson. 1995. Small mammals in managed, naturally young, and old-growth forests. *Ecological Applications* 5:336–352.

Carey, A.B., M.M. Hardt, S.P. Horton, and B.L. Biswell. 1991. Spring bird communities in the Oregon Coast Range. *In*: L.F. Ruggiero, K.B. Aubry, A.B. Carey, and M.F. Huff (tech. coords.). *Wildlife and vegetation of unmanaged Douglas-fir forests.* USDA Forest Service Gen. Tech. Rep. PNW-GTR-285. Portland, OR. pp. 123–144.

Corn, P.S. and R.B. Bury. 1991a. Terrestrial amphibians in the Oregon Coast Range. *In*: L.F. Ruggiero, K.B. Aubry, A.B. Carey, and M.F. Huff (tech. coords.). *Wildlife and vegetation of unmanaged Douglas-fir forests.* USDA Forest Service Gen. Tech. Rep. PNW-GTR-285, Portland, OR. pp. 305–317.

Corn, P. S., and R. B. Bury. 1991b. Small mammal communities in the Oregon Coast Range. *In*: L.F. Ruggiero, K.B. Aubry, A.B. Carey, and M.F. Huff (tech. coords.). *Wildlife and vegetation of unmanaged Douglas-fir forests.* USDA Forest Service Gen. Tech. Rep. PNW-GTR-285, Portland, OR. pp. 241–254.

Corn, P.S., R.B.Bury, and T.A. Spies. 1988. Douglas-fir forests in the Cascade

Mountains of Oregon and Washington: is the abundance of small mammals related to stand age and moisture? *In*: R.C. Szaro, K.E. Severson, and D.R. Patton (tech. coords.). *Management of amphibians, reptiles, and small mammals in North America*. Proceedings of a symposium. USDA Forest Service Gen. Tech. Rep. RM-166, Fort Collins, CO. pp. 340–352.

CSP (Scientific Panel for Sustainable Forest Practices in Clayoquot Sound). 1995. *Sustainable ecosystem management in Clayoquot Sound. Planning and practices*. Victoria, B.C.

Culver, D.C. 1970. Analysis of simple cave communities. I. Caves as islands. *Evolution* 27:689-695.

Darlington, P.J., Jr. 1943. Carabidae of mountains and islands: data on the evolution of isolated faunas, and on atrophy of wings. *Ecological Monographs* 13(1).

Darlington, P.J., Jr. 1957. *Zoogeography The geographical distribution of animals*. John Wiley and Sons, New York, NY.

deMaynadier, P.G., and M.L. Hunter, Jr. 1995. The relationship between forest management and amphibian ecology: a review of the North American literature. *Environmental Review* 3:320-261.

Diamond, J.M. 1975. The island dilemma: lessons of modern biogeographic studies for the design of nature reserves. *Biological Conservation* 7: 129-146

Dmowski, K., and M. Kozakiewicz. 1990. Influence of a shrub corridor on movements of passerine birds to a lake littoral zone. *Landscape Ecology* 4:99-108

Dunning, J.B., R. Borgella, K. Clements, and G.K. Meffe. 1995. Patch isolation, corridor effects, and colonization by a resident sparrow in a managed pine woodland. *Conservation Biology* 9:542-550.

Dupuis, L.A., J.N.M. Smith, and F.L. Bunnell. 1995. Relation of terrestrial-breeding amphibian abundance to tree-stand age. *Conservation Biology* 9:645-653.

Eldredge, N. 1992. *The miner's canary. Unraveling the mysteries of extinction*. Virgin Books, London, U.K.

Fahrig, L. 1997. Relative effects of habitat loss and fragmentation on population extinction. *Journal of Wildlife Management* 61:603-610.

Fahrig, L., and G. Merriam. 1985. Habitat patch connectivity and population survival. *Ecology* 66:1762-1768.

Fahrig, L., and G. Merriam. 1994. Conservation of fragmented populations. *Conservation Biology* 8:50-59.

Fahrig, L., and J. Paloheimo. 1988. Determinants of local population size in patchy habitats. *Theoretical Population Biology* 34:194-213.

Fenger, M.A., and A.P. Harcombe. 1990. *Biodiversity, old-growth forests and wildlife in British Columbia*. B.C. Ministry of Environment, Victoria, B.C.

(FEMAT) Forest Ecosystem Management Assessment Team. 1993. *Forest ecosystem management: an ecological, economic, and social assessment*. Report of the Forest Ecosystem Management Assessment Team. US Government Printing Office, Washington, DC.

Forster, J.R. 1778. *Observations made during a voyage round the world, on physical geography, natural history, and ethical philosophy*. G. Robinson, London, UK.

Freemark, K. 1990. Landscape ecology of forest birds in the Northeast. *In*: R.M. DeGraaf and W.M. Healy (eds.). *Is forest fragmentation a management issue in the Northeast?*. USDA Forest. Service Gen. Tech. Rep. NE-GTR-140, Saint Paul, MN. pp. 7-12

Freemark, K.E., and G. Merriam. 1986. Importance of area and habitat heterogeneity to bird assemblages in temperate forest fragments. *Biological Conservation* 36:115-41.

Gashwiler, J.S. 1970. Plant and mammal changes on a clearcut in west-central Oregon. *Ecology* 51:1018-1026.

Gause, G.F. 1934. *The struggle for existence*. Williams & Wilkins, Baltimore, MD.

Gilbert, F.F., and R. Allwine. 1991a. Small mammal communities in the Oregon Cascade Range. *In*: L.F. Ruggiero, K.B. Aubry, A.B. Carey, and M.F. Huff (tech. coords.). *Wildlife and vegetation of unmanaged Douglas-fir forests*. USDA Forest Service Gen. Tech. Rep. PNW-GTR-285, Portland, OR. pp. 257–267.

Gilbert, F. F., and R. Allwine. 1991b. Terrestrial amphibian communities in the Oregon Cascade Range. *In*: L.F. Ruggiero, K.B. Aubry, A.B. Carey, and M.F. Huff (tech. coords.). *Wildlife and vegetation of unmanaged Douglas-fir forests*. USDA Forest Service Gen. Tech. Rep. PNW-GTR-285, Portland, OR. pp. 319–324.

Gilpin, M.E., and M.E. Soulé. 1986. Minimum viable populations: the processes of species extinctions. *In*: M.E. Soulé (ed.). *Conservation biology: the science of scarcity and diversity*. Sinauer Associates, Sunderland, MA. pp. 13–34.

Gould, S.J. 1997. The paradox of the visibly irrelevant. *Natural History* 106(11):12–18.

Grant, P.R. 1986. *Ecology and evolution of Darwin's finches*. Princeton University Press, Princeton, NJ.

Haas, C. A. 1995. Dispersal and use of corridors by birds in wooded patches on an agricultural landscape. *Conservation Biology* 9:845-854.

Hagar, J.C., W.C. McComb, and C.C. Chambers. 1995. Effects of forest practices on wildlife, Chapter 9 *In*: R.L. Beschta, J.R. Boyle, C.C. Chambers [and others]. *Cumulative effects of forest practices in Oregon: literature and synthesis*. Oregon State University, Corvallis, OR.

Hansen, A.J., T.A. Spies, F.J. Swanson, and J.L. Ohmann. 1991. Conserving biodiversity in managed forests. Lessons from natural stands. *BioScience* 41:382-392.

Hansen, A.J., W.C. McComb, R. Vega, [and others]. 1995. Bird habitat relationships in natural and managed forests in the west Cascades of Oregon. *Ecological Applications* 5:555-569.

Hanski, I., and M. Gilpin. 1991. Metapopulation dynamics: brief history and conceptual domain. *Biological Journal of the Linnaean Society* 42(1-2):3-16.

Harrison, R. 1992. Toward a theory of inter-refuge corridor design. *Conservation Biology* 6:293-295.

Harris, L.D. 1984. *The fragmented forest*. University of Chicago Press, Chicago, IL.

Harris, L. D., and K. Atkins. 1991. Faunal movement corridors in Florida. *In*: W. E. Hudson (ed.) *Landscape linkages and biodiversity*. Island Press, Washington, D.C. pp. 117–134.

Henderson, M. T., G. Merriam, and J. Wegner. 1985. Patchy environments and species survival: chipmunks in an agricultural mosaic. *Biological Conservation* 31:95-105.

Hess, G. 1994. Conservation corridors and contagious disease: a cautionary note. *Conservation Biology* 8:256-262.

Hobbs, R. 1992. The role of corridors in conservation: solution or bandwagon. *Trends in Ecology & Evolution* 7:389-392.

Hooven, E.F., and H.C. Black. 1976. Effects of some clearcutting practices on small-mammal populations in western Oregon. *Northwest Science* 50:189-208.

Huffaker, C.B. 1958. Experimental studies on predation: dispersion factors and predator-prey oscillations. *Hilgardia* 27:343-383.

Janzen, D.H. 1968. Host plants in evolutionary and contemporary time. *American Naturalist* 102:592-595.

Keitt, T. H., D.L. Urban, and B.T. Milne. 1997. Detecting critical scales in fragmented landscapes. *Consevation Ecology* [online] 1(1):4 http://www.consecol.org/vol1/iss1/art4.

Knopf, F.L. 1986. Changing landscapes and the cosmopolitism of the eastern Colorado avifauna. *Wildlife Society Bulletin* 14:132-142.

Knopf, F.L., and F.B. Samson. 1994. Scale perspectives in avian diversity in western riparian ecosystems. *Conservation Biology* 8:669-676.

Lande, R. 1988. Genetics and demography of biological conservation. *Science* 241:1455-60

Lande, R., and G.F. Barrowclough. 1987. Effective population size, genetic variation, and their use in population management. *In*: M.E. Soulé (ed.). *Viable populations for management*. Cambridge University Press, Cambridge, UK. pp. 87–144.

LaPolla, V.N., and G. Barrett. 1993. Effects of corridor width and presence on the population dynamics of the meadow vole (*Microtus pennsylvanicus*). *Landscape Ecology* 8:25-37.

Levins, R., 1970. Extinction. *In*: M. Gerstenhaber (ed.). *Lectures on mathematics in the life sciences. Vol. 2*. American Mathematical Society, Providence, RI pp. 77–107.

Lindenmayer, D. B. 1994. Wildlife corridors and the mitigation of logging impacts on fauna in wood-production forests in southeastern Australia: a review. *Wildlife Research* 21:323-340.

Lindenmayer, D., and H. Nix. 1993. Ecological principles for the design of wildlife corridors. *Conservation Biology* 7:627-630.

Lundquist, R.W., and J.M. Mariani. 1991. Nesting habitat and abundance of snag-dependent birds in the southern Washington Cascade Range. *In*: L.F. Ruggiero, K.B. Aubry, A.B. Carey, and M.F. Huff (tech. coords.). *Wildlife and vegetation of unmanaged Douglas-fir forests*. USDA Forest Service Gen. Tech. Rep. PNW-GTR-285, Portland, OR. pp. 221–240.

MacArthur, R.H., and E.O. Wilson. 1963. An equilibrium theory of insular zoogeography. *Evolution* 17:373-387.

MacArthur, R.H., and E.O. Wilson. 1967. *The theory of island biogeography*. Princeton University Press, Princeton, NJ.

MacClintock, L., R.F. Whitcomb, and B.L. Whitcomb. 1977. Island biogeography and "habitat islands" of eastern forest II. evidence for the value of corridors and minimization of isolation in preservation of biotic diversity. *American Birds* 31:6-16.

Machtans, C., M. Villard, and S.J. Hannon. 1996. Use of riparian buffer strips as movement corridors by forest birds. *Conservation Biology* 5:1366-1379.

Main, A.R., and M. Yadov. 1971. Conservation of macropods in reserves in Western Australia. *Biological Conservation* 3:123-133.

Manley, I. A. 1999. Behaviour and habitat selection of marbled murrelets nesting on the Sunshine Coast. MSc thesis, Simon Fraser University, Burnaby, BC.

Manuwal, D.A. 1991. Spring bird communities in the southern Washington Cascade Range. *In*: L.F. Ruggiero, K.B. Aubry, A.B. Carey, and M.F. Huff (tech. coords.). *Wildlife and vegetation of unmanaged Douglas-fir forests*. USDA Forest Service Gen. Tech. Rep. PNW-GTR-285, Portland, OR. pp. 161–175.

Manuwal, D.A., and M.H. Huff. 1987. Spring and winter bird populations in a Douglas-fir forest sere. *Journal of Wildlife Management* 51:586-595.

Marcot, B.G. 1986. The use of expert systems in wildlife-habitat modeling. *In*: J. Verner, M.L. Morrison, and C.J. Ralph (eds.). *Wildlife 2000: modeling habitat relationships of terrestrial vertebrates*. University of Wisconsin Press, Madison, WI. pp. 145–150.

Marshall, D. B., M. Chilcote, and H. Weeks. 1996. *Species at risk. Sensitive, threatened, and endangered vertebrates of Oregon*. Oregon Department of Fish and Wildlife, Portland, OR.

Mayr, E. 1963. *Animal species and evolution*. Harvard University Press, Cambridge, MA.

McComb, W.C., R.G. Anthony, and M. Newton. 1993a. Small mammal and amphibian abundance in streamside and upslope habitats of mature Douglas-fir stands, western Oregon. *Northwest Science* 76:7-15.

McComb, W.C., T.A. Spies, and W.H. Emmingham. 1993b. Douglas-fir forests. Managing for timber and mature-forest habitat. *Journal of Forestry* 91(12):31-42.

McGarigal, K. and W.C. McComb. 1995. Relationships between landscape structure and breeding birds in the Oregon Coast Range. *Ecological Monographs* 65:235-260.

Merriam, G., and A. Lanoue. 1990. Corridor use by small mammals: field measurement for three experimental types of *Peromyscus leucopus*. Landscape Ecology 4:123-131.

Naiman, R.J., H. Décamps, and M. Pollock. 1993. The role of riparian corridors in maintaining regional biodiversity. *Ecological Applications* 3:209-212.

Nelson, S.K. 1988. *Habitat use and densities of cavity-nesting birds in the Oregon coast ranges.* MSc thesis, Oregon State University, Corvallis, OR.

Noss, R.F. 1987. Corridors in real landscapes: a reply to Simberloff and Cox. *Conservation Biology* 1:159-164.

Noss, R. F., and L.D. Harris. 1986. Nodes, networks, and MUMs: preserving diversity at all scales. *Environmental Management* 10:299-309.

O'Neill, R.V., B.T. Milne, M.G. Turner, and R.H. Gardner. 1988. Resource utilization scales and landscape pattern. *Landscape Ecology* 2:63-69.

Planty-Tabacchi, A., E. Tabacchi, R.J. Naiman [and others]. 1996. Invasibility of species-rich communities in riparian zones. *Conservation Biology* 10:598-607.

Preston, F. 1962a. The canonical distribution of commonness and rarity: Part I. *Ecology* 43:185-215.

Preston, F. 1962b. The canonical distribution of commonness and rarity: Part II. *Ecology* 43:410-432.

Quammen, D. 1997. *The song of the dodo. Island biogeography in an age of extinctions.* A Touchstone Book, New York, NY.

Raedeke, K.J. (ed.) 1988. *Streamside management: riparian wildlife and forestry interactions.* Proceedings of a symposium, 11-13 February 1987, University of Washington, Seattle. Institute of Forest Resources, Contribution No. 59,University of Washington, Seattle, WA.

Ralph, C.J., P.W.C. Paton, and C.A. Taylor. 1991. Habitat association patterns of breeding birds and small mammals in Douglas-fir/hardwood stands in northwestern California and southwestern Oregon. *In*: L.F. Ruggiero, K.B. Aubry, A.B. Carey, and M.F. Huff (tech. coords.). *Wildlife and vegetation of unmanaged Douglas-fir forests.* USDA Forest Service Gen. Tech. Rep. PNW-GTR-285, Portland, OR. pp. 379–393.

Raphael, M.G. 1984. Wildlife diversity and abundance in relation to stand age and area in Douglas-fir forests of Northwestern California. *In*: W.R. Meehan, T.R. Merrell, and T.A. Hanley (eds.). *Proceedings of the Symposium on Fish and Wildlife Relationships in Old-growth Forests.* American Institute of Fishery Research Biologists. Morehead City, NC. pp. 259–274.

Raphael, M.G. 1988a. Long-term trends in abundance of amphibians, reptiles, and mammals in Douglas-fir forests of northwestern California. *In*: R.C. Szaro, K.E. Severson, and D.R. Patton (tech. eds.). *Management of amphibians, reptiles, and small mammals in North America.* Proceedings of a symposium. 19-21 July 1988, Flagstaff, AZ. USDA Forest Service Gen. Tech. Rep. RM-166, Fort Collins, CO. pp. 23–31.

Raphael, M.G. 1988b. Habitat associations of small mammals in a subalpine forest, southeastern Wyoming. *In*: R.C. Szaro, K.E. Severson, and D.R. Patton (tech. eds.). *Management of amphibians, reptiles, and small mammals in North America.* Proceedings of a symposium. 19-21 July 1988, Flagstaff, AZ. USDA Forest Service Gen. Tech. Rep. RM-166, Fort Collins, CO. pp. 359-367.

Raphael, M.G. 1991. Vertebrate species richness within and among seral stages of Douglas-fir/hardwood forest in northwestern California. *In*: L.F. Ruggiero, K.B. Aubry, A.B. Carey, and M.F. Huff (tech. coords.). *Wildlife and vegetation of unmanaged Douglas-fir forests.* . USDA Forest Service Gen. Tech. Rep. PNW-GTR-285, Portland, OR. pp. 415–423.

Rice, J., R.D. Ohmart, and B.W. Anderson. 1983. Turnovers in species composition of avian communities in contiguous riparian habitats. *Ecology* 64:1444-1455.

Rodrick, E., and R. Milner. (tech. eds.). 1991. *Management recommendations for Washington's priority habitats and species*. Washington Dept. Wildlife, Olympia, WA.

Rosenberg, K.V., and M.G. Raphael. 1986. Effects of forest fragmentation on vertebrates in Douglas-fir forests. *In:* J. Verner, M. Morrison, and C. Ralph, (eds.). *Wildlife 2000: modeling habitat relationships of terrestrial vertebrates*. Univ. of Wisconsin Press. Madison, WI. pp. 263–272.

Rosenberg, D.K., K.A. Swindle, and R.G. Anthony. 1994. Habitat associations of California red-backed voles in young and old-growth forests in western Oregon. *Northwest Science* 68:266-272.

Rosenberg, D.K., B.R. Noon, and E.C. Meslow. 1997. Biological corridors: forms, function, and efficacy. *BioScience*. 47:677-687.

Ruefenacht, B., and R.L. Knight. 1995. Influences of corridor continuity and width on survival and movement of deer mice, *Peromyscus maniculatus*. *Biological Conservation* 71:269-274.

Shaffer, M.L. 1981. Minimum population sizes for species conservation. *BioScience* 31(2):131-134.

Shaffer, M.L. 1987. Minumum viable populations: coping with uncertainty. *In:* M.E. Soulé (ed.) *Viable populations for conservation*. Cambridge University Press, Cambridge, UK.

Shaffer, M.L. 1990. Population viability analysis. *Conservation Biology* 4:39-40.

Simberloff, D.S. 1988. The contribution of population and community biology to conservation science. *Annual Review of Ecology & Systematics* 19:473-511.

Simberloff, D.S. and J. Cox. 1987. Consequences and costs of conservation corridors. *Conservation Biology* 1:63-71.

Simberloff, J., J. Cox, and D. Mehlman. 1992. Movement corridors: conservation bargains or poor investments. *Conservation Biology* 4:493-504.

Simpson, G.G. 1944. *Tempo and mode in evolution*. Columbia University Press, New York, NY.

Slatkin, M. 1985. Gene flow in natural populations. *Annual Review of Ecology & Systematics* 16:292-430.

Sneath, P.H.A. and R. Sokal. 1973. *Numerical taxonomy*. W.H. Freeman, San Francisco, CA.

Small, M. F., and M.L. Hunter. 1988. Forest fragmentation and avian nest predation in forested landscapes. *Oecologia* 76:62-64.

Sokal, R.R. and E.J. Rohlf. 1995. *Biometry: the principles and practice of statistics in biological research*. W.H. Freeman Co., New York, NY.

Spackman, S.C. and J.W. Hughes. 1995. Assessment of minimum strean corridors width for biological conservation: species richness and distribution along mid-order streams in Vermont, USA *Biological Conservation* 71:325-332.

Spies, T.A. and J.F. Franklin. 1991. The structure of natural young, mature, and old-growth Douglas-fir forests. *In:* L.F. Ruggiero, K.B. Aubry, A.B. Carey, and M.F. Huff (tech. coords.). *Wildlife and vegetation of unmanaged Douglas-fir forests*. USDA Forest Service Gen. Tech. Rep. PNW-GTR-285, Portland, OR. pp. 99–110.

Stauffer, D.F. and W. Aharony. 1985. *Introduction to percolation theory*. Taylor and Francis, London, UK.

Taylor, C.A., C.J. Ralph, and A.T. Doyle. 1988. Differences in the ability of vegetation models to predict small mammal abundance in different aged Douglas-fir forests. *In:* R.C. Szaro, K.E. Severson, and D.R. Patton (eds.). *Management of amphibians, reptiles, and small mammals in North America*. USDA Forest Service Gen. Tech. Rep. RM-166, Ft. Collins, CO. pp. 368–374.

Taylor, P.D., L. Fahrig, K. Henein, and G. Merriam. 1993. Connectivity is a vital element of landscape structure. *Oikos* 68:571-573.

Thomas, D.W. and S.D. West. 1991. Forest age associations of bats in the southern Washington Cascade and Oregon Coast

Ranges. *In*: L.F. Ruggiero, K.B. Aubry, A.B. Carey, and M.F. Huff (tech. coords.). *Wildlife and vegetation of unmanaged Douglas-fir forests*. USDA Forest Service Gen. Tech. Rep. PNW-GTR-285, Portland, OR. pp. 295–303.

Thompson, R.L. 1996. *Home range and habitat use of western red-backed voles in mature coniferous forests in the Oregon Cascades*. MSc thesis, Oregon State Univ, Corvallis, OR

Tiebout, H.M., and R.A. Anderson. 1996. A comparison of corridors and intrinsic connectivity to promote dispersal in transient successional landscapes. *Conservation Biology* 11:620-627.

Vander Haegen, W.M, and R.M. Degraaf. 1996. Predation on artificial nests in forested riparian buffer strips. *Journal of Wildlife Management* 60:542-50.

Vega, R.M.S. 1993. *Bird communities in managed conifer stands in the Oregon Cascades: habitat associations and nest predation*. MSc thesis, Oregon State University, Corvallis, OR.

Vuilleumier, F. 1970. Insular biogeography in continental regions. I. The northern Andes of South America. *American Naturalist* 104:373-388.

Wegner, J.F., and H.G. Merriam. 1979. Movements by birds and small mammals between a wood and adjoining farmland habitats. *Journal of Applied Ecology* 16:349-357

Welsh, H.H., Jr., and A. Lind. 1988. Old growth forests and the distribution of the terrestrial herpetofauna. *In*: R.C. Szaro, K.E. Severson, and D.R. Patton (tech. eds.). *Management of amphibians, reptiles, and small mammals in North America*. Proceedings of a symposium. 19-21 July 1988, Flagstaff, AZ. USDA Forest Service Gen. Tech. Rep. RM-166, Fort Collins, CO. pp. 439–458.

Welsh, H.H., Jr., and A.J. Lind. 1991. The structure of the herpetofaunal assemblage in the Douglas-fir/hardwood forests of northwestern California and southwestern Oregon. *In*: L.F. Ruggiero, K.B. Aubry, A.B. Carey, and M.F. Huff (tech. coords.). *Wildlife and vegetation of unmanaged Douglas-fir forests*. USDA Forest Service Gen. Tech. Rep. PNW-GTR-285, Portland, OR. pp. 395–413.

West, S.D. 1991. Small mammal communities in the southern Washington Cascade Range. *In*: L.F. Ruggiero, K.B. Aubry, A.B. Carey, and M.F. Huff (tech. coords.). *Wildlife and vegetation of unmanaged Douglas-fir forests*. USDA Forest Service, Gen. Tech. Rep. PNW-GTR-285, Portland, OR. pp. 269–283.

Wetmore, S.P., R.A. Keller, and G.E.J. Smith. 1985. Effects of logging on bird populations in British Columbia as determined by a modified point-count method. *Canadian Field-Naturalist* 99:224-233.

Willson, M.F., and T.A. Comet. 1996. Bird communities of northern forests: patterns of diversity and abundance. *Condor* 98:337-349.

Wolff, J.O., E.M. Schauber, and W.D. Edge. 1997. Effects of habitat loss and fragmentation on the behaviour and demography of gray-tailed voles. *Conservation Biology* 11:945-956.

Vulnerability of Forested Ecosystems in the Pacific Northwest To Loss of Area

J. Michael Scott

The forests of Idaho, Oregon, and Washington have been subjected to a wide variety of disturbances, anthropogenic and natural, and exhibit substantial variation. Recent anthropogenic disturbances (e.g. farming, urbanization, silviculture practices etc.) have dramatically modified the composition, structure, and ecological processes associated with historical landscapes. In this paper, I review available information on the management status of Idaho, Oregon, and Washington forests. Forests at low elevations and higher soil productivity have suffered the greatest losses of area and ecological integrity. These are also the areas with the least acreage in biological reserves. Ponderosa pine, riparian habitats, and oak woodlands are among those forest types that are most vulnerable to further loss in area and/or ecological integrity, whether through fragmentation or conversion to earlier successional stages. Activities that tend to further fragment these habitats should be avoided.

Key words: ecosystem, fragmentation, conservation, reserve

1. INTRODUCTION

The west was viewed by many as the last frontier, settlement of which would be the final expression of our nation's manifest destiny. The great push to populate the West began in the middle of the 19th century and accelerated after that. Many of those western migrants came to the Pacific Northwest motivated in large part by the rich soils and bountiful supplies of water, timber, fish, and mineral resources. Populations are an order of magnitude greater today in Idaho and Oregon and 15 times greater in Washington than at the time of the first census in 1890. Concurrent with the increase in numbers of people in the Pacific Northwest came major changes in land cover and consequently, loss and fragmentation of natural landscapes. Rich bottomlands were converted to croplands and then increasingly to urban landscapes. Today, these same areas are primarily agricultural, urban, and industrial lands (Oregon Biodiversity Project 1998 and Cassidy et al. 1997). The large tracts of forest of the coast range and Cascade mountains served as seemingly endless sources of building material and jobs. But concern at the turn of the century regarding the use of those forested lands and the possible loss of future options resulted in the creation of the U.S. Forest Service and forest reserves (Pinchot 1947). The slopes of the coast range and the Cascade Mountains are still dominated by forests, but far less of it is the old-growth that dominated prior to 1800 (Garman et al. 1999). Historically, vegetation maps show the Willamette Valley and Puget Sound Trough as being dominated by prairie, oak woodlands, and bottomland hardwoods. The mountain areas from the crest of the Cascades to the coast were dominated by Douglas-fir (*Pseudotsuga menziesii*). These forests have a history of natural disturbance (Agee, 1999, Garman, et al., 1999) and in recent years, have been greatly modified by timber harvest.

U.S. Geological Survey, Biological Resources Division, Idaho Cooperative Fish and Wildlife Research Unit, University of Idaho, Moscow Idaho

The historical patterns of patch size, stand structure, and species composition resulting from natural disturbances are different than those resulting from today's anthropogenic disturbances. The overall effect has been a reduction in area of contiguous old-growth forest types in both public, but especially private, ownership (Garman et al., 1999).

Concerns about effects of management actions on stand structure, composition, associated ecological processes, and ecosystem services (Noss 1990, Noss et al. 1995, Daily 1997) were a factor in the decision to hold this conference. A clear need existed to bring together the most current information available.

In this paper, I use habitat fragmentation to mean "a breaking apart or subdividing of a formerly contiguous tract of forest." The Forest Ecosystem Management Assessment Team (FEMAT 1993) defined fragmentation as "the process of reducing size and connectivity of stands that compose a forest" and forest fragmentation as "the change in the forest landscape, from extensive and continuous forests of old-growth to a mosaic of younger stand conditions." All fragmentation results in loss of habitat (McGarigal and McComb 1999). In this chapter, I emphasize the importance of avoiding additional loss of natural forest habitat types through fragmentation and therefore additional loss of those forested ecosystems in Washington, Oregon, and Idaho that have already suffered major loss of area and/or ecological integrity. The emphasis of this paper will be on habitat loss regardless of its cause. While some forest types (e.g., Douglas fir) may actually have greater acreage today because of plantings and cultivation, I distinguish between these anthropogenic stands and intact "natural" ecosystems. These modified forests differ in their composition of associated species, stand structure, as well as ecological processes from the natural occurring old-growth ecosystems they replaced. They differ in ways that we are only beginning to understand. Recently, concerns about habitat loss resulting from harvest of old-growth forests spawned the often acrimonious debate (Cover Story 1992, Bainbridge 1993) around species like the Northern Spotted Owl (*Strix occidentalis*) and the Marbled Murrelet (*Brachyramphus marmoratus*) and in part contributed to the adoption of the Northwest Forest Plan (FEMAT 1993).

2. ECOLOGICAL CONSEQUENCES OF FRAGMENTATION AND HABITAT LOSS

Geographical (eastern vs. western forests) differences in species responses to fragmentation and influence of ecological context are just beginning to be identified period. Uncertainty exists (Mills and Tallmon 1999) as to whether species responses in Western managed forests are the same as have been observed in the eastern United States (Bunnell 1999, J. Marzluff pers. comm.). The forest edges so often found in Pacific Northwest forests were discussed by Leopold (1933) in his text *Game Management* as a desired condition; they enhanced numbers of game species and access to them. However, as the ascendancy of landscape ecology and conservation biology as scientific disciplines brought increased attention, concerns have arisen regarding possible negative impacts of fragmentation (Rosenberg and Raphael 1986, Terborgh 1989). As a result, fragmentation has been the focus of an increasing number of research efforts. A search of *Ecology*, *The Journal of Wildlife Management*, and the major ornithological journals of North America found 11 articles with "fragment" as a key word for the period 1945-1960 and 151 since 1990. Several major symposia have been convened on the topic, and two were held in Portland, Oregon, within 6 months of each other (J. Rochelle, L. Lehmann and J. Wisniewski this volume and D. Dobkin and L. George, in prep.).

Perhaps the most obvious consequence of habitat loss is the loss of species that were restricted in their occurrence to the patches of destroyed habitat. Endemic species (those species found in very limited areas) are most vulnerable to this effect. Those species about which we know the least, invertebrates and plants, have the highest percentage of endemism. Their vulnerability to the effects of habitat loss would be compounded relative to that of vertebrates by their generally diminished capabilities of movement. Loss of species from fragmented landscapes in a local area, and even extinction, have been well documented in tropical forests and coastal shrub environments (Gentry 1986, Bolger et al. 1997). This phenomena is best observed in small island settings and in the tropics where species ranges are much smaller and where it is not unusual to have a species distribution restricted to a single valley or mountain top (Howarth and Mull 1992). However, similar patterns of limited mountain top distributions can also be found in North America (Wilcove 1987).

A second consequence of habitat loss which may result from fragmentation is creation of barriers to movement and dispersal. Barriers may be as small as a road 30 meters wide (Adams and Geis 1983, Oxley et al. 1974). The nature of barriers will vary among species. A 30 meter wide little-traveled, dirt road may serve as a dispersal barrier to many nonvolant invertebrates but not to a black bear (*Ursus americanus*). However, a four-lane freeway can be a major source of mortality for black bears (Brody and Pelton 1989). What constitutes a movement barrier for a non-volant invertebrate (Mader 1984) would be no barrier at all for a bird capable of flight. Barriers to movement between populations can lead to changes in the genetic composition of those remaining individuals (e.g., reduced heterozygosity, inbreeding depression, genetic drift). These issues are discussed in detail elsewhere in this volume (Mills and Tallmon 1999).

Crowding effects are another consequence of habitat loss associated with fragmentation (Leck 1979). With loss of habitat, displaced individuals move into the remaining suitable habitat patches. This may result in increased competition with resident animals and reduced fitness in individual animals. Fragmentation of forested areas by road construction increases vulnerability of large mammal species to disturbance and take by hunters (Basile and Lonner 1970, Rost and Bailey 1979). Other biological consequences of fragmentation include decreased pollination by native pollinators (Aizen and Feinzinger 1994a,b), increased nest predation (Andren 1992), and increased brood parasitism (Donovan et al. 1997, Brittingham and Temple 1983.

Wilcove and his colleagues (1998) found that 85% of the threatened and endangered species in the United States were imperiled by habitat destruction and loss of ecological integrity, both of which may result from fragmentation. Examples of forest dwelling species in the Pacific Northwest for which this may be true include the previously cited Spotted Owl and Marbled Murrelet.

Conversion of old-growth forests to urban or agricultural cover types has a much greater impact on species composition than does conversion to early successional stages. Conversion of one habitat to another results in improved habitat for some species and loss of habitat or reduction in habitat quality for others. For example, McCarigal and McComb (1999) found that an equal number of species were negatively and positively associated with late seral forests and early successional forests and that species abundances were generally greater in more fragmented landscapes. However, it is not those species that associate with early successional stages of forests that are presently considered at risk. Providing habitat for species like Dark-eyed juncos (*Junco hyemalis*) or Yellow-rumped warblers (*Dendroica cornata*) that thrive in early successional forests is currently not a challenge. It is the decrease in number or absence of species primarily associated with old-growth forests when these stands are converted to early successional stages about which attention has been focused. With human populations predicted to increased significantly in the Pacific Northwest by 2050, additional conversions of forested landscapes to non-forested landscapes and unharvested forests to harvested forests can be expected as desire and demand for larger homes, bigger industries, and more roadways increase. With hundreds of species and many human activities dependent on these forested landscapes, it will be a challenge to maintain the full range of biodiversity in the living landscapes of the Pacific Northwest, especially those ecosystems that are most vulnerable to additional loss of area and ecological integrity.

3. GEOGRAPHY OF PROTECTED AREAS

Recent studies reporting the occurrence of mapped vegetation types in the western United States suggest that perhaps 40% of the mapped cover types have 10% or less of their area included in areas dedicated for the long-term maintenance of natural systems (Stoms et al. 1998, Davis et al. 1995, Caicco et al. 1995, Edwards et al. 1995, Cassidy et al. 1997, Davis and Stoms 1996, R. G. Wright et al. In prep.). These numbers suggest that if we are to achieve the goal of protecting the full range of biodiversity for future generations, we must be more disciplined in our selection of future reserve areas than we have been in the past. Traditionally, conservation areas are viewed as doing an excellent job of protecting rocks and ice and other areas with low potential for economic return. This perception is only reinforced when one examines the distribution of existing protected areas in the conterminous United States. Results of this assessment revealed that bioreserves occur most frequently in areas of high elevation and/or low soil productivity (J. M. Scott et al. Unpub. Manuscript), not necessarily the areas of greatest biodiversity. Similar patterns of under-protection of low elevation, high soil productivity, and flat terrain areas, those most desirable for agriculture, urban development or silviculture, have also been documented in New South Wales (Pressey 1994) and Nepal (Hunter and Yonzon 1993). The establishment of large acreages of old-growth forest in ecological reserves and additional acreage managed to create late successional stage forests on federal lands by the Northwest Forest Plan (FEMAT 1993) was a positive step towards maintaining the ecological integrity and species composition of forested ecosystems in Oregon, Washington, and Northern California.

If we are to minimize future losses of biodiversity, we must focus our attention on those ecosystems, species, and areas that are most vulnerable. Ideally, this would require complete information on the compositional, structural, and functional elements of biodiversity (Noss 1990, Noss and Cooperrider 1994), something we do not have.

Table 1. Endangered ecosystems in the Pacific Northwest (after Noss et al. 1995)

Old growth Ponderosa pine (*Pinus ponderosa* in Idaho)

Maritime-like forests in the Clearwater Basin of Idaho

Old-growth forests in Douglas-fir region of Oregon and Washington

Coastal temperate rainforests in Oregon

Old-growth Ponderosa pine forests in Oregon

Oak savannas

However, I do not believe we need to wait for the final answers before we can make informed decisions of which forested ecosystems in which to minimize additional losses of areas and ecological integrity. We can increase chances of saving the full range of forest ecosystems and associated species by using that incomplete information to make better informed management decisions or reassessing research priorities.

In the remainder of this chapter, I will discuss those forest ecosystems in the Pacific Northwest that have been identified as being at greatest risk. These ecosystems have been identified through examination of the literature and from my personal research in this area. These ecosystems are considered at risk because they are rare, have suffered large loss of area or degradation in their structure composition or function, or are under represented in bioreserves (12% or less of their distributional area in reserves, Bruntland 1987).

4. FORESTED ECOSYSTEMS OF CONCERN

Noss, LaRoe, and Scott (1995) identified 165 ecosystems that they considered to be endangered. Sixty-nine of those were forest ecosystems, and 6 of those ecosystems occurred in the Pacific Northwest (Table 1). Included were oak woodlands, old-growth, riparian, and ponderosa pine (*Pinus ponderosa*) forests. Several of these same forest ecosystems (e.g., oak savanna and woodlands, riparian forests, ponderosa pine, and old-growth forests as well as bottomland hardwoods) were identified by the Oregon Biodiversity Project (1998) as ecosystems of concern.

In a separate study, The Nature Conservancy identified 63 forests, 42 woodland, and 8 sparse woodland plant communities in the West as rare (Grossman et al. 1994). This was 30, 19.8, and 3.9% respectively of the rare communities identified by those authors in the western United States. Of those forest communities occurring in Idaho, Oregon, and/or Washington (Table 2), ponderosa pine com-

Table 2. Forest and woodland associations and their plant alliances that occur in Washington, Oregon, and/or Idaho that have been identified as being rare. Association – the finest level of the classification system, a physiognomically uniform group of vegetation stands that share one or more diagnostic (dominant, differential, indicator, or character) overstory and understory species. These elements occur as repeatable patterns of assemblages across the landscapes and are generally found under similar habitat conditions. The Association refers to existing vegetation, not to a potential vegetation type. A plant alliance is a physiognomically uniform group of Associations sharing one or more diagnostic (dominant, differential, indicator, or character) species, that, as a rule, are found in the uppermost stratum of the vegetation) (Grossman et al. 1994, 1998)

Forest Alliances	Forest associations
Abies grandis Forest Alliance	4
Alnus rhombifolia Forest Alliance	3
ChamaecypAris lawsoniana Forest Alliance	6
Picea breweriana Forest Alliance	3
Picea sitchensis Forest Alliance	1
Populus trichocarpa Forest Alliance	3
Pseudotsuga menziesii Forest Alliance	5
Thuja plicata Forest Alliance	6
Tsuga heterophylla Forest Alliance	4
Woodland Alliances	
Abies grandis	1
Juniperus osteosperma	1
Pinus contorta	1
Pinus flexilis	1
Pinus monophylla	4
Woodland Alliances	**Woodland associations**
Pinus ponderosa	5
Pinus ponderosa-Pseudotsuga menziesii	1
Pinus ponderosa-Quercus garryana	1
Populus angustifolia-Picea pungens	1
Pseudotsuga menziesii	1
Quercus garryana	1
Sparse Woodland Alliances	**Sparse Woodland associations**
Juniperus occidentalis	2
Juniperus osteosperma	2
Pinus ponderosa	1
Quercus garryana	1

Table 3. Plant Communities with 10% or less of their mapped distribution in protected areas in Idaho (After Caicco et al. 1995. Conservation Biology Vol. 9)

Subalpine Park	Montane Forests – Steppe Transitions
Subalpine Fir – Mountain Hemlock	Douglas-Fir – Limber Pine/Montane Brush Mosaic
Subalpine Forests	Douglas-Fir – Aspen/Montane Brush or Sagebrush
Subalpine fir- Douglas-Fir – Quaking Aspen	Lodgepole Pine – Quaking Aspen Mountain Brush
Mountain Hemlock – Subalpine Fir	Ponderosa Pine- Grassland or Montane Sagebrush
Lodgepole Pine – Quaking Aspen – Subalpine Fir	Mountain sagebrush and forest mosaic
Montane Forests	Foothills and Plains Woodlands
Western Larch – Douglas-Fir	Limber Pine/Gressewood
Douglas-Fir – Western Larch	Western Juniper/Low Sagebrush Mosaic
Douglas-Fir – Englemann Spruce	Western Juniper/Mountain Sagebrush
Western Red Cedar - Western Hemlock	Utah Juniper/Big Sagebrush
Grand Fir – Western Red Cedar	Utah Rockmountain Juniper/Big Sagebrush
Western Red Cedar – Grand Fir – Douglas-Fir	Riparian
Grand Fir - Douglas-Fir	Lodgepole Pine Floodplain Riparian
Douglas-Fir – Limberpine – Whitebark Pine	Willow Floodplain Riparian
Ponderosa Pine – Lodgepole Pine	
Lodgepole Pine Mixed Conifer	

munities were the single largest grouping of rare communities identified. In a separate study in Idaho, 35 of the 46 mapped forest cover types had 12% or less of their area in bioreserves (Caicco et al. 1995; Table 3). In Washington, 3 of 9 mapped eastside forest zones (Table 4 Cassidy et al. 1997) had less than 10%, often cited as the standard to be met, of their areas protected (Miller 1984). These same authors found 6 of 8 of the identified forest zones on the west side (Table 5) as having less than 12% of their area in bioreserves. With few exceptions, vegetation types that occurred in alpine or subalpine areas had very large percentages of their acreage in protected areas.

5. CONCLUSIONS

The ecosystems identified above represent a first cut at assessing those forest ecosystems in the Pacific Northwest that can least afford to suffer habitat loss. These ecosystems represent

Table 4. Conservation Status of areas in east-side forest zone in Washington.. Figures presented are percent of mapped are in each of four management categories. Status three indicates public lands not specially designated as conservation lands and status 4 lands are private lands not designated as conservation lands. Status 2 lands are considered to be lands that are managed primarily for the long term maintenance of self sustaining natural ecosystems. Examples of area classified as status 1 or 2 include national parks, wilderness areas, national wildlife refuges, and nature conservancy reserves (after Cassidy et al. 1997).

East-side forest zone	Status 1	Status 2	Status 3	Status 4
Oak	0.02	3.36	13.88	82.75
Ponderosa Pine	1.36	2.53	24.69	71.41
Interior Douglas-fir	4.51	3.14	41.46	50.89
Grand Fir	9.52	1.60	54.97	33.92

Table 5. Conservation Status of a west-side forest zones 1 status designations are the same as in Table 4 (after Cassidy et al. 1997). Figures given are percentage of each area in each management status

West-side forest zone	Status 1	Status 2	Status 3	Status 4
Willamette Valley	0.00	2.22	2.79	94.99
Woodland/Prairie Mosaic	0.00	1.54	3.05	95.41
Puget Sound Douglas-Fir	0.03	1.10	7.36	91.51
Sitka Spruce	4.34	1.18	10.60	83.89
Olympic Douglas-fir	33.96	0.00	58.83	7.21
Western Hemlock	4.28	0.61	32.93	62.18

various levels of classifications and biological organization. Orians (1992) and Franklin (1992) have pointed out the importance of the scale of biological organization used in determining outcomes of any conservation assessment. Thus, a conservation assessment (e.g., A gap analysis Scott et al. 1993) of the management status of vegetation associations, the finest level of organization recognized by the Nature Conservancy's natural vegetation classification system, would undoubtedly result in identification of more types at risk than would an analysis at the alliance level in the same standardized vegetation classification system (see table 2 for definition of association and alliances). However, it will be some time before we map the vegetation of the Pacific Northwest at that level of detail. Future assessment of the management status of plant communities will benefit from use of a standard classification system that is now possible with the recent publication of a national vegetation classification system (Grossman et al. 1998).

All studies I reviewed identified ponderosa pine, riparian habitats, and oak woodlands as being in need of further conservation actions on their behalf. Additionally, ancient stands (e.g., old-growth) were considered in greater need of conservation than were earlier successional stages of the same type. Further, converging lines of evidence repeatedly identified the lower elevation areas and those with more productive soils as having the most change from historical conditions and the least area in bioreserves. Thus, I suggest that a prudent management strategy would be to minimize habitat loss and fragmentation in ecosystems identified in Tables 1-5, particularly those acreages occurring on highly productivity soils at low elevation sites.

If we are to avoid future losses or at risk forest ecosystems, we must be: 1) aware of which these forested ecosystems are most vulnerable to additional loss and fragmentation (different but very related processes); 2) willing to use best available information on their occurrence to work together to effect land management practices that are ecologically defensible, and 3) sensitive to committing Type II errors (A type II error is made when if is stated that an oberved treatment has no significant impact on a species or ecosystem when in fact it does.) when assessing impacts of management actions in at-risk ecosystems.

LITERATURE CITED

Agee, J. K. 1999. Fire effects on landscape fragmentation in interior west forests. In press. J. Rochelle, L. Lehman, and J. Wisniewski (editors). Forest fragmentation. Brill Academic Publishers. Leiden, Netherlands.

Adams, L. W., and A. D. Geis. 1983. Effects of roads on small mammals. Journal of Applied Ecology 20:403-415.

Aizen, M. A., and P. Feinsinger. 1994a. Forest and fragmentation, pollination, and plant reproduction in a Chaco dry forest, Argentina. Ecology 75:330-351.

Aizen, M. A., and P. Feinsinger 1994b. Habitat fragmentation, native insect pollinators, and feral honeybees in Argentina. Ecological Applications 4:378-392.

Andren, H. 1992. Corvid density and nest predation in relation to forest fragmentation: a landscape perspective. Ecology 73:794-804.

Bainbridge, C. 1993. Timber, Owls, and Communities: The history of a northwest controversy. Masters Thesis. University of Idaho, Moscow, ID.

Basile, J. V., and T. N. Lonner. 1970. Vehicle restrictions influence elk and hunter distribution in Montana. Journal of Forestry 77:155-159

Bolger, D. T., A. C. Alberts, R. M. Sauvajot, P. Potenza, C. McCalvin, D. Tran, S. Mazzoni, and M. E. Soule. 1997. Response of rodents to habitat fragmentation in coastal southern California. Ecological Applications 7:552-563.

Brody, A. J., and M. P. Pelton. 1989. Effects of roads on black bear movements in western North Carolina. Wildlife Society Bulletin 17:5-10.

Brittingham, M., and S. Temple. 1983. Have cowbirds caused forest songbirds to decline? Bioscience 33:31-35.

Bruntland, G. H. 1987. Our Common Future. Oxford University Press. New York, New York, NY.

Bunnell, F. L. 1999. What habitat is an island? In press. J. Rochelle, L. Lehmann, and J. Wisniewski (editors). Forest fragmentation: wildlife and management implications. Brill Academic Publishers. Leiden, The Netherlands.

Caicco, S., B. Butterfield, and B. Csuti. 1995. A gap analysis of management status of the vegetation of Idaho. Conservation Biology 9:498-511.

Cassidy, K. M., M. R. Smith, C. E. Grue, K. M. Dvornich, J. E Cassady, K. R. McAllister, R. E. Johnson. 1997. Gap Analysis of Washington State. Washington State Gap Analysis Project Final Report Volume 5. Cooperative Research Unit University of Washington, Seattle, WA.

Cover Story. 1992. "Earth summit at Rio, see the Earth Summit: No More Hot Air: It's time to talk sense about the Environment." Newsweek. June 1, 1992.

Daily, G. 1997. Natures Services. Island Press, Washington, D.C.

Davis, F. W., P. A. Stine, D. M. Stoms, M. I. Borchert, and A. D. D. Hollander. 1995. Gap analysis of the actual vegetation of California: The southwest Region. Madrono 42:40-78.

Davis, F. W., and D. M. Stoms. 1996. Sierran vegetation: A gap analysis. Pp. 671-689. Sierra Nevada Ecosystem Project: Final Report to Congress. Vol. 11. Assessments and scientific basis for management options. Wildland Resources Center Tech. Rep. No. 37. University of California, Davis, CA.

Donovan, T. M., P. W. Jones, E. M. Ann and F. R. Thompson III. 1997. Variation in local scale edge effects: mechanisms and landscape context. Ecology 78:2064-2070.

Edwards, T. E., C. A. Homer, S. D. Bassett, A. Falconer, R. D. Ramsey, and D. W. Wright. 1995. Utah Gap Analysis: An environmental information system. Final Project Report 951. Utah State University, Logan, UT.

Famighetti, R. (editor). 1997. The world almanac of book of facts. World almanac books K-111. Reference Corp. Mahwah, NJ.

Fahrig, L. 1999. Forest loss and fragmentation: which has the greater effect on persistence on forest dwelling animals? In press. J. Rochelle, L. Lehman, and J. Wisniewski (editors). Forest fragmentation: wildlife and management implications. Brill Academic Publishers. Leiden, The Netherlands.

Forest Ecosystem Management Assessment Team (FEMAT). 1993. Forest ecosystem management: an ecological, economic, and social assessment. U.S. Government Printing Office, Washington, D.C.

Franklin, J. F. 1992. Preserving biodiversity: Species, ecosystems, or landscapes? Ecological Applications 3(2):202-205.

Garman, S. L., F. J. Swanson, and T. A. Spies. 1999. Past, present, and future landscape patterns in the Douglas-fir region of The Pacific Northwest. In press. J. Rochelle, L. Lehman, and J. Wisniewski (editors). Forest fragmentation. Brill Academic Publishers. Leiden, Netherlands.

Gentry, A. W. 1986. Endemism in tropical versus temperate plant communities. Pp.

153-181 in M. E. Soule (editor). Conservation Biology: The Science of Scarcity and Diversity. Sinauer, Sunderland, MA.

Grossman, D. H., K. L. Goodin, and C. L. Ruess (editors). 1994. Rare plant communities of the conterminous United States: An initial survey. The Nature Conservancy, Arlington, VA.

Grossman, D. H., D. Faber, Langendoen, A. S. Weakley, McAnderson, P. Bourgeron, R. Crawford, K. Goodin, S. Landaal, K. Metzler, K. Patterson, M. Pyne, M. Reid, and L. Snedden. 1998. International classification of ecological communities. Terrestrial vegetation of the United States Volume 1. The Nature Conservancy, Arlington, VA.

Howarth, F. G., and W. P. Mull. 1992. Hawaiian insects and their kin. University of Hawaii Press, Honolulu, HI.

Hunter, M. L., and P. Yonzon. 1993. Altitudinal distributions of birds, mammals, people, forests, and parks in Nepal. Conservation Biology 7:420-423.

Leck, C. F. 1979. Avian extinctions in an isolated tropical wet-forest preserve, Ecuador. Auk 96:343-359.

Leopold, A. 1933. Game management. Charles Scribner Sons, New York, NY.

Mader, H. J. 1984. Animal habitat isolation by roads and agriculture fields. Biological Conservation 29:81-96.

McGarigal, K., and W. C. McComb. 1999. Forest fragmentation effects on breeding bird communities in the Oregon coast ranges. In press. J. Rochelle, L. Lehman, and J. Wisniewski (editors). Forest fragmentation. Brill Academic Publishers. Leiden, Netherlands.

Miller, K. R. 1984. The Bali Action Plan: A framework for the future of protected areas. Pp. 756-764 in J. A. McNeely and K. R. Miller (editors). National Parks, Conservation and Development. Smithsonian Institution Press, Washington, D.C.

Mills, L. S., and D. A. Tallmon. 1999. The role of genetics in understory forest fragmentation. In press. J. Rochelle, L. Lehman, and J. Wisniewski (editors). Forest fragmentation. Brill Academic Publishers. Leiden, Netherlands.

Noss, R.F., E.T. LaRoe III, and J.M. Scott. 1995. Endangered ecosystems of the United States: A preliminary assessment of loss and degradation. Biological Report 28, U.S. Department of Interior, National Biological Service, Washington, D.C.

Noss, R. F. 1990. Indicators for monitoring biodiversity: A hierarchical approach. Conservation Biology 4:355-364.

Noss, R.F., and A.Y Cooperrider 1994. Saving nature's legacy; protecting and restoring biodiversity. Island Press, Washington, D.C.

Oregon Biodiversity Project. 1998. Oregon's living landscapes. Defenders of Wildlife Washington, D.C.

Orians, G. H. 1992. Endangered at what level? Ecological Applications 3:206-208.

Oxley, D J., B. Fenton, and G. R. Carmody. 1974. The effects of roads in populations of small mammals. Journal of Applied Ecology 11:51-59.

Pinchot, G. 1947. Breaking New Ground. Harcourt Brace and Co, New York, NY.

Pressey, R. L. 1994. Ad hoc reservations: forward or backward steps in developing representative reserve systems? Conservation Biology 8: 662-668.

Rost, G. R., and J. A. Bailey. 1979. Distribution of mule deer and elk in relation to roads. Journal of Wildlife Management 43:634-641.

Scott, J. M., F. Davis, B. Csuti, R. Noss, B. Butterfield, C. Groves, H. Anderson, S. Caicco, F. D'Erchia, T. Edwards Jr., J. Ulliman, and R. G. Wright. 1993. Gap analysis: A geographic approach to protection of biological diversity. Wildlife Monographs 123:1-41.

Stoms, D., F.W. Davis, K.L. Driese, K.M. Cassidy, and M.P. Murray. 1998. Gap analysis of the vegetation of the intermountain semi-desert ecoregion. The Great Basin Naturalist. 58(3):199-216.

Terborgh, J. W. 1989. Where have all the birds gone? Princeton University Press, Princeton, NJ.

Wilcove, D. S., D. Rothstein, J. Dubow, A. Phillips, and E. Losos. 1998. Quantifying the threats to imperiled species in the United States. Bioscience 48(8):607-615.

Wilcove, D. S. 1987. From fragmentation to extinction. Natural Areas Journal 7:23-29.

Fire Effects on Landscape Fragmentation in Interior West Forests

James K. Agee

Interior West forests include a wide variety of forest types with unique characteristics. Disturbances, primarily by fire, tended to be both cyclic and stochastic, and in some cases equilibrium landscapes were the result. The strongest case for equilibrium landscapes is in the low-severity fire regimes, where fire was frequent. Fuels were fragmented and forests were not. The weakest case is in the high-severity fire regimes, because major disturbance was infrequent and might overlap long-term climatic changes. Patch sizes were smallest in low-severity fire regimes and highest in high-severity fire regimes, with edge maximized in moderate-severity fire regimes. Modern human management has homogenized fuels at high levels, converting low-severity fire regimes to high-severity, and changed the landscape patch size and edge character. Special forest types including old-growth and riparian forests in the interior West need to be evaluated differently than those in coastal maritime environments, because of higher past, present, and future probability of disturbance.

Key words: Forest fire, fire regimes, landscape ecology, forest fragmentation, western United States, Interior West forests

1. INTRODUCTION

Fragmentation has had a notorious context for Western forests since Larry Harris published "The Fragmented Forest" in 1984. This book applied island biogeography theory to Douglas-fir (*Pseudotsuga menziesii*) forests on the west side of the Cascade Range that were being fragmented by clearcutting practices. Old-growth forest was being compartmentalized into "islands" separated by an "ocean" of heavily manipulated young-growth forest. Fragmentation in westside forests is being discussed in this book by Garman et al., but I would like to revive in this paper a positive view of fragmentation as applied to historical forests of the interior West. I have adopted the definition of fragmentation as "the process of reducing size and connectivity of stands that compose a forest," and will argue that in the disturbance-prone forests of the interior West, that fragmentation was created and maintained by disturbance, often in subtle ways. This fragmentation resulted in both stable and unstable forests, and in both equilibrium and non-equilibrium systems. Fragmentation may be an appropriate goal in coarse-filter conservation strategies for interior West forest types. This chapter summarizes the landscape character of interior West forests, the changes that have occurred over the past century, the unique character of old growth and riparian zones in these forests, and challenges for the future.

College of Forest Resources, University of Washington, Seattle, Washington 98195

Table 1. Common forest types and major species of the interior West forests. M = major species; m = minor species. Forest series or zone refers to the species that is the potential vegetation dominant, and will always be labeled with an "M". Seral species of importance will also have an "M" label

Species	Ponderosa pine	Douglas-fir	Douglas-fir/hard-wood	White fir	Grand fir	Red fir	Subalpine fir	Whitebark fir
Ponderosa pine	M	M	M	M	M	-	-	-
Lodgepole pine	m	M	-	m	M	M	M	-
Knobcone pine	-	m	m	-	-	-	-	-
Sugar pine	-	m	m	m	-	-	-	-
Western white pine	-	-	-	m	m	M	-	-
Whitebark pine	-	-	-	-	-	-	-	M
Douglas-fir	-	M	M	M	M	-	m	-
White fir	-	-	-	-	M	-	m	-
Grand fir	-	-	-	-	M	m	-	-
Red fir	-	-	-	-	-	M	-	-
Subalpine fir	-	-	-	-	-	-	M	m
Incense-cedar	-	m	m	m	-	m	-	-
Western larch	-	m	-	-	M	-	-	-
Subalpine larch	-	-	-	-	-	-	m	-
Engelmann spruce	-	-	-	-	m	-	M	-

Latin names not in text: knobcone pine = *Pinus attentuata*; subalpine fir = *Abies lasiocarpa*; incense-cedar = *Calocedrus decurrens*; subalpine larch = *Larix lyallii*; Engelmann spruce = *Picea engelmannii*

2. THE CHARACTER OF INTERIOR WEST FORESTS

The most important character of interior West forests is their variability. An entire paper could be devoted to the character of interior West forests. It is no more possible to generalize about "interior West forests" as a whole than it is to generalize about the music tastes of the human cultures of the world. These forests are not exclusively coniferous, they occur across wide ranges in elevation and precipitation, and possess a wide range of adaptations and susceptibilities to disturbance.

Gradients of Forest Types

- Alpine
- Whitebark pine
- Subalpine fir
- Grand fir
- Douglas-fir
- Ponderosa pine

- Alpine
- Mountain hemlock
- Red fir
- White fir
- Douglas-fir
- Ponderosa pine
- Oak

Gradients of Fire Regimes

- Moderate Severity
- High Severity
- Weather Driven
- Moderate Severity
- Low Severity
- Fuel Driven

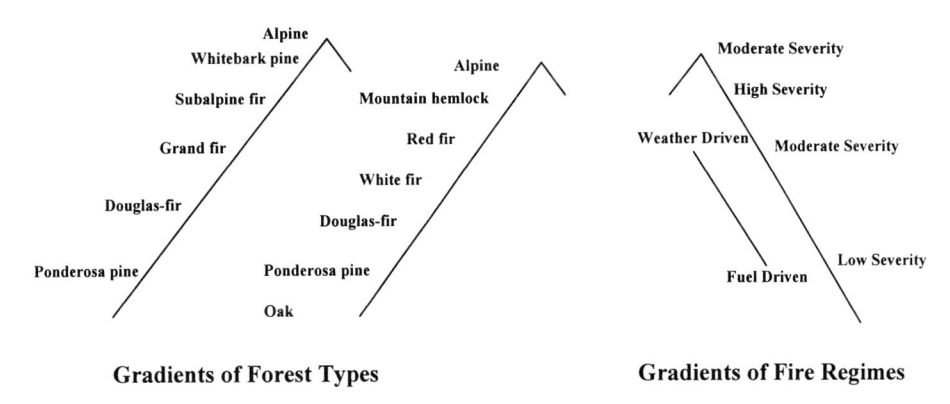

Figure 1. Common interior forest types with associated fire regimes and factors driving the fire regimes. The gradient of forest types shows major species responses to changing temperature and moisture conditions with changes in elevation. The historical fire regimes show a corresponding change. Refer to Table 1 and text for Latin names.

When forests are classified by zones (Franklin and Dyrness, 1973) or forest series or plant associations (Daubenmire, 1968) generalizations become more possible (Agee, 1993). Many of the major forest types of the interior West are widely distributed, have similar species within them, and similar patterns of disturbance (Table 1, Figure 1). They share in common some traits: most have a cool to cold winter (depending on elevation), and a warm, dry summer, and most have a significant history of natural disturbance, particularly from forest fires.

However, the juxtaposition of landscape types may be important for understanding one forest type in the context of adjacent types (Agee et al., 1990), and species near the edge of their range may have unique characteristics of composition, structure, and disturbance (Shinneman and Baker, 1997).

3. DISTURBANCE AS AN ECOLOGICAL FACTOR

Fire has been the most pervasive natural disturbance factor across Western forest landscapes (Barnes et al., 1998), but it did not work independently of other disturbances. Fire has had both fine and coarse scale effects on the forests of western North America (Agee, 1993), but these effects differ considerably by fire regime.

3.1 The Fire Regime

Natural disturbances range from benign to catastrophic, and can be generated from within or outside of the ecosystem (White, 1987). The disturbance effects in either case are due in part to current pattern or structure and to the nature of the disturbance. Disturbance is usually characterized by a combination of factors: type, frequency, variability, magnitude, extent, seasonality, and synergism with other disturbances (White and Pickett, 1985). Western forest fires have a wide range of historic frequencies from less than 10 to over 500 years that vary considerably by forest type. Predictability is associated with variability, and either very short or very long fire return intervals compared to the average interval can have major ecological effects. Non-sprouting species killed by one fire can be locally extirpated by a second closely-spaced fire; when fire intervals are unusually long, fire-sensitive species may pass through the critical fire-sensitive period of their life history. Magni-

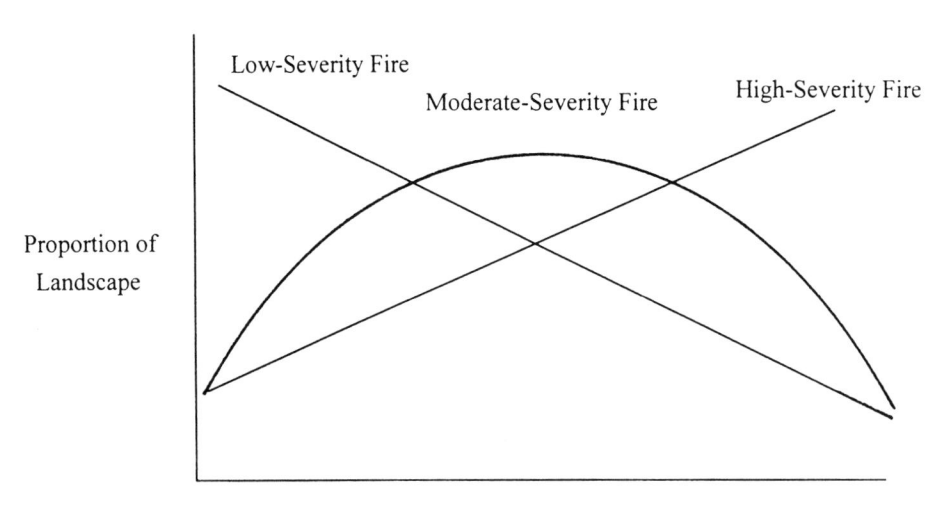

Figure 2. Proportions of severity by fire regime. The low-severity (nonlethal) fire regime has primarily low-severity fire, the high-severity (lethal) fire regime has primarily high-severity fire, and the moderate-severity (mixed) fire regime has complex mixes of both (adapted from Agee, 1993). The moderate-severity "arc" indicates that "thinning-like" severity with substantial mortality and survival is least common in the low- and high-severity fire regimes, and modal in the moderate-severity fire regime.

tude is often described as fireline intensity, a measure of energy output related to flame length, although other factors such as total fuel consumption or duration of smoldering can also be important. Extent describes the scale of the fire, but is generally poorly related to fire effects without knowledge of magnitude. Seasonality describes when fires occur in the year. In the American Southwest, spring months are inferred to be the most common fire season (Swetnam and Betancourt, 1990), while in the Pacific Northwest, mid- to late-summer appears to be the most common season (Wright, 1996). Synergism, or the interaction of fire with other disturbances, is poorly understood and generally unpredictable. Insects, disease, and wind may follow fire events with more than endemic background effects, and conversely, accelerated fire effects may follow other disturbances. Many secondary effects such as soil mass movement may follow intense fires (Swanson, 1981).

The fire regimes of western forests are usually described in terms of historical fires (defined here as pre-European settlement), and interpreted much the same way as potential vegetation (e.g., Daubenmire, 1968), namely what occurred historically and what the trajectories of change may be with or without management (Agee, 1993). The effect of Native Americans on fire ignitions is inextricably confounded with lightning ignitions, and the historical fire regime therefore includes native ignitions. Fire regimes based on fire severity (Agee, 1993) are defined by effects on dominant organisms, such as trees, and although broadly described in three classes, can be disaggregated to the forest type or plant association level if desired. The approach below is to use these broader classes as an organizing paradigm within which individual forest types are discussed. The high-severity fire regimes were those in which the effect of a fire was usually a stand replacement event (Figure 1). Fire return intervals were generally 100+ yrs. The low-severity fire regimes were those in which the typical fire was benign to dominant organisms across much of the area it burned, and fire return intervals were generally 5-25 yrs. The moderate-severity fire regimes had a complex mix of severity levels (Figure 1, Figure 2), with fire return intervals usually 25-75 yrs. The Interior Columbia Basin Ecosystem Management Project (ICBEMP) renamed these fire regimes lethal, nonlethal, and mixed, respectively (Quigley et al., 1996), and these terms can be interpreted synonymously in this paper.

The basic factors influencing fire behavior — fuels, weather, and topography — are the drivers of these fire regimes. While some suggest that weather is the dominating factor in all natural fire regimes (Bessie and Johnson, 1995), this argument makes best sense in the historic high-severity fire regimes, and those historic and low- and moderate-severity fire regimes altered by fire exclusion into high-severity fire regimes (Agee, 1997). Fuels were historically important as limiting factors to both fire spread and extent, and as driving factors for fragmentation, in the historic low- and moderate-severity fire regimes. As weather, or more precisely, climate became wetter and cooler, the fire regime is less fuel-dominated and driven more by extreme weather.

4. LANDSCAPE EFFECTS OF HISTORICAL FIRE REGIMES

4.1. Concepts in Ranges of Variability

The concept of ranges of variability in forest structural stages has been suggested as a framework for coarse filter conservation strategies. If it is possible to produce or mimic the historic ranges in stand size, composition, and connectivity by forest type on current and future landscapes, then much of the habitat for native flora and fauna might well be present. Fine-filter strategies, such as individual species plans or snag retention, might still be needed, but most species and ecosystem elements should be present (Haufler et al., 1996).

The *process* of fire created variability in forest *pattern*, and it clearly varied in frequency and intensity by forest type in the interior West (Agee, 1994). The spatial distribution of variously-aged stands, and those of varying species composition or structure, was a result of this process, but it is not always clear whether these distributions are temporally unique or stable over time. The two major types of assumptions necessary to produce estimates of the distribution of historic stand ages or structures are whether the processes are *cyclic* or *stochastic*, and whether *equilibrium* or *non-equilibrium* systems were, are, or can ever be expected.

Any forest stand has a probability that a disturbance will enter and kill the trees in the stand. The nature of the probabilistic process can either be *cyclic* or *stochastic*, independent of the actual probability of disturbance. Consider a coin with heads (H) and tails (T) as the only possible outcomes. If you turned it alternatively from side to side, it would create a pattern of HTHTHTHTHT etc. The probability, p, would be 0.5 for a head or a tail. This is a cyclic process, as it is very regular and predictable. If the coin were flipped in the air, the probability would remain 0.5 for a head or tail, but the pattern might be HHTTHTTHHT, so that subsequent tosses might produce runs of heads or tails. This is a stochastic process, even though the probability of head or tail is exactly the same as the cyclic process.

Now consider the distribution of forest stand ages as a cyclic process (Figure 3). The even-aged, area-regulated (equal area in each age class) managed forest is a prime example of a cyclic system (assuming that natural disturbances are eliminated). The probability of age class X moving to age class X+1 is close to one (p ~ 1.0), and the probability of being "recycled" to age 0 is close to zero (p ~ 0), until the rotation age is reached, at which time the stand is cut and it moves to age class 0 with p = 1. Forest stands operating under a cyclic process will have a rectangular age class distribution, truncated at the rotation age (R). The average stand age is R/2, and there are no stands older than R.

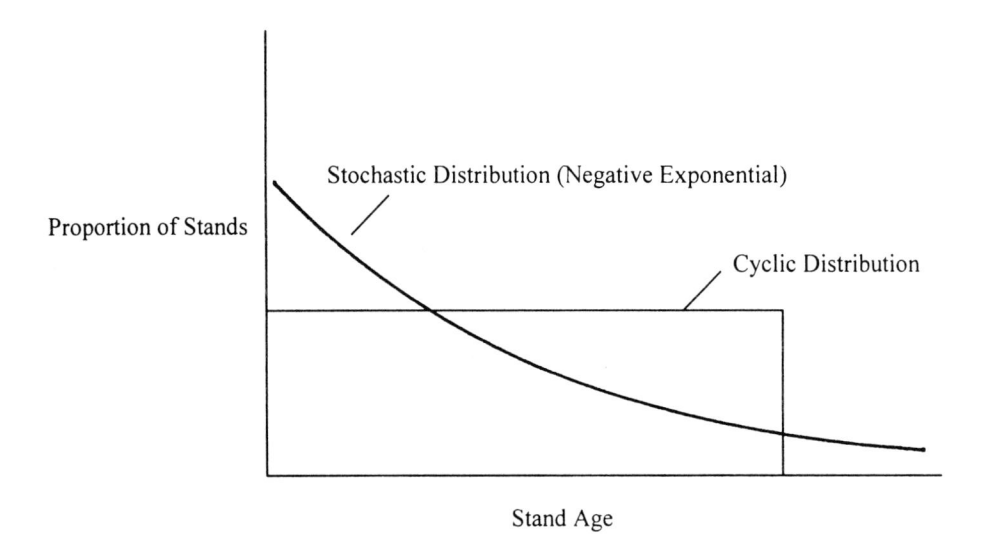

Figure 3. Proportions of the landscape in various age classes depend on the nature of disturbance. Cyclic processes result in rectangular-shaped distributions, while stochastic processes may be represented by (in this case) a negative exponential distribution.

Forests can also be "regulated" by stochastic processes. Boreal forests are the classic example of a natural forest regulated by fire acting in a stochastic manner (Johnson and Van Wagner, 1985). The probability of any stand being burned is low, described by p. The probability of a stand of age X moving to age X+1 is then 1-p. If this process is carried out for a long time, then a negative exponential distribution like that shown in Figure 1 will result. The frequency of any age class x is

$$f(x) = pe^{-px}$$

The average stand age is 1/p, or C, and this is commonly called the fire cycle. Roughly 1/3 of the stands are older than the average stand age, and there is no maximum age for a stand, in contrast to the distribution created by cyclic processes. The number of older stands is small, and they are likely to be affected by disease or insects if protected from fire, but this is disturbance at a much finer grain than the landscape effects of fire. Alternatives to the negative exponential distribution include the Weibull, a more complex form of the negative exponential and equal to it when flammability (the c parameter) is assumed to remain constant ($c = 1$). With the Weibull, flammability can be assumed to change with age, although the change must be assumed to be monotonic (linear) and is usually assumed to increase over time ($c > 1$).

4.2. Equilibrium or Non-equilibrium Landscapes

If a given cyclic or stochastic process remains constant over time, then an equilibrium landscape is eventually produced, and the shape of the age-class distribution will remain constant over time, although quite different between the cyclic and stochastic models (Figure 3). While no stand is in equilibrium, the distribution of stand ages that make up the landscape will be stable under the assumption of a constant set of disturbance probabilities. However, if the probabilities of disturbance change over time, then a non-equilibrium state is introduced which may "ripple" through the age class distribution as a "blip", or if the probabilities of disturbance are continually changing, there may be no predictable range of variability for any age class or the age class distribution as a whole.

Equilibrium is in part a function of scale: as scale decreases (in a cartographic sense), the total area becomes larger, more fires will occur, and the average fire size as a proportion of the total landscape declines. As this occurs, the assumptions necessary to apply a model such as the negative exponential become more reasonable, and equilibrium landscape conditions are therefore more plausible. What is an appropriate scale? The scale must be such that fire occurrence is regular rather than episodic, and does not burn "large" areas of the landscape when it occurs, and can require very large areas (Pickett and Thompson, 1978; Lertzman and Fall, 1998). The minimum is the area required to contain a balance of patches in every successional state over time. The Yellowstone landscape (~1 million ha) is a good example of a non-equilibrium landscape. It is a broad subalpine plateau that rarely burns, but much of it burned in the early 1700's (Romme, 1982). The forest age classes created by those early fires created a very large age class that moved as a "wave" through the age class distribution over time, so that much of the park's forest in the 1980's was old-growth, with lodgepole pine being replaced by subalpine fir on many sites. In 1988, the large fires that burned the national park killed much of the old-growth age class, as well as stands of younger ages, and a large proportion of the forest there is now (1999) in the 10-year old age class, which may move as a similar age class "wave" in the future or be broken up by other non-equilibrium fires. Obviously, for the past 300 years, and now for centuries to come, the Yellowstone forests have not and will not possess any equilibrium age class structure. If large subalpine areas of the Rockies are considered, fire may well be more an equilibrium-maintaining process at that scale, particularly if topography is broken (cliffs, rocky areas, etc.) so that individual fires cannot burn large proportions of the landscape in any year.

The application of a concept like historic range of variability (Morgan et al., 1994) implies that within fairly narrow limits, the forest landscape does possess equilibrium properties. If it did not, then we could substitute a range of 0-100% in each structural stage of each forest type and be satisfied that we have encompassed the possible ranges of all non-equilibrium landscapes. Ecological theory today is rife with non-equilibrium evidence (Baker, 1989; Lesica, 1996), from which one might infer that equilibrium landscapes do not exist and are probably not a useful target for management. I think this is partially the result of backlash against a century of ecological theory that postulated, with little evidence, that convergence of a wide variety of regional vegetation types would occur, that later successional communities were always more stable and diverse, and that natural disturbance was rare. All of these postulates have now been discarded, and the notions of chaos and complexity theory (Peterson and Parker, 1998) are new ecological paradigms, built into theory and practice, such as adaptive management (Walters, 1986). More than any other thing we have learned in the last two decades is that in plant ecology, there is no "one size fits all" solution when it comes to ecosystems, and that trumps the other issues. In the following section, I argue that there are, in some forest types and at appropriate scales, cyclic and stochastic processes that result in equilibrium landscapes.

4.3. Landscape metrics

There are myriad metrics that may be generated for landscapes, and the important ones may differ depending on the problem (McGarigal and Marks, 1994). Patch size, edge, shape, core area, nearest neighbor, and diversity metrics are among the most common. For historical disturbance regimes, few of these metrics are rarely available in quantifiable form, and because the pattern of scale is so variable, metrics are not easily compared across fire regimes. The "grain" size may be much less than 1 ha in historic ponderosa pine (*Pinus ponderosa*) forests (White, 1985), but thousands of hectares in subalpine or boreal forests (Bessie and Johnson, 1995). Two landscape metrics are compared below across the spectrum of Western forest fire regimes: patch size and patch edge (Table 2, Figure 4). Patch

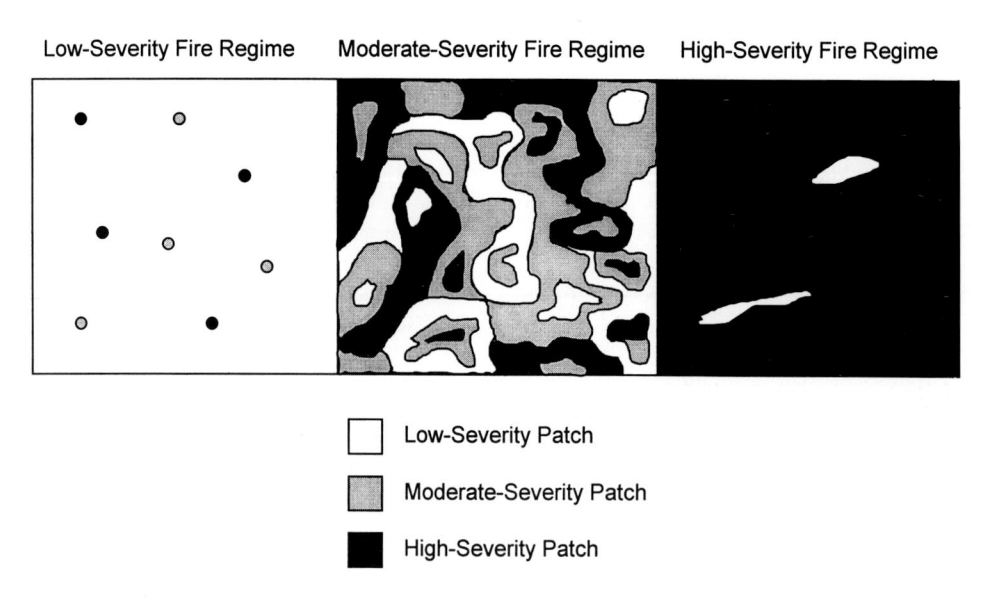

Figure 4. A schematic landscape pattern of fire regimes. Black dots in low-severity fire regimes are very old patches of large, old trees being killed by insects and decomposed by fire, and gray dots are emerging small-sized stands that have less-defined edge with older forest than the recently killed patches. The moderate-severity fire regime is typically a complex mosaic of larger patches of the three fire severity levels, while the high-severity fire regime has large, stand replacement patches. From Agee (1998).

size is quite different than fire extent. A low-severity fire may be quite large, but create conditions for a new age class of trees only in selected, small locales, resulting in a very small patch size from a large event. Similarly, edge is defined by the changes in species and structure, not by the perimeter of the fire event. Fragmentation may be measured by these parameters and also by parameters that limit future disturbance, such as fuel amounts and arrangements.

4.4. Low-Severity (Nonlethal) Fire Regimes

Low-severity fire regimes typically had large fires (Wright 1996) but small patch sizes. Fires burned frequently in these forests, and by regularly consuming fuels, killing small trees,

Table 2. Relative landscape characters of Western forest fire regimes.

Landscape character	Fire Regime		
	Low-severity	Moderate-severity	High-severity
Patch Size[1]	Small(~ 1 ha)	Medium(1-300+ ha)	Large (1-10000+ ha)
Edge	Low Amount	High Amount	Moderate Amount
Pre-Post Fire Similarity[2]	High	Moderate	Low

[1] The average patch within which tree regeneration will be open-grown.

[2] Of the total area burned, the proportion resembling the pre-fire forest structure.

and pruning the boles of residual trees, maintained a relatively fire-resistant landscape. Forests with significant components of ponderosa pine had very small patch sizes, ranging from 0.02-0.35 ha (Cooper, 1960; West, 1965; Bonnicksen and Stone, 1981; Morrow, 1985; White, 1985; and Agee, 1998). Patches were often created by group kill of pines by bark beetles (*Dendroctonus* spp.), and subsequent consumption of the debris by several fires. Because these dead patches were small and limited in extent (Figure 3), patch edge was also limited. Most of the forest was a fairly uniform mosaic of mature tree clusters and grassy understories. Coarse woody debris under such a regime would have been limited and clustered, with most of it in the dead clusters created by beetles and being consumed by fire. Over the landscape was a highly connected system of older forest, dotted with regenerating patches and clusters of coarse woody debris. This was a classic example of a cyclic equilibrium system (Figure 3), as defined by forest structure, and one in which fragmentation was more in subtle aspects of fuel structure (needle and grass biomass) than in tree size or structure. Varying fire return intervals around some frequent mean obviously introduced some variability, but fire tended to maintain more than disrupt equilibrium structure. There would have been very little visual evidence to separate 250 yr old stands from 350 yr old stands, and because fires were very large in extent compared to the patch sizes of trees (often >1000 times), the fuel fragmentation would have been at quite a different scale than the forest structure. Periodic fire minimized forest fragmentation by maintaining fuel fragmentation.

Before getting carried away by the idyllic vision of the equilibrium forest, it is important to note that exceptions did occur, even in ponderosa pine forest. Defoliation by pine butterfly (*Neophasia menapia*) created much larger patches, at least in some areas of south-central Washington (Weaver, 1961), and the Black Hills of South Dakota had pine stands that appear to have had less frequent but more severe fires, and much large patch sizes (Shinneman and Baker, 1997). These latter stands are transitional to boreal forest, containing some white spruce (*Picea glauca*).

Clearly, though, the wide range of studies in low-severity fire regimes support the notion of an equilibrium system, including at least some of the Black Hills (Shinneman and Baker, 1997).

The system link between fuels and fire is that of negative feedback: fire consumes fuels, and thus limits it own ability to spread, at least for a time. In eastern Washington, Wright (1996) reconstructed historic fire patterns over past centuries in dry Douglas-fir and dry grand fir plant association groups (PAGs). Where fires were closely spaced in time (<3 years), fire boundaries appear not to overlap (Figure 5A). The cured grass and pine needle fuels apparently require several years to recover to levels that will support fire spread. In these ecosystems, there is a strong argument that fires fragmented fuels such that fire spread became self-limiting. Stand structure was too open in itself to carry crown fire, and fires were limited not by stand fragmentation, but more subtly by the fuels structure. Fires occurred often enough that available fuel and fire intensity were limited, helping to maintain an unfragmented forest structure dominated by fire-tolerant tree species.

4.5. The Moderate-Severity (Mixed) Fire Regimes

Moderate-severity fire regimes had larger patch sizes and considerable edge (Figure 4). Fires here maintained both a naturally fragmented forest structure and fuel structure. Patch size in the moderate-severity fire regimes is typically larger than for the low-severity fire regimes. Patch size for moderate-severity forests (including some drier westside Cascade forests) range from 2.5 to 250 ha (Agee 1998). Patch edge is typically much higher for moderate-severity fire regimes than for high-severity fire regimes (although the methods for defining a patch will significantly influence any edge metric). The result, both from fire and other disturbances, was considerable patchiness on the landscape (Taylor and Halpern, 1991). Low-severity patches had surface fuels removed and only understory trees killed. Moderate-severity patches had some overstory removed (similar to the first entry on a shelterwood), and a favorable envi-

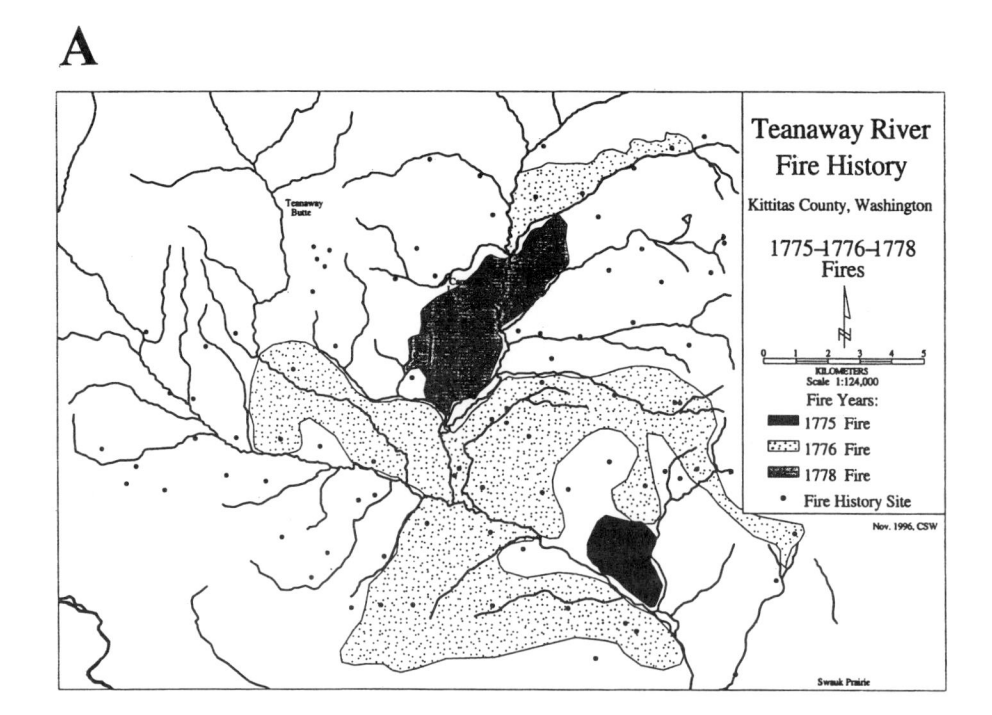

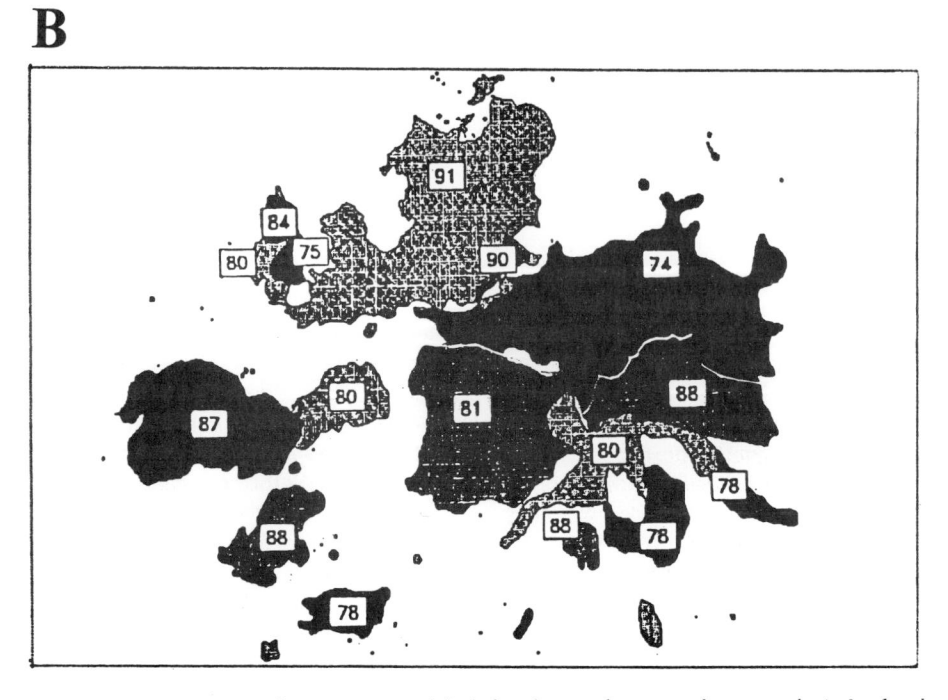

Figure 5. Fires can create fragmentation of fuel that limit subsequent fire spread. A. In the dry grand fir plant association groups (PAG) of the eastern Cascades of Washington, Wright (1996) found that historic fires stop at boundaries of very recent fires. B. In red fir forests, van Wagtendonk (1995) found a similar pattern in recent natural fires allowed to burn in Yosemite National Park.

ronment for regeneration of a new age class of trees (generally shade-tolerant species: white [*Abies concolor*] or grand fir [*Abies grandis*], Douglas-fir on dry sites, red fir [*Abies magnifica*], sugar pine [*Pinus lambertiana*]). High-severity patches within the moderate-severity fire regime had all the overstory killed and created an environment for shrubfields or new shade-intolerant tree species (western larch [*Larix occidentalis*], lodgepole pine [*Pinus contorta*]) (Antos and Habeck, 1981; Cobb, 1988). Landscape position in part explains differential severity: lower slope positions had the least amount of severe fire, while upper slopes, particularly of west or south aspect, and ridgetops experienced more severe fire (Taylor and Skinner, 1998).

Both the structure and species composition of the forest stands were defined by disturbance; the fragmentation of stands by fire altered fuel structure, too, and this limited subsequent fire spread, such as in red fir forests (van Wagtendonk, 1995; Figure 5B). As red fir stands age, they become more fire tolerant and less likely to result in a stand-replacement patch in the next fire (Chappell and Agee, 1996).

Coarse woody debris would likely have been less clustered than for the low-severity fire regimes. It would have experienced a net loss in the low-severity portions of a fire, because debris would have largely been consumed (logs are generally drier than in coastal forests) while little new debris of any size was created. In the moderate-severity portions of a fire, possibly a net gain would have occurred, as overstory trees would have been killed even though existing log biomass was consumed. In the high-severity portions of fires, a net gain would occur because almost all of the live tree component was converted to snags. The moderate patch size suggests that a coarse woody debris resource was almost constantly available across much of the landscape, although the spatial distribution of debris might vary over time.

4.6. The High-Severity (Lethal) Fire Regimes

Disturbance events in high-severity fire regimes often have large patch sizes. Although the large majority of fires historically remained quite small, the vast majority of area affected by fire is from the few large events that cover thousands of hectares (Romme, 1982; Romme and Despain, 1989; Bessie and Johnson, 1995). Small fires tend to have little edge, while larger events tend to be more patchy and leave more residual islands (unburned stringers) (Eberhart and Woodard, 1987). Generally, the edge created in the high-severity fire regimes is less than in the moderate-severity fire regimes (Agee, 1998).

The distribution of stand ages in the high-severity fire regimes is not clear. Even if an assumption about the nature of disturbance allows a fit of age classes to a distribution such as the Weibull or negative exponential, there are still assumptions about long-term stability (e.g., stable climate over centuries) that may alter the age class structure. Graphical solutions to these changes in assumptions, based on shapes of cumulative age class distributions (e.g., Johnson and Gutsell, 1994), are usually flawed (see Huggard and Arsenault, in review). Several characters of the high-severity fire regimes of interior West forests are unquestionable: (1) the fire return intervals were long (usually >100 yrs); (2) some stands are much older than the average fire return interval, suggesting that either the concept of "refugia" (Camp et al., 1997) or just random chance are operating; and (3) fires often impose a new landscape mosaic by burning stands of various ages (Bessie and Johnson, 1995).

Coarse woody debris dynamics in the high-severity fire regimes typically follow a "boom and bust" cycle. After a large fire event, coarse woody debris is at a high, as the live trees are all converted to snags and then logs (Harmon et al., 1986). As the new stand develops, coarse woody debris levels drop, and snags created by self-thinning of the new stand are too small to add much biomass. In mid-succession, thinning by disease and insects can create pulses of coarse woody debris that are slowly increased by additions of individual trees as succession proceeds. A new fire event will start the cycle over. If a fire occurs during early succession existing coarse woody debris is consumed and little new biomass is added to this component, creating a long-term minimum in coarse woody debris (Brown, 1975; Agee, 1998).

5. MODERN HUMAN IMPACT ON HISTORICAL PATTERNS

The influence of modern humans on interior West forests has been to radically alter the basic patterns of landscape created by natural disturbance. In the low elevation forests, a naturally fragmented forest has been homogenized by impacts of grazing, fire exclusion, and timber harvest, creating ecosystems with larger patch sizes and converting historically low-severity fire regimes to high-severity fire regimes. In the historical high-severity fire regimes, application of forest regulation schemes through timber harvest have attempted to move a stochastic system to a cyclic one.

Low-severity fire regimes have been greatly affected by grazing, fire exclusion, and timber harvest. Grazing removed the herbaceous fuel that was partly responsible for spread of historic fire through these forests. A reduction in fire frequency was the result of introduction of large numbers of livestock (Savage and Swetnam, 1990). The lack of fire, due to lack of grassy fuels and early fire suppression efforts, together with exposure of surface soil due to livestock trampling, encouraged tree regeneration and even further declines in herbaceous understory (Irwin et al., 1994). When timber harvest began, the large, fire-tolerant pines were generally removed, and other species and smaller trees were left. In general, a more homogeneous forest resulted, with uniform canopy, small tree diameter, and continuous fuels from the forest floor to the crown (Figure 6). Tree density increased, herbaceous and low shrub understory decreased, and multi-layered forests emerged (Everett et al., 1997). Increases in both the amount of surface fuel and "ladder fuels" of suppressed understory trees have converted these historic low-severity fire regimes to high-severity fire regimes. Photographs such as Figure 6, and age class analyses (McNeil and Zobel, 1980) suggest that by 1950 most forests in low-severity fire regimes had such high stem density that further tree regeneration declined. Natural forest openings have declined in size (Skinner, 1995).

The stress that these forests have endured since then, and the changes in species composition towards more shade-tolerant species, have encouraged bark beetle attacks on the pines, and defoliator attacks, such as spruce budworm (*Choristoneura occidentalis*), on Douglas-fir and true fir in these stands (Anderson et al., 1987; Gast et al., 1991; Hessburg et al., 1994; Covington and Moore, 1994; Quigley et al., 1996). Forest fire problems have steadily increased since the 1960's (Agee, 1993). Fires may not be larger in extent than historically, but are more severe in effect (Brown and Arno, 1990). Stand replacement burning is common, and the recovery of the stand to trees of any species may be very slow. The resulting large patch size (often dominated by fire-sensitive species like hardwoods or lodgepole pine) limits opportunities to use prescribed fire for fuel reduction and tree density control in the future.

The high-severity fire regime forests have been less affected by the selective harvest so widespread at lower elevation. Patch clearcuts have been more common in these types, which tend to be at higher elevation and are low productivity ecosystems. Ironically, the patches are usually smaller in size than the natural patch sizes found from historic forest fires (DeLong, 1998). Because of the dense nature of these higher elevation forests, less herb and shrub understory was present, except in early stages of succession. Grazing impacts have been concentrated in adjacent meadow environments, and many have not recovered from heavy grazing that peaked a century ago.

White pine blister rust, caused by *Cronartium ribicola*, has a deleterious effect on white pines in general, and specifically whitebark pine (*Pinus albicaulis*) at timberline in the interior West. The effect of blister rust at lower elevation on western white pine (*Pinus monticola*) and sugar pine is more subtle than at higher elevation, because these lower elevation pines grow with a mix of other species and their absence is less striking than the absence of whitebark pine at high elevation. Where other disturbances, such as fires, continue to occur, regeneration of whitebark pine in the face of this introduced disease is unlikely to occur. Although much more subtle than other human-induced disturbance, blister rust may have even more significant effects than roads, grazing, or logging for the persistence of a species, whitebark pine.

6. SPECIAL STANDS: OLD GROWTH AND RIPARIAN AREAS

Two issues in interior West forests seem to garner special attention beyond the general issues discussed above. One is the issue of old growth forest: how it is defined and how best to manage it. The other is the riparian zone, defined as critical both for its inherent aquatic habitat and as corridor habitat for terrestrial animals. These issues, from the perspective of fragmentation, need to be dealt with differently than for westside coastal forests.

6.1 Old-Growth Forest

Old-growth forest is defined as forest of antiquity that has species composition, structure, and function similar to what the forest type had in historical time. This definition is sometimes different than that of "late-successional forest," which is a forest that has species composition, structure, and function of late-seral successional states. In the low- to moderate-severity fire regimes of the interior West, old-growth forest had a significant to predominant proportion of early successional tree species that were large and had open understories with a healthy grass/low shrub component (Figure 6A). Late successional forest in these same types is a multi-layered forest with predominantly late successional species dominance (Figure 6B), and has been produced by fire-exclusion policies and land-use activities across much of the remaining unmanaged interior West forests in the low- to moderate-severity fire regimes. Much of it is at significant risk of stand-replacing fire, as the late-successional structure has moved the historical fire regime to a human-induced high-severity fire regime, and the associated landscape ecology implications (larger patch sizes, abrupt edge) of that shift. Entry into such areas by prescribed fire or low thinning (removing the smallest tree sizes) may be quite appropriate for restoration of old-growth forests in the interior West (Agee, 1997).

In the high-severity fire regimes of the interior West, the correspondence between old-growth forest and late-successional forest is much stronger, because unmanaged forests were disturbed much less frequently and tended to naturally develop a multi-layered structure of late-successional species in the old-growth condition. Modern indirect management practices (such as fire exclusion) have had much less impact on these forest types, and the need for restorative action is less.

6.2 Riparian Forests

With the increased emphasis on threatened fish species and watershed condition, riparian forests have been targeted for special protection. From a disturbance perspective, riparian areas differ from the associated uplands in several respects. Riparian areas may have a higher component of deciduous trees and shrubs than uplands, and are thought to have higher levels of foliar moisture through the dry season. Riparian areas generally have gentler slopes than uplands, and may have extensive terraces in some locations, so fire intensity often declines in riparian zones. Conversely, headwater areas may funnel winds and riparian areas may experience increases in fire intensity under those conditions. Riparian zones generally are of higher productivity than uplands, with tendencies towards higher tree density and cover and a multi-layered structure (more fuel). Riparian trees should be under less stress than upland trees because of additional access to moisture, and therefore be less susceptible to insect attack and to fire (higher fuel moisture). These tradeoffs require a site-specific evaluation.

Is it appropriate to consider riparian zones of interior West forests "stable" corridors that can connect old-growth patches? Is "restoration" of these areas by harvest entry needed? These are questions for which we have a poor information base to develop answers. However, we can document observations that suggest any strategies should again be stratified by forest type and disturbance

Figure 6. Repeat photography of a low-elevation forest in western Montana from 1908 to 1948 show an increase in tree density, a change in species dominance from ponderosa pine to Douglas-fir, a reduction of understory herbs and shrub biomass, and much higher potential for crown fire (from Gruell et al., 1982; photos courtesy USDA Forest Service Archives).

regime. In low-severity fire regimes, trees with multiple fire scars are often seen adjacent to the stream channel, suggesting historical disturbance may not have discriminated between the riparian and upland. In high-severity fire regimes, stand replacement burns often stopped at riparian zones. Fire "refugia" in these fire regimes are often found adjacent to perennial stream channels (Camp et al., 1997) but they were not always fire barriers. In fact, there are records of riparian zones acting as crown-fire corridors because of heavy fuel loads while the adjacent uplands would not burn at all (Agee, 1998), although in earlier intense fire events the riparian zones apparently did not burn. Given this scenario, old-growth riparian corridors could in some cases carry fire from one old-growth upland patch to another (e.g., Simberloff and Cox, 1987), a negative consequence of connectivity. Whatever the outcome, it is unlikely that there will be a "one size fits all" solution for riparian zone management: not every zone needs treatment and not every zone needs total protection.

7. CONCLUSIONS

Fragmentation is a difficult concept to apply to interior West forests using the typical attributes of island biogeography theory. That theory, as applied to moist, westside forests, suggested that large, undisturbed patches of old-growth forest connected by corridors is a successful coarse-filter conservation strategy (Harris 1984). For interior forests, that were and are highly mediated by disturbance, the coarse-filter was a mosaic of generally smaller patch sizes, with less abrupt edge than westside forests, except at high elevation where high-severity fire regimes existed. The forests with low- and moderate-severity fire regimes naturally had more open structure that in the "westside" sense would be called fragmented. The fragmentation was much more subtle than the contrast between old-growth and adjacent clearcuts. Negative feedback by frequent fire fragmented the fuel structure of these historical forests. Twentieth century management has tended to homogenize the interior forests, at least at low elevation. Pick and pluck harvest of old-growth trees, and ingrowth of shade-tolerant understory, has removed much of the clumped structure of the older forest. Exclusion of fire has increased insect and disease problems, and altered historical fire regimes almost exclusively toward high-severity fire regimes, with large patches and abrupt edge.

The challenge faced by interior West forest managers and interested publics is to first recognize that fragmentation is not always something to avoid. In fact, most proposed forest restoration strategies are specifically designed to fragment the continuous fuel across the interior West landscapes. The second part of that challenge is to begin to implement these new conservation strategies. Currently, for example, there is great debate about the value of fuel breaks, or defensible fuel profile zones, compared to more area-wide treatments such as prescribed fire (Agee et al., in review). How will they affect fire behavior? What will be the ecological effects of using fire compared to other treatments, such as thinning? If fire is the tool of choice, how much smoke will local communities tolerate? What is financially practical for society to consider at a sub-regional scale? None of these are easy questions to answer, and the answers are likely to be different in various areas of the West, not only because forest types differ but because the social and economic environment will also be unique.

Some of the questions can only be answered with large-scale manipulations. Ecological modeling can help with this effort (Peterson and Parker 1998) but the real test is the production of sustainable landscapes for the future.

Acknowledgments. The discussion of cyclic and stochastic processes was adapted from a fire training lesson plan developed by K. McKelvey. Thoughtful reviews were provided by S. Stephens and D.L. Peterson.

LITERATURE CITED

Agee, J.K. 1993. Fire Ecology of Pacific Northwest Forests. Island Press. Washington, D.C.

Agee, J.K. 1994. Fire and weather disturbances in terrestrial ecosystems of the eastern Cascades. Gen. Tech. Rep. PNW-GTR-320. Portland, OR: U.S. Department of Agriculture, Forest Service. Pacific Northwest Research Station. 52 pp.

Agee, J.K. 1997. The severe weather wildfire: too hot to handle? *Northwest Science* 71: 153-156.

Agee, J.K. 1998. The landscape ecology of western forest fire regimes. *Northwest Science* 72 (special issue): 24-34.

Agee, J.K., Finney, M., and deGouvenain, R. 1990. Forest fire history of Desolation Peak, Washington. *Canadian Journal of Forest Research* 20: 350-356.

Agee, J.K., Bahro, B., Finney, M.A., Omi, P.N., Sapsis, D.B., Skinner, C.N., van Wagtendonk, J.W., and Weatherspoon, C.P. (in press). The use of fuelbreaks in landscape fire management. *Forest Ecology and Management*.

Anderson, L., Carlson, C.E., and Wakimoto, R.H. 1987. Forest fire frequency and western spruce budworm outbreaks in western Montana. *Forest Ecology and Management* 22: 251-260.

Antos, J.A., and Habeck, J.R. 1981. Successional development in *Abies grandis* (Dougl.) Forbes forests in the Swan Valley, western Montana. *Northwest Science* 55: 26-39.

Baker, W.L. 1989. Effect of scale and spatial heterogeneity on fire-interval distributions. *Canadian Journal of Forest Research* 19: 700-706.

Barnes, B.V., Zak, D.R., Denton, S.R., and Spurr, S.H. 1998. Forest Ecology. Fourth Ed. John Wiley and Sons. New York.

Bessie, W.C. and Johnson, E.A. 1995. The relative importance of fuels and weather on fire behavior in subalpine forests. *Ecology* 76: 747-762.

Bonnicksen, T.M., and Stone, E.C. 1981. The giant sequoia-mixed conifer forest community characterized through pattern analysis as a mosaic of aggregations. *Forest Ecology and Management* 3: 307-328.

Brown, J.K. 1975. Fire cycles and community dynamics in lodgepole pine forests. pp. 429-456 In: Baumgartner, D.M. (ed) Management of lodgepole pine ecosystems: Symposium proceedings. Washington State Univ Extension Service, Pullman, WA.

Brown, J.K., and Arno, S.F. 1990. The paradox of wildland fire. *Western Wildlands* (Spring): 40-46.

Camp, A., Oliver, C., Hessburg, P., and Everett, R. 1997. Predicting late-successional fire refugia pre-dating European settlement in the Wenatchee Mountains. *Forest Ecology and Management* 95: 63-77.

Chappell, C.B., and Agee, J.K. 1996. Fire severity and tree seedling establishment in *Abies magnifica* forests, southern Cascades, Oregon. *Ecological Applications* 6: 628-640.

Cobb, D.F. 1988. Development of mixed western larch, lodgepole pine, Douglas-fir, and grand fir stands in eastern Washington. M.S. thesis. University of Washington. Seattle, WA.

Cooper, C.F. 1960. Changes in vegetation, structure, and growth of southwestern pine forests since white settlement. *Ecological Monographs* 30: 129-164.

Covington, W.W., and Moore, M.M. 1994. Southwestern ponderosa pine forest structure: changes since Euro-American settlement. *Journal of Forestry* 92: 39-47.

Daubenmire, R.F. 1968. Plant Communities: a Textbook of Plant Synecology. Harper and Row. New York.

DeLong, S.C. 1998. Natural disturbance rate and patch size distribution of forests in northern British Columbia: Implications for forest management. *Northwest Science* 72 (special issue): 35-48.

Eberhart, K.E., and Woodard, P.M. 1987. Distribution of residual vegetation associated with large fires in Alberta. *Canadian Journal of Forest Research* 17: 1207-1212.

Everett, R., Schellhaas, D., Spurbeck, D., Ohlson, P., Keenum, D., and Anderson, T. 1997. Structure of northern spotted owl nest stands and their historical conditions on the eastern slope of the Pacific Northwest Cascades, USA. *Forest Ecology and Management* 94: 1-14.

Franklin, J.F., and Dyrness, C.T. 1973. Natural vegetation of Oregon and Washington. USDA Forest Service Gen. Tech. Rep. PNW-8.

Gast, Jr., W.R., Scott, D.W., Schmitt, C. [and others]. 1991. Blue Mountains forest health report: New perspectives in forest health. Portland, OR: U.S. Department of Agriculture, Forest Service. Pacific Northwest Region, Malheur, Umatilla, and Wallowa-Whitman National Forests.

Gruell, G.E., Schmidt, W.C., Arno, S.F., and Reich, W.J. 1982. Seventy years of vegetative change in a managed ponderosa pine forest in western Montana: Implications for resource management. USDA Forest Service Gen. Tech. Rep. INT-130.

Harmon, M.E., Franklin, J.F., Swanson, F.J., Sollins, P., Gregory, S.V., Lattin, J.D., Anderson, N.H., Cline, S.P., Aumen, N.G., Sedell, J.R., Liemkaemper, G.W., Cromack, Jr., K., and Cummins, K.W. 1986. Ecology of coarse woody debris in temperate ecosystems. *Advances in Ecological Research* 15: 133-302.

Haufler, J.B., Mehl, C.A., and Roloff, G.J. 1996. Using a coarse-filter approach with species assessment for ecosystem management. *Wildlife Society Bulletin* 24,2: 200-208.

Harris. L. 1984. The Fragmented Forest. University of Chicago Press.

Hessburg, P.F., Mitchell, R.G., and Filip, G.M. 1994. Historical and current roles of insects and pathogens in eastern Oregon and Washington forested landscapes. USDA Forest Service Gen. Tech. Rep. PNW-GTR-327.

Huggard, D. and Arsenault, A. (in review). Mistakes in fire frequency analysis.

Irwin, L.L., Cook, J.G., Riggs, R.A., and Skovlin, J.M. 1994. Effects of long-term grazing by big game and livestock in the Blue Mountains forested ecosystems. USDA Forest Service Gen. Tech. Rep. PNW-GTR-325.

Johnson, E.A. 1992. Fire and Vegetation Dynamics: Studies from the North American Boreal Forest. Cambridge University Press. Cambridge.

Johnson, E.A., and Gutsell, S. 1994. Fire frequency models, methods, and interpretation. *Advances in Ecological Research* 25: 329-287.

Johnson, E.A., and Van Wagner, C.E. 1985. The theory and use of two fire history models. *Canadian Journal of Forest Research* 15: 214-220.

Lertzman, K., and Fall, J. 1998. From forest stands to landscapes: Spatial scales and the role of disturbances. Chapter 16 In: Peterson, D.L., and Parker, V.T. (eds) Ecological Scale. Columbia University Press. New York.

Lesica, P. 1996. Using fire history models to estimate proportions of old growth forest in northwest Montana, USA. *Biological Conservation* 77: 33-39.

McGarigal, K., and Marks, B. 1994. FRAGSTATS: spatial pattern analysis program for quantifying landscape character. Oregon State University, Corvallis, OR.

McNeil, R.C., and Zobel, D.B. 1980. Vegetation and fire history of a ponderosa pine-white fir forest in Crater Lake National Park. *Northwest Science* 54:30-46.

Morgan, P., Aplet, G.H., Haufler, J.B., Humphries, H.C., Moore, M.M., and Wilson, W.D. 1994. Historical range of variability: a useful tool for evaluating ecosystem change. *J. Sustainable Forestry* 2: 87-112.

Morrow, R.J. 1985. Age structure and spatial pattern of old-growth ponderosa pine in Pringle Falls Experimental Forest, central Oregon. Oregon State University, Corvallis. M.S. Thesis.

Peterson, D.L., and Parker, V.T. 1998. Ecological Scale: Theory and Applications. Columbia University Press. New York.

Pickett, S.T.A., and Thompson, J.N. 1978. Patch dynamics and the design of nature reserves. *Biological Conservation* 13: 27-37.

Quigley, T., Haynes, R.W., and Graham, R.T. 1996. Integrated scientific assessment for ecosystem management in the interior Columbia Basin. USDA Forest Service Gen. Tech. Rep. PNW-GTR-382.

Romme, W.H. 1982. Fire and landscape diversity in subalpine forests of Yellowstone National Park. *Ecological Monographs* 52: 199-221.

Romme, W.H., and Despain, D.G. 1989. Historical perspective on the Yellowstone fires of 1988. *Bioscience* 39: 695-699.

Savage, M. and Swetnam, T.W. 1990. Early nineteenth-century fire decline following sheep pasturing in a Navajo ponderosa pine forest. *Ecology* 71: 2374-78.

Shinneman, D.J., and Baker, W.L. 1997. Nonequilibrium dynamics between catastrophic disturbance and old-growth forests in ponderosa pine landscapes of the Black Hills. *Conservation Biology* 11: 1276-1289.

Simberloff, D., and Cox, J. 1987. Consequences and costs of conservation corridors. *Conservation Biology* 1: 63-71.

Skinner, C.N. 1995. Change in spatial characteristics of forest openings in the Klamath Mountains of northwestern California, USA. *Landscape Ecology* 10: 219-228.

Swanson, F.J., Jones, J.A., and Grant, G.E. 1997. The physical environment as a basis for managing ecosystems. Chap. 15 In: Kohm, K.A. and Franklin, J.F. (eds) Creating a Forestry for the 21st Century. Island Press. Washington, D.C.

Swetnam, T.W., and Betancourt, J.L. 1990. Fire-Southern Oscillation relations in the southwestern United States. *Science* 249: 1017-1020.

Taylor, A.R, and Skinner, C.N. 1998. Fire history and landscape dynamics in a Douglas-fir late-successional reserve, Klamath Mountains, California, USA. *Forest Ecology and Management 111*: 285-301.

Taylor, A.R., and Halpern, C.B. 1991. The structure and dynamics of *Abies magnifica* forests in the southern Cascade Range, USA. *Journal of Vegetation Science* 2: 189-200.

van Wagtendonk, J.W. 1995. Large fires in wilderness areas. pp. 113-116 In: Brown, J.K., Mutch, R.W., Spoon, C.W., and Wakimoto, R.H. (tech coord) Proceedings: symposium on fire in wilderness and park management. USDA Forest Service Gen. Tech. Rep. INT-GTR-320.

Walters, C. 1986. Adaptive management of renewable natural resources. Macmillan. New York.

Weaver, H. 1961. Ecological changes in the ponderosa pine forest of Cedar Valley in southern Washington. *Ecology* 42: 416-420.

West, N.E. 1969. Tree patterns in central Oregon ponderosa pine forests. *American Midland Naturalist* 81: 584-590.

White, A.S. 1985. Presettlement regeneration patterns in a southwestern ponderosa pine stand. *Ecology* 66: 589-594.

White, P.S. 1987. Natural disturbance, patch dynamics, and landscape pattern in natural areas. *Natural Areas Journal* 7: 14-22.

White, P.S., and Pickett, S.T.A. 1985. Natural disturbance and patch dynamics: An introduction. Chapter 1 In: Pickett, S.T.A., and White, P.S. (eds) The Ecology of Natural Disturbance and Patch Dynamics. Academic Press, New York.

Wright, C.S. 1996. Fire history of the Teanaway River drainage, Washington. Univ of Washington, Seattle. M.S. thesis.

Past, Present, and Future Landscape Patterns in the Douglas-fir Region of the Pacific Northwest

Steven L. Garman *, Frederick J. Swanson**, Thomas A. Spies**

Landscape patterns in the Douglas-fir region of the western Pacific Northwest (PNW) have been strongly influenced by both natural and anthropogenic disturbances. Historically, landscape patterns were driven primarily by wildfire, which varied in frequency and intensity along regional moisture gradients and in response to variable climatic trends. At the time of settlement, however, old-growth forests covered about half of the forest land base. Over the past 150 years, settlement activities and timber harvesting have resulted in extensive fragmentation of the pre-settlement old-growth forest matrix, producing more regular patterns with more predictable trajectories. Fifty years of dispersed clearcutting have created a checkerboard-like pattern of young and older forest patches throughout federally-managed lands. Use of short-rotation, large aggregated clearcuts has produced large patches of young seral forests on non-federal timber lands.

Future landscape pattern will be driven by recent changes in forest-management policies. The ecosystem-management based approach adopted for federally-managed lands uses a series of land allocations to achieve biodiversity and commodity production objectives. Forest-management guidelines for state and private industrial lands impose structural retention requirements for aquatic and upslope areas. Based on a qualitative assessment of future large-scale patterns under these plans, it becomes apparent that late-successional forests largely will be restricted to large blocks of federal reserves distributed throughout the region and along aquatic systems on all federally-managed lands. Also apparent is that large-scale connectivity (both physical and functional) of these reserves will be highly dependent on site-specific management prescriptions on intervening land allocations and ownerships. Understanding how to manage these intervening lands to enhance connectivity of late-successional forests within and among landscapes is an important management and research question.

Key words: forest fragmentation, forest management, ecosystem management, landscape pattern, natural disturbance, Pacific Northwest

---

* Dept. of Forest Science, Oregon State University, Corvallis, OR
** USDA Forest Service, Pacific Northwest Research Stn., Corvallis, OR

1. INTRODUCTION

Landscape patterns in the Douglas-fir region of the western Pacific Northwest (PNW) have been strongly influenced by both natural and anthropogenic disturbances. Prior to Euro-American settlement, landscape patterns were driven by wildfire and other natural disturbances, which in turn were influenced by a variable climatic regime. Forest patterns created by historical disturbances changed slowly with settlement in the mid 1800's, then more rapidly as timber harvesting became widespread. A significant consequence of land-use activities over the past 150 years has been the gradual fragmentation of the old-growth forest matrix. Today, only 50-60% of the pre-settlement old-growth Douglas-fir (*Pseudotsuga menziesii*) forests remain (Bolsinger and Waddell, 1993), and connectivity of these forests is greatly reduced by extensive areas of managed plantations.

Future landscape patterns will be guided by the recent Pacific Northwest Forest Management Plan (NWFP) (FEMAT, 1993) for federal ownerships, and state forest practices regulations (Oregon Department of Forestry, 1997; Washington State Department of Natural Resources, 1997) for state and private industrial timber lands. Although goals and objectives differ between federal and state forest policies, both impose guidelines for stand or landscape structure to better balance ecological diversity and commodity production. Under these plans, large-scale trends in landscape pattern will depend on a variety of factors. These include current conditions, the management strategies of adjacent ownerships, and the dispersion of ownerships. Viewing future conditions in the context of these factors is important to understand the potential benefits and limitations of current policies to reduce fragmentation of older forests throughout the region.

The purpose of this chapter is to provide a synopsis of historical and future landscape pattern dynamics in the western PNW region. Specifically, we review the influence of natural disturbance, primarily wildfire, and historical timber harvesting on forest-pattern development, and provide case studies of contemporary landscape conditions for western Oregon. We also provide a qualitative prognosis of future landscape conditions under current federal and state forest-management policies.

2. THE REGION AND FOREST COMMUNITIES

The Douglas-fir region of the PNW comprises 14.1 million hectares west of the crest of the Oregon and Washington Cascade mountains to the Pacific Ocean (Figure 1). About 38% of the total land base is administered by federal agencies (26% USDA Forest Service, 7% USDI Bureau of Land Management, >5% National Park and other agencies). Seventy-seven percent of this region is designated as forest land with 66% classified as commercial timber land. Of the designated timber lands, 48% is in public ownership (38% USDA Forest Service and USDI Bureau of Land Management, remainder state and county) and 32% is classified as private industrial forest lands.

Nine forest zones are recognized in this region (Franklin and Dyrness, 1988). Of these, the western hemlock zone is the most extensive and includes the majority of the Douglas-fir forests. At the time of Euro-American settlement, an estimated 60-70% of the forested land base was covered by old-growth Douglas-fir (Franklin and Spies, 1984; Booth, 1991). General characteristics of these forests include large (>100-cm dbh) overstory stems; multiple sub-canopy layers comprised of shade-tolerant species such as western hemlock (*Tsuga heterophylla*) and western redcedar (*Thuja plicata*); and large snags and large amounts of downed logs (Franklin and Spies, 1991). These and additional features provide unique habitat conditions for a range of plant and animal species relative to other seral stages and forest communities. Other dominant forest zones in the region include Sitka spruce (*Picea sitchensis*), which occurs along a narrow (<16 km) coastal zone, and Pacific silver fir (*Abies amabilis*) and mountain hemlock (*Tsuga mertensiana*) which occupy high elevation sites. Oregon white oak (*Quercus garryana*) savannahs occur in the valley fringes and bottoms, and were more extensive prior to urban and agricultural development and wildfire suppression. Red alder (*Alnus rubra*) and big-leaf maple (*Acer macrophyllum*) are the most common and wide-

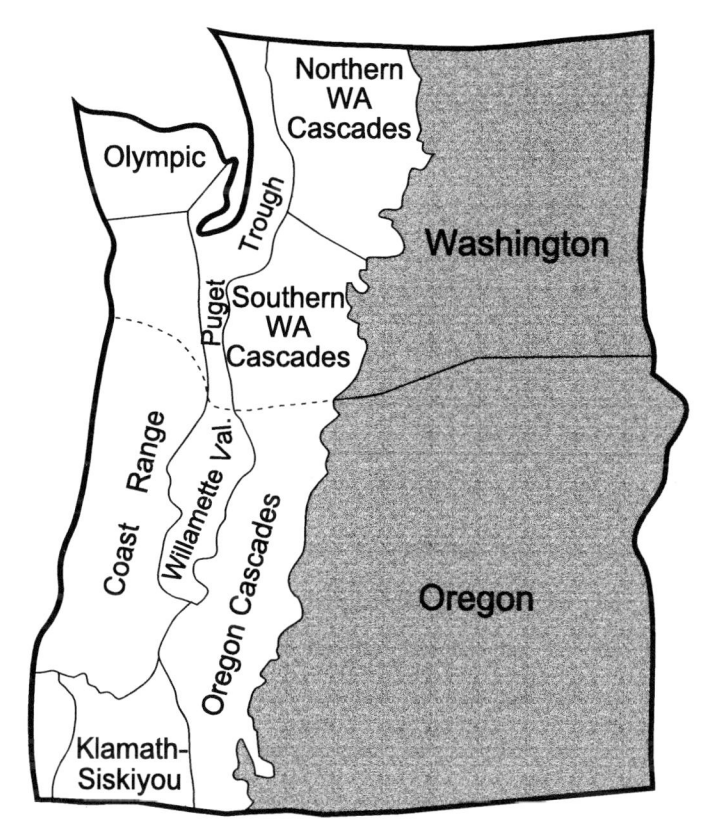

Figure 1. Provinces of the Douglas-fir region in the Pacific Northwest.

spread hardwood species, but thrive at lower elevations in riparian areas and on highly disturbed sites.

3. NATURAL LANDSCAPE PATTERNS

Wildfire has been the primary natural disturbance process in the region (Agee, 1990), although windthrow (Ruth and Yoder, 1953), insect- and disease-induced mortality (Rudinsky, 1962; Powers, 1995), and other disturbances were significant in specific subregions. Prior to Euro-American settlement, lightning was the primary ignition source (Agee, 1993), which increased with distance from the ocean and with increasing elevation (Morris, 1934). Native Americans had a profound, rather well-documented role in the burning of the Willamette Valley (Agee, 1993; Zyback, 1993), and along waterways in the Oregon Coast Range (Sauter and Johnson, 1974). Their use of upper elevation, mountainous terrain and of valley floor interiors is indicated by widespread artifacts (e.g., obsidian flakes and points); however, there is little evidence that they had a wide-spread influence on the fire regime of these areas (Agee, 1990).

Understanding pre-historic wildfire patterns and landscape conditions is problematic due to the spatial limitations of lake-sediment paleoecological studies and erasure of tree-ring records by more recent disturbances. However, fire-history reconstruction studies covering the last 500 yrs suggest broad geographic patterns in frequency, severity, and patch size decreasing severity of fire from the mesic..... (Agee, 1993; Heyerdahl et al., 1995). The general pattern from the 1450-1850 AD period appears to be one of increasing frequency and decreasing severity from the mesic northern Cascades to the drier southern portions of the range. Wetter conditions of the northern Cascades reduced the occurrence of fires, which resulted in accumulation of high fuel loadings. When fires did occur, they tended to be large stand-replacement events. Warmer and drier conditions in the southern portion of the range resulted in frequent, low severity underburns, which created and maintained fine-scale patterns (i.e., small, dispersed canopy openings) (Agee, 1998).

A latitudinal difference in fire regimes is also evident. Work by Impara (1997) and Weisberg (1998) in central western Oregon (latitude of Eugene, OR) indicate an interior Coast Range fire regime characterized by infrequent but high severity, large stand-replacement events. Fires in the moderate-severity regime of the west-central Oregon Cascades were more frequent but less severe. Fires in this regime are characterized as producing a complex mosaic of patches experiencing different severities due to the range of weather conditions during burning, and variable topographic and fuel conditions (Agee, 1998). Low severity but relatively frequent fires occurred along the east and west fringe of the Willamette Valley. Estimates of mean fire return intervals range from 230 years for the entire Coastal region (Fahnestock and Agee, 1983) to 237-242 years for the Oregon Coast Range (Ripple 1994). Teensma et al. (1991) estimated 150-300 years between stand-replacement fires in the Oregon Coast Range. Based on charcoal distribution within sediments of a Coastal Oregon lake, Long (1995) estimated a local fire-return interval of 175 years. Estimates of mean fire return intervals for the west-central Oregon Cascades range from 95 to 145 years (Morrison and Swanson, 1990; Teensma, 1987; Means, 1982; Weisberg, 1998); Weisberg (1998) estimated a return interval of 197 years for just stand-replacement events.

General temporal trends in landscape patterns over the past 500 years also are evident from fire-history reconstruction studies. The period 400-500 years before present was one of extensive, high severity fires which initiated many of today's old-growth stands. Studies in the west-central Oregon Cascades (Wallin et al., 1996) and the Oregon Coast Range (Impara, 1997) suggest that few but large fires occurred during this period. Following this period was several centuries (1600-mid 1800's) with low wildfire activity, which likely resulted from the relatively cooler climate of the little ice age. Comparisons of fire sizes (i.e., proportion of study area burned by a fire) reported in Rasmussen and Ripple (1998), Garza (1995), and Wallin et al. (1996) suggest that individual fire events were substantially smaller during this period compared to those of the previous period. Relatively fewer and smaller fires promoted more extensive development of forests. Examining spatial patterns of wildfire in two west-central Oregon Cascade landscapes, Wallin et al. (1996) estimated that closed canopy forests covered >80% of his study areas from 1600 until time of settlement. It is estimated that by 1840, >61% of the forests in the Oregon Coast Range were old-growth (>200 years old) and only 3.5% of the forests were <100 years old (Teensma et al., 1991; Ripple, 1994).

The mid and late 1800's were characterized by rather widespread fire (Wallin et al., 1996; Impara, 1997; Van Norman, 1998; Weisberg, 1998), reflecting both a climatic warming trend and ignition by Euro-American settlers. The impact of human-caused wildfires was fairly extensive. Fires linked to settlers burned over 34% of the Oregon Coast Range in the mid 1800's (Teensma et al., 1991). Morris (1934) estimated that in western Oregon seven times as much land was deforested by human-caused fires in the mid 1800's than by natural wildfire in the three previous decades. The past century has been one of relatively little wildfire due to increasingly effective fire suppression efforts (Agee, 1990).

4. FOREST MANAGEMENT AND LANDSCAPE PATTERNS

4.1 Historical Trends

Timber harvesting has been an important regional industry since the mid to late 1800's. By 1900, log production was estimated to be ca. 1.5 and 0.4 billion (i.e., 10^9) board feet in Washington and Oregon, respectively (Wall, 1972). A growing population and increasing demand for wood products resulted in an almost linear

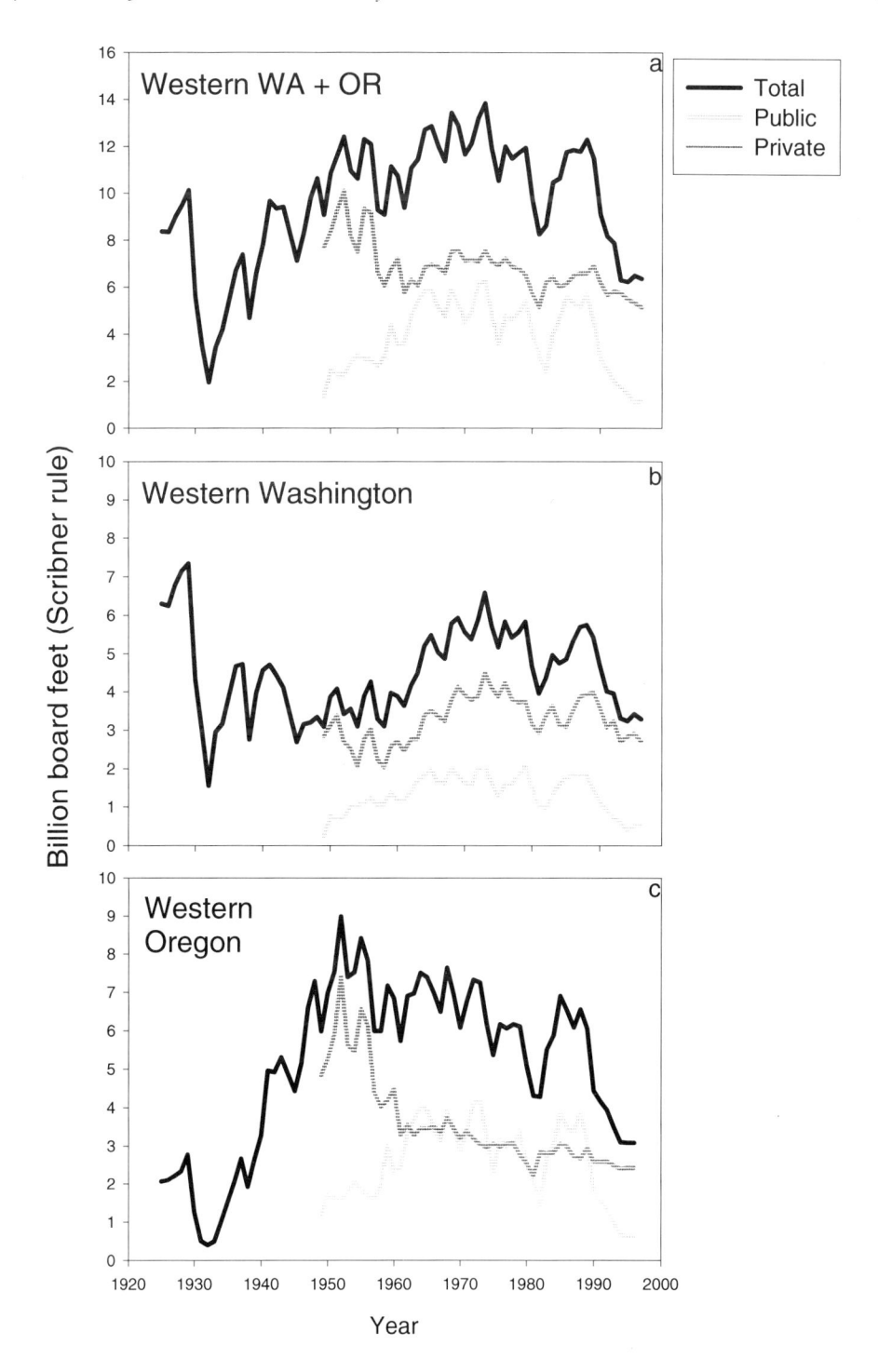

Figure 2. Historical trends in log production in western Washington and western Oregon by general ownership categories. Data from Wall (1972), Gedney et al. (1986a,b; 1987), MacLean et al. (1992), Bourhill (1994), and Larsen (1997). Ownership data only shown for 1948-96.

increase in regional log production until the 1930 depression (Figure 2a). Log production in Washington reached an all time high as early as 1929 (Figure 2b). By the mid 1930's, 32% of the forested lands in Washington were logged at least once with most of the cutting occurring in low elevation, old-growth Douglas-fir forests on private lands (Andrews and Cowlin, 1940). A general migration of harvesting operations into Oregon after 1940 in combination with post-war timber needs resulted in peak production in the state in 1952 (Figure 2c). Since 1952, log production has been slightly declining in western Oregon and the source of logs has shifted. As prime old-growth stock became limiting on private lands, public lands became a more important source of timber. From 1960 until recently, federally-managed public lands in Oregon provided about half of the state's timber each year (Figure 2c). Regional production has varied between about 11-14 billion board feet since the mid 1950's and peaked in 1972 (Figure 2a). Recent decline in production on public lands reflects legal constraints on timber harvesting in the early 1990's and changes in forest management practices. Production on private lands also has declined, but less so than on public land.

The influence of forest management on landscape pattern up to the mid 1930's is described by written accounts and forest maps produced by Andrews and Cowlin (1940) and Cowlin and Moravets (1940). Ease of logging and access to water-transportation networks concentrated the earliest timber operations in Puget Sound and coastal Washington. By 1933, the landscapes of these earlier timber operations were described as "vast expanses of cut-over land largely barren of conifer growth" (Andrews and Cowlin, 1940). Large tracts of old-growth by this time still remained along the coast of Washington and on the upper slopes of the Olympic Mountains and Washington Cascades. Early harvesting in the northern part of the Oregon Coast Range had removed much of the original forest. However, forest inventory maps indicated extensive tracts of old-growth on the slopes and foothills of the Oregon Cascade Range, and in the southwest region of the Oregon Coast Range. Of the estimated 5.7 million ha of Douglas-fir old-growth in the region prior to logging, only 2.8 million remained by 1933, with about 75% of the old-growth located in western Oregon. Deforested burns, old non-restocked and recent cut-overs occurred as large patches across the landscape and totaled more than 1.7 million hectares. Large burns and recent cut-overs created a notable landscape mosaic in the northern coastal region of Oregon and in the southern coastal region of Washington in the mid 1930's.

4.2 Recent Trends

Harvest strategies of federal and private industrial ownerships have notably differed over the past 50 years. Since 1940, private industrial lands have used clearcut harvesting with the removal of nearly all live and dead material and reforestation with primarily Douglas-fir, short rotations (40-60 yrs), and until recently, large aggregated clearcuts. Over the past five decades, federally-managed forests have employed a dispersed clearcut system with rotation intervals of 70-80 years. This system dispersed relatively small cutting units (5-25 ha) evenly across large areas of older forests to facilitate development of road networks, disperse the effects of clearcutting on watersheds, and provide edge and open habitat for game species (Franklin and Forman, 1987).

Recent trends in landscape patterns reflect two important consequences of these different harvest strategies. First, the higher rate of harvesting on private industrial lands promoted a more rapid reduction in amount and connectivity of older forests than on public lands. Using classified MSS (multi-spectral scanner) satellite imagery of a 259,00-ha section of the west-central Oregon Cascades, Spies et al. (1994) estimated a 45% decrease in closed canopy forests on private industrial lands compared to 13% on adjacent federally-managed timber lands between 1972-88. They also found private industrial lands experienced a more rapid decrease in closed-canopy interior habitat relative to public lands. However by 1988, even harvest patterns on federal lands had eliminated large tracts of interior forests outside of special land allocations (i.e., wilderness areas, experimental forest and natural areas, river corridors).

Second, although rates of fragmentation of older forests have been lower on federally-managed lands, the dispersed cutting scheme produced a template for accelerated fragmentation of the old-growth matrix. By the mid 1980's, 40 years of dispersing small patch cuts across the landscape was beginning to create a checkerboard pattern of old-growth and younger forests. The potential for continued dispersed cutting to accentuate fragmentation of late-successional forests was suggested by Franklin and Forman (1987). Simulation studies since have demonstrated the dispersed cutting system to substantially decrease extent and patch size of interior forests compared to an aggregated, long-rotation strategy (Li et al., 1993; Wallin et al., 1994), and compared to natural disturbance patterns (Wallin et al., 1994). Using subsamples of classified MSS satellite imagery, Spies et al. (1994) empirically demonstrated the effects of the dispersed scheme by comparisons of closed-canopy forest patterns between public and private ownerships. For comparable proportions of area cut, the dispersed cutting patterns on public lands resulted in ca. 10-30% less closed-canopy interior and twice as much edge habitat (closed - open-canopy forest interface) than the aggregated cutting scheme used on private industrial lands. At the landscape level, however, private lands had much less interior forest and more edge than public lands because of the historical high rate of cutting on private lands.

4.3 Contemporary Patterns

Assessments of landscape conditions using satellite imagery offer a comprehensive picture of contemporary forest patterns. Results of ongoing assessments for the Oregon Cascade and Coastal Provinces using 1988 Thematic Mapper (TM) satellite imagery are summarized here to illustrate important ownership and geographic differences.

4.3.1 Oregon Cascade Range
To illustrate landscape patterns of the Cascade Range, we used a 0.5-million ha section of the land-cover map produced by Cohen et al. (1995) (Figure 3). In this sample, differences in land-use practices among ownerships are clearly evident. Extensive timber harvesting on private industrial (PI) and land-clearing on private nonindustrial (PNI) ownerships have resulted in large, extensive patches of regeneration forest (semi-closed, closed mixed, and young conifer forests combined) (Figures 3,4). Mature and old conifer forest, combined, accounted for <23% of the land base of either ownership (Figure 4). Forest Service (USFS) lands had more mature and old conifer forest compared to PI lands, reflecting differences in harvesting histories (Figure 4). Bureau of Land Management (BLM) lands had a higher proportion of regeneration forest and lower proportions of mature and old conifer compared to USFS lands, but higher or similar proportions compared to the PI ownership (Figure 4).

Distributions of forest patch sizes further illustrate differences in landscape conditions among ownerships. About 89% of regeneration forest occurred in patches >1000 ha on PNI and PI ownerships (Figure 5a), and these large patches comprised >67% of the land base of an ownership (Figure 5b). In contrast, 60% of regeneration forest on USFS lands occurred in patches >1000 ha (Figure 5a), but only comprised 23% of this ownership (Figure 5b). The relatively small size of the BLM parcels restricted maximum patch sizes. Most of the regeneration forest in this ownership occurred in patches 10-1,000 ha in size (Figure 5a), although these patches encompassed about 55% of the BLM land base (Figure 5b). Connectivity of old conifer was highest on USFS lands. Large (>1000 ha) patches of old conifer comprised 30% of the USFS land base (Figure 5e,f). BLM lands had noticeably smaller patches of old conifer compared to USFS lands. Most (>77%) of the old conifer on private lands occurred in small (<100 ha) patches (Figure 5e), and comprised only about 10% of the land base of an ownership (Figure 5f). Mature conifer forest was limited across all ownerships and only occurred in patches >100 ha on USFS lands (Figure 5d).

The extent to which these patterns are representative of other portions of the Cascade Range varies with geographic location. Federal ownership dominates this province (e.g., Figure 6). Wilderness areas and National Parks, which generally occupy the higher elevations, contain more contiguous tracts of older forest and extensive amounts of nonforest (i.e., alpine

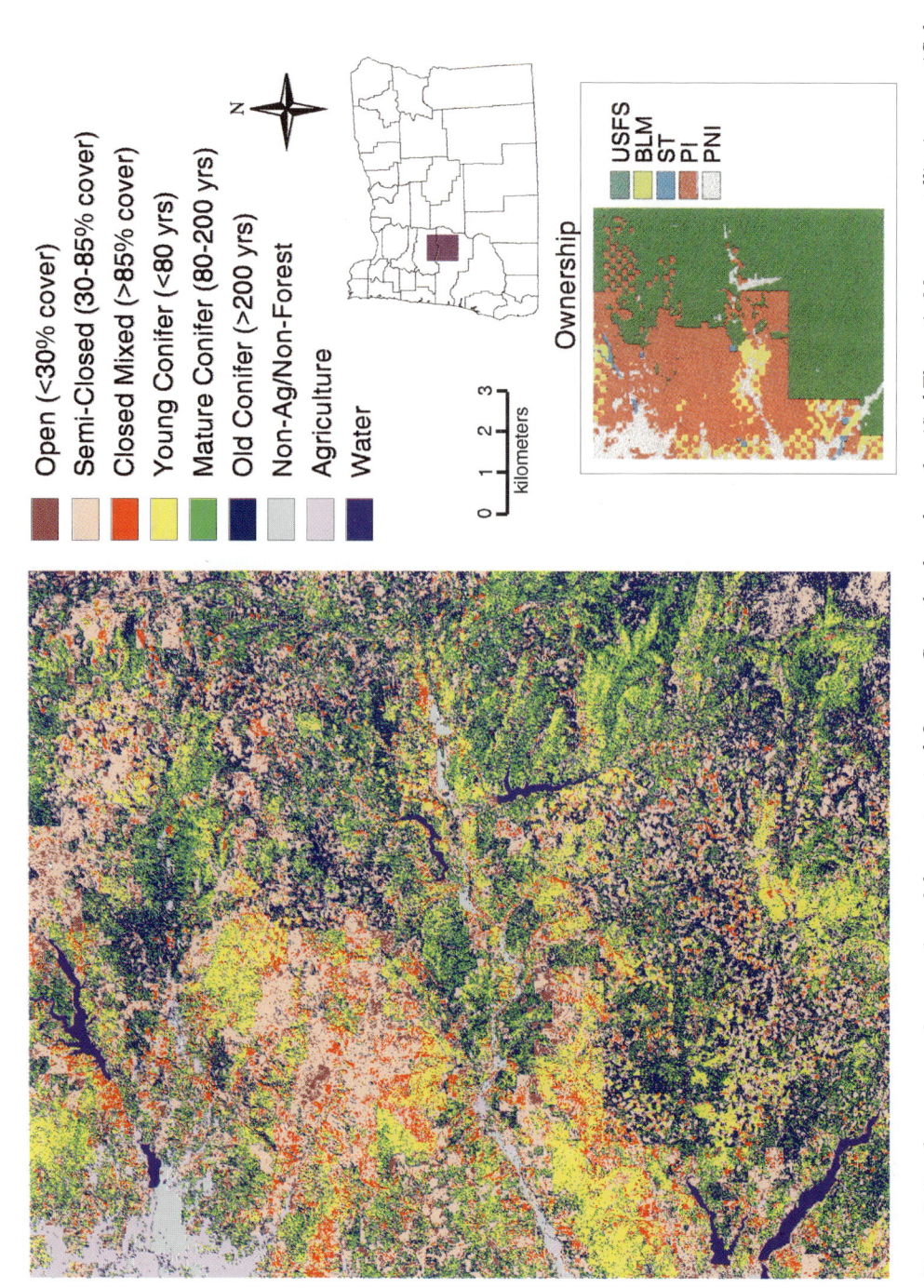

Figure 3. Land-cover map for a section of the west-central Oregon Cascades, based on classified Thematic Mapper satellite imagery (Cohen et al., 1995). Open, semi-closed, and closed-mixed types have <86% canopy cover of conifer species. PNI - private nonindustrial forests; PI - private industrial forests; STATE - all State lands; BLM - USDI Bureau of Land Management; USFS - USDA Forest Service.

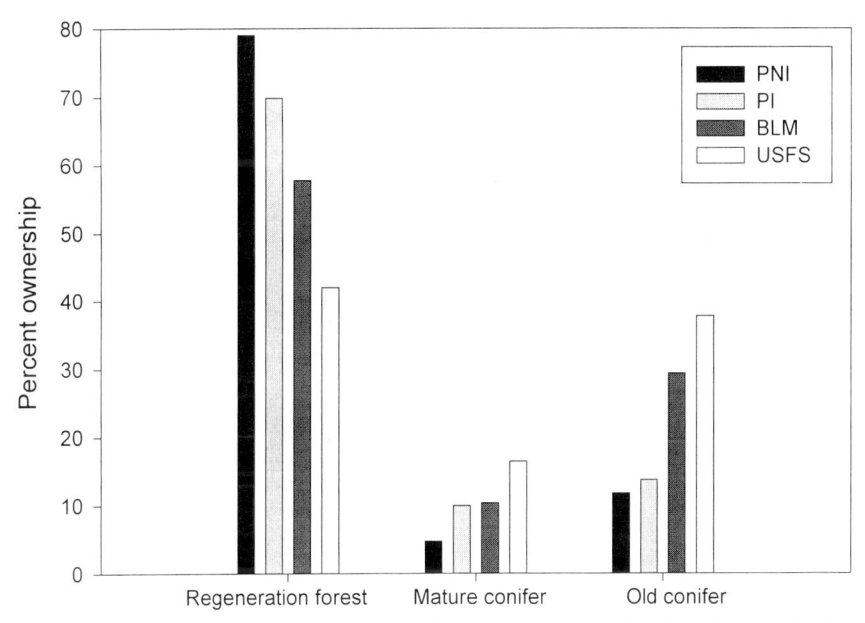

Figure 4. Distribution of forest types by ownership category for a 0.5 million-ha sample of the west-central Oregon Cascade landscape (see Figure 3). State lands not included due to small sample size. Regeneration forest - semi-closed, closed-mixed, and young conifer combined.

meadows, boulder fields, snow capped mountain peaks). Outside of these areas, however, managed federal and private industrial landscapes are qualitatively similar to our sample. Studies to further quantify landscape conditions across this province are in progress (Cohen, pers. comm.)

4.3.2 Oregon Coast Range

Assessment of the ecological effects of the ownership mosaic of the Oregon Coast Range is an ongoing effort of the Coastal Landscape Analysis and Modeling Study (CLAMS) (Spies et al., in prep.). Covering 5-million ha, the CLAMS study area extends from the Pacific Ocean to the Willamette Valley fringe (see gray area on locator map, Figure 7). The ownership mosaic of this region has substantially influenced overall landscape pattern and has significant implications for future patterns. Ownership is dominated by private lands (24% PNI, 38% PI), with federal lands comprising 25% (11% USFS, 14% BLM) and Oregon State lands comprising 12% of the land base. Simply based on the geometry of ownership parcels, Forest Service and State lands can support the largest contiguous patches of forests; the checkerboard pattern of BLM lands and intervening PI lands limits potential patch sizes on these ownerships (e.g., see Figure 6).

Similar to the Oregon Cascade example, landscape patterns of federal and private ownerships in the Oregon Coast Range can be quite distinctive. The sample of the Coast Range land-cover map in Figure 7 illustrates pattern differences among ownerships. In this sample, PI lands are dominated by younger forest (i.e., semi-closed, small/medium classes) and USFS lands by older forests (i.e., large, very large). Also apparent in this sample is the staggered clearcutting on federal lands (e.g., lower left, Figure 7).

Assessments of riparian ($\leq$ 100 m from water) and upslope (>100 m from water) forest patterns have illustrated differences in historical land-use strategies. Riparian zones across all ownerships were dominated by the early-successional class (open, semi-closed, broadleaf classes combined), with the large class (comprised of mostly conifer species) the second most dominant (Figure 8a). The high proportion of the early-successional class on PNI

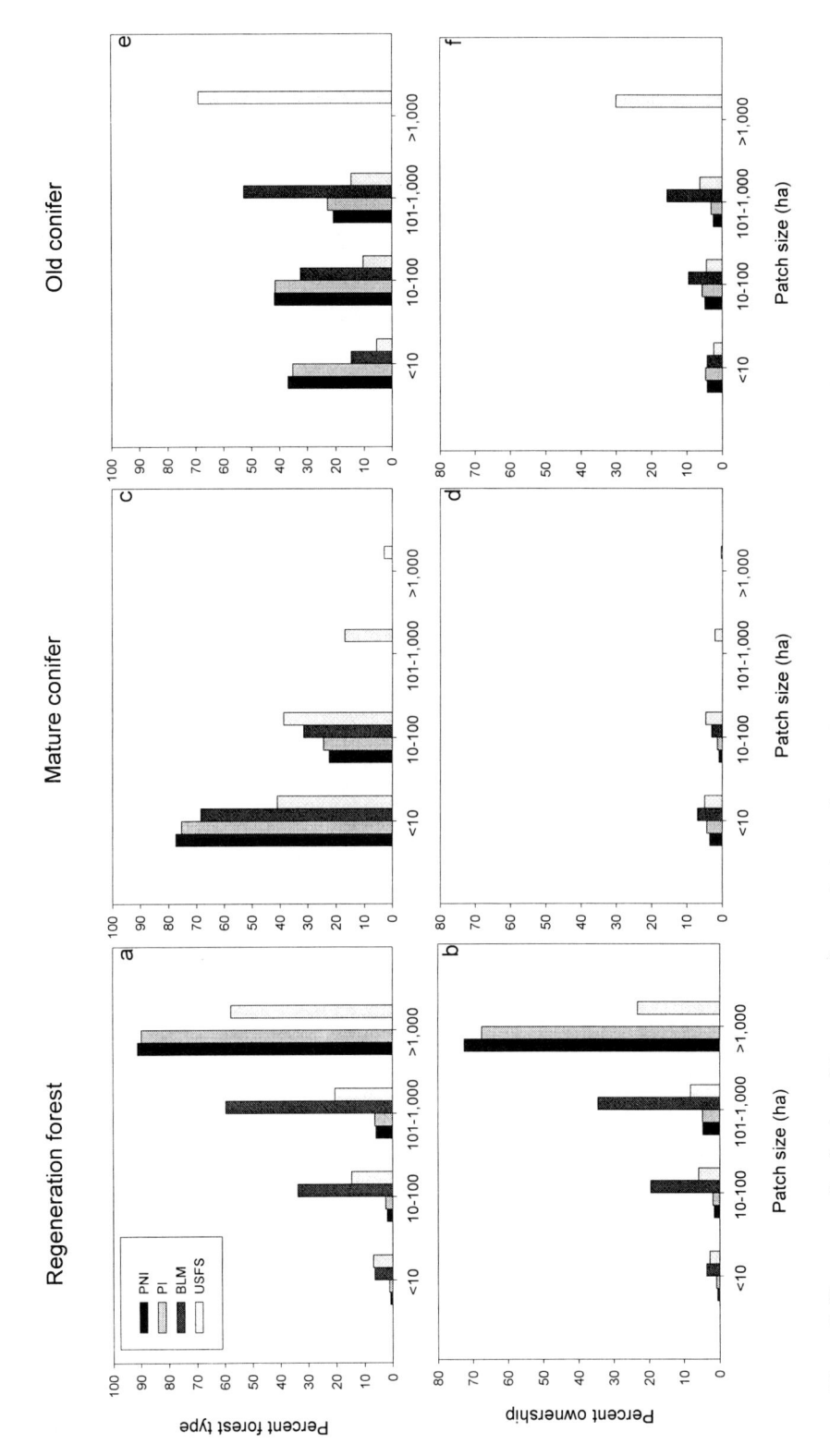

Figure 5. Frequency distribution of forest types by patch size and ownership category for the west-central Oregon Cascades landscape sample (see Figure 3). For percent forest type, percentages of a forest type sum to 100 for an ownership category. For percent ownership, percentages across all three forest types sum to 100 for an ownership. Regeneration forest - semi-closed, closed-mixed, and young conifer classes combined (classes defined in Figure 3).

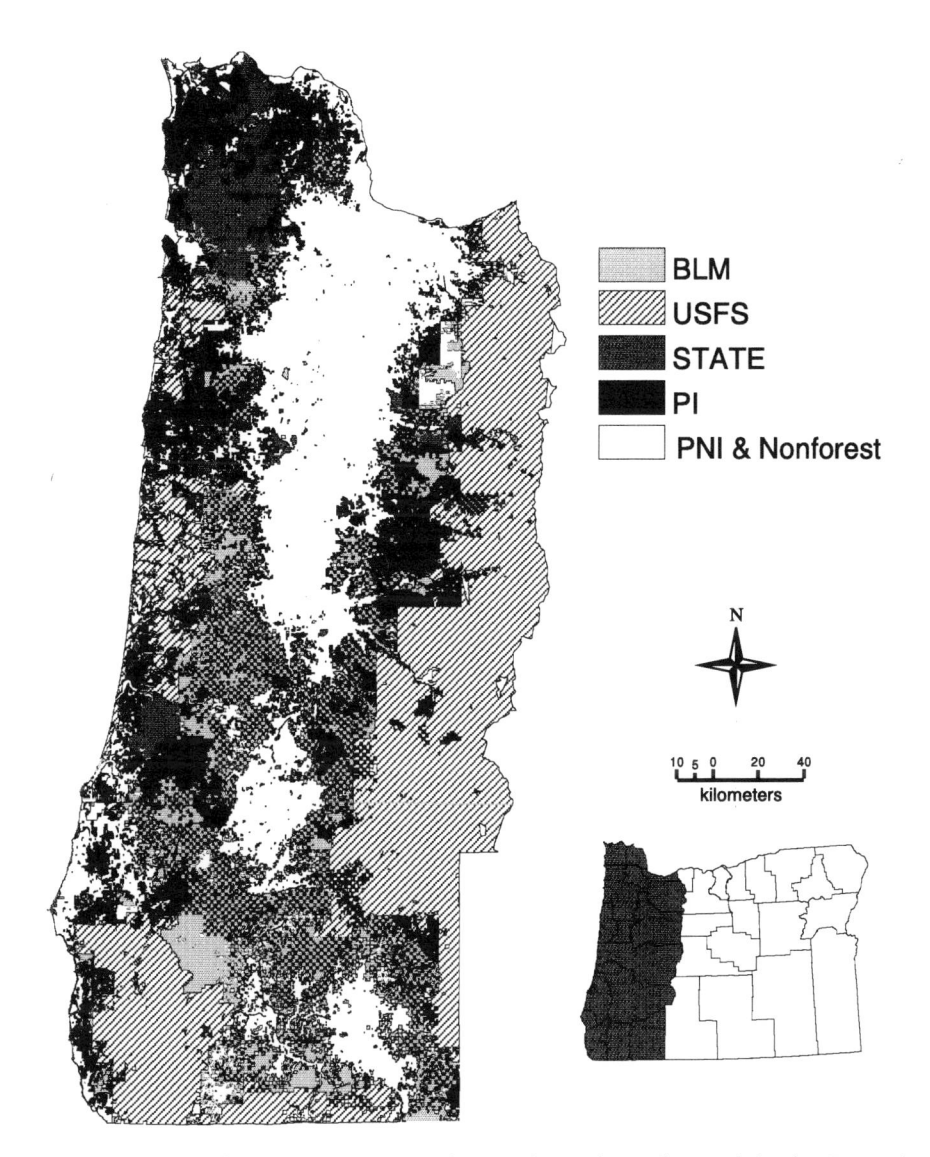

Landscape Pattern Dynamics in the Western Pacific Northwest 71

Figure 6. Ownership patterns in western Oregon. Ownership codes are defined in Figure 3.

lands reflected the extent of land-clearing and agriculture. Greater protection of aquatic systems on public lands, however, has resulted in lower proportions of this class than on PI lands. Riparian areas on federally-managed lands were comprised of almost equal proportions of the large and early-successional classes. Similar to riparian areas, upslope areas on private lands were dominated by the early-successional class (Figure 8b). However, large or medium classes were slightly more promi-nent than the early-successional class on federal and state lands. BLM lands had the highest proportion of old-growht in both riparian and uplsope areas, followed by USFS lands. Even adjusting for differences in land and area, BLM lands supported the highest amount of old-growth of all ownerships in 1988.

Pattern differences among ownerships were also evident. For the PNI ownership, land clearing, agriculture, and natural regrowth of disturbed sites have resulted in the predomi-

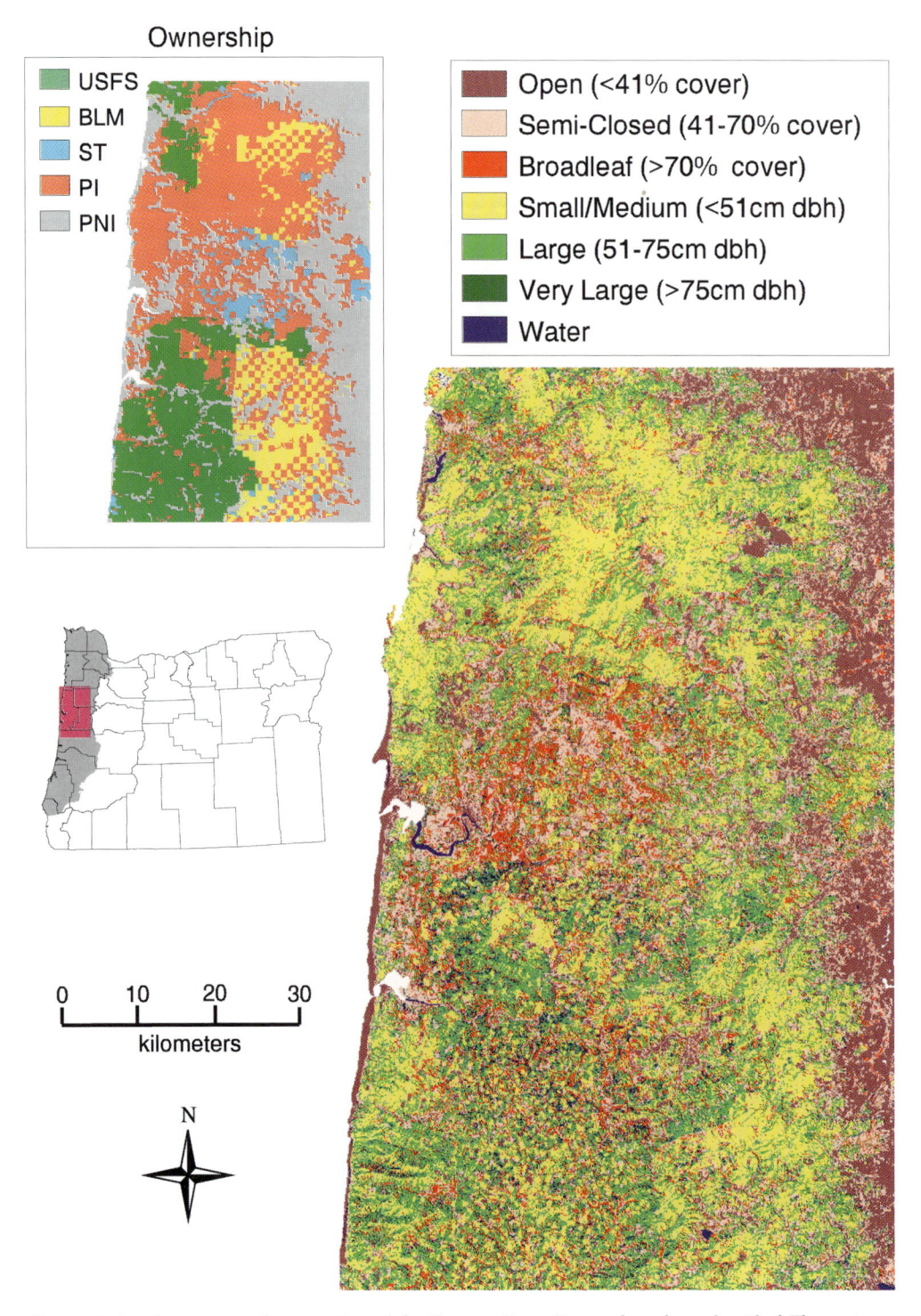

Figure 7. Land-cover map for a section of the Oregon Coast Range, based on classified Thematic Mapper satellite imagery (Spies et al., in prep.). Forest types are defined in Figure 8. Ownership codes are defined in Figure 3 (ST - STATE).

Figure 8. Distribution of vegetation classes by ownership category for the Oregon Coastal Landscape Analysis and Modeling Study (CLAMS) study area. Cover classes derived from Thematic Mapper satellite imagery. Small, medium, large, and very large classes are based on average of overstory trees using the diameter model of Cohen et al. (1995). Old Growth based on a canopy mosaic algorithm (Spies et al., in prep.). Ownership codes are defined in Figure 3.

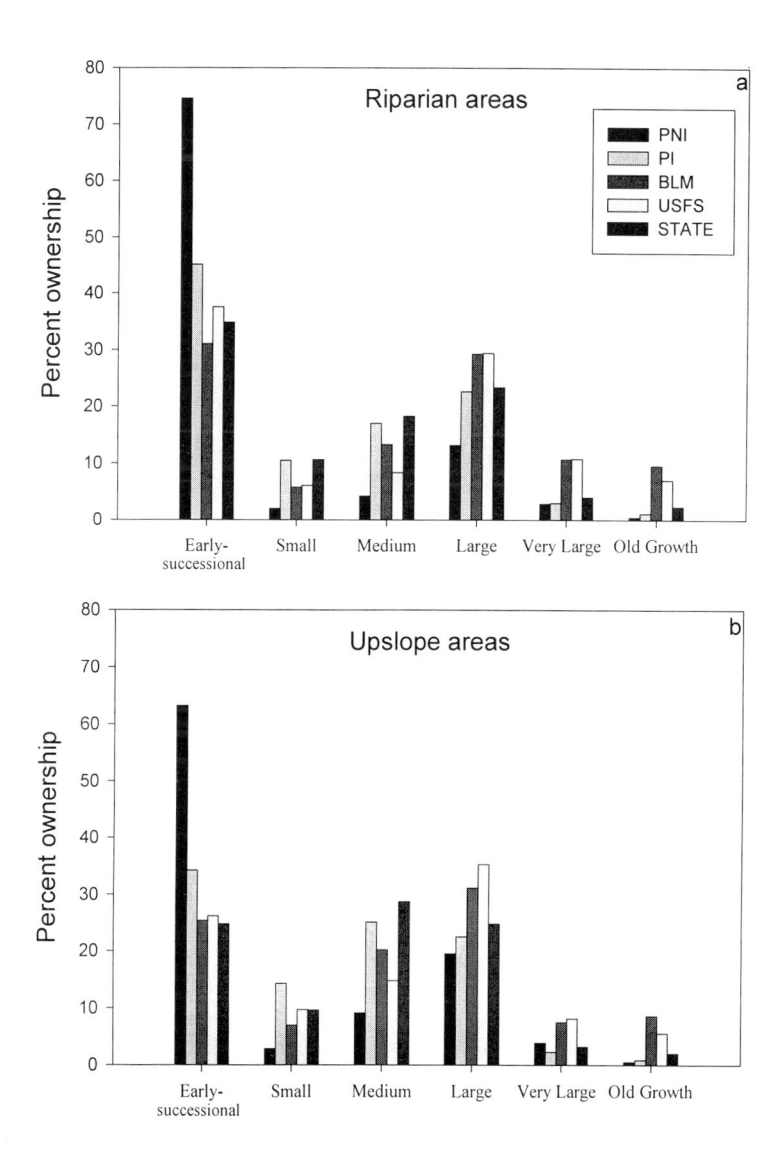

nance of large patches (>1000 ha) of early-successional forest (Figure 9a) over almost half of the land base (Figure 9b). This contrasts with PI lands where only 10% of the land base was comprised of early-successional patches of this size. On other ownerships, early-successional forest was distributed among smaller patches and only comprised <25% of the land base of an ownership. Patches sizes of small/medium class (mostly young plantation forests) reflected the aggregated clearcutting scheme of non-federal lands. Sixty to eighty percent of patches of this class on PI and STATE lands were large (>1000 ha) (Figure 9c) and occupied 30-40% of an ownership (Figure 9d). On USFS lands, only 45% of this class occurred in large patches and these patches only comprised 18% of this ownership. For BLM and PI ownerships, the small/medium class was generally distributed among smaller patch sizes.

Patch-size distributions of late-successional forest (large, very large classes combined) illustrate how extensive the forest matrix has been fragmented within individual ownerships. A large proportion of this forest type occurred in patches <100 ha across all ownerships (Figure 9e). Federal lands had a higher proportion of patches >100 ha and a higher proportion of their land base comprised by these patch sizes than other ownerships combined (Figure 9f). Large patches (>1000 ha) were most prevalent on public lands, but they only represented <5% of the total area of an ownership (Figure 9f). This contrasts with the Oregon Cascade example presented above where large patches of old conifer covered about 30% of the USFS ownership (Figure 5f). Private lands in the Coast Range noticeably lacked patches of late-successional forests >100 ha in size (Figure 9f).

5. FUTURE LANDSCAPE PATTERNS

5.1 Forest Management Regulations

Future landscape patterns will be largely influenced by current federal and state forest-management guidelines. Recent concerns for sustaining biological diversity and late-successional ecosystems have led to significant changes in forest policy for federal lands. The Northwest Forest Plan (NWFP) (FEMAT, 1993), which applies to all federal lands within the range of the northern spotted owl (*Strix occidentalis*), designates specific land allocations and management guidelines (Table 1) with the primary intent of creating an interconnected system of forest reserves capable of maintaining viable populations of old-growth associated species. Late-successional reserves (LSRs) constitute the backbone of the regional conservation strategy (Figure 10). LSRs were designated to provide a wide distribution of reserves, and currently encompass a range of forest-age classes. Management intervention is permitted only to accelerate structural development of younger stands; otherwise, LSRs are to remain untouched in perpetuity. Ten Adaptive Management Areas (AMAs) (2 in northern CA., 4 each in WA and in OR) were set aside for demonstration, implementation, and evaluation of monitoring programs and innovative management practices that integrate ecological and economic values. Matrix lands are the primary source of timber, and have variable requirements for retention of live and dead material during harvest to enhance the ecological diversity of managed forests. Transcending all allocations is the Aquatic Conservation Strategy. In addition to general watershed restoration requirements, this strategy requires development and protection of riparian reserves (45-90 m on each side of a stream, pond, or wetland) to protect aquatic ecosystems and to provide an important linkage between riparian and upslope habitats.

Commercial timber harvesting on non-federal lands is regulated by WA and OR State Forest Practice Acts (Oregon Department of Forestry, 1997; Washington State Department of Natural Resources 1997). Collectively, these acts limit the size of regeneration harvests (e.g., <48 ha in Oregon), set minimum re-stocking guidelines, specify retention levels for green-tree (e.g., 5/ha >25-30cm diameter at breast height) and coarse woody debris (e.g., 5-7/ha for snags and for logs), and limit harvest activities in riparian zones. Recent enhancements over the past decade include increased structural, compositional, and width requirements for riparian buffers (Lorensen et al., 1994). In Oregon, riparian reserves are 6 m on each side of most streams. Riparian management zones

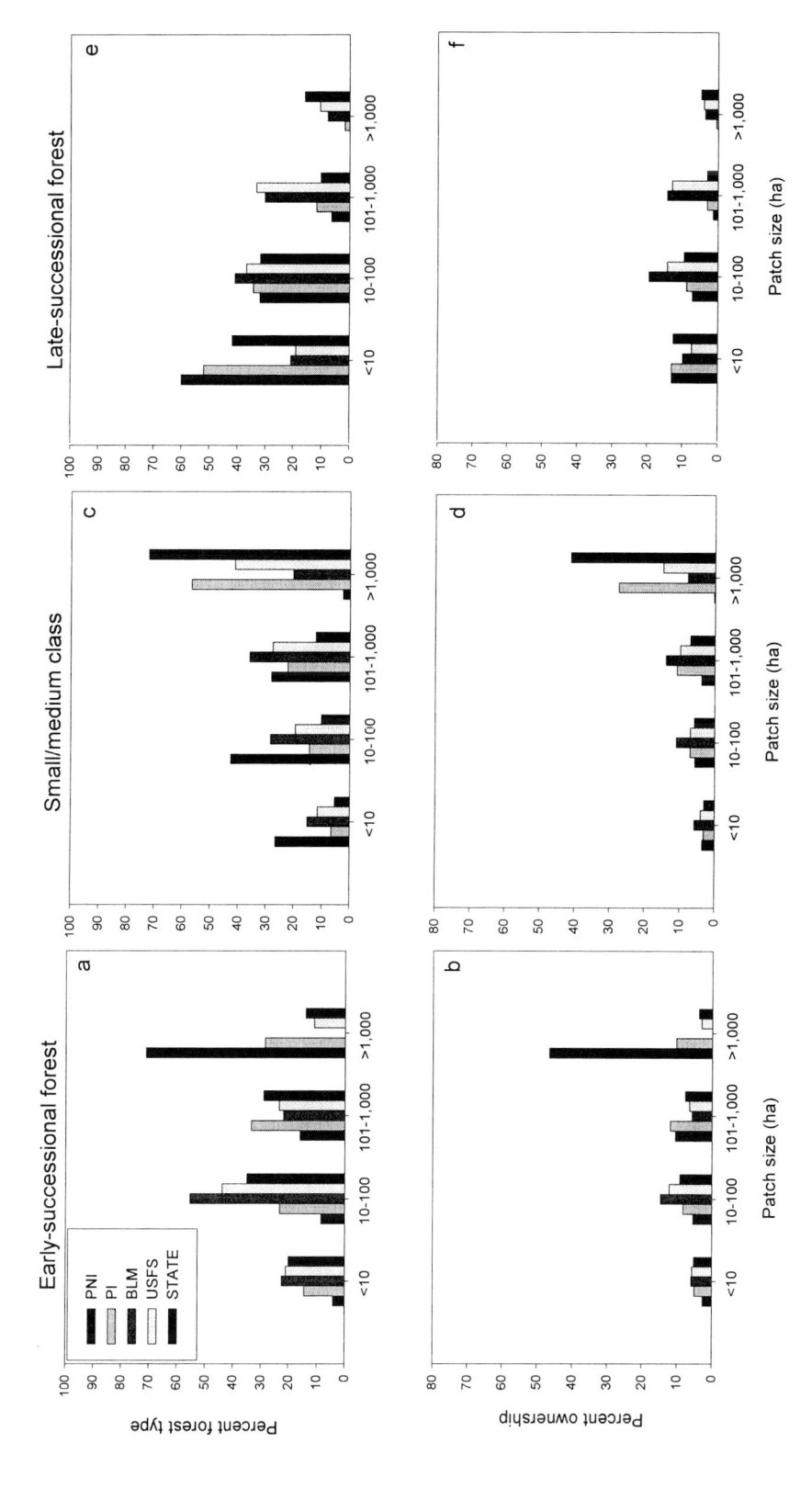

Figure 9. Frequency distribution of forest types by patch size and ownership category for the Oregon Coastal Landscape Analysis and Modeling Study (CLAMS) study area. For percent forest type, percentages of a forest type sum to 100 for an ownership category. For percent ownership, percentages across all three forest types sum to 100 for an ownership. Early-successional - open, semi-closed and broadleaf classes combined; Late-successional - large and very large classes combined (classes defined in Figure 8).

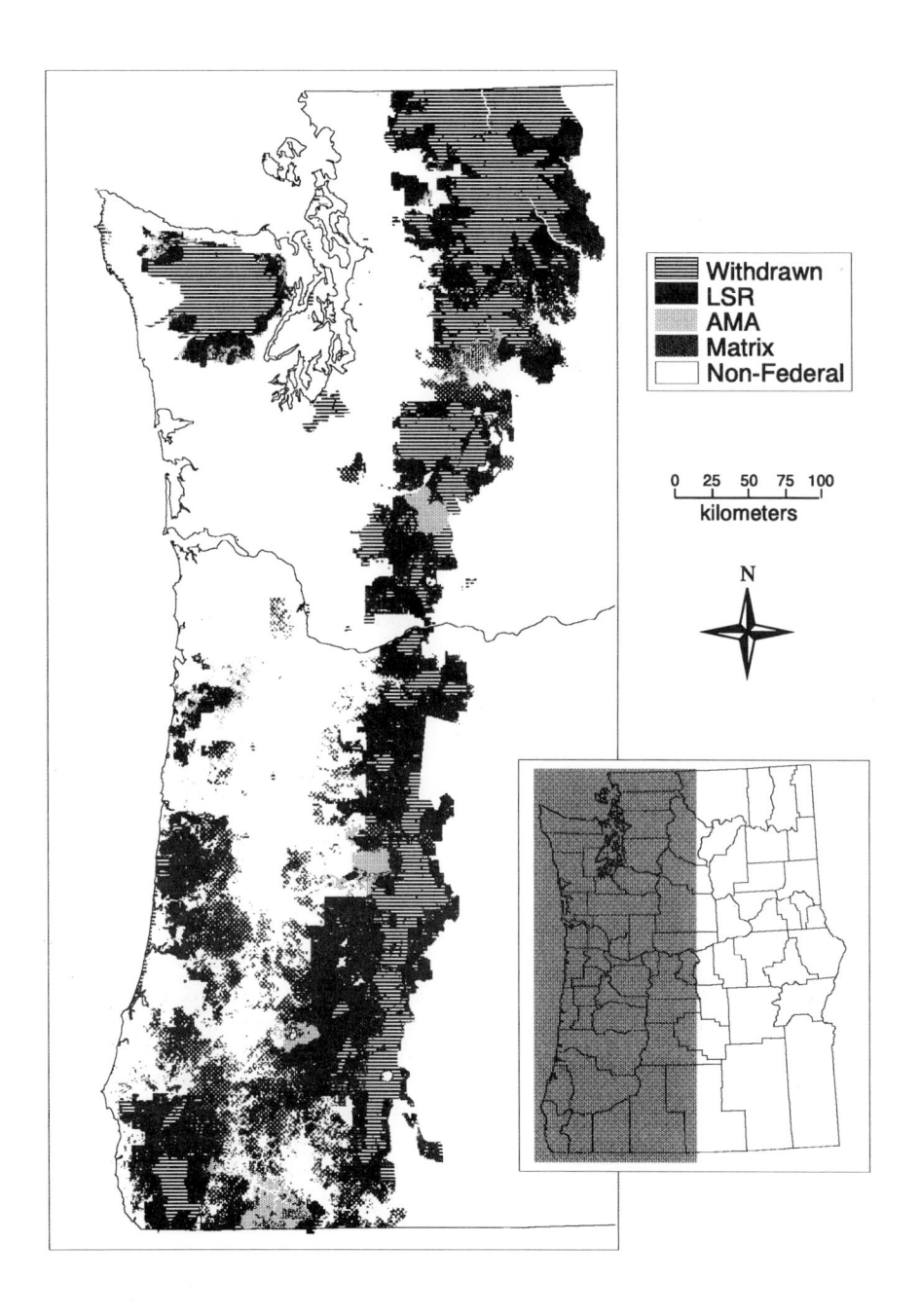

Figure 10. Federal land allocations in western Washington and western Oregon (FEMAT, 1993). See Table 1 for definitions of land allocations.

Table 1. Summary of land allocations and management standards and guidelines for federal ownerships within the range of the northern spotted owl (Interagency ROD-S&G Team, 1994)

Land Allocations	Standards and Guidelines (S&G)
Congressionally Reserved	No alteration of congressional mandates for these areas (National Parks and Monuments, Wilderness Areas, Wild and Scenic Rivers, National Wildlife Refuges, Dept. of Defense Lands).
Administratively Withdrawn Areas	The most restrictive S&G or existing plans apply (recreational and visual areas, back country and other areas not scheduled for timber harvest).
Late-Successional Reserves (LSR)	Thinning or other silvicultural treatments which will promote late-successional forest conditions may occur in stands <80-100 years old.
Managed Late-Successional Areas	Suitable owl habitat surrounding owl activity centers will be maintained with various management methods.
Adaptive Management Areas (AMA)	Potential to demonstrate and test alternative silvicultural and landscape designs (may be subject to LSR polcies).
Matrix	The more restrictive S&G apply where appropriate. Late-successional habitat maintained around owl activity centers. Manage for renewable supply of large down logs and for snags. Variable retention requirements depending on geographic location (e.g., 15%, $\geq$15/ha). 25-30% of Oregon BLM lands north of Grants Pass managed in late-successional conditions (connectivity/diversity blocks).
Riparian Reserves (40% of all allocations)	Promote, maintain buffers: width depends on aquatic category (i.e., fish bearing, domestic water supply). Silvicultural prescriptions permitted which promote renewable supply of large trees to streams.

(RMZ) can extend out to 30 m on a side in which 10-23m^2/ha of mostly conifer basal area must be retained during a harvest. In Washington, required leave-tree densities in RMZs (8-30 m on each side) range from 61-247/ha, but there are also strict shade requirements which can lead to higher retention levels.

5.2 Prognosis of Future Patterns

Portraying future trends in landscape patterns under current management guidelines is difficult due to uncertainty of management intentions and other factors. However, broad generalizations can be made based on current conditions, and ownership and land allocation patterns. For instance, development of large tracts of older forests will likely be confined to federal reserves (LSRs, riparian reserves, spotted owl activity centers, marbled murrelet (*Brachyramphus marmoratus*) sites, and portions of Administratively Withdrawn areas) and on portions of BLM matrix lands (Table 1). Because of the range of forest conditions currently present within these areas, the proposed network of late-successional forest reserves will not be fully realized for some time. Currently, 30-40% of the stands in LSRs are young with a substantive proportion being managed Douglas-fir plantations. Results from computer simulations suggest that in the absence of catastrophic natural disturbances and management intervention, these stands may require >140 yrs to develop late-successional characteristics; even with thinning, stands may still require >80 yrs to begin to resemble late-successional forests (Garman, 1999). The range of current

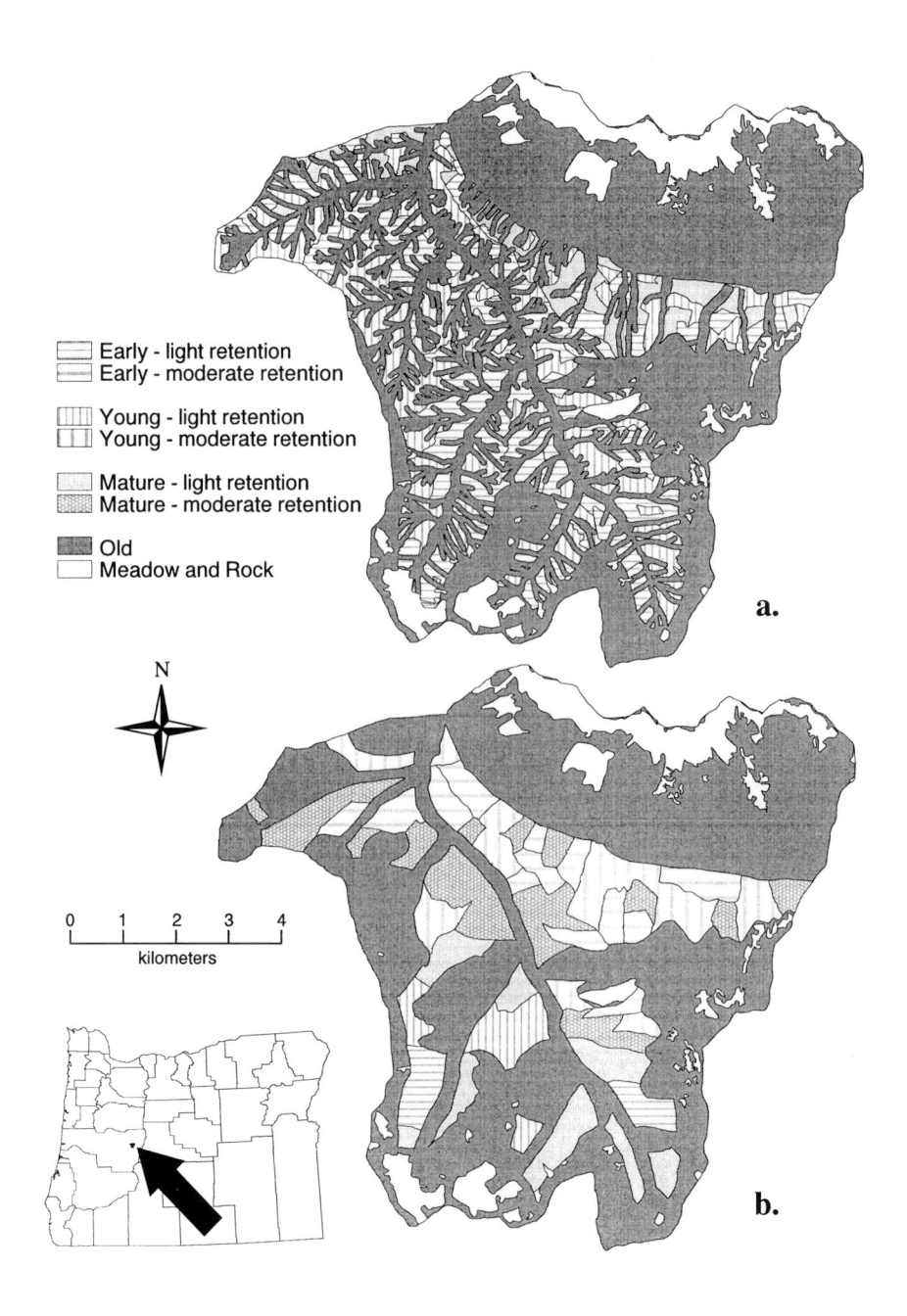

Figure 11. Projected landscape conditions (200 yrs from the present) of the Augusta Creek watershed under an interim forest plan (a) and a landscape design based on historical fire regimes (b) (Cissel et al., 1998). The interim forest plan was based on the matrix and riparian reserve standards and guidelines of the Pacific Northwest Forest Management Plan (FEMAT, 1993). Cover types were based on years since harvest and canopy retention levels. Early - ≤40 yrs, Young - 41-80 yrs, Mature - 81-200 yrs, Old >200 yrs; light retention - 15% canopy cover, moderate retention - 30-50% canopy cover.

forest conditions within designated riparian reserves also will determine developmental rate of these late-successional reserves. The Oregon Coast Range case study presented above illustrated that 30-35% of the riparian area on federally-managed lands was comprised of early-successional forests (open, semi-closed, broadleaf classes). Even with active restoration efforts, it may require several centuries to develop conifer-dominated forests in coastal riparian areas currently occupied by hardwoods or shrubs. In some areas, conifer-dominated forests may not develop due to competitive abilities of coastal shrub species (Nierenberg, 1996).

Landscape patterns on matrix lands will depend on several factors. An important consequence of the NWFP is that it limits late-successional forests on matrix lands to streamside areas. Riparian reserves may form a large single patch, depending on the stream network, but will be narrow and provide little interior habitat (e.g., Figure 11a). Structurally diverse forests may develop within the harvested portions of matrix lands, but that will depend on how required levels of green-tree retention are distributed. For non-coastal and non-BLM matrix lands, 15% of the area to be harvested must be retained, with 70% of this retention aggregated (>0.2 ha patches) and the remainder dispersed. Use of small, widely spaced retention patches with the remaining retention also widely distributed would promote greater vertical structure over a harvest unit than aggregating retention in large patches. However, large retention patches distributed in relation to retention patches on adjacent harvest units could provide important refugia or dispersal habitat for certain species across the managed area. An additional consideration is whether retention patches are retained or harvested in subsequent harvest entries. Retaining some of these patches through multiple rotations (80 yrs on most matrix lands) would provide overall greater spatial diversity of forest structure than always establishing new retention patches comprised of the most recent cohort.

Other matrix lands with special patch-retention requirements are the connectivity/diversity blocks (263 ha in size) of the BLM checkerboard ownership. These lands are to be managed on a long rotation (150 yrs) with 20-30% of the block managed as late-successional forest at any point in time. Pattern development on these lands similarly will be influenced by both the dispersion of the late-successional forest patches within and among blocks and the turnover rate of these patches. Where late-successional stands within riparian reserves satisfy the percentage requirement, patches of forest reserves will likely not be established on upslope areas.

Management requirements for other matrix lands include retention of 15-20 trees/ha at harvest for non-coastal BLM matrix lands south of Grants Pass, OR, and protection of stands occupied by marbled murrelets (0.8 km around an occupied site) for coastal matrix lands. Patterns on these lands will consist of riparian and other reserves, and young (<80 yrs) managed stands of varying structural characteristics.

Landscape patterns in AMAs will be determined by the type and amount of other land allocations they contain and the types of management experiments implemented. Management experiments are currently being designed. An example of the landscape experiment being conducted in the Central Cascades AMA is presented below.

Landscape patterns on non-federal commercial forest lands largely will be driven by state forest-management policies, and goals and objectives of individual land owners. Harvesting rates will be influenced by market prices. Based on historical cutting rates, however, it would be expected that about 20% of this land base will be clearcut harvested each decade. The current retention requirements for clearcut sites have the potential to enhance the structural and functional diversity of managed stands, but do not promote the development and maintenance of late-successional forests. Tracts of late-successional forests may be established by Habitat Conservation Plans to provide habitat for threatened and endangered species (e.g., Oregon Dept. of Forestry, 1998). Aquatic zone regulations in both Washington and Oregon provide protection of stream-side forests. However, estimates suggest that these areas are only a small portion of the land base. For example, <5% of the state and private industrial land base in Coastal Oregon is estimated to be within riparian management zones (RMZs), and only 20% of this area is protected from thinning of any kind. Minimum requirements for retained stand structures can be exceeded, and exceptions to regulations are possible if a land owner can demonstrate increased effectiveness of alternative strategies. In general, however, future patterns on these lands likely will consist of a mosaic of <60-yr old Douglas-fir stands managed primarily for high volume production.

Potential landscapes patterns in combination with the dispersion of land allocations will lead to varying levels of connectivity of late-successional forests throughout the region. LSRs in the Cascade Range will likely produce a more complete network of interacting late-successional forests than in the Coast Range where reserves are more dispersed because of federal ownership patterns (e.g., Figure 10). However even in the Cascade Range, patterns on matrix lands separating LSRs will have an important role in determining connectivity. Specifically, stream densities and management approaches to green-tree retention on harvested portions of matrix lands will determine the degree to which late-successional forests are physically connected among LSRs. For the connectivity/diversity blocks of the BLM ownership to provide large, contiguous patches of late-successional habitat, placement of older stands must be coordinated among adjacent blocks. Gap-crossing abilities and patch-size requirements of species determine the degree to which a landscape is functionally connected (With, 1999). For certain species associated with late-successional forests, riparian reserves on matrix lands and aquatic buffers on non-federal timber lands may be sufficient for them to readily disperse among reserves. For other species, the extent of physically connected, large patches of older forests will determine their ability to move throughout the landscape.

The interspersion of multiple ownerships has important implications for future landscape diversity and connectivity at basin scales. Multiple ownerships with contrasting management objectives can potentially create a greater range of forest and habitat conditions than a single ownership. For instance, the hypothetical example of future patterns of a Coastal Oregon watershed shown in Figure 12 illustrates the importance of private lands in maintaining open areas and broadleaf forests between large tracts of late-successional forests (i.e., the Large class in Figure 12) on federal lands. This mixture of conditions has the potential to support both late-successional species and those requiring open and closed forests in proximity for feeding and nesting. This example also illustrates how connectivity of late-successional forests within a watershed will depend on management objectives of intervening ownerships as well as the configuration of reserve parcels (e.g., BLM lands in Figure 12). Both the spatial configuration of intervening parcels and the amount of time forests on these parcels differ structurally and functionally from late-successional forests will influence connectivity of surrounding reserves. A more subtle point illustrated in this watershed example is the potential for undesired interactions among contrasting forest patterns. For instance, exposure of forested edges by clearcutting on adjacent ownerships could induce edge-related compositional changes due to climatic factors and increase susceptibility of stands to natural disturbances such as windthrow. An effect of this interaction would be a reduction in the effective size of a forested patch. Long-term conditions and connectivity of late-successional reserves on the checkerboard portion of the BLM ownership especially will be sensitive to management actions on adjacent lands.

5.2.1 Natural Disturbance

Adding to the uncertainty of long-term landscape patterns is the potential influence of natural disturbances. Regardless of preventive measures, landscapes will be affected by outbreaks of insects and pathogens, landslides, floods, and windthrow. Even with suppression efforts, wildfires burn thousands of hectares of forests annually (Agee, 1990). Patterns produced by natural disturbances will vary with frequency and intensity of the disturbance, and restoration efforts. Small non-stand replacing events have the potential to increase the overall spatial and compositional diversity of forested stands. Stand-replacement events have a similar role when viewed over a larger spatial extent. Predicting when and where disturbances will occur is nearly impossible. How

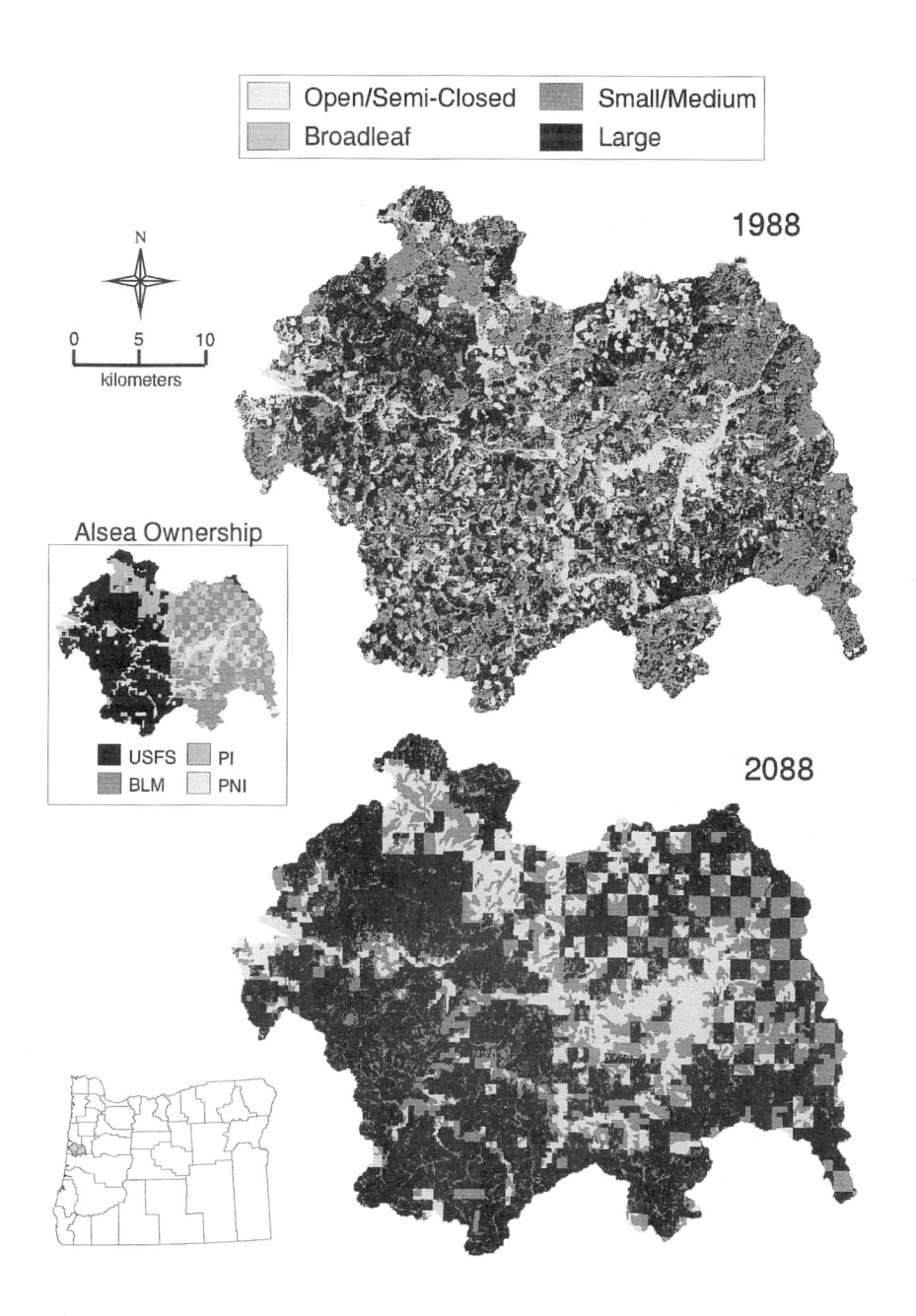

Landscape Pattern Dynamics in the Western Pacific Northwest 81

Figure 12. Recent and hypothetical future landscape patterns for the 118,000-ha Alsea Basin, Oregon (Johnson, unpubl). Future conditions simulated using LSR objectives for federal lands, current management practices for PI lands, and assuming land-clearing to dominate PNI lands. Riparian management zones on PI lands were not considered. See Figure 8 for definition of vegetation classes.

ever, studies have shown that spatial pattern of previous disturbances can significantly influence initiation and propagation of subsequent disturbances (e.g., Bradshaw and Garman, 1994; Jones et al., 1998). For instance, forests adjacent to large clearcut patches are more susceptible to windthrow, which in turn can influence bark beetle outbreaks. Clearcutting on steep sites increases the probability of slope failures, which can have long-lasting impacts on down-slope landscape patterns. Leaving large amounts of slash in harvested areas can increase the likelihood of wildfire ignition, and subsequent spreading of fire into surrounding forested areas. In general, areas where management practices create contrasting conditions among adjacent forests or adversely impact the stability of steep slopes will be more susceptible to certain natural disturbances, and thus have greater variability in landscape patterns over time.

5.2.2 Alternative Landscape Management Designs

Confounding any assessment of future landscape pattern is the inevitable evolution of forest policies. Largely untested, current state and federal forest-management plans for the PNW region will ultimately change as we better understand their ability to meet desired goals and as alternative management prescriptions are shown to be better. Assessments of forest policies using computer modeling are continuing to show how current management approaches can be modified to provide greater protection of biodiversity and long-term economic returns (e.g., Carey et al., 1996). One of the largest ongoing efforts to assess forest policies is the Coastal Landscape Analysis and Modeling Study (CLAMS) (Spies et al., in prep.). This study is designed to quantitatively test the assumptions of current state and federal policies, and to evaluate consistency between projected future outcomes and policy goals for Coastal Oregon. Additionally, tools and methods developed in this effort will facilitate the design and testing of alternative forest-management policies. Because of the spatial extent and scope of proposed assessments, results of the CLAMS project have a significant potential to influence landscape policies of different ownerships in the Coastal Oregon province.

Use of historical natural patterns and disturbance regimes as a guide to landscape management has been proposed as an alternative to the reserve-matrix approach on federally-managed lands (Swanson et al., 1993). Earlier applications of this historical range of natural variability concept can be found in the management of National Parks, where disturbances (e.g., prescribed wildfire) have been introduced to restore and maintain historical conditions (Agee, 1993). Its use in the management of commercial timber lands is fairly novel, although of increasing interest throughout the United States (Baker, 1992; Hunter, 1993) and western Canada (Stuart-Smith and Hebert, 1996). The premise of this concept is that ecological processes and native species are adapted to the temporal and spatial range of landscape patterns resulting from natural disturbance regimes; operating outside of this range may negatively affect ecological processes and species' populations. Managed landscapes designed from natural patterns thus have a greater potential to provide habitat and pattern dynamics necessary to sustain indigenous species and processes.

The natural variability concept has been applied to two watersheds in west-central Oregon (Cissel et al., 1998; in press), with the most recent application extending over a large portion of the Central Cascades Adaptive Management Area. Using wildfire regimes as reference points for land-management prescriptions, these applications have illustrated important differences between matrix prescriptions of the NWFP and a natural variability design. Compared to the NWFP, management approaches based on historical patterns produced larger harvest-unit sizes, longer harvest rotation intervals, higher structural retention requirements, and aggregated aquatic reserves in headwater areas. Consequences of the NWFP and natural variability landscape designs have been examined by long-term (200 yr) projections of patterns (e.g., Figure 11). Over this projection, the natural variability designs maintained large, spatially connected tracts of structurally-diverse older forests in addition to a range of other seral stages on the matrix allocation (e.g., Figure 11b). In contrast,

prescriptions under the NWFP produced a landscape quickly dominated by young, structurally simple forests, with late-successional forests on matrix lands limited to riparian areas (e.g., Figure 11a). Additional benefits of the natural variability design were greater habitat protection for most species with only a nominal reduction in timber production. Given the inherent adaptive nature of landscape management, the ability to quickly modify landscape patterns to meet evolving resource objectives is essential. In comparison with the NWFP design, the distribution of mature and older forests maintained by the natural variability designs provided increased flexibility to add or redistribute reserve locations, modify sizes and shapes of closed-canopy forests patches, and to increase amounts and dispersion of open-canopy forests.

The benefits and limitations of using historical patterns to design future landscapes will continue to be assessed as these alternative landscape designs are implemented and monitored. However, the currently perceived benefits make the natural variability approach an appealing alternative to the matrix-riparian reserve design of the NWFP. Wide-spread adoption of the natural variability concept, in total or in part, on other allocations or ownerships has the potential to produce landscape patterns very different from what would be expected under current forest-management guidelines.

6. CONCLUSIONS

Landscape patterns of the western PNW region have changed considerably since Euro-American settlement, yet the region still retains the largest blocks of temperate old-growth forests in North America. The mosaic of older and younger forests produced by variable natural disturbance regimes has been simplified by settlement activities and clearcut timber harvesting over the past 150 years. Land clearing and agriculture on private nonindustrial lands has produced large patches of open and early-successional habitat. The high rate and method of timber harvesting on non-federal timber lands over the past century have resulted in large patches of young plantation forests and small fragments of older forests on these ownerships. Because of the shorter history of harvesting, federally-managed lands currently contain more and larger patches of the pre-settlement forest compared to other ownerships, but dispersed cutting practices over the past 50 years has also promoted the development of a checkerboard of plantation and older forests.

Current forest-management policies have a mixed potential to reverse recent trends in fragmentation of older forests. The structural retention requirements for upslope and riparian areas on non-federal commercial forest lands have a limited potential to promote the development of diverse, managed forests, but do not address the development of late-successional forests or interior forest condition. Under the NWFP, federally-managed forests will support the majority of late-successional forests which will be distributed across the landscape mostly as large blocks of forest reserves and along riparian systems within landscapes managed for timber production. Large-scale connectivity (both physical and functional) of late-successional reserves will be highly dependent on management prescriptions in intervening land allocations and ownerships. Understanding how to manage these intervening lands to enhance connectivity of late-successional forests within and among landscapes is an important management and research question. At least for federally-managed lands, experimentation with silvicultural prescriptions and landscape designs to improve on ecological diversity and commodity production is an integral part of future management. In general, testing and refinement of forest policies over time will be essential to ensure the continued evolution of management practices which best satisfy multiple resource objectives. Also, continual adjustment of management practices will be important as we better understand the effects of landscape pattern on species' populations and ecological processes.

Acknowledgments. We thank J. Cissel for the use of Figure 11, K. N. Johnson for the use of Figure 12, W. Cohen for the use of the Oregon Cascade TM satellite imagery, D. Oetter and A. Moses for help in spatial data analysis, and K. Ronnenberg for preparing the figures. In addition, we thank P. Weisberg and K. McGarigal for reviews of an earlier version of this manuscript. Financial support for data analyses and preparation of this paper was provided by the USDA Forest Service-Pacific Northwest Research Station, the Coastal Landscape Analysis and Modeling Study, the National Science Foundation Long-term Ecological Research (LTER) program, and the Oregon Forest Resource Institute.

LITERATURE CITED

Agee, J. K., 1990, The historical role of fire in Pacific Northwest forests. Pages 25-38 in Walstad, J. D., Radosevich, S. R., and Sandberg, D. V., editors. *Natural and Prescribed Fire in Pacific Northwest Forests*. Oregon State University Press, Corvallis, OR.

Agee, J.K., 1993, *Fire Ecology of Pacific Northwest Forests*. Island Press, Washington, DC.

Agee, J. K., 1998, The landscape ecology of western forest fire regimes. *Northwest Science* 72, 24-34.

Andrews, H. J., and Cowlin, R. W., 1940, Forest resources of the Douglas-fir region. *USDA Misc. Publ. No. 389*.

Baker, W. L., 1992, The landscape ecology of large disturbances in the design and management of nature reserves. *Landscape Ecology* 7, 181-194.

Bolsinger, C. L., and Waddell, K. L., 1993, Area of old-growth forests in California, Oregon, and Washington. *USDA Forest Service, Pacific Northwest Research Station, Resource Bulletin, PNW-RB-197*.

Booth, D. E., 1991, Estimating prelogging old-growth in the Pacific Northwest. *Journal of Forestry* 89, 25-29.

Bourhill, B., 1994, *History of Oregon's Timber Harvest and Lumber Production*. Oregon Dept. of Forestry, Salem, OR.

Bradshaw, G. A., and Garman, S. L., 1994, Detecting fine-scale disturbances in forested ecosystems as measured by large-scale landscape patterns. Pages 535-550 in Michner, W. K., Brunt, J. W., and Stafford, S.G., editors. *Environmental Information Management and Analysis: Ecosystem to Global Scales*. Taylor & Francis LTD, Bristol, PA.

Carey, A. B., Elliott, C., Lippke, B. R., Sessions, J., Chambers, C. J., Oliver, C. D., Franklin, J. F., and Raphael, M. G., 1996, *Washington forest landscape management project–a pragmatic, ecological approach to small-landscape management*. Washington State Dept of Natural Resources, Olympia, WA.

Cissel, J. H., Swanson, F. J., and Weisberg, P. J., In press, Landscape management using historical fire regimes: Blue River Oregon. *Ecological Applications*.

Cissel, J. H., Swanson, F. J., Grant, G. E., Olson, D. H., Gregory, S. V., Garman, S. L., Ashkenas, L. R., Hunter, M. G., Kertis, J. A., Mayo, J. H., McSwain, M. D., Swetland. S. G., Swindle, K. A., and Wallin, D. O., 1998, A landscape plan based on historical fire regimes for a managed forest ecosystem: the Augusta Creek Study. *USDA Forest Service, Pacific Northwest Research Station, General Tech. Report, PNW-GTR-422*.

Cohen, W. B., Spies, T. A., and Fiorella, M., 1995, Estimating the age and structure of forests in a multi-ownership landscape of western Oregon, USA. *International Journal of Remote Sensing* 16, 721-746.

Cowlin, R. W., and Moravets, F. L., 1940, Forest resources of Washington. *Pacific Northwest Forest and Range Experiment Station, Forest Survey Report 101*.

Fahnestock, G. R., and Agee, J. K., 1983, Biomass consumption and smoke production by prehistoric and modern forest fires in western Washington. *Journal of Forestry* 81, 653-657.

FEMAT, 1993, *Forest Ecosystem Management: An Ecological, Economic and Social Assessment. Report of the Forest Ecosystem Management Assessment Team*. July 1993, Portland, OR. U.S. Government Printing Office.

Franklin, J. R., and Dyrness, C. T., 1988, *Natural vegetation of Oregon and Washington*. Oregon State University Press, Corvallis, OR.

Franklin, J. F., and Forman, R. T. T., 1987, Creating landscape patterns by forest cutting: ecological consequences and principles. *Landscape Ecology* 1, 5-18.

Franklin, J. F., and Spies, T. A., 1984, Characteristics of old-growth Douglas-fir forests. Pages 212-229, in *New Forests for a Changing World*, Society of American Foresters, Bethesda, MD.

Franklin, J. F., and Spies, T. A., 1991, Ecological definitions of old-growth Douglas-fir forests. Pages 61-69 in Ruggiero, L. F., Aubrey, K. A., Carey, A. B., and Huff, M. H., editors. *Wildlife and Vegetation of Unmanaged Douglas-fir Forests*. USDA Forest Service, Pacific Northwest Research Station, General Tech. Report, PNW-GTR-285.

Garman, S. L., 1999, *Accelerating Development of Late-successional Conditions From Young Managed Douglas-fir Stands: A Simulation Study*. Administrative Report, Blue River Ranger District, Willamette National Forest, Blue River, OR.

Garza, E. S., 1995, Fire History and Fire Regimes of East Humbug and Scorpion Creeks and Their Relation to the Range of *Pinus lambertiana* Dougl. M.F. Thesis, Oregon State University, Corvallis, OR.

Gedney, D. R., Bassett, P. M., and Mei, M. A., 1986a, Timber resource statistics for non-federal forest land in southwest Oregon. *USDA Forest Service, Pacific Northwest Research Station, Resource Bulletin*, PNW-138.

Gedney, D. R., Bassett, P. M., and Mei, M. A., 1986b, Timber resource statistics for non-federal forest land in northwest Oregon. *USDA Forest Service, Pacific Northwest Research Station, Resource Bulletin*, PNW-140.

Gedney, D. R., Bassett, P. M., and Mei, M. A., 1987, Timber resource statistics for non-federal forest land in west-central Oregon. *USDA Forest Service, Pacific Northwest Research Station, Resource Bulletin*, PNW-143.

Heyerdahl, E. K., Berry, D., and Agee, J. K., 1995, *Fire History Database of the Western United States*. Interagency Agreement DW12934530. University of Washington: U.S. Environmental Protection Agency/USDA Forest Service.

Hunter, M. L., Jr., 1993, Natural fire regimes as spatial models for managing boreal forests. *Biological Conservation* 65, 115-120.

Impara, P., 1997, Spatial and Temporal Patterns of Fire in the Forests of the Central Oregon Coast Range. Ph.D. dissert., Oregon State University, Corvallis, OR.

Interagency ROD-S&G Team, 1994, Record of Decision for amendments to Forest Service and Bureau of Land Management planning documents within the range of the northern spotted owl: Standards and Guidelines for management of habitat for late-successional and old-growth forest related species within the range of the northern spotted owl. U.S. Government Printing Office, 589-111/00001.

Jones, J. A., Swanson, F. J., and Sinton, D. S., 1998, A history of windthrow, wildfire, and forest cutting in the Bull Run watershed, Oregon. *Report to City of Portland Water Bureau and the Mt. Hood National Forest*.

Larsen, D. N., 1997, *Washington Timber Harvest 1996*. Washington State Department of Natural Resources, Resource Planning and Asset Management.

Li, H., Franklin, J. F., Swanson, F. J., and Spies, T. A., 1993, Developing alternative forest cutting patterns: a simulation approach. *Landscape Ecology* 8, 63-75.

Long, C. J., 1995, Fire history of the central coast range, Oregon: a circa 9000 year record from Little Lake. M.A. Thesis, University of Oregon, Eugene, OR.

Loresen, T., Andrus, C., and Runyon, J., 1994, *The Oregon Forest Practices Act Water Protection Rules*. Forest Practices Policy Unit, Oregon Department of Forestry, Salem, OR.

MacLean, C. D., Bassett, R. M., and Yeary, G., 1992, Timber resource statistics for western Washington. *USDA Forest Service, PNW Research Station, Resource Bulletin*, PNW-RB-191.

Means, J. E., 1982, Developmental history of dry coniferous forests in the central western Cascade Range of Oregon. Pages 142-158 in Means, J. E., editor. *Forest Succession and Stand Development Research in the Northwest*. Forest Resources Laboratory, Oregon State University, Corvallis, OR.

Morris, W. G., 1934, Forest fires in Oregon and Washington. *Oregon His. Quart.* 35, 313-339.

Morrison, P. H., and Swanson, F. J., 1990, Fire history and pattern in a Cascade Range landscape. *USDA Forest Service, Pacific Northwest Research Station, General Tech. Report*, PNW-GTR-254.

Nierenberg, T. R., 1996, Characterization of unmanaged riparian overstories in the central Oregon Coast Range. M.S. Thesis, Oregon State University, Corvallis, OR.

Oregon Department of Forestry, 1997, *Oregon Department of Forestry Forest Practice Administrative Rules*. Oregon Department of Forestry, Salem, OR.

Oregon Department of Forestry, 1998, *Western Oregon State Forests Habitat Conservation Plan*. Oregon Department of Forestry, Salem, OR.

Powers, J. S., 1995, Spatial and Temporal Dynamics of the Douglas-fir Bark Beetle (*Dendroctonus pseudotsugae*, Hopk.) In the Detroit Range District, Oregon: a Landscape Ecology Perspective. M.S. Thesis, Oregon State University, Corvallis, OR.

Rasmussen, M. C., and Ripple, W. J., 1998, Retrospective analysis of forest landscape patterns in western Oregon. *Natural Areas Journal* 18, 151-163.

Ripple, W. J., 1994, Historic spatial patterns of old forests in western Oregon. *Journal of Forestry* 92, 45-49.

Rudinsky, J. A., 1962, Ecology of Scolytidae. *Annual Review of Entomology* 7, 327-347.

Ruth, R. H., and Yoder, R. A., 1953, Reducing wind damage in the forests of the Oregon coast range. *USDA Forest Service, Pacific Northwest Forest and Range Experiment Station, Research Paper 7*, Portland, OR.

Sauter, J., and Johnson, B., 1974, *Tillamook Indians of the Oregon Coast*. Binfords & Mort, Portland, OR.

Spies, T. A., Ripple, W. J., and Bradshaw, G. A., 1994, Dynamics and pattern of a managed coniferous forest landscape in Oregon. *Ecological Applications* 4, 555-568.

Stuart-Smith, K., and Hebert, D., 1996, Putting sustainable forestry into practice at Alberta Pacific. *Canadian Forest Industries* April/May.

Swanson, F. J., Jones, J. A., Wallin, D. O., and Cissel, J. H., 1993, Natural variability-implications for ecosystem management. Pages 89-103 in Jensen, M. E., and Bourgeron, P. S., editors *Eastside Forest Ecosystem Health Assessment - Volume II: Ecosystem management: principles and applications*. USDA Forest Service, Pacific Northwest Research Station.

Teensma, P. D. A., 1987, Fire History and Fire Regimes of the Central Western Cascades of Oregon. Ph.D. dissert., University of Oregon, Eugene.

Teensma, P. D. A., Rienstra, J. T., and Yeiter, M. A., 1991, Preliminary Reconstruction and Analysis of Change in Forest Stand Age Classes of the Oregon Coast Range from 1850 to 1940. USDI, Bureau of Land Management, Tech. Note T/N OR-9.

VanNorman, K. J., 1998, Historical Fire Regime in the Little River Watershed, Southwestern Oregon. M.S. Thesis, Oregon State University, Corvallis, OR.

Wall, B. R., 1972, Log production in Washington and Oregon: an historical perspective. *USDA Forest Service, Pacific Northwest Research Station, Resource Bulletin*, PNW-42.

Wallin, D. O., Swanson, F. J., and Marks, B., 1994, Landscape pattern response to changes in pattern generation rules: land-use legacies in forestry. *Ecological Applications* 4, 569-580.

Wallin, D. O., Swanson, F. J., Marks, B., Cissel, J. H., and Kertis, J., 1996, Comparison of managed and pre-settlement landscape dynamics in forests of the Pacific Northwest, USA. *Forest Ecology and Management* 85, 291-309.

Washington State Department of Natural Resources, 1997, *Forest practices illustrated*. Washington State Department of Natural Resources. Olympia, WA.

Weisburg, P. J., 1998, Fire History, Fire Regimes, and Development of Forest Structure in the Central Western Oregon Cascades. Ph.D. dissert., Oregon State University, Corvallis, OR.

With, K. A., 1999, Is landscape connectivity necessary and sufficient for wildlife management? in Rochelle, J.A., Lehmann, L.A., and Wisniewski, J., editors. *Forest Fragmentation: Implications for Wildlife Management*. Brill Academic Publishers, Leiden, The Netherlands. (This volume).

Zybach, R., 1993, Native forests of the Northwest: 1788-1856 American Indians, culture fire, and wildlife habitat. *Northwest Woodland* 9, 14-15.

Forest Loss and Fragmentation: Which has the Greater Effect on Persistence of Forest-dwelling Animals?

Lenore Fahrig

Because habitat loss and fragmentation typically co-occur, the term "fragmentation" is often used as a short-hand for both processes. However, the loss of habitat has a far greater effect than the fragmentation, or "breaking apart", of habitat on population survival, and many or most species may exhibit a threshold response to habitat loss. Use of the word "fragmentation" as a short-form for loss and fragmentation or, in some cases, for loss only, shifts research focus away from habitat loss and toward habitat pattern. This has the unfortunate consequence of shifting conservation focus away from habitat conservation and restoration. Since "forest" and "habitat" are not synonymous for forest-interior species, the breaking apart of forest habitat entails additional loss of habitat (i.e., interior forest) for these species. I suggest that there is currently an over-emphasis on habitat configuration and an under-emphasis on habitat amount. Since habitat loss is the main factor causing species extinctions, this shift in focus is detrimental to species conservation.

Key words: habitat loss, habitat destruction, habitat fragmentation, habitat configuration, landscape scale, forest-interior, forest-edge, threshold, conservation, habitat restoration

1. DEFINITION OF HABITAT FRAGMENTATION: JUST SEMANTICS?

Habitat loss is the most important factor causing the current species extinction event (Groombridge 1992, Bibby 1995, Ehrlich 1995, Thomas and Morris 1995). This fact was, however, not reflected in a speech by Dr. Robert Watson of the World Bank at the 1996 annual meeting of the Ecological Society of America. Dr. Watson, head of environmental policy for the Bank, did not even include habitat loss on his list of factors causing species extinctions. His list did, however, include "habitat fragmentation". Just what did Dr. Watson mean by "habitat fragmentation"? More importantly, what do the directors of the World Bank think Dr. Watson means when he tells them that "habitat fragmentation" is an important threat to biodiversity?

According to the dictionary, "fragmentation" means "the breaking apart or up into pieces" (Merriam-Webster Inc. 1987). Presumably, then, "habitat fragmentation" means the breaking apart of habitat. Unfortunately, the dictionary definition of fragmentation does not apply perfectly to habitat fragmentation. When a porcelain vase is "fragmented", the amount of porcelain remains constant. Habitat fragmentation, on the other hand, generally occurs through habitat removal. Therefore, habitat loss and habitat fragmentation are inextricably linked.

Because of the co-occurrence of habitat loss and fragmentation, the term "fragmentation" is often used as a short-hand for the

---

Ottawa-Carleton Institute of Biology, Carleton University, Ottawa, Canada K1S 5B6

combined processes of habitat loss and fragmentation (e.g., Diffendorfer et al. 1995, Schumaker 1996). However, replacing the words "habitat loss" with "habitat fragmentation" has negative consequences for species conservation. If species go extinct mainly because of habitat loss, the solution is straightforward: habitat conservation and restoration. On the other hand, if species go extinct mainly because of the fragmentation or "breaking apart" of habitat, the problem may appear less severe and at the same time more complex, and the solution seems less obvious. One might even conclude that: (i) loss of habitat is not a serious threat to species survival, as long as the remaining habitat is not broken apart; and (ii) to restore endangered populations we need not restore large tracts of habitat, but just enough to connect up the "broken apart" pieces of remaining habitat.

Such conclusions would be erroneous. Available evidence (reviewed below) suggests that the effect of habitat loss on population persistence is much greater than the effect of "breaking apart" habitat.

The definition of "habitat fragmentation" is therefore not simply an issue of semantics. Use of the word "fragmentation" as a shortform for habitat loss and fragmentation or, in some cases, for loss only (Andrén 1996, Taylor and Merriam 1996), shifts research focus away from habitat loss and toward habitat pattern. This has the unfortunate consequence of shifting conservation focus away from habitat conservation and restoration.

2. FRAGMENTATION EFFECTS MUST BE STUDIED AT A LANDSCAPE SCALE

One of the reasons that habitat loss and fragmentation have been used synonymously is that "fragmentation" studies are typically conducted at a scale too small to differentiate between the effects of loss and breaking apart of habitat. Almost all studies that report effects of "habitat fragmentation" on population density or distribution use the patch as the unit of observation. Evidence of patch size effects or patch isolation effects are cited as "fragmentation effects" (e.g., Dodd 1990, Robinson et al. 1992, van Apeldoorn et al. 1992, Celada et al. 1994, Hunter et al. 1995, Andrén 1994, Hinsley et al. 1996, Schmiegelow et al. 1997, Vos and Chardon 1998). As stated by McGarigal and McComb (1995), "nearly all of the studies on fragmentation have employed a patch-centered sampling scheme in which independent forest patches, not landscapes, were sampled. Based on...a variety of "patch" characteristics, such as patch size and isolation, inferences often have been made about how "landscape" structure affects wildlife populations. Yet it is unclear whether relationships derived at the patch level can be extrapolated to the landscape level."

In fact, relationships derived at the patch scale can not be extrapolated to infer effects of habitat fragmentation. Decreases in patch size and increases in patch isolation do not necessarily imply that the habitat has become more broken apart (fragmented). Habitat loss alone at the landscape scale can account for these changes (Figure 1).

Documentation of the effects of habitat fragmentation therefore requires a truly landscape-scale study. In such a study the landscape is the unit of observation (e.g., McGarigal and McComb 1995, Miller et al. 1997, Trzcinski et al., in press). The structures of many independent landscapes are quantified and then the measures of landscape structure are related to density, diversity, or distribution of the focal organism(s). In such studies, each data point is an individual landscape (not a patch).

3. RELATIVE IMPORTANCE OF HABITAT LOSS AND FRAGMENTATION: SIMULATIONS AND DATA

Several studies have documented the effects of habitat fragmentation while holding habitat amount constant (e.g., Andreassen et al. 1998). In such studies, different treatments contain the same amount of habitat broken up into different numbers of pieces. These studies have revealed negative (Burkey 1989, 1995, Atmar and Patterson 1993, Adler and Nuernberger 1994, Irlandi

Forest Loss and Fragmentation

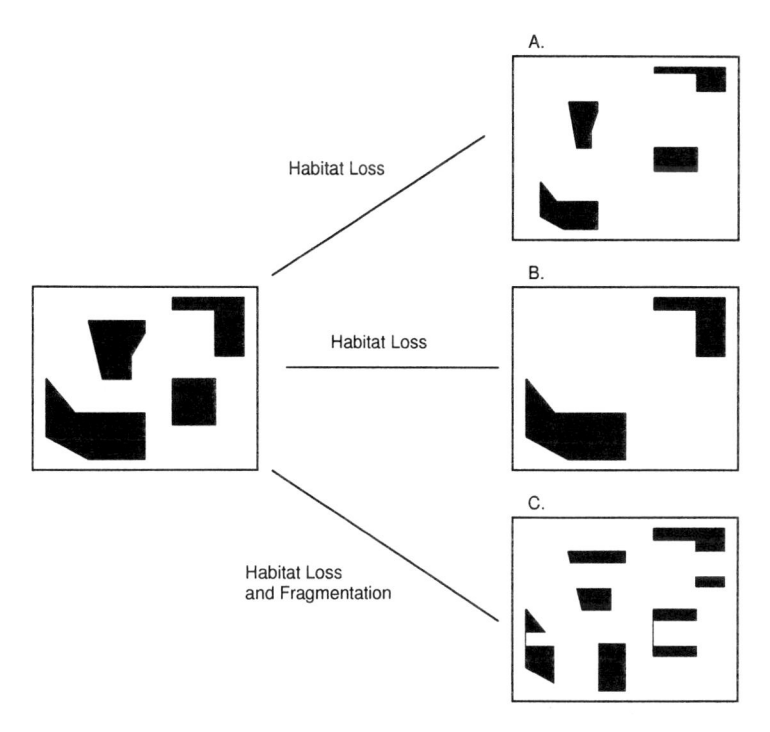

Figure 1. Definition of habitat loss and fragmentation. Increasing patch size (A) and patch isolation(B) can result from habitat loss alone. From Fahrig (1997).

1994, Dytham 1995), positive (McGarigal and McComb 1995, Collins and Barrett 1997) and neutral (Middleton and Merriam 1983, Hamel et al. 1993) effects of fragmentation.

To date, 1 simulation study (Fahrig 1997) and 3 empirical studies (McGarigal and McComb 1995, Miller et al. 1997, Trzcinski et al., in press) have documented the relative effects of habitat amount and fragmentation. In all cases, the effects of habitat amount were found to greatly outweigh the independent effects of breaking apart or fragmentation of the habitat. For example, Trzcinski et al. (in press) determined the independent effects of forest cover and forest fragmentation (i.e., configuration after controlling for cover), on the probability of presence of forest-breeding birds in 94 10 km x 10 km landscapes in southern and eastern Ontario, Canada. Of the 31 species studied, 25 showed statistically significant positive effects of forest cover (i.e., negative effects of forest loss). Only 6 species showed significant effects of forest fragmentation; 4 of these fragmentation effects were negative and 2 were positive.

While both McGarigal and McComb (1995) and Trzcinski et al. (in press) looked at responses of birds to landscape structure, McGarigal and McComb worked in mainly forested landscapes of the Oregon Coast Range, while Trzcinski et al. used data from agricultural landscapes in southern Ontario. The similar conclusion produced by these studies, that effects of forest loss far outweigh effects of forest fragmentation, in very different regions, suggests that the conclusion may be quite general.

These results suggest that conservation emphasis should be placed primarily on conservation and restoration of habitat. While there are certainly some circumstances in which habitat configuration affects species (McGarigal and McComb 1995, Fahrig 1998), very little benefit will accrue to most species of concern through manipulations or judicious planning of habitat configuration alone. Emphasis on habitat configuration appears largely misguided if the objective is species conservation.

Why then has there been so much emphasis on habitat configuration relative to habitat loss? First, as argued above, the use of the term "habitat fragmentation" in place of habitat loss has led to an over-estimate of the importance of habitat configuration, because the term "fragmentation" itself implies configuration (not loss).

Second, while many researchers have investigated the potential effects of habitat spatial pattern, very few have actually compared the relative effects of habitat pattern and habitat amount. In a conservation context, it is not enough to know that a certain factor (e.g., fragmentation) can have an effect on population survival. We must also know how large this effect is in comparison with other factors. In the absence of such comparison, and in light of the large number of studies of the effects of spatial pattern, it has been tempting to conclude that habitat configuration is at least as important as habitat amount (e.g., Kareiva and Wennergren 1995). This conclusion is unfounded.

A final reason for the misplaced emphasis on pattern over amount of habitat is that habitat loss appears to be inevitable. One could argue that, if habitat loss cannot be stopped, then we must make the best of a bad situation and at least ensure that the remaining habitat is not too fragmented. This would be a reasonable stance if habitat configuration could in fact mitigate habitat loss to some extent. However, the studies to date indicate that this is highly unlikely. In fact, if insufficient habitat remains for a population's survival, that population will not persist, no matter how that habitat is arranged. Focussing on habitat arrangement in this case creates a false hope. Efforts would be much better spent on determining how much habitat is necessary to ensure population survival and working toward restoring habitat to that level.

4. HOW MUCH IS ENOUGH?: THRESHOLD RESPONSE OF POPULATION EXTINCTION TO HABITAT LOSS

If the effects of habitat loss cannot be mitigated by alterations in habitat configuration, the most important question conservation biologists must answer is: how much habitat is enough? For example, how much mature forest of various types must be maintained on the landscape to ensure persistence of species that depend on these forest types? Several studies (Lande 1987, Bascompte and Solé 1996, Fahrig, unpublished data) suggest that the answer to this depends mainly on 2 factors: (i) the demographic potential of the organism; and (ii) the survival rate of the organism while travelling through non-habitat or "matrix" areas. The greater the demographic potential and the less hostile the matrix habitat, the less habitat the organism needs to ensure its survival on the landscape.

It also seems likely that the relationship between habitat loss and population survival is a threshold phenomenon (Bascompte and Solé 1996). Figure 2 illustrates the relationship between habitat amount and regional population survival probability, from simulations using the model described in Fahrig (1997, 1998). These simulations indicate a very sharp threshold in the relationship. The occurrence of such a threshold suggests, again, that it would be catastrophic for conservation efforts to ignore habitat amount while emphasizing pattern. It would be easy to inadvertently cross the threshold in habitat amount, thus ensuring population extinction, no matter what the configuration of the remaining habitat.

5. FOREST FRAGMENTATION VS. HABITAT FRAGMENTATION: INTERIOR SPECIES AND EDGE EFFECTS

While this discussion has been framed in terms of habitat loss and fragmentation, the focus of the conference on which this book is based was forest fragmentation. The difference between habitat and forest depends on the focal organism. For species for which

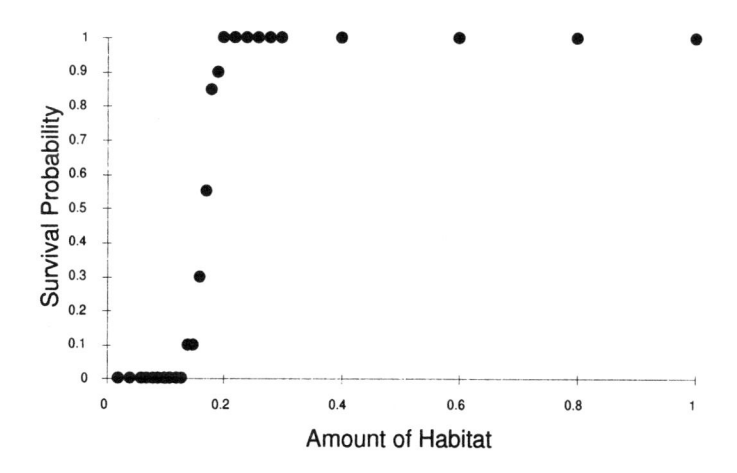

Figure 2. Population survival probability vs. habitat amount from model simulations. Each point represents the proportion of 20 runs in which the population survived for 500 time steps. Note the sharp threshold.

forest and habitat are synonymous, I conclude that forest amount is the main issue. However, for forest-edge species and forest-interior species, the mere breaking apart of forest (with no forest loss) changes the amount of habitat. The more broken apart the forest, the less habitat there is available to forest-interior species, and the more habitat there is available to forest-edge species (Figure 3).

Bender et al. (1998) conducted a meta-analysis of all patch size-density studies and found that population density was positively related to forest patch size for forest-interior species and negatively related to forest patch size for edge species. In such studies, density is computed as number of individuals divided by forest patch area. Since interior species do not use the edge (by definition), as the patch size decreases the population density decreases, since it is calculated over the whole patch. The reverse is true for edge species. Note that this is true even if the density of each species in its actual habitat (edge or interior) remains constant with forest patch size.

Therefore, forest fragmentation can be expected to have a negative effect on forest-interior species because it represents an additional loss of forest interior habitat. The predictions for forest-edge species, at the landscape scale, however, will depend on the total amount of edge in the landscape as forest is removed. Forest removal can reduce the total amount of edge at the landscape scale (e.g., Trzcinski et al., in press) which should have an overall negative effect on edge species. However, if the remaining forest is very broken apart, forest removal may increase the total amount of edge, which would have a positive effect on edge species.

6. MORE ON THE DEFINITION OF FRAGMENTATION

To understand the relationship between alterations of forest structure and population responses of focal species it is therefore critical to: (i) view the problem from a landscape (not a patch) scale; (ii) differentiate between "forest" and "habitat"; and (iii) differentiate between "loss" and "fragmentation" of forest or habitat. To illustrate the confusion over the use of the term "fragmentation", I end with some examples from the literature. These fall into 3 categories: (i) fragmentation = loss; (ii) fragmentation = loss and breaking apart; and (iii) fragmentation = breaking apart only.

6.1 Examples of Studies that Use Fragmentation Synonymously with Loss

The oft-cited paper by Andrén (1994) is the most important example of inconsistency in use of the term "fragmentation". Andrén defines habitat fragmentation as "the process of subdividing a continuous habitat into smaller pieces..." While this definition does imply "breaking apart" of habitat, Andrén goes on to state, "Habitat fragmentation has three major components, namely loss of the original habitat, reduction in habitat patch size, and increasing isolation of habitat patches..." Clearly, however, habitat loss alone leads to a reduction in habitat patch size, and increasing isolation of habitat-patches (Figure 1). Therefore, Andrén's use of the term fragmentation is inconsistent. While the definition itself implies breaking apart, the proposed components of fragmentation do not.

Taylor and Merriam (1996) state: "Habitat fragmentation occurs when parts of a continuously distributed habitat are replaced by new habitat that differs from the original... An example is forest fragmentation - a landscape where parts of a continuous forest have been replaced by non-forest." This definition does not involve breaking apart of habitat. In fact, a few clear-cuts in a continuous forest would meet this definition of fragmentation, i.e., replacement of part of a continuous habitat with a new habitat. Since the forest remains contiguous, this use of fragmentation clearly involves only loss of habitat.

6.2 Examples of Studies that Use Fragmentation Synonymously with Loss and Breaking Apart

In his memo to the conference speakers, Jim Rochelle defined forest fragmentation as "the process of reducing size and connectivity of stands that compose a forest." This is similar to Andrén's definition above, in that it focuses at the patch scale (forest stands). At the landscape scale, the main process reducing size of stands is removal of forest (Figure 1). However, by including connectivity in the definition, Rochelle goes further than loss alone to include breaking apart of forest. This definition of fragmentation is identical to that used by Dooley and Bowers (1998): "Habitat fragmentation, the process by which relatively continuous areas of habitat are broken into smaller parcels or fragments..."

Similarly, With and Crist (1995) state that through "habitat fragmentation...the landscape becomes dissected into smaller and smaller parcels." Again, reduction of the size of individual parcels can occur through habitat loss alone. However, the word "dissected" implies that the habitat also becomes broken apart. In a later paper, With et al. (1997) define habitat fragmentation as "a disruption in landscape connectivity", where landscape connectivity is "the functional linkage among habitat patches, either because habitat is physically adjacent or because the dispersal abilities of the organisms effectively connect patches across the landscape." This definition focuses more on "breaking apart" and less on loss of habitat, although the disruption of connectivity presumably occurs mainly through habitat loss.

As a final example, Diffendorfer et al. (1995) refer to "habitat fragmentation, an anthropogenic process that increases heterogeneity across space by degrading once-continuous natural habitats into remnant pieces." Again, this definition implies both loss and breaking apart of habitat.

6.3 Examples of Studies that Use Fragmentation Synonymously with Breaking Apart

The final set of examples illustrates the use of habitat fragmentation to refer only to the breaking apart of habitat. As stated above, the dictionary definition of fragmentation refers only to "breaking apart". McGarigal and McComb (1995) state: "Habitat fragmentation alters the spatial configuration of habitats, leading to population subdivision." This definition focuses solely on subdivision or breaking apart, and their distinction between breaking apart and loss of habitat is

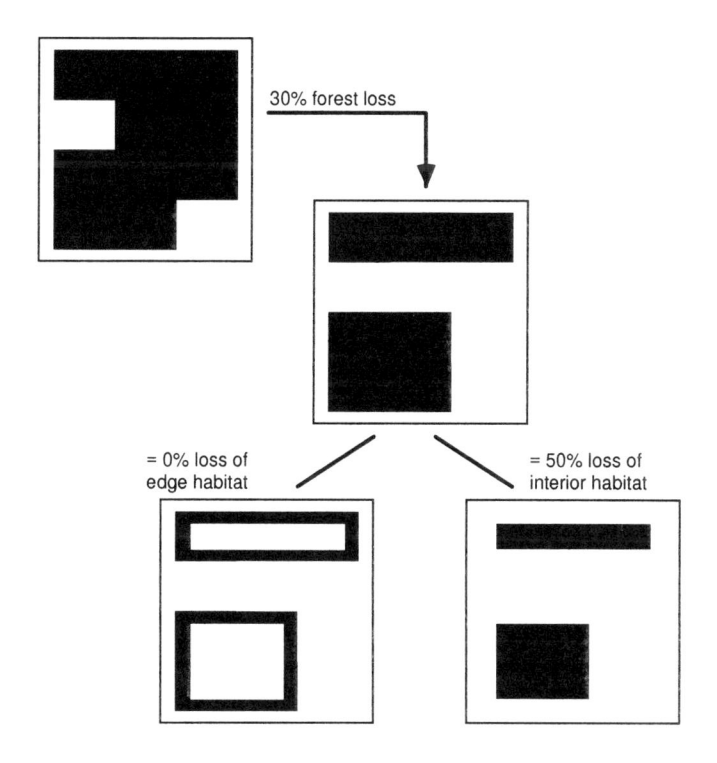

Figure 3. Effect of forest loss on habitat amount for forest edge and interior species. From Fahrig (1997).

reflected in the methods they used to analyze their data.

Several authors imply that habitat loss and fragmentation are separate factors through their explicit use of both terms. For example, Nève et al. (1996) state: "Habitat loss and fragmentation are the main threats to biodiversity...", and Lindenmayer and Possingham (1996) state: "The destruction of temperate and tropical forests and the fragmentation of the remaining forested areas is a major problem..." Finally, Bascompte and Solé (1996) express the difference very explicitly: "...we can define habitat destruction..., a reduction of the fraction of available sites. Habitat fragmentation, on the other hand, is a consequence of such destruction."

7. CONCLUSIONS

(i) Habitat loss has a much greater effect on population persistence than changes in habitat configuration.

(ii) Population survival may show a threshold response to habitat loss.

(iii) Conservation efforts should focus primarily on habitat conservation and restoration. Alteration of habitat configuration cannot compensate for habitat loss.

(iv) Patch-scale data such as patch size or patch isolation cannot provide evidence for landscape-scale fragmentation effects. Such effects can only be observed through landscape-scale studies.

(v) The term "fragmentation" should not be used if habitat loss is the main factor being considered. Focus on "fragmentation effects" leads to the erroneous conclusion that negative effects of habitat loss can be compensated by alteration of habitat configuration.

LITERATURE CITED

Adler, F.R. and B. Nuernberger. 1994. Persistence in patchy irregular landscapes. Theoretical Population Biology 45: 41-75.

Andreassen, H.P., K. Hertzberg and R.A. Ims. 1998. Space-use responses to habitat fragmentation and connectivity in the root vole Microtus oeconomus. Ecology 79: 1223-1235.

Andrén, H. 1994. Effects of habitat fragmentation on birds and mammals in landscapes with different proportions of suitable habitat: a review. Oikos 71: 355-366.

Andrén, H. 1996. Population responses to habitat fragmentation: statistical power and the random sample hypothesis. Oikos 76: 235-242.

Atmar, W. and B.D. Patterson. 1993. The measure of order and disorder in the distribution of species in fragmented habitat. Oecologia 96: 373-382.

Bascompte, J. and R.V. Solé. 1996. Habitat fragmentation and extinction thresholds in spatially explicit models. Journal of Animal Ecology 65: 465-473.

Bender, D.J., T.A. Contreras and L. Fahrig. 1998. Habitat loss and population decline: A meta-analysis of the patch size effect. Ecology 79: 517-533.

Bibby, C.J. 1995. Recent past and future extinctions in birds. Pages 98-110 in: J.H. Lawton and R.M. May. Extinction rates. Oxford University Press, Oxford.

Burkey, T.V. 1989. Extinction in nature reserves: the effect of fragmentation and the importance of migration between reserve fragments. Oikos 55: 85-81.

Burkey, T.V. 1995. Extinction rates in archipelagos: implications for populations in fragmented habitats: Conservation Biology 9: 527-541.

Celada, C., G. Bogliani, A. Gariboldi and A. Maracci. 1994. Breeding bird communities in fragmented wetlands. Biological Conservation 69: 177-183.

Collins, R.J. and G.W. Barrett. 1997. Effects of habitat fragmentation on meadow vole (Microtus pennsylvanicus) population dynamics in experiment landscape patches. Landscape Ecology 12: 63-76.

Diffendorfer, J.E., M.S. Gaines and R.D. Holt. 1995. Habitat fragmentation and movements of three small mammals (Sigmodon, Microtus, and Peromyscus). Ecology 76: 827-839.

Dodd, C.K. 1990. Effects of habitat fragmentation on a stream-dwelling species, the flattened musk turtle Sternotherus depressus. Biological Conservation 54: 33-45.

Dooley, J.L. and M.A. Bowers. 1998. Demographic responses to habitat fragmentation: experimental tests at the landscape and patch scale. Ecology 79: 969-980.

Dytham, C. 1995. The effect of habitat destruction pattern on species persistence: a cellular model. Oikos 74: 340-344.

Ehrlich, P.R. 1995. The scale of the human enterprise and biodiversity loss. Pages 214-226 in: J.H. Lawton and R.M. May. Extinction rates. Oxford University Press, Oxford.

Fahrig, L. 1997. Relative effects of habitat loss and fragmentation on species extinction. Journal of Wildlife Management. 61: 603-610.

Fahrig, L. 1998. When does fragmentation of breeding habitat affect population survival? Ecological Modelling 105: 273-292.

Freemark, K. and B. Collins. 1989. Landscape ecology of birds breeding in temperate forest fragments. Pages 443-454 in: J.M. Hagan and D.W. Johnston (eds.). Ecology and conservation of neotropical migrant landbirds. Smithsonian Institution Press, Washington, DC, USA.

Groombridge, B. (ed.). 1992. Global biodiversity: state of the earth's living resources. Chapman and Hall, New York, NY. 585 pp.

Hamel, P.W., W.P. Smith and J.W. Wahl. 1993. Wintering bird populations of fragmented forest habitat in the central basin, Tennessee. Biological Conservation. 64: 107-115.

Hinsley, S.A., R. Pakeman, P.E. Bellamy and I. Newton. 1996. Influences of habitat fragmentation on bird species distributions and regional population sizes. Proceedings of the Royal Society of London B 263: 307-313.

Hunter, J.E., R.J. Gutiérrez and A.B. Franklin. 1995. Habitat configuration around spotted owl sites in northwestern California. Condor 97: 684-693.

Irlandi, E.A. 1994. Large- and small-scale effects of habitat structure on rates of predation: how percent coverage of seagrass affects rates of predation and siphon nipping on an infaunal bivalve. Oecologia 98: 176-183.

Kareiva, P. and U. Wennergren. 1995. Connecting landscape patterns to ecosystem and population processes. Nature 373: 299-302.

Lande, R. 1987. Extinction thresholds in demographic models of territorial populations. Am. Nat. 130: 624-635.

Lindenmayer, D.B. and H.P. Possingham. 1996. Modelling the inter-relationships between habitat patchiness, dispersal capability and metapopulation persistence of the endangered species, Leadbeater's possum, in south-eastern Australia. Landscape Ecology 11: 79-105.

McGarigal, K. and W.C. McComb. 1995. Relationships between landscape structure and breeding birds in the Oregon Coast Range. Ecological Monographs 65: 235-260.

Merriam-Webster, Inc. 1987. Webster's ninth new collegiate dictionary. Thomas Allen and Son Limited, Markham, Ontario.

Middleton, J. and G. Merriam. 1983. Distribution of woodland species in farmland woods. Journal of Applied Ecology 20: 625-644.

Miller, J.N., R.P. Brooks and M.J. Croonquist. 1997. Effects of landscape patterns on biotic communities. Landscape Ecology 12: 137-153.

Neve, G., L. Mousson and M. Baguette. 1996. Adult dispersal and genetic structure of butterfly populations in a fragmented landscape. Acta Oecologica 17: 621-626.

Robinson, G.R., R.D. Holt, M.S. Gaines, S.P. Hamburg, M.L. Johnson, H.S. Fitch and E.A. Marinko. 1992. Diverse and contrasting effects of habitat fragmentation. Science 257: 524-526.

Schmiegelow, F.K.A., C.S. Machtans and S.J. Hannon. 1997. Are boreal birds resilient to forest fragmentation? An experimental study of short-term community responses. Ecology 78: 1914-1932.

Schumaker, N.H. 1996. Using landscape indices to predict habitat connectivity. Ecology 77: 1210-1225.

Taylor, P.D. and G. Merriam. 1996. Habitat fragmentation and parasitism of a forest damselfly. Landscape Ecology 11: 181-189.

Thomas, J.A. and M.G. Morris. 1995. Rates and patterns of extinction among British invertebrates. Pages 111-130 in: J.H. Lawton and R.M. May. Extinction rates. Oxford University Press, Oxford.

Trzcinski, M.K., L. Fahrig and G. Merriam. Independent effects of forest cover and fragmentation on the distribution of forest breeding birds. Ecological Applications, in press.

Van Apeldoorn, R.C., W.T. Oostenbrink, A. van Winden and F.F. van der Zee. 1992. Effect of habitat fragmentation on the bank vole, Clethrionomys glareolus, in an agricultural landscape. Oikos 65: 265-274.

Vos, C.C. and J.P. Chardon. 1998. Effects of habitat fragmentation and road density on the distribution pattern of the moor frog Rana arvalis. Journal of Applied Ecology 35: 44-56.

With, K.A. and Crist, T.O. 1995. Critical thresholds in species responses to landscape structure. Ecology 76: 2446-2459.

With, K.A., R.H. Gardner and M.G. Turner. 1997. Landscape connectivity and population distributions in heterogeneous environments. Oikos 78: 151-169.

Is Landscape Connectivity Necessary and Sufficient for Wildlife Management?

Kimberly A. With

Landscape connectivity is a central issue in conservation and wildlife management. Nevertheless, connectivity has eluded a precise definition and has been difficult to quantify and implement in practice. Neutral landscape models, derived from percolation theory, have been presented as a quantitative tool for defining landscape connectivity. In this framework, landscape connectivity is defined by the likelihood of having a connected cluster of habitat that spans the landscape (the *percolation cluster*). Connectivity emerges as a consequence of the scale at which organisms interact with the scale of fragmentation (e.g., a species' gap-crossing abilities). Thus, habitat need not be adjacent to be considered connected if the organism has good gap-crossing abilities. I analyzed species' perceptions of landscape connectivity for different forest management scenarios modeled with fractal neutral landscapes. Although all landscapes with >40% forest were perceived to be connected, gap-sensitive species (incapable of crossing non-forest gaps) actually perceived fragmented landscapes (timber harvest conducted at fine scales) to be more connected than clumped landscapes (timber harvest conducted at a coarse scale) when half (40-60%) of the landscape was forested. A percolation cluster was more likely to form on fragmented landscapes because of the more dispersed habitat distribution. Nevertheless, these landscapes had less suitable habitat (connected habitat meeting the MAR of the species) and supported smaller populations than clumped landscapes. Landscape connectivity (or habitat connectivity at some scale) may thus be a necessary, but not sufficient condition, for wildlife management. Assessment of connectivity is further complicated by the fact that species may exhibit different scales of movement in response to different scales of patchiness on the landscape. Maintenance of connectivity at a variety of scales will ultimately be necessary for successful conservation and wildlife management.

Key words: connectivity, dispersal, forest fragmentation, fractal landscapes, gaps, habitat corridors, hierarchy, movement, neutral landscape models, perceptual resolution, percolation theory, scale

1. INTRODUCTION

With the recognition that forestry and wildlife management must be implemented at a broader ecosystem or landscape scale

Department of Biological Sciences, Bowling Green State University, Bowling Green, OH

(Salwasser, 1991; Franklin, 1993; Petit et al., 1995; Beattie, 1996; Thomas, 1996) has come the challenge of maintaining or restoring the integrity of landscapes and their associated ecosystems. A disruption in the structural integrity of landscapes, owing to disturbances such as anthropogenic habitat loss and fragmentation, is expected to impede ecological flows of water, nutrients, energy, and the movement (dispersal) of organisms across the landscape (Gardner et al., 1993). Subsequently, a disruption in landscape structure may compromise its functional integrity, by interfering with the critical ecological processes necessary for population persistence and the maintenance of biodiversity and ecosystem health. *Landscape connectivity* thus involves both a structural component (the amount and spatial distribution of habitat on the landscape) and a functional component (the interaction of ecological flows with landscape pattern). A landscape that is structurally connected is assumed to be functionally connected. For example, dispersal should occur freely within contiguous habitat or among patches connected by corridors, and as a result, populations are well connected and the species is able to persist on the landscape. The principal concern facing wildlife and land managers in today's increasingly fragmented forests, however, is whether a landscape can maintain its functional integrity even if it is not structurally connected.

In this chapter, I review issues related to the definition and quantification of landscape connectivity, particularly as these apply to the management of the structural and functional integrity of forested landscapes. I present an analysis of different forest management scenarios to evaluate how species that differ in their minimum area requirements and gap-crossing abilities are affected by different intensities and patterns of timber harvest that fragment the landscape. This permits a species-centered assessment of the suitability and connectivity of landscapes, which is a necessary perspective if we are to implement successful strategies for wildlife and conservation management in fragmented forests.

2. PERSPECTIVES ON LANDSCAPE CONNECTIVITY

Although connectivity is considered a "vital element of landscape structure" (Taylor et al., 1993), it has eluded a precise definition and has been difficult to quantify and implement in practice. Two perspectives have emerged, however, which reflect different conceptualizations of landscape structure (Wiens, 1994; 1995). The first derives from island biogeographic theory, in which habitat fragments are viewed as analogues of oceanic islands in an inhospitable sea or ecologically neutral matrix (e.g., forest fragments in an agricultural matrix). The second is the landscape mosaic perspective, in which landscapes are viewed as spatially complex, heterogeneous assemblages of habitats, which can not be simply categorized into discrete elements such as patches, matrix and corridors.

2.1 Habitat Corridors: Lifelines or Landmines?

Strategies for conservation management and reserve design tend to consider connectivity in a very literal sense, by advocating the creation of habitat corridors between isolated habitat fragments or between reserves in a regional reserve network (Noss, 1991). Corridors are intended to facilitate movement or dispersal of organisms between habitat patches in the hopes that metapopulation dynamics will be maintained on the landscape (Hansson, 1991). Corridors are clearly a legacy of an island biogeographic or patch-based view of the world. Landscape structure is viewed in simple terms of habitat patches embedded within an inhospitable matrix, such that habitat corridors are required to facilitate dispersal among isolated fragments (Figure 1). Emphasis is placed on the structural component of connectivity, but the expectation (or hope) is that corridors will also restore the functional integrity of landscapes.

Although intuitively appealing, evidence in support of corridors is equivocal (Simberloff and Cox, 1987; Simberloff et al., 1992; Hobbs, 1992; Mann and Plummer, 1995). Some species use them, others do not. Species documented to use corridors are generally those that are unable or unwilling to cross gaps of unsuitable habitat (i.e., the inhospitable matrix). Small mammals, such as chipmunks (*Tamias striatus*), white-footed mice (*Peromyscus leucopus*) and voles (*Microtus pennsylvanicus*), use corridors in the form of fencerows connecting woodlots (chipmunks, deer mice) or experimentally created vegetative strips among old-field patches (voles), presumably because corridors provided increased cover and safety from

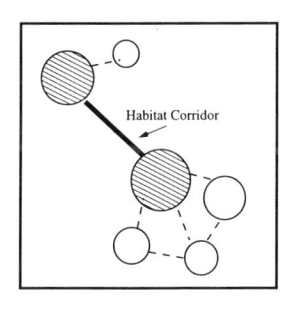

Island Biogeographic/Metapopulation	Landscape Mosaic	Neutral Landscape
Habitat Corridors	Relative Habitat Viscosity	Percolation Cluster

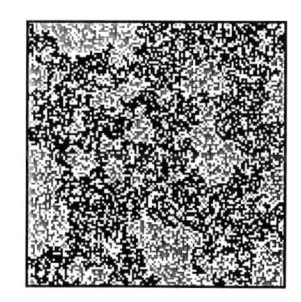

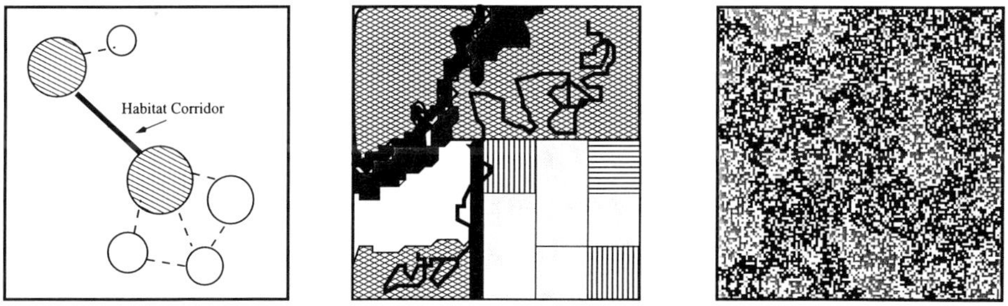

Figure 1. Different perspectives on landscape connectivity. The Island Biogeographic/ Metapopulation perspective views landscape structure in terms of discrete elements such as patches, habitat corridors, and an "inhospitable matrix." Habitat corridors are deemed necessary to facilitate dispersal between isolated fragments to maintain metapopulation dynamics [dashed lines indicate dispersal between adjacent populations, some of which act as sources (shaded) or sinks]. The Landscape Mosaic perspective represents landscape structure as a spatially complex mosaic of different habitat types and landscape elements (rivers, roads). Connectivity is defined on the basis of the differential movement of organisms through habitat. Habitat may differ in its "viscosity" or resistance to movement, facilitating movement through certain elements of the landscape and impeding it in others. The solid, convoluted line represents the movement pathway of a hypothetical organism through this landscape mosaic. Note that it is deflected by certain habitat boundaries (e.g., agricultural fields). Neutral Landscapes offer a means of quantifying landscape connectivity, based on the movement responses (gap-crossing abilities) of species. Transitions in landscape connectivity can be determined by the presence or absence of a single cluster of habitat that spans the entire landscape (percolation cluster shown in black; additional habitat not part of the percolation cluster shown in gray).

predators relative to the surrounding matrix (Wegner and Merriam, 1979; Henderson et al., 1985; La Polla and Barrett, 1993).

Even highly vagile organisms, such as birds and large carnivores, which can easily traverse areas of unsuitable habitat may nevertheless be reluctant to do so. Birds, with their prodigious flight capabilities, may nevertheless utilize corridors to locate patchily distributed food resources or isolated patches of breeding habitat on a landscape (Haas, 1995; Machtans et al., 1996). Cougars (*Felis concolor*) may also benefit from well-placed corridors on the landscape (Beier, 1993; 1996). Urban sprawl has resulted in the increasing isolation of cougar populations among the mountain ranges of southern California. Persistence of this metapopulation is critically dependent upon maintaining dispersal among populations, which is facilitated by riparian corridors or by freeway underpasses which minimize traffic-related mortality.

Wolves (*Canis lupus*), however, traverse fragmented landscapes without benefit of corridors, as evidenced by the recolonization in recent years of areas in Montana where they had been extirpated (Forbes and Boyd, 1997). Wolves are such effective dispersers, barring human persecution or interference, that discrete corridors connecting populations (e.g., reintroduced wolf packs in Idaho and Yellowstone National Park) may not be necessary as long as the overall structural integrity of the landscape is maintained ("broad landscape linkages already in use by wolves;" Forbes and Boyd, 1997).

Reliance upon corridors may be influenced by the surrounding landscape context (Anderson and Danielson, 1997). Bachman's Sparrow (*Aimophila aestivalis*) was primarily a forest species prior to the logging of open, mature pine forest in the southeast, but now nests in open habitats of early pine regeneration that contain the open shrub and understory conditions preferred by this species (4-7 years post-harvest; Dunning et al. 1995). Subsequently, most of the forest now represents unsuitable habitat and colonization of clearcuts of suitable age, which are distributed as isolated patches, is facilitated by dispersal along corridors of early-successional habitat such as utility right-of-ways.

Habitat specialists do not necessarily require corridors to locate suitable breeding habitat in fragmented landscapes, however. The northern Spotted Owl (*Strix occidentalis caurina*) requires old-growth forest for breeding, but is able to traverse nonforest habitat during juvenile dispersal (Gutiérrez and Harrison, 1996). Other studies have demonstrated, both theoretically and empirically, that habitat quality of the corridor itself may influence whether the corridor is ultimately used (Henein and Merriam, 1990; Bennett et al., 1994).

As these examples illustrate, it is difficult to predict whether species will utilize corridors, because this is not simply a function of body size or vagility. We lack basic information on species' gap-crossing abilities and thus their potential reliance on corridors for most vertebrates, especially amphibians and reptiles (Lidicker and Koenig, 1996). Direct observation or "tenacious tracking" of radio-tagged individuals has provided the strongest evidence of corridor use in some species, but is obviously time intensive, especially considering that corridor use may be infrequent (Haas, 1995; Beier, 1996). If a species can cross gaps of unsuitable habitat, corridors may not be necessary. Innovative approaches to measuring species' responses to gaps have recently been conducted in birds by quantifying the extent to which individuals can be lured into different matrix habitats or across gaps of varying sizes using taped playbacks of territorial songs or mobbing calls (Sieving et al., 1996; Rail et al., 1997; Desrochers and Hannon, 1997).

Thus, the jury is still out on the efficacy of corridors. Corridors are costly to construct if they are deemed necessary to maintain patch connectivity, and potentially costly to conservation efforts if they also serve as conduits for predators, competitors or disease (e.g., Hess, 1994). The debate over the utility of corridors will probably never be resolved because it depends upon the nature of the target species and the structure of the landscape. In the meantime, it is assumed that a landscape connected by corridors is better than a fragmented landscape that is not (Beier and Noss, 1998). Landscape connectivity can be assessed and realized in other ways, however.

2.2 The Landscape Mosaic Perspective

A newly emerging perspective on landscape connectivity represents a departure from the traditional island biogeographic/metapopulation view of patches, matrix and corridors and instead characterizes landscapes as complex, heterogeneous mosaics comprised of multiple habitat types (Wiens, 1994; Wiens, 1995). This approach focuses on the differential responses of organisms to landscape structure. Landscape connectivity is assessed by the extent to which movement is facilitated or impeded through different habitat types across the landscape (Taylor et al., 1993; With et al., 1997; Pither and Taylor, 1998). Highly connected landscapes are those in which organisms can move easily among habitats, and landscapes with low connectivity contain habitats or configurations of habitat that resist the movement of organisms across the landscape (Figure 1).

Apart from its representation of the landscape, this perspective on landscape connectivity differs from the island biogeographic/metapopulation view in another important way: it explicitly incorporates an organismal or species-centered view of landscape structure (Pearson et al., 1996). We must guard against viewing landscapes purely in anthropocentric terms, or as a level of organization in the conventional ecological hierarchy that lies beyond the ecosystem level, but below that of the biome, both of which imply an area of broad spatial extent (King, 1997). Given that landscapes must be defined relative to the scale at which the organism or ecological process of interest operates (Wiens, 1989; With, 1994), it follows that connectivity of a given landscape must also be determined by how organisms perceive and respond to landscape structure. Boundaries (e.g., habitat edges) that deflect some species will be readily traversed by others (Lidicker and Koenig, 1996). Thus, a landscape is not inherently connected or fragmented; the same landscape may in fact be both from the perspective of two different species that differ in their gap-crossing abilities or movement responses to landscape structure.

This perspective represents a more holistic view of landscapes, in that connectivity is an emergent property of landscapes resulting from the interaction of species with landscape structure. Note that this approach does not require the a priori dissection of the landscape into discrete elements, such as patches and corridors, because patch structure emerges from how, and at what scales, organisms interact with landscape heterogeneity (the abundance and distribution of habitat types). Patches are seldom discrete in nature, and the hierarchical patch structure characteristic of environmental heterogeneity is resolved differently by species depending upon the spatial grain and extent of their activities (Kotliar and Wiens, 1990). The perceptual resolution of an organism is defined by its grain, the finest scale of patchiness to which an organism responds, and its extent, the broadest scale of patchiness that it perceives and to which it responds (Kotliar and Wiens, 1990; With, 1994). This is analogous to the concept of an "ecological neighborhood" (Addicott et al., 1987) or "ambit" (Hutchinson, 1953), which define the appropriate spatial and temporal scales that bound a particular process (e.g., movement). Subsequently, corridors need not be discrete structures on the landscape (linear elements connecting patches), but may emerge from particular configurations of habitat that facilitate movement (Gustafson and Gardner, 1996).

How can this species-centered definition of landscape connectivity be operationalized and put into practice? How can we assess whether a given landscape is connected from the perspective of species that differ in their scale of interaction with landscape structure, in terms of their minimum area requirements and gap-crossing abilities? Neutral landscape models, developed in the field of landscape ecology, provide a means of quantifying landscape connectivity based on species-specific movement rules.

3. A PRIMER TO NEUTRAL LANDSCAPE MODELS

Neutral landscape models are grid-based (raster) maps in which complex landscape patterns are created using theoretical spatial distributions (e.g., With, 1997; With and King, 1997). They are thus "neutral" to the biophysical processes that structure real landscapes. Neutral landscape models were inspired by *percolation theory*, which concerns the flow of liquids through material aggregates (percolation) and provides a quantitative means for describing connectivity in heterogeneous systems (Stauffer and Aharony, 1991). Robert Gardner and his colleagues (Gardner et al., 1987; Gardner and O'Neill, 1991) were quick to see the potential application of percolation theory for modeling ecological flows in heterogeneous landscapes. Neutral landscapes provide a useful null model for assessing the effect of spatial pattern on ecological processes, and for quantifying when landscape connectivity becomes disrupted (With and King, 1997; Pearson and Gardner, 1997).

In the framework of neutral landscape models, habitat connectivity is first determined by the ability of organisms to move among habitat cells, which is a function of the ability or willingness of a species to cross gaps of unsuitable or less-preferred habitat. The willingness to cross gaps may be further modified by the nature of the intervening matrix habitat; an organism may be willing to cross certain habitats and not others. Species' gap-crossing abilities are then used to identify whether a cluster of habitat spans the entire landscape (Figure 1). Such a landscape is considered to be connected because organisms moving among habitat cells within this percolation cluster would at least be able to disperse freely. Thus, the functional response of species to landscape pattern (gap-crossing ability) is used to identify a structural feature of the landscape that confers overall connectivity (percolation cluster). The use of percolation theory to identify regions of the landscape, such as percolation clusters, that facilitate dispersal for a given species is a potentially powerful tool in the management and restoration of landscape connectivity.

At what point does the landscape become "disconnected?" Percolation theory predicts that this will occur abruptly, as a critical threshold, over a small range of habitat loss. Imagine a completely forested landscape. As habitat is destroyed, there is a decreasing probability that a landscape will remain connected with each subsequent loss (e.g., 10%) of forest cover. Habitat loss compromises the structural integrity of the percolation cluster, until finally only one or a few cells of habitat are holding the entire structure intact. These cells serve as critical links between regions of the percolation cluster, and once severed, abruptly fragment the percolation cluster into two or more segments; the landscape is no longer connected. As the threshold level of habitat loss is reached, the probability of having a connected landscape rapidly goes to zero when the percolation cluster is fragmented (i.e., the relationship between habitat loss and landscape connectivity is not linear). Thus, while landscape connectivity is a probabilistic function of habitat loss, the non-linear nature of this relationship makes it convenient to refer to landscapes as being either "connected" or "disconnected" (fragmented) with respect to this threshold.

The threshold level of habitat loss at which landscapes become disconnected depends on the scale at which organisms interact with the scale of patchiness on the landscape (i.e., a species' gap-crossing or dispersal ability relative to the gap-size distribution on the landscape; Dale et al., 1994; With and Crist, 1995; Pearson et al., 1996; With et al., 1997). For example, Pearson et al. (1996) used hierarchical random neutral landscapes, in which habitat abundance is varied at different levels in a nested fashion (the habitat distribution at the broader scale constrains the distribution at the next, finer scale) to model the hierarchical patch structure that exists in nature (Kotliar and Wiens, 1990; Figure 8.3 in Pearson and Gardner, 1997). A hypothetical species that lacked gap-crossing abilities (movement through adjacent habitat cells only; "Rule 1," Plotnick and Gardner, 1993) was very sensitive to the pattern of habitat loss. Fine-scale habitat loss was more disruptive to dispersal in this species than if habitat loss occurred at a coarser scale. Although coarse-scale habitat loss created large gaps (e.g., clearcuts in a forested landscape), the remaining habitat was also distributed as large blocks which facilitated dispersal (at least within the forest stand). Thus, landscape connectivity was maintained across a greater degree of habitat loss under this scenario for a gap-sensitive species; that is, the critical threshold occurred at lower levels of remaining habitat when habitat was destroyed in large tracts than if fine-scale habitat loss (e.g., selective logging) occurred throughout the landscape.

Neutral landscape models have the advantage of permitting the unlimited exploration of how spatial pattern (e.g., owing to habitat loss and fragmentation) affects dispersal or some other ecological process across a wide range of landscape scenarios. Although most neutral landscape applications have been binary maps of habitat vs. non-habitat, heterogeneous landscape mosaics comprised of multiple habitat types can be generated (With and Crist, 1995; With et al., 1997). Real landscape maps (e.g., land-cover map layer within a GIS) can also be used. An example of such an application (albeit with binary maps of forest and agriculture) is an analysis of how current and projected patterns of land-use change in the central Amazon are expected to impact rainforest species that differ in their gap-crossing abilities and area requirements (minimum amount of connected habitat an organism requires; Dale et al., 1994). Species in this analysis were as diverse as jaguars (*Felis onca*; large area requirements and high gap-crossing ability), three-toed sloths (*Bradypus variagatus*; small area requirements and low gap-crossing ability), and insects (e.g., Euglossine bees, which forage over a broad area and thus have large area requirements, but very limited gap-crossing abilities); all of these species need to be considered in the development of a comprehensive management strategy. Gap-sensitive species with large area requirements (i.e., insects) were most sensitive to the different scenarios of deforestation; suitable habitat for such species was perceived to be lost at a much faster rate than the actual rate of forest loss. This study did not quantify landscape connectivity per se, but the authors recommended maintaining connectivity of forest patches particularly to prevent the extirpation of species with low gap-crossing abilities and large area requirements.

Although maintenance of landscape connectivity has become a central issue in conservation and wildlife management, organisms clearly require a certain amount of connected habitat to meet their minimum area requirements, as the Dale et al. (1994) study illustrates. Is habitat connectivity necessary and sufficient for assessing landscape suitability for a particular species? How is the amount of suitable habitat available to the organism (connected habitat meeting the minimum area requirements of the species) affected by habitat fragmentation? How does this affect assessment of landscape connectivity? To address these questions, I performed a spatial analysis similar to that conducted by Dale et al. (1994), but with fractal landscapes to provide a more general assessment of how habitat fragmentation affects the connectivity and suitability of landscapes for different species.

4. CONNECTIVITY OF SPECIES-DEFINED LANDSCAPES

The procedure used to assess the suitability and connectivity of species-defined landscapes follows that of Dale et al. (1994), but a general overview is presented here and detailed below for completeness. The gap-crossing ability of the species is used to determine whether individual cells of habitat are connected. The minimum area requirement (MAR) of the species imposes a filter on the landscape by identifying habitat that is not only connected, but also of sufficient area to support individual territories. The resulting *species-defined landscape* can then be analyzed for connectivity, based on the species' gap-crossing abilities.

4.1 Spatial Analyses

4.1.1 Generation of Forested Landscapes
Neutral landscape maps were generated as 256 x 256-cell grids (i.e., raster maps) using the midpoint displacement algorithm to produce fractal distributions of habitat (see With et al., 1997, With 1997 for details). Spatial contagion (clumping) of habitat can be adjusted by altering the value of a single parameter (H). For a given level of habitat abundance, landscape patterns can be produced that are clumped ($H = 1.0$) or extremely fragmented ($H = 0.0$; Figure 2). For this application, it will be assumed that these fractal landscapes are forested landscapes undergoing different intensities and patterns of timber harvest. Clumped fractal landscapes represent timber harvest conducted at a coarse scale by clearcutting extensive forested areas, whereas fragmented fractal landscapes represent fine-scale timber harvest (selective logging). While the absolute size of neutral landscapes is arbitrary, it is necessary to assign units of measurement in this application because the availability of habitat on the landscape will ultimately be based on the MAR (ha) of a given species. I will therefore consider the resolution of individual grid cells (pixels) to be 30 m, which is consistent with the resolution of widely available remotely sensed data (e.g., Landsat Thematic Mapper Imagery). Thus, each pixel is 0.09 ha (30 x 30 m) and the overall landscape is ~5900 ha or ~60 km². Again, the size of the landscape is not important for the objectives of this analysis; larger landscapes can be generated easily. My primary interest here is in understanding how the scale(s) at which species operate *relative* to the scale of landscape pattern affects assessment of habitat suitability and landscape connectivity. The results of this analysis should therefore not be taken literally, as applying to a particular species in a given landscape, but is useful to wildlife and land managers by analogy.

A total of 16 management scenarios were created by generating fractal neutral landscapes across eight levels of forest cover (p = 0.01, 0.1, 0.2...0.7) and two levels of fragmentation (Clumped vs. Fragmented; Figure 2). Forest cover represents the intensity of harvest or degree of land transformation. Previous studies have demonstrated that landscapes with >70% habitat are well connected regardless of the spatial configuration of habitat or the gap-crossing abilities of the species (Gardner et al., 1989; With and Crist, 1995; Pearson et al., 1996; With et al., 1997; With and King, 1999a). Therefore, habitat levels above this were not analyzed. Forested landscapes with <10% cover have obviously undergone extensive transfor-

Clumped

Fragmented

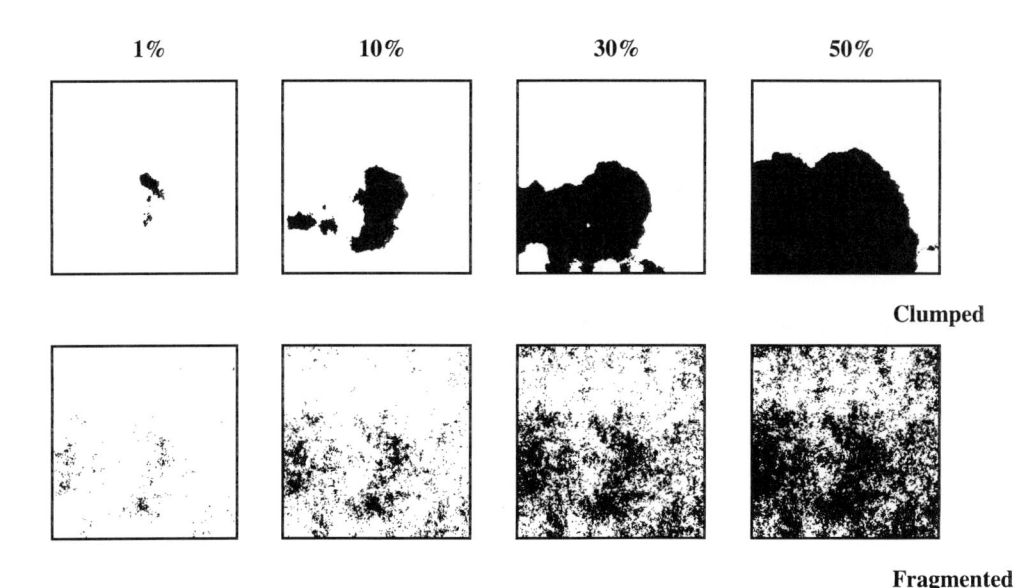

Figure 2. Fractal neutral landscapes at different levels of habitat abundance. These landscapes can be used to model different forest management scenarios. Timber harvest is conducted at either a coarse scale to create extensive clearcut areas so as to preserve large forest stands (Clumped), or at a fine scale via selective logging, producing an extensively fragmented forest (Fragmented).

mation, such as has occurred in the central hardwoods forest of the Midwest. While such landscapes may no longer be a concern from a forest management perspective, they nevertheless are a land management concern and serve to illustrate the extreme case. Five replicate landscapes were generated for each of the 16 management scenarios, resulting in a total of 80 landscape maps which served as the basis for this analysis of habitat suitability and connectivity in species-defined landscapes.

4.1.2 Species Characteristics
To explore a wide range of species' responses to landscape structure, I analyzed habitat suitability and connectivity from the standpoint of a suite of hypothetical species that differed in their gap-crossing abilities and MAR. Species differed in their MAR by four orders of magnitude (1, 10, 100, 1000 ha). Most forest passerines have territory sizes <1 ha, and small forest-dwelling mammals, such as deer mice and red-backed voles (*Clethrionomys californicus*) likewise have home ranges on the order of 1 ha or less (e.g., Nowak, 1991). Porcupines (*Erethizon dorsatum*) and opossums (*Didelphis virginiatus*) are examples of species with home range sizes around 10 ha, and mesopredators such as racoons (*Procyon lotor*) and striped skunks (*Mephitis mephitis*) may have home ranges in the vicinity of 100 ha (Nowak, 1991). Northern spotted owls require unusually large tracts of old-growth forest for breeding habitat (800 ha; Gutiérrez and Harrison, 1996). Pine martens (*Martes americana*) may likewise require suitable forested habitat on the order of 1000 ha (Nowak, 1991).

Gap-crossing abilities indicate the size of the non-forest gap the species is able or willing to cross. Species were either unable to cross gaps (gap-sensitive species) or could cross gaps of 1 cell (30 m) or 3 cells (90 m) in this analysis. In their search for forested cells, gap-sensitive (Gap = 0 cells) species were able to move among the eight neighboring cells (four adjacent cells, plus diagonals; "Rule 2" movement; Plotnick and Gardner, 1993). A species capable of crossing 1 cell of unsuitable habitat had a movement neighborhood of 24 cells (a 5 x 5-cell block centered on the individual's current location), and a species capable of crossing 3 non-forested cells had a movement neighborhood of 80 cells (a 9 x 9-cell block).

The combinations of MAR and gap-crossing abilities give rise to 12 species-types (e.g., species with small area requirements and high gap-crossing abilities). Although terrestrial vertebrates with large area requirements also tend to be highly vagile and have good gap-crossing abilities, certain forest insects may need to cover large areas to find suitable food resources but are strongly influenced by microclimatic changes of forest openings which would prevent them from crossing gaps (large area requirements and limited gap-crossing ability; Dale et al., 1994).

4.1.3 Analytical Procedure
Analysis of species-defined landscapes for each management scenario (n = 80 maps) entailed the following steps:
1) Identify clusters of forested cells that are connected. A "cluster" is defined as a group of forested cells that can be accessed by an individual, based on its gap-crossing abilities. Thus, forested cells need not be adjacent within a cluster if the species can cross gaps;
2) Measure the size of each cluster on the landscape. The number of forested cells comprising each cluster identified in (1) is counted;
3) Assess suitability of clusters. Suitable forest clusters are those large enough to meet the area requirements of the species. Thus, forest clusters considered suitable for a species with an area requirement of 10 ha are > 111 forested cells in size in this application;
4) Produce species-defined landscape map. Only the suitable forest clusters (connected habitat meeting the MAR of the species) that were identified in (3) are retained. The amount of forest habitat remaining on the landscape and the total number of territories (units of minimum area) that can be supported (suitable habitat area/minimum area) are calculated; and,
5) Analysis of connectivity for species-defined landscapes. The species-defined landscape map generated in (4) was then surveyed for the presence of a percolation cluster (forest cluster that spans the landscape) using the species' gap-crossing abilities to quantify whether landscapes are connected.

To illustrate the implementation of this approach, consider a landscape with 10% forest cover (Figure 2). For a species with a MAR of 100 ha, most (99.6%) of the available habitat is suitable for this species if the forest cover is clumped (total suitable forest cover on species-defined landscape = 9.9%; Figure 3). Three of the patches on the original landscape are sufficiently large to meet the area requirements of this species; only a few isolated cells of forest on the periphery of the larger patches and one small patch are not suitable (compare Figures 2 and 3). In fragmented landscapes, however, less than half (46.2%) of the available habitat is suitable if the species lacks gap-crossing abilities (total forest cover on species-defined landscape = 4.6%; Figure 3). If the species has gap-crossing abilities, however, nearly twice as much forest cover is now suitable (total forest cover on species-defined landscape = 8.9%; Gap = 3, Figure 3). For gap-sensitive species, the clumped forested landscape is clearly the preferable management scenario, given that it supports nearly twice as many territories (or units of minimum area) as the fragmented landscape and 40% more territories than expected for a species with good gap-crossing abilities on the same fragmented landscape (Figure 3). Population sizes are thus likely to be greater on the clumped forested landscape. This does not indicate whether landscapes with such limited forest cover will actually be able to support viable populations of this species, however.

4.1.4 Overview of analyses
Analyses were conducted for gap-sensitive species (Gap = 0 cells) on clumped and fragmented landscapes, because differences were expected to be greatest between forest management scenarios for these species (80 landscapes x 4 levels of MAR = 320 species-defined landscapes analyzed). Initial analyses revealed that gap-crossing ability had a greater effect on the suitability and connectivity of fragmented landscapes than on clumped landscapes. Therefore, subsequent analyses focused on fragmented landscapes (40 landscapes x 4 levels of MAR x 3 levels of gap-crossing ability = 480 species-defined landscapes analyzed).

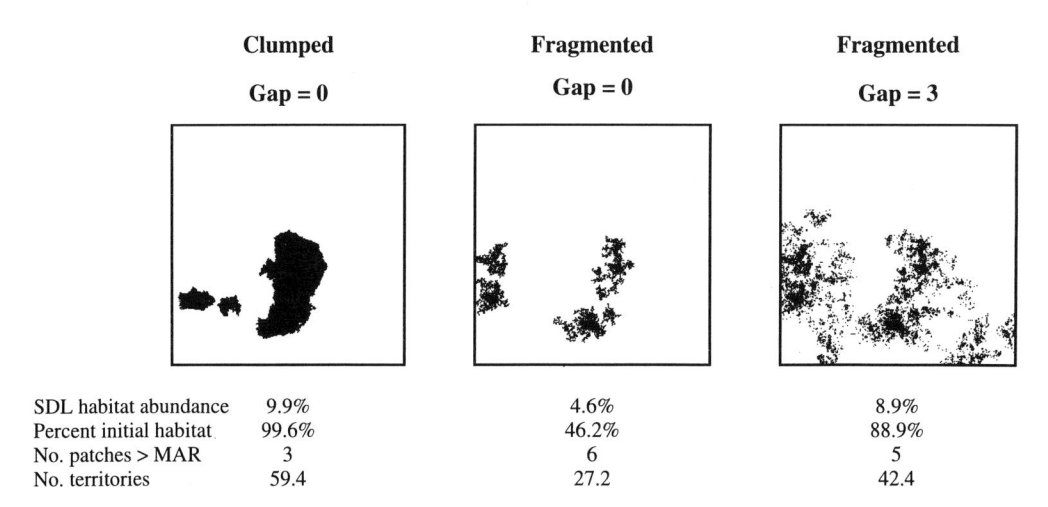

	Clumped Gap = 0	Fragmented Gap = 0	Fragmented Gap = 3
SDL habitat abundance	9.9%	4.6%	8.9%
Percent initial habitat	99.6%	46.2%	88.9%
No. patches > MAR	3	6	5
No. territories	59.4	27.2	42.4

Figure 3. Species-defined landscapes. Landscapes are either 10% clumped or fragmented landscapes (see Figure 2). The suitability of forest cover is affected by the scale at which the species interacts with landscape pattern (i.e., its gap-crossing abilities). For a species requiring a minimum area of 100 ha, most of the forest cover on the clumped landscape is available even to a species incapable of crossing gaps of unsuitable habitat ("gap-sensitive species"). Only half of the available forest cover is suitable for gap-sensitive species on the fragmented landscape. If the species has good gap-crossing abilities, however, most of the forest can be utilized, although fewer territories can be supported on the fragmented landscape.

4.2 Results

Gap-sensitive species (Gap = 0 cells) were able to utilize nearly all of the available habitat when forest management preserved large tracts of forest (Clumped landscapes, Figure 4). The proportion of suitable habitat on the landscape exhibited a precipitous decline for gap-sensitive species with large area requirements (MAR = 100, 1000 ha), however. This is because the total amount of habitat on the landscape was less than the MAR of the species ($\leq$1% for species with MAR = 100 ha; <10% for species with MAR = 1000 ha). In contrast, gap-sensitive species on fragmented landscapes perceived a much faster rate of habitat loss relative to the actual percentage of forest cleared (Figure 4). Fragmentation significantly reduced the amount of suitable forest for gap-sensitive species if more than 40% (MAR = 100, 1000 ha) or 50% (MAR = 1, 10 ha) of the forest was cleared. The difference between management scenarios (effect of fragmentation) is greatest when habitat is limiting (e.g, 1-20%). Note, for example, that a species with large area requirements (MAR = 1000 ha) could not be supported on a fragmented landscape with only 20% forest cover, but nearly 80% of available habitat would be deemed suitable (0.8 × 20% available = 16% suitable) if the landscape was managed to create large forest tracts (Figure 4). This is not to suggest that this will be sufficient habitat to support viable populations of this species, however.

Gap-crossing abilities clearly mitigated the effects of fragmentation (Figure 5). Even if a species was only able to cross a single cell of non-forest (30 m; Gap = 1 cell), a greater proportion of available forest cover was perceived to be suitable than for gap-sensitive species on fragmented landscapes. Species with good gap-crossing abilities (Gap = 3 cells) were generally able to use most of the available habitat on fragmented landscapes (Figure 5). At low levels of habitat abundance (<30%), however, less of the available habitat is suitable for these species on fragmented landscapes then for gap-sensitive species on clumped landscapes

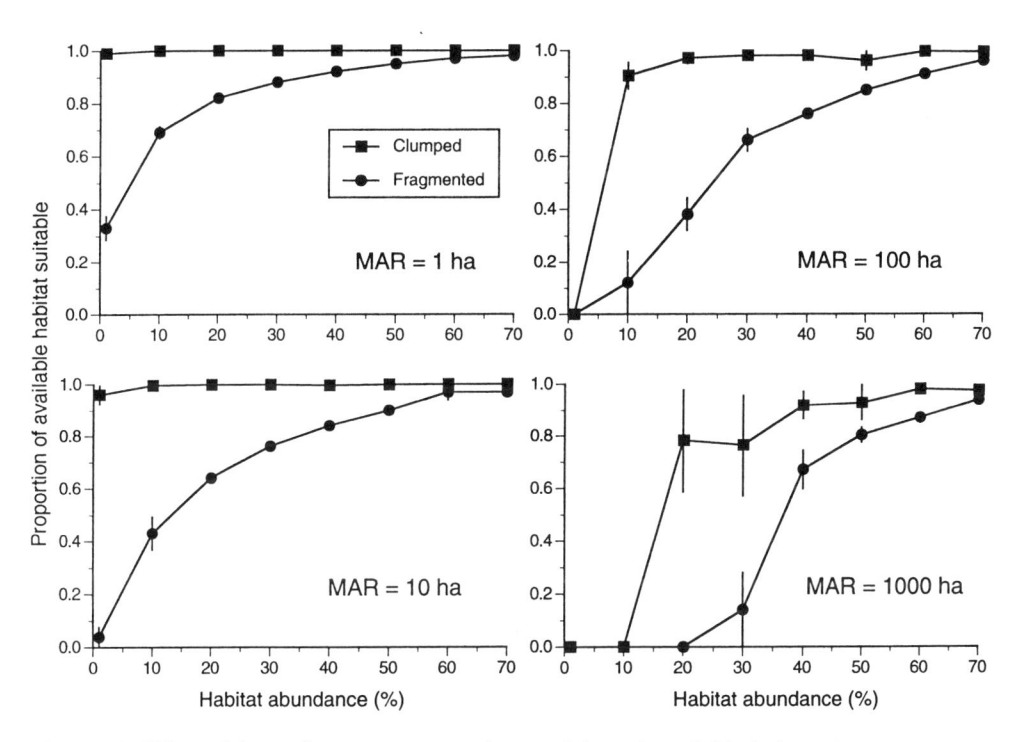

Figure 4. Effect of forest fragmentation on the suitability of available habitat for gap-sensitive species (incapable of crossing gaps of unsuitable habitat) that differ in minimum area requirements. If all of the available habitat on the landscape can be utilized by the species, then the proportion of available habitat that is suitable is 1.0. "Clumped" and "Fragmented" refer to the different forest management scenarios depicted in Figure 2. Error bars are ± 1 SE.

(compare Figures 4 and 5). Still, the number of territories that could be supported on a clumped and fragmented landscape at a given level of habitat abundance was about the same if the species has good gap-crossing abilities (Gap = 3 cells; Figure 6). Fragmented landscapes support smaller populations of gap-sensitive species, however, especially for species with moderate to large area requirements (Gap = 0 cells; Figure 6).

All landscapes containing at least 40% forest cover were connected, if we adopt the convention of defining connectivity as a >50% probability that a percolation cluster occurs on the landscape (Figure 7; Pearson and Gardner, 1997). The critical threshold at which landscapes became disconnected was influenced by the degree of forest fragmentation and the species' gap-crossing abilities. Gap-sensitive species on fragmented landscapes perceived a disconnected landscape when forest cover decreased to about 40%, whereas such species on clumped landscapes perceived a connected landscape until 30% forest cover (Figure 7A). Note, however, that there is a higher probability that fragmented landscapes will be connected than clumped landscapes in the range of 40-60% habitat cover (Figure 7A). Habitat is more dispersed on fragmented fractal landscapes, and thus it is more likely that a single cluster of forest cover that spans the landscape will emerge at this range of habitat abundance. If the organism possesses gap-crossing abilities, all landscapes with >20% cover are connected (e.g., Fragmented landscape, Gap = 3 cells; Figure 7B). Thus, fragmented landscapes are functionally connected, if not physically connected, for species with gap-crossing abilities. Such species are able to integrate forest cover across areas of unsuitable habitat to fulfill their MAR.

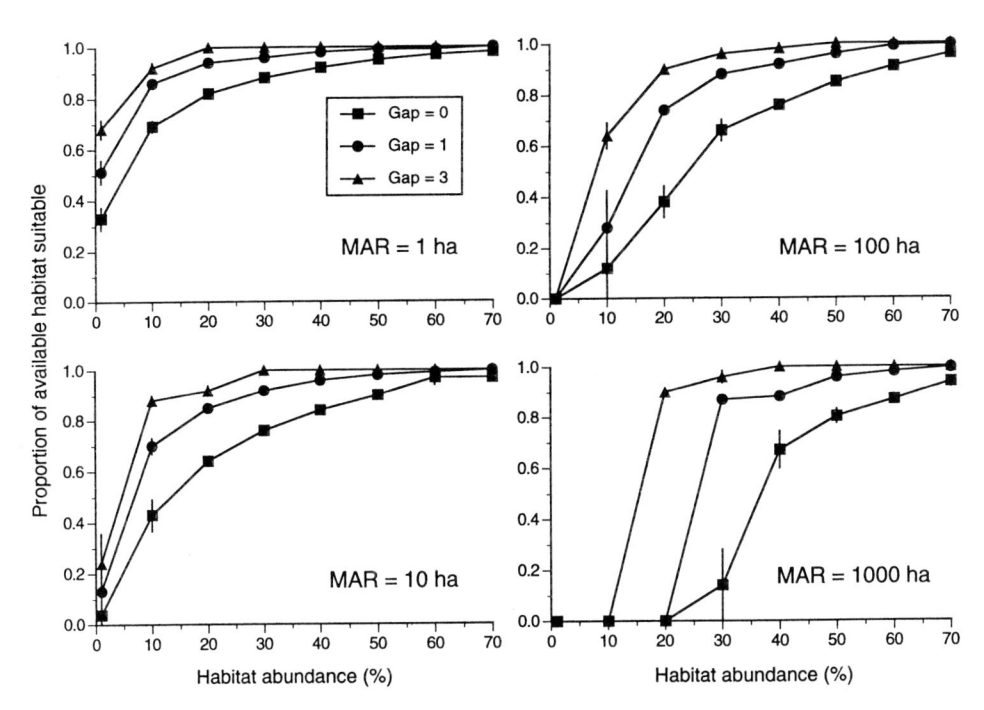

Figure 5. Effect of gap-crossing abilities on the suitability of available habitat on fragmented landscapes for species that differ in minimum area requirements. Gap-crossing ability refers to the "movement neighborhood" of the organism; that is, the ability or willingness to cross gaps of unsuitable habitat (e.g., Gap = 1 is a 24-cell neighborhood in which the species is able to cross gaps 1-cell wide). Error bars are ± 1 SE.

4.3 Discussion

It may seem surprising at first that a fragmented landscape can be connected, or that it could have a higher degree of connectivity than landscapes with more intact habitat distributions. Our surprise should serve as a reminder that landscapes are neither inherently fragmented nor connected. Assessment of landscape connectivity is based on the scale at which organisms interact with the scale of fragmentation. In this analysis, species with good gap-crossing abilities, which therefore operate at a broader spatial scale than the fine-scale pattern of timber harvest depicted in the fragmented fractal landscapes, perceived all fragmented landscapes as connected until only about 10-20% forest remained. Such landscapes are considered to be *functionally connected*, because the organism is able to connect disjunct habitat patches via dispersal. Gap-sensitive species, however, were more affected by the scale of fragmentation, which complicated the assessment of landscape connectivity. Fragmented landscapes (timber harvest conducted at fine spatial scales) were *more likely* to be perceived as connected than clumped landscapes (timber harvest conducted at coarse spatial scales) if about half (40-60%) of the forested landscape was cleared (Gap = 0, Clumped vs. Fragmented; Figure 7).

Habitat clusters appear to coalesce sooner, forming larger patches, on fragmented fractal landscapes than on clumped fractal landscapes within this particular range of available habitat (Figure 5 in With and King, 1999a). This is a consequence of the more dispersed habitat distribution on fragmented landscapes, which makes it easier for patches to coalesce into a percolation cluster (the defining characteristic of a connected landscape in the neutral landscape/percolation framework). At 30% habitat, however, gap-sensitive species no longer perceived fragmented landscapes to be connected, although most clumped landscapes were still

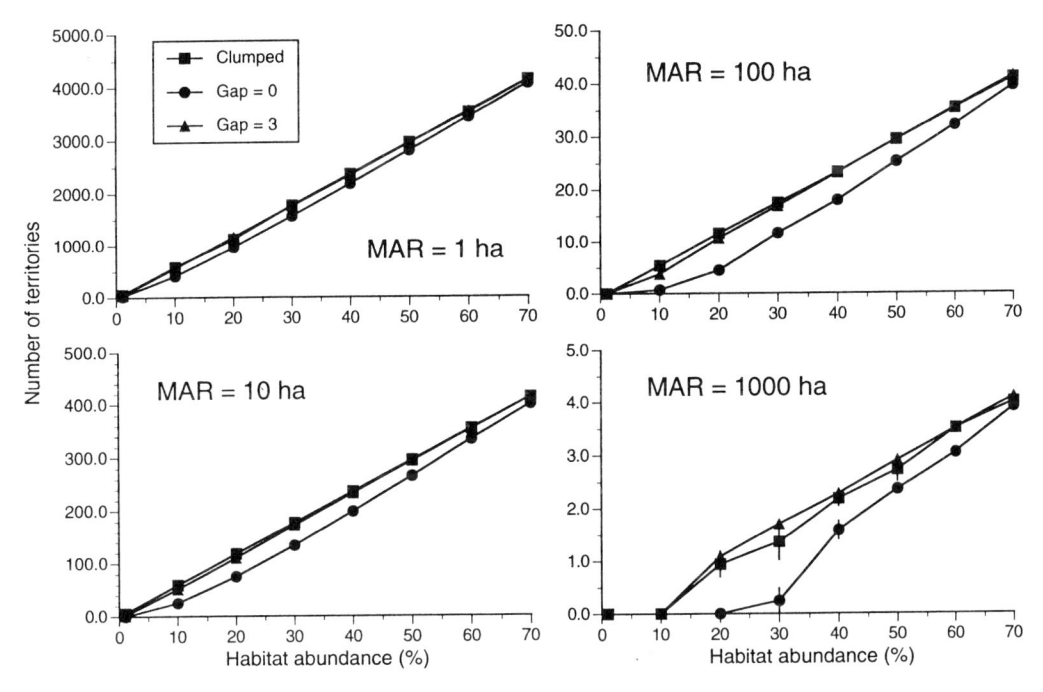

Figure 6. Effect of forest fragmentation and gap-crossing abilities on the number of territories that can be supported on the landscape. Population sizes are expected to be lower for gap-sensitive species (Gap = 0) on fragmented landscapes, then for such species on "Clumped" landscapes. Species with good gap-crossing abilities (Gap = 3) are able to utilize most of the available habitat, and thus fragmented landscapes support as many territories as for gap-sensitive species on clumped landscapes. Error bars are ± 1 SE.

connected at this level. In addition, the disruption of landscape connectivity occurred more precipitously (i.e., as a threshold) in fragmented landscapes than in clumped landscapes (Figure 7).

While landscape connectivity (or habitat connectivity at some scale) is a necessary condition for wildlife management, is it sufficient? Consider a scenario where half of the landscape is slated for timber harvest. If forest management is concerned solely with the maintenance of connected landscapes, then one might interpret the results of this analysis to suggest that fine-scale fragmentation would be preferable to a coarser scale of deforestation. Gap-sensitive species would then be assured of a connected landscape. [Note that this is a different recommendation than presented by Pearson et al. (1996) because a different neutral landscape model (fractal vs. hierarchical random) and movement rule (Rule 2 vs. Rule 1) is being used to assess landscape connectivity for gap-sensitive species.] However, the amount of suitable habitat (connected habitat meeting individual area requirements), and thus the number of territories that can be supported, is substantially reduced on fragmented landscapes relative to clumped landscapes, particularly for gap-sensitive species with large area requirements (MAR = 100, 1000 ha; Figures 5-7). Fewer territories presumably equates to smaller populations, and smaller population sizes should enhance extinction risk for species in fragmented landscapes. This was borne out in a recent modeling synthesis that coupled a generalized metapopulation model with fractal neutral landscapes to predict extinction thresholds for species that differ in dispersal ability and life-history traits (With and King, 1999b). Species on clumped fractal landscapes either did not go extinct, or persisted longer across a greater range of habitat loss, than species on fragmented fractal landscapes.

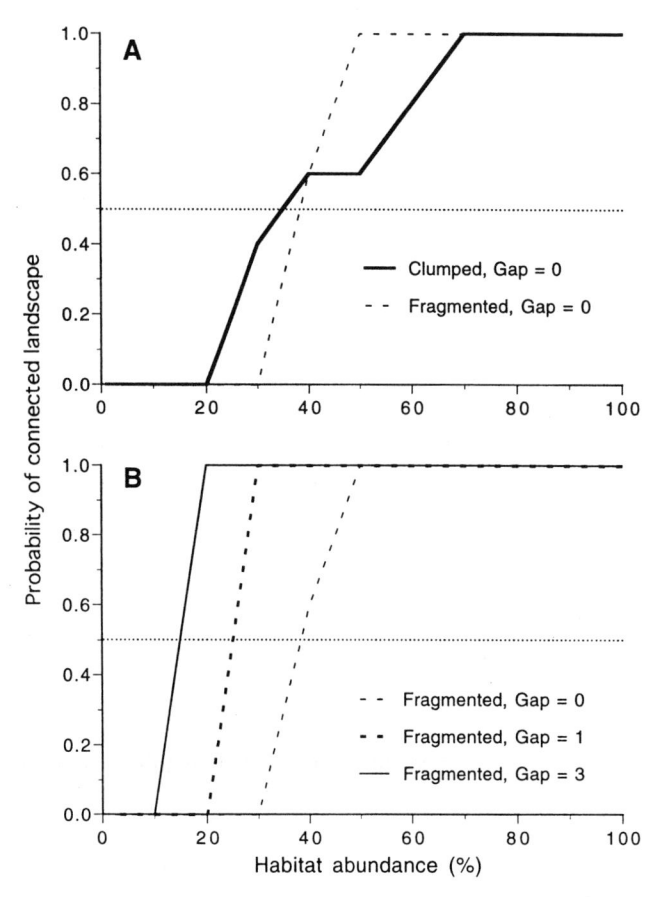

Figure 7. Landscape connectivity in different forested landscapes (Clumped vs. Fragmented), as assessed by species with different gap-crossing abilities. Connectivity is defined as a >50% probability that a percolation cluster (a single cluster of habitat that spans the landscape) occurs on the landscape (see Figure 1). Thus, all landscapes are connected if forest cover >40%. A. Comparison of landscape connectivity in clumped and fragmented forests for species without gap-crossing abilities (Gap = 0). B. Comparison of landscape connectivity in fragmented forests for species that differ in gap-crossing ability.

Thus, landscape connectivity may not be sufficient for wildlife management. Fragmented landscapes may be functionally connected, but support smaller, less viable populations than clumped landscapes that lack overall connectivity. This apparent paradox can be resolved by considering the scale at which connectivity is assessed. In this analysis, landscape connectivity is being inferred from the scale at which organisms are able to integrate forest cover (based on gap-crossing ability) to meet their MAR. Thus, the suitability of available habitat on the landscape for a given species is defined as the amount of connected habitat that meets the area requirements for individuals of that species. This is accomplished by identifying patches of connected habitat that are at least as large as the MAR for that species. If a patch of connected habitat spans the landscape, the overall landscape is considered to be connected (i.e., habitat connectivity implies overall landscape connectivity).

Is this an appropriate scale at which to assess landscape connectivity, however? Strictly speaking, percolation measures the resistance of the overall landscape to movement. Theoretically, an individual could traverse the entire landscape on the percolation cluster of a connected landscape (Gardner et al., 1989). This definition of landscape connectivity would have little relevance to a territorial vertebrate or to organisms bounded by physiological and energetic constraints that limit their foraging range, however. Such organisms are unlikely to operate at the spatial extent of the entire landscape, which is the scale at which connectivity is being defined. Such scale mismatches in how connectivity is defined relative to the process being studied may explain why thresholds in landscape connectivity (i.e., percolation thresholds) often do not coincide with thresholds in ecological responses to habitat fragmentation (e.g., Wiens et al. 1997, With and King 1999a, b).

Does this represent a failure of percolation theory to define landscape connectivity in a meaningful way for wildlife management? Not necessarily. The key is to define connectivity at a level appropriate to the scale at which movement or dispersal occurs. However, assessment of connectivity is complicated further by the fact that many species may exhibit different scales of movement in response to different scales of patchiness on the landscape (Kotliar and Wiens, 1990; Pearson et al., 1996). For example, birds are able to fly across extensive regions of the landscape in search of suitable habitat, but are often very specialized in their foraging requirements such that habitat edges, roads and waterways may be barriers to movement at a finer scale within territories (Lidicker and Koenig, 1996). Thus, the scale at which connectivity is assessed at one level (connected habitat that will fulfill breeding requirements) may be inappropriate for assessing connectivity at another level (connected patches of breeding habitat required for population persistence).

Connectivity thus needs to be assessed at multiple scales. At the finest scale, gap-crossing abilities could be used to quantify the amount of suitable connected habitat that is capable of fulfilling the individual area requirements of the species (this study). Dispersal distances, the ability and propensity of individuals to move between suitable habitat patches, might then be used to assess the connectivity of the overall landscape, and also to evaluate the need for habitat corridors or "stepping stones" to connect isolated fragments. Connectivity should be evaluated within a hierarchical framework (e.g., King, 1997). For example, Reed Noss (1991) identified three scales of connectivity that might be necessary in species management: 1) fencerow scale, which is the connection of close habitat patches by corridors, such as hedgerows connecting isolated woodlots; 2) landscape mosaic scale, which are larger corridors connecting major landscape features, such as riparian vegetation along rivers; and 3) regional scale, in which broad-scale landscape corridors connect nature reserves. Although these recommendations are based on a very literal definition of connectivity (i.e., corridors), they nevertheless demonstrate how connectivity can be assessed and maintained at different scales. Maintenance of connectivity at a variety of scales will ultimately be necessary for the successful conservation and management of wildlife.

4.4 Management Considerations

The analysis of species-defined landscapes presented in this chapter should not be misconstrued as endorsement for a particular landscape management scenario as being superior over another in maintaining habitat or landscape connectivity. Such was not the intent of this analysis. Instead, this analysis illustrated the application of neutral landscape models to demonstrate how connectivity (defined by percolation clusters) can be quantified from the perspective of species that operate at different scales on the landscape, highlighted some of the difficulties inherent in assessing landscape connectivity, and provided some general guidelines regarding the definition and management of landscape connectivity. Some general recommendations that emerge from this analysis are as follows:

1) *Landscape connectivity needs to be assessed from an organismal perspective.* A landscape is not inherently connected or fragmented. Assessment of connectivity is based on the scale at which organisms interact with the scale of fragmentation on the landscape;

2) *Connectivity is a necessary, but not sufficient, condition for evaluating landscape suitability.* A fragmented landscape may be connected, and yet have less suitable habitat (connected

habitat meeting the individual area requirements for the species) and support smaller populations than a more clumped habitat distribution;

3) *Connectivity needs to be assessed across multiple scales and species.* A species may exhibit different scales of movement which may influence its response to different scales of patchiness that exist on the landscape. Thus, connectivity perceived at a broad scale does not guarantee that patch structure at a fine scale will also be perceived as connected, and vice versa. Furthermore, various species on the same landscape will likely have different perceptions as to whether the landscape is connected at a particular scale;

4) *Connectivity must be maintained across a variety of scales and for a variety of species.* A hierarchical approach to the maintenance and management of landscape connectivity must be adopted to ensure connectivity at the different levels necessary to sustain minimum viable populations and to ensure the regional persistence of the species (e.g, connected habitat to fulfill MAR at the level of the individual, connected habitat patches at the level of the population, and connected populations at the level of the metapopulation); and,

5) *Habitat corridors may not be strictly necessary to achieve connectivity.* Percolation theory may facilitate the identification of critical regions on the landscape that facilitate dispersal (percolation clusters), and which need to be maintained or restored to promote landscape connectivity (e.g., Keitt et al., 1997). Because habitat patches within the percolation cluster may be functionally connected by dispersal for species with gap-crossing abilities, this argues more for the protection of habitat "stepping stones" rather than continuous strips of habitat (corridors). Nevertheless, corridors may enhance dispersal success for even vagile species by reducing mortality incurred during dispersal (e.g., cougars; Beier 1993, 1996).

The analytical approach presented here, founded on percolation theory, can be implemented for real species on actual landscapes to assess connectivity, to identify regions of the landscape essential in preserving connectivity, or to evaluate how different scenarios of forest management may disrupt connectivity. Nevertheless, it should be obvious that the assessment of landscape connectivity, from whatever perspective and at whatever scale, is dependent upon information on species' movement responses to patch structure, the gap sizes they are willing to cross, or the distances they can disperse. Such basic information is unavailable for most vertebrates (Lidicker and Koenig, 1996). While innovative approaches have recently been developed to obtain these data (e.g., Sieving et al., 1996; Rail et al., 1997; Desrochers and Hannon, 1997; Zollner and Lima, 1997), this continues to be a research priority. In the meantime, neutral landscape models may at least permit a general assessment of how patterns of deforestation affect landscape connectivity from the perspective of different species, which is a necessary vista if we are to adopt meaningful conservation strategies (Hansen and Urban, 1992).

Acknowledgments. My research on species' responses to landscape structure and the application of neutral landscape models in conservation biology has been supported by the Conservation and Restoration Biology Program of the National Science Foundation through grants DEB-9532079 and DEB-9610159. I thank A. W. King for developing the script files that permitted the rapid analysis of "species-defined landscapes" and R. H. Gardner for the use of RULE, the Fortran program that generated the fractal neutral landscapes. Lenore Fahrig, R. K. Bangert, and S. J. Cadaret provided constructive comments on this chapter, which have hopefully enabled me to clarify some of the ideas presented herein.

LITERATURE CITED

Addicott, J. F., Aho, J. M., Antolin, M. F., Padilla, M. F., Richardson, J. S., and Soluk, D. A. 1987. Ecological neighborhoods: scaling environmental patterns. *Oikos* 49, 340-346.

Anderson, G. S., and Danielson, B. J. 1997. The effects of landscape composition and physiognomy on metapopulation size: the role of corridors. *Landscape Ecology* 12: 261-271.

Beattie, M. 1996. An ecosystem approach to fish and wildlife conservation. *Ecological Applications* 6, 696-699.

Beier, P. 1993. Determining minimum habitat areas and habitat corridors for cougars. *Conservation Biology* 7, 94-108.

Beier, P. 1996. Metapopulation models, tenacious tracking, and cougar conservation. Pages 293-323 in McCullough, D. R., editor. *Metapopulations and Wildlife Conservation.* Island Press, Washington, D. C.

Beier, P., and Noss, R. F. 1998. Do habitat corridors provide connectivity? *Conservation Biology* 12, 1241-1252.

Bennett, A. F., Henein, K., and Merriam, G. 1994. Corridor use and the elements of corridor quality: chipmunks and fencerows in a farmland mosaic. *Biological Conservation* 68, 155-165.

Dale, V. H., Pearson, S. M., Oferman, H. L., and O'Neill, R. V. 1994. Relating patterns of land-use change to faunal biodiversity in the central Amazon. *Conservation Biology* 8, 1027- 1036.

Desrochers, A., and Hannon, S. J. 1997. Gap crossing decisions by forest songbirds during the post-fledging period. *Conservation Biology* 11, 1204-1210.

Dunning, J. B., Jr., Borgella, Jr., R., Clements, K., and Meffe, G. K. 1995. Patch isolation, corridor effects, and colonization by a resident sparrow in a managed pine woodland. *Conservation Biology* 9, 542-550.

Franklin, J. F. 1993. Preserving biodiversity: species, ecosystems, or landscapes? *Ecological Applications* 3, 202-205.

Forbes, S. H., and Boyd, D. K. 1997. Genetic structure and migration in native and reintroduced Rocky Mountain wolf populations. *Conservation Biology* 11, 1226-1234.

Gardner, R. H., and O'Neill, R. V. 1991. Pattern, process and predictability: the use of neutral models for landscape analysis. Pages 289-307 in Turner, M. G., and Gardner, R. H., editors. *Quantitative Methods in Landscape Ecology.* Springer-Verlag, New York.

Gardner, R. H., O'Neill, R. V., and Turner, M. G. 1993. Ecological implications of landscape fragmentation. Pages 208-226 in Pickett, S. T. A., and McDonnell, editors. *Humans as Components of Ecosystems: Subtle Human Effects and the Ecology of Populated Areas.* Springer-Verlag, New York.

Gardner, R. H., Milne, B. T., Turner, M. G., and O'Neill, R. V. 1987. Neutral models for the analysis of broad-scale landscape pattern. *Landscape Ecology* 1, 19-28.

Gardner, R. H., O'Neill, R. V., Turner, M. G., and Dale, V. H. 1989. Quantifying scale-dependent effects of animal movement with simple percolation models. *Landscape Ecology* 3, 217-227.

Gustafson, E. J., and Gardner, R. H. 1996. The effect of landscape heterogeneity on the probability of patch colonization. *Ecology* 77, 94-107.

Gutiérrez, R. J., and S. Harrison. 1996. Applying metapopulation theory to spotted owl management: a history and critique. Pages 167-185 in McCullough, D. R., editor. *Metapopulations and Wildlife Conservation.* Island Press, Washington, D. C.

Haas, C. M. 1995. Dispersal and use of corridors by birds in wooded patches on an agricultural landscape. *Conservation Biology* 9, 845-854.

Hansen, A. J., and Urban, D. L. 1992. Avian response to landscape pattern: the role of species' life histories. *Landscape Ecology* 7, 163-180.

Hansson, L. 1991. Dispersal and connectivity in metapopulations. *Biological Journal of the Linnean Society* 42, 89-103.

Henein, K., and Merriam, G. 1990. The elements of connectivity where corridor quality is variable. *Landscape Ecology* 4, 157-170.

Henderson, M. T., Merriam, G., and Wegner, J. 1985. Patchy environments and species survival: chipmunks in an agricultural mosaic. *Biological Conservation* 31, 95-105.

Hess, G. R. 1994. Conservation corridors and contagious disease: a cautionary note. *Conservation Biology* 8, 256-262.

Hobbs, R. J. 1992. The role of corridors in conservation: solution or bandwagon? *Trends in Ecology and Evolution* 7, 389-392.

Hutchinson, G. E. 1953. The concept of pattern in ecology. *Proceedings of the National Academy of Science of the USA* 105, 1-12.

Keitt, T. H., Urban, D. L., and Milne, B. T. 1997. Detecting critical scales in fragmented landscapes. *Conservation Ecology* [online] 1, 4. Available from the Internet. URL: http://www.consecol.org/vol1/iss1/art4.

King, A. W. 1997. Hierarchy theory: a guide to system structure for wildlife biologists. Pages 185-212 in Bissonette, J. A., editor. *Wildlife and Landscape Ecology: Effects of Pattern and Scale.* Springer-Verlag, New York.

Kotliar, N. B., and Wiens, J. A. 1990. Multiple scales of patchiness and patch structure: a hierarchical framework for the study of heterogeneity. *Oikos* 59, 253-260.

LaPolla, V. N., and Barrett, G. W. 1993. Effects of corridor width and presence on the population dynamics of the meadow vole (*Microtus pennsylvanicus*). *Landscape Ecology* 8, 25-37.

Lidicker, Jr., W. Z., and Koenig, W. D. 1996. Responses of terrestrial vertebrates to habitat edges and corridors. Pages 85-109 in McCullough, D. R., editor. *Metapopulations and Wildlife Conservation.* Island Press, Washington, D. C.

Machtans, C. S., Villard, M.-A., and Hannon, S. J. 1996. Use of riparian buffer strips as movement corridors by forest birds. *Conservation Biology* 10, 1366-1379.

Mann, C. C., and Plummer, M. L. 1995. Are wildlife corridors the right path? *Science* 270, 1428-1430.

Noss, R. F. 1991. Landscape connectivity: different functions at different scales. Pages 27-39 in Hudson, W., editor. *Landscape Linkages and Biodiversity,* Island Press, Washington, D. C.

Nowak, R. M. 1991. *Walker's Mammals of the World.* Fifth edition. Volumes 1 and 2. Johns Hopkins University Press, Baltimore, Maryland.

Pearson, S. M., and Gardner, R. H. 1997. Neutral models: useful tools for understanding landscape patterns. Pages 215-230 in Bissonette, J. A., editor. *Wildlife and Landscape Ecology: Effects of Pattern and Scale.* Springer-Verlag, New York.

Pearson, S. M., Turner, M. G., Gardner, R. H., and O'Neill, R. V. 1996. Pages 77-95 in Szaro, R. C., and Johnston, D. W., editors. *Biodiversity in Managed Landscapes: Theory and Practice.* Oxford University Press, Oxford.

Petit, L. J., Petit, D. R., and Martin, T. E. 1995. Landscape-level management of migratory birds: looking past the trees to see the forest. *Wildlife Society Bulletin* 23, 420-429.

Pither, J., and Taylor, P. D. 1998. An experimental assessment of landscape connectivity. *Oikos* 83, 166-174.

Plotnick, R. E., and Gardner, R. H. 1993. Lattices and landscapes. Pages 129-157 in Gardner, R. H., editor. *Lectures on Mathematics in the Life Sciences: Predicting Spatial Effects in Ecological Systems.* Volume 23. American Mathematical Society, Providence, Rhode Island.

Rail, J.-F., Darveau, M., Desrochers, A., and Huot, J. 1997. Territorial responses of boreal forest birds to habitat gaps. *Condor* 99, 976-980.

Salwasser, H. 1991. New perspectives for sustaining diversity in U.S. National Forest ecosystems. *Conservation Biology* 5, 567-569.

Sieving, K. E., Willson, M. F., and DeSanto, T. L. 1996. Habitat barriers to movement of understory birds in fragmented south-temperate rainforest. *Auk* 113, 944-949.

Simberloff, D., and Cox, J. 1987. Consequences and costs of conservation corridors. *Conservation Biology* 1, 63-71.

Simberloff, D., Farr, J. A.., Cox, J., and Mehlman, D. W. 1992. Movement corridors: conservation bargains or poor investments? *Conservation Biology* 6, 493-504.

Stauffer, D., and Aharony, A. 1991. *Introduction to Percolation Theory.* Second edition. Taylor and Francis, London.

Taylor, P. D., Fahrig, L., Henein, K., and Merriam, G. 1993. Connectivity is a vital element of landscape structure. *Oikos* 68, 571-573.

Thomas, J. W. 1996. Forest service perspective on ecosystem management. *Ecological Applications* 6, 703-705.

Wegner, J. F., and Merriam, F. 1979. Movement by birds and small mammals between a wood and surrounding farmland habitat. *Journal of Applied Ecology* 16, 349-358.

Wiens, J. A. 1989. Spatial scaling in ecology. *Functional Ecology* 3, 385-397.

Wiens, J. A. 1994. Habitat fragmentation: island vs landscape perspectives on bird conservation. *Ibis* 137, S97-S104.

Wiens, J. A. 1995. Landscape mosaics and ecological theory. Pages 1-26 in Hansson, L., Fahrig, L., and Merriam, G., editors. *Mosaic*

Landscapes and Ecological Processes. Chapman and Hall, London.

Wiens, J. A., Schooley, R. L., and Weeks, Jr., R. D. 1997. Patchy landscapes and animal movement: do beetles percolate? *Oikos* 78, 257-264.

With, K. A. 1994. Using fractal analysis to assess how species perceive landscape structure. *Landscape Ecology* 9, 25-36.

With, K. A. 1997. The application of neutral landscape models in conservation biology. *Conservation Biology* 11, 1069-1080.

With, K. A., and Crist, T. O. 1995. Critical thresholds in species' responses to landscape structure. *Ecology* 76, 2446-2459.

With, K. A., and King, A. W. 1997. The use and misuse of neutral landscape models in ecology. *Oikos* 79, 219-229.

With, K. A., and King, A. W. 1999a. Dispersal success on fractal landscapes: a consequence of lacunarity thresholds. *Landscape Ecology* 14, 73-82.

With, K. A., and King, A. W. 1999b. Extinction thresholds for species in fractal landscapes. *Conservation Biology* 13, *in press.*

With, K. A., Gardner, R. H., and Turner, M. G. 1997. Landscape connectivity and population distributions in heterogeneous environments. *Oikos* 78, 151-169.

Zollner, P. A., and Lima, S. L. 1997. Landscape-level perceptual abilities in white-footed mice: perceptual range and the detection of forested habitat. *Oikos* 80, 51-60.

Edge effects: Theory, evidence and implications to management of western North American forests

Laurie Kremsater and Fred L. Bunnell

"Edge effects" in forests range from the beneficial effects for which the term was first coined, to potential negative effects that are more frequently examined today. We first examine microclimatic and vegetation distributions around edges, then summarize plant and animal distributions near edges. We expose probable reasons for the distributions where we can, and briefly review studies of processes that could influence distribution and are often concentrated around edges (e.g., predation and nest parasitism). There is tremendous variability in responses around edges, both for abiotic variables and biotic responses. Findings from studies in a non-forested matrix (most of the current literature) do not transfer appropriately to conditions within an area of forest stands of different ages. Evidence for positive edge effects is similar in all studies (species richness generally increases with edge and more species are positively associated with edges than are negatively associated with edges). Evidence of negative edge effects differs starkly between eastern or mid-western North America and coastal montane forests. In the east and midwest many studies document increased predation and parasitism near edges; in the Pacific Northwest researchers have found little effect of patch area or negative edge effects, and no species consistently associated with forest interior. Reported edge effects range from 0 to several hundred meters into forests. Microclimatic effects generally extend about 1 to 3 tree heights into the forest, but edge influences on distributions of organisms or factors affecting organisms (such as predation and nest parasitism) are concentrated within 50 m of the edge. Effects of disturbance from road traffic can extend farther.

Key words: forest edges, edge effects, forest management, biological diversity, microclimate, predation, nest parasitism

1. INTRODUCTION

In theory, forest edges are easy to define. In practice, they are more complex and their features and effects are elusive. Edges are places where plant communities meet or where successional stages or vegetation conditions within plant communities come together (Giles 1978; Thomas *et al.* 1979; Forman and Godron 1986). Conceptually, edges have environments significantly different from the interior of adjacent patches and hence also typically differ in biomass, soil characteristics, and species composition and abundance (Forman 1986). In reality, many edges are regions of continual change. Clements (1907) described an ecotone (edge) as a tension zone where principal species from adjacent environments reach their limits. For practical purposes we can identify

Centre for Applied Conservation Biology, University of British Columbia, Vancouver, Canada

four kinds of forest edges: natural, (relatively) permanent edges (e.g., forest next to lakes), permanent anthropogenic (e.g., forests next to agricultural or urban area), natural successional (e.g., edges caused by fires or windthrow), and anthropogenic successional (e.g., edges caused by timber harvesting). Edges have distinct features. Thomas et al. (1979) described edge structure in terms of edge contrast, which refers to the degree of structural difference along the boundary between two habitats. Recent concepts of edge extend beyond vegetation structure to include microclimatic attributes and ecological processes (e.g., succession, tree mortality, competition, predation, nest parasitism). When first investigated, edges were documented as positive features of the landscape for wildlife (Leopold 1933; Odum 1971). Edges were seen as important habitats rich in food and cover, either because they abutted two quite different habitats or because they were discrete habitats of themselves. Documented edge effects included increased abundance of game birds and mammals. More recently potential negative effects of edges, such as increased predation and parasitism, promotion of habitat generalist species, and reduction in forest interior species have attracted interest (Laudenslayer 1986; Soulé 1986; Reese and Ratti 1988; Yahner 1988; Harris 1989; Laurance and Yensen 1991; Start 1991).

Studies of edges can be separated into those describing species abundance and richness near edge habitats without evaluating underlying causes for those distributions, and those that attempt to discover processes affecting species' distributions and abundance near edges. Our goal is to describe vertebrate distributions around edges, exposing probable reasons for the distributions where we can. In the sections following we first examine microclimatic and vegetation distributions around edges, then summarize vertebrate species' distributions near edges, briefly review studies of processes that could influence distribution and are often concentrated around edges (e.g., predation and nest parasitism), and finally summarize the major findings relevant to forest practices.

2. EDGES AND MICROCLIMATE

Microclimates at edges are different from those in the adjacent forest or opening, with measures at edges often intermediate between the two habitats (Figure 1). Measures of depth of edge have been documented for solar radiation, incident light, humidity, temperature, and wind speed (Forman and Baudry 1984; Forman and Godron 1986; Young 1988; Chen et al. 1993). Generally the ground in clearcuts receives more direct solar radiation and loses more outgoing long wave radiation than does ground under a forest canopy (Matlack 1993; Dunsworth and Arnott 1995). Radiation strongly influences temperature regimes. In cleared areas daytime air temperatures are usually higher and nighttime temperatures usually lower than in more heavily vegetated areas (Geiger 1965; Bunnell et al. 1985; Dunsworth and Arnott 1995). Open areas and edges usually have greater temperature ranges at the soil surface and upper soil layers, and greater incidence of frost than do forested areas (Geiger 1965; Saunders et al. 1991; Chen et al. 1993; Scougall et al. 1993).

Edges may experience different rainfall, interception, and evapotranspiration regimes and wind dynamics than do surrounding habitats (e.g., Swanson et al. 1988). Chen et al. (1993) reported that mean daily average relative humidity increased from the clearcut into the forest, although sometimes edges had the least humidity. The distance over which wind speed declines as air flows from open areas into adjacent forests (Geiger 1965; Saunders et al. 1991) depends on the structure and density of vegetation. Sometimes a dense growth of herbaceous and woody vegetation near the edge reduces the distance to which microclimatic effects extend into forests (Wales 1972; Ranney et al. 1981; Forman and Godron 1986). Chen et al. (1993) noted that wind velocities were consistently highest in clearcuts, intermediate in edges, and lowest in forest areas.

Deviations from these generalizations occur depending on the weather and aspect, and forest cover in the immediate vicinity. The form of the function describing the decline of microclimate variables near edges depends on several factors including the orientation of the edge with respect to solar angle or wind direc-

Edge effects: theory, evidence and implications 119

| 200 | 150 | 100 | 50 | 0 | 50 | 100 | 150 | 200 | 250 |

Air temperature
Concannon 1995[b]
Chen et. al. 1990[a]
Young & Mitchell 1994
Matlack 1993

Wind
Chen et al.
Concannon

Humidity/ VPD
Concannon[c]
Chen et al.[c]
Young & Mitchell[d]
Matlack[c]

Radiation
Matlack
Young & Mitchell

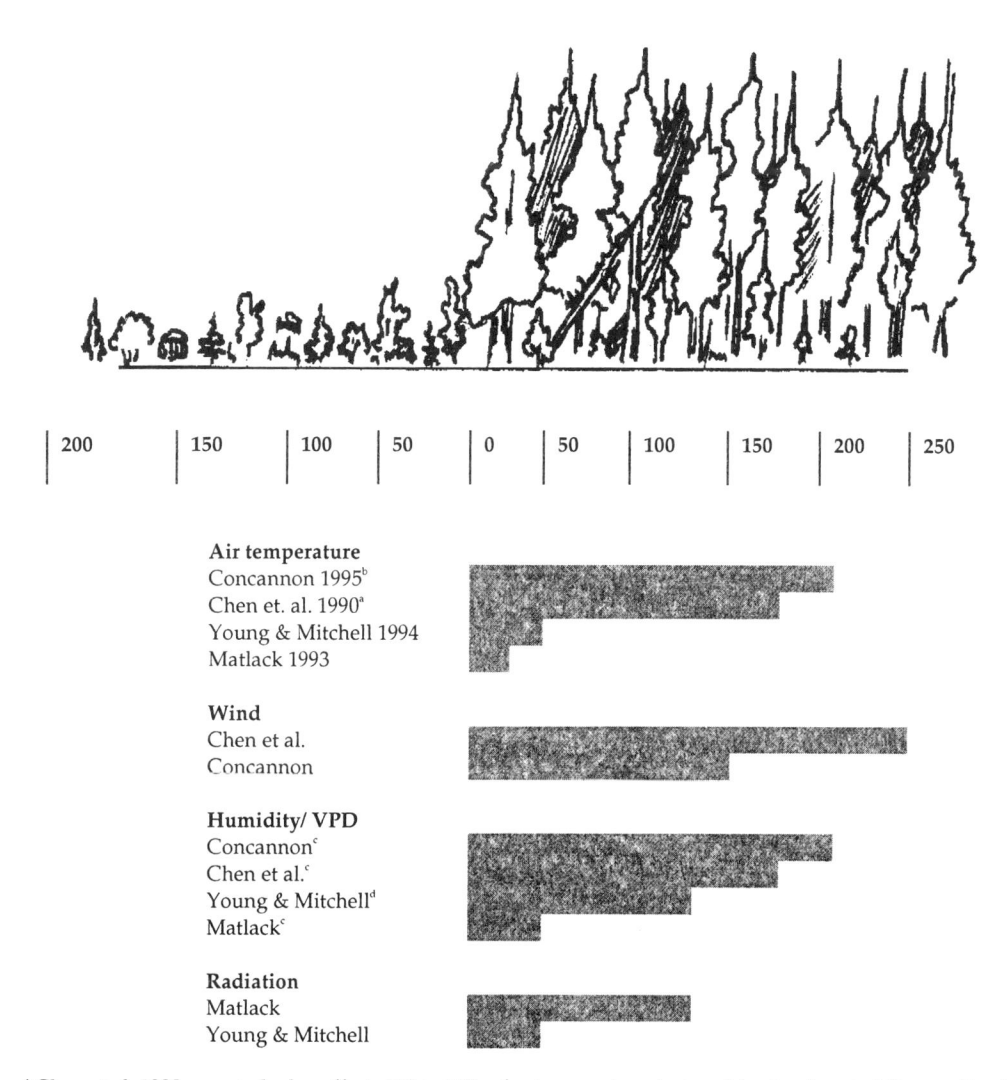

a/ Chen et al. 1990 reported edge effects 120 to 180m for temperature, but could extend up to 240m on South aspects and very hot days.
b/ We have only Concannon's abstract which reports only the value of 200m.
c/ Relative humidity reported.
d/ Vapour pressure deficit reported.

Figure 1. Reported depths of microclimate effects from edges into forests. These values are summaries. Refer to the text for detail.

tion. The magnitude of these effects differs with time of year and latitude as a function of solar angle (Geiger 1965; Ranney *et al.* 1981; Koppenaal and Mitchell 1992; Chen *et al.* 1993; Young and Mitchell 1994), and with prevailing weather.

Because physical variables are more easily and precisely measured than biological variables, microclimatic edge effects are the most quantified forest edge effects. Generalizations for the Pacific Northwest suggest that microclimatic effects extend up to 2 to 3 tree heights (about 100 m to 150 m) inside a patch (Figure 1; discussions of Franklin and Forman 1987; Fritschen et al. 1971 in Harris 1984; Lehmkuhl and Ruggiero 1991). A distance of 2–3 tree heights is consistent with review of the silvicultural literature. Smith (1962:206) reported that the silvicultural literature suggests that microclimatic conditions at the centre of an opening with a diameter 2–3 times the height of the surrounding trees are similar to those of larger openings.

This general pattern changes with aspect -- warmer aspects (usually southern and western aspects) may have two or three times the depth of edge effects as compared to northern aspects (Whitney and Runkle 1981; Ranney et al. 1981; Forman and Godron 1986; Young 1988 in Bennett 1990a; Matlack 1993; Young and Mitchell 1994). Depths of effects can also be reduced by shrubby vegetation at edges that may reduce penetration of wind into forest patches and reduce light, humidity, and temperature changes inside forested patches (Gysel 1951; Ranney et al. 1981; Concannon 1995).

Figure 1 illustrates that specific microclimatic measures may be less than or greater than the generalization of 150 m to 160 m. In the eastern U.S. Matlack (1993) found most depths of edge for microclimate to be less than 50 m (but he looked only 50 m into the forest). Concannon (1995) found winds penetrated 120 m into forests from beach edges, but temperature and relative humidity penetrated only 30 m; effects on radiation and temperature penetrated 200 m into the forest. Chen et al. (1990) found below-canopy wind effects extended 240 m into the forest. They reported that temperature and humidity returned to forest interior conditions within 120 m to 180 m of the opening, but that edge effects for temperature and humidity could extend up to 240 m into the forest in extreme cases of hot, windy days and southern exposures, or could extend only a few meters on a cloudy day. Soil temperature and moisture approached 'interior' forest levels within 60 m to 120 m of the edge.

3. EDGES AND PLANT DISTRIBUTIONS

Plant responses to edges are usually a function of microclimate conditions. Changes in microclimate, wind, moisture and light regimes affect seedling germination, survival, and growth (Wales 1972; Bennett 1990b; Gates and Mosher 1981; Levenson 1981; Ranney et al. 1981; Ng 1983 in Laurance and Yensen 1991; Laurance 1989 in Laurance and Yensen 1991; de Casenave et al. 1995). For the Pacific Northwest Chen et al. (1992) and Zen (1995) reported changes in seedling survival and tree growth rates near edges—some species benefitted from edges, others did not. Wind from edges may transfer dust, detritus, and seeds into the forest interior (Geiger 1965; Cale and Hobbs 1991). In some cases this transfer may increase the movement of exotic plants into forests (Janzen 1983). The implications of changes in turbulent transfer for plant gas exchange and growth have not been examined. Moreover, nutrient cycling near edges is related to microclimatic conditions. Decomposition rates of litter near the edge may be greater than in forest interiors due to the high availability of soil moisture and higher temperatures that accelerate activity of fungi and other decomposer organisms (e.g., Edmonds and Bigger 1984).

Plant species common at successional edges include species that benefit from disturbance. Edges often have higher light intensity, greater foliage density, and more foliage layers than areas deeper inside the forest (Table 1). That pattern appears common around edges as less shade-tolerant herbs and shrubs from openings mingle with forest overstory. Shade-intolerant, mid- or early-successional vegetation or exotics often thrive along margins of forest patches (Ranney et al. 1981; Lovejoy et al. 1986; Alverson et al. 1988; Frost 1992; and Table 1). Table 1 collates reported distributions of vegetation around successional edges. Many of the species are shrubs or taller deciduous trees, supporting the concept that edges favour less shade-tolerant and more invasive or early seral species. The generality is weakened by the fact that many of these studies are somewhat anecdotal and were conducted in eastern forests where deciduous tree species are more common. If the pattern of deciduous and shrubby species being more common at edges is general, that should be reflected in the distribution of animals around edges (section following).

Radiation may be the most significant variable affecting plants at edges, but wind also exerts considerable influence. Increased exposure to wind at edges may increase damage to trees and alter forest structure and composition by pruning, windthrow, and desiccation (Ruth and Yoder 1953; Alexander 1964; Moen 1974; Ranney et al. 1981; Janzen 1986; Lovejoy et al. 1986; Brett 1989; Reville et al. 1990; Chen et al. 1992). Increased wind-shear forces near edges can cause elevated rates of treefalls and tree mortality that alter forest structure and tree species composition (Chen et al. 1990). Several authors have noted that edges may have more stressed, dead, and downed trees than do adjacent forests (Geiger 1965; Chen et al. 1992) and so may support a rich insect fauna. That condition provides forage for many birds, but may also encourage insect infestations of nearby timber. In warm edges, for example, a timber pest, Vienna moth (*Ocneria monacha*) may emerge earlier than its parasites so that populations build up and do more damage than under typical emergence patterns (Geiger 1965). Other insects (e.g., some bark beetles) are killed by exposure to heat and dryness.

Finally, because vegetation responds to radiation and other aspects of microclimate, one might expect edge effects for plants to extend inward for distances similar to those of microclimatic effects. In fact, edge effects on vegetation seem to penetrate less deeply into forests than do effects on microclimate. Perhaps many statistically significant changes in microclimate are not meaningful biologically, or perhaps microclimate variables are measured with greater accuracy and precision so that effects appear to extend farther than are possible to measure for biological organisms or processes. Most researchers report effects on vegetation composition of less than 50 m, many are less than 25 m, but values over 100 m have been reported, depending on the vegetation variables (e.g., Wales 1972; Caruso 1973; Wagner 1980; Franklin et al. 1981; Gates and Mosher 1981; Ranney et al. 1981; Chen et al. 1990; Palik and Murphy 1990; Williams-Linera 1990 in Chen et al. 1992; Angelstam 1992; Chen et al. 1992; Fraver 1994; Zen 1995; P. Burton pers. comm.). Responses of organisms presumably integrate the large variation in microclimatic influences noted above. The fact that changes in plant composition and abundance extend less far into forests than do microclimatic effects (animals also integrate effects similarly to plants and demonstrate narrower edge effects than those exhibited by abiotic variables, see section 4) demonstrates that extreme penetrations of microclimatic effects may be unreliable indicators of biological responses.

4. EDGES AND ANIMAL DISTRIBUTIONS

Many studies document trends in species abundance near edges, but do not evaluate reasons for the patterns (Table 2). Given the variability in both microclimatic changes and plant responses, we expect only the broadest patterns to be consistent. Implications of these patterns for vertebrate distributions can be summarized simply. Unlike permanent edges, as around lake shores or wetlands, successional edges undergo more rapid change and should experience high rates of vertebrate turnover. Because the plant species mix is unstable and changing, vertebrate affinities with edge likely will be more strongly related to structural features than to individual plant species (e.g., Lovejoy et al. 1986; Buechner 1987; Laurance 1987).

If vertebrates primarily respond to broad differences in vegetation (e.g., structure), the effective edge width should be narrower than it is for microclimatic variables (about 150 m) and closer to that exhibited by vegetation (about 15 m to 50 m, but see section 5). We can predict four types of species that we expect to respond positively to edges. First, small organisms which are relatively responsive to microclimatic changes and find the edge microclimate favourable (e.g., many insects and other arthropods) should be more abundant at edges than in the forest. The second group includes species feeding primarily on insects (some birds, amphibians, reptiles, and perhaps bats). Among insectivorous birds, those that exploit structural differences, as do flycatchers, should be well represented. A third group includes species that find openings more favourable for foraging and forests more favourable for cover. We expect more large than small species in this group simply because small species are more likely to find favourable

Table 1. Plants of edge and interior forests

Edge	Edge width and study	Interior[a]
More abundant salal	Chen et al. 1996. Oregon; Edge effect 10 m-15 m for˘ salal; rattlesnake plantain peaked at 250 m inside edge. Unclear how much these are idealized distributions for model exercise and what is based on data	Rattlesnake plantain
Higher number of dead trees Higher growth rate Higher density thin stemmed trees	Chen et al. 1992. S. Washington, C. Oregon forest patches adjacent to clearcut. Edge effect penetrated about 60 m into forest; interior generally greater than 120 m into forest	Higher canopy cover Higher growth rates
Bedstraw Blue cohosh Blue-stemmed golden rod Canada bluegrass Cleaver's bedstraw Common wood sorrel Hawthorns Motherwort Queen Anne's Lace Rough cinquefoil Staghorn sumac *Torilus anthriscus* White avens White trillium Wild rye	Gysel 1951. Michigan. Beech/maple forest surrounded by pasture. These species restricted to edge (not found in open or forest) or interior (not found in open or edge). Edge width from 0 m to about 6 m; interior plots were undocumented distance into woodlands that ranged from about 5 ha to 14 ha	Maple-leaved Viburnum Running strawberry bush

Higher old growth canopy density
Higher 2nd growth canopy density
More abundant:
  Hop hornbeam (ironwood)
  Butternut hickory
  White ash
	Palik and Murphy 1990. Eastern United States 10 m edge effect on north side; 20 m on south side; forest interior >25 m into forest	More abundant:
		American beech and sugar maple

More shrubs	LaRue et el. 1995. Quebec, birch-balsam fir. Edge plot was 350 m by 250 m on riparian edge; interior was at least 300 m away; stated shrubs were a "narrow" edge effect	

More shrubs.
Higher density and basal area of:
  Oaks (alba, rubra, velutina, coccinea)
  American blackhaw
  Sweet cherry
  Red maple
  White ash
  Shagbark hickory
	Wales 1972. New Jersey oak hickory forest adjacent to field. Edge effect for shrubs 10 m on north side, none on south side.	

Table 1. Continuing.

Edge	Edge width and study	Interior[a]
More non-native species [a]	Scougall et al. 1993. Australian eucalyptus forest. Non-native species declined sharply over 10 m into the forest, then continued to decline more gradually	

Higher tree basal area
Higher tree stem density
Strongly edge-oriented species:
  Black walnut
  Burr oak
  European buckthorn
  Hawthorns
  Hickory
  Manitoba maple
  Panicled dogwood
  Trembling aspen
  Wild plum
  Willow spp.
  Yellow birch
	Ranney et al. 1981. Wisconsin, deciduous forest adjacent to field. Edge effect penetrated about 15 m	Beech
		Red maple
		Sugar maple

Higher density of thin stems
Higher density & basal area of:
  Basswood
  Ironwood
  Musclewood
  Red oak
  Shagbark hickory
  White ash
	Whitney and Runkle 1981. Ohio, beech-maple. 4 ha–16 ha forests. In their study only sugar maple density or basal area greater in interior than edge (edge 10 m–20 m; unclear where interior is, but study plots 4 ha and 16 ha)	Sugar maple
		Beech

Higher density thin stemmed trees	Young and Mitchell 1994. Argentina semi-arid forest next to natural grassland. 10 m edge; interior was 50 m to 250 m inside forest depending on patch size	Higher basal area of trees. Higher density of thick stems

[a] The definition varies by study.
[b] This finding is widespread and occurs around permanent as well as successional edges (e.g., Naiman et al. 1993; Spackman and Huges 1995; Planty-Tabacchi et al. 1996)

cover in the opening. These species (e.g., black-tailed deer) would use both the open and forest habitat, but because they move back and forth their movements or sightings should be concentrated around edges. The fourth group is the predators of those species that concentrate use around edges. Predators, especially the smaller forms, also may be responding to the structural juxtaposition and exploiting concealment near an opening. It is more difficult to predict, from first principles, the kinds of species that should avoid edges. Edges are favourable to the four broad groups of species noted above, but are also dangerous places. Predation rates and nest parasitism rates are often elevated near edges (see section 5.2 and Marzluff and Restani this volume). Clearly, species seeking forest interiors should include those species for which edges confer no advantage (e.g., bark gleaners and conifer seed eaters). This broad group of forest interior species is thus very loosely defined as those that have become sufficiently proficient at foraging in forest interiors that they need not venture near dangerous edges. A second group of forest interior species may be those that benefit from microclimate conditions away from edges. Although microclimate measures are very variable near edges, conditions at edges and in forest interiors will be quite different during extreme weather events. During those conditions forest interiors could provide moderate conditions (e.g., provide moisture during drought for species such as bryophytes; provide protection for vertebrates during deep snow and strong winds).

4.1 Invertebrates

Warmer temperatures near edges have positive effects on some invertebrates and negative effects on others. For example, Edmonds and Bigger (1984) suggested that the microclimate at edges may affect abundance of insect species, and linked increased abundance of ants to increases in temperature. Similarly, Usher et al. (1993) reported that edges may have greater species richness of ground-dwelling arthropods as a result of favourable higher temperatures. Fuller and Warren (1991) argued that external south-facing woodland edges provide warmer habitats for insects than those found either in the open or within the woodland. Light, as well as temperature, affects invertebrate distributions near edges. For example, in British woodlands deleterious effects of shading on butterflies within narrow "rides" through forests is documented (Pollard 1982; Warren 1985; Warren et al. 1986).

Many studies do not document reasons for invertebrate responses to edge. Sisk and Margules (1993) studied ground-dwelling beetles in small plots (0.25 ha to 3.06 ha) of Eucalyptus forest. They sampled 641 "morphospecies" of which 22 were abundant enough to permit statistical tests; 3 of the 22 showed a significant response to edges (always positive). As with birds, many species were habitat specialists that occupied one habitat or the other, but did not change in density near the edge. Scougall et al. (1993), however, found little difference in ant abundance or species composition across forest:paddock edges not subject to livestock grazing. Wood and Samways (1991) noted that edges may serve as insect pathways; Huggard (pers. comm.) documented that *Gryllobats* seem to prefer edges and edge-rich treatments (0.1 ha patch cut arrays, perhaps because of invertebrate accumulations due to wind disturbance.

Distances of edge effects on invertebrates are poorly documented. Scougall et al. (1993) reported changes in ant species composition within 10 m of edges subject to livestock grazing. The response reflected reduced vegetative cover and associated changes in soil temperature and moisture. Brown (in Lovejoy et al. 1986) reported that "light-loving" butterflies typical of younger forest penetrated 200 m–300 m into older tropical forests.

4.2 Amphibians and reptiles

Reptiles and amphibians should respond differently to microclimatic conditions near edges—higher temperatures would favor reptiles but encourage desiccation among amphibians. Species of both groups, however, would benefit from elevated arthropod richness and abundance. For a few amphibian species, DeMaynadier and Hunter (1997) reported more captures beyond 25 to 35 m from edges. They noted plethodonid salamanders showed greater negative responses to edges than did anurans and also reported greater responses around high contrast edges. Amphibian richness in northwestern California was positively correlated with length of edge at both stand-level (their 'plot' scale) and landscape-level (their 1 000 ha block scale) scales (Rosenberg and Raphael 1986). Ensatina was negatively correlated with edge within plots; rough-skinned newt and Pacific tree frog were positively correlated with total edge over 1 000 ha blocks. Toads and Pacific tree frogs are more resistant to dessication than most amphibians, and the both have been reported as more abundant in clearcuts than in natural stands (Raphael 1988). Gates (1991) reported salamanders, lizards, and toads were more active along edges than inside forests or open areas, but advanced no reasons. Data of Welsh and Lind (1991) indicate greater abundance of reptiles in more open, older forest. Reptiles may also be seeking basking sites, and openings will serve as well as an edge. Gregory (1978 in Sadoway 1986) observed that garter snakes and northern alligator lizards use edge habitats.

4.3 Mammals

Early studies were stimulated because edges promote abundant game mammals and birds (Leopold 1933; Whitcomb et al. 1976). For example, edges created by clearcutting were documented as beneficial to deer and elk (Julander and Jeffery 1964; Reynolds 1962, 1966; Harper 1969; McCaffery and Creed 1969; Willms 1972; Weger 1977). In most studies, deer or elk use of open habitats (usually clearcuts) decreased with increasing distance from the edge. Further study has revealed that deer, elk, and sometimes moose (e.g., Eastman 1976; Kirchhoff and Schoen 1983; Nyberg and Janz 1990) fall into that group of species that concentrate their use around edges when forage is not sufficient within the forest. Interpretation of findings for such species depends on the habitat mosaic and scale of measurement, because edge effects vary with juxtaposition of openings (Reynolds 1962, 1966; Wallmo 1969; Lyon and Jensen 1980; Hanley 1983; Kremsater and Bunnell 1992). Contrary to early studies that showed benefits of edge, Kirchhoff and Schoen (1983) concluded that successional edges conferred no value over old growth to deer in southeast Alaska. They noted (as did Weger 1977 for coastal BC) that old-growth forest had the most beneficial mosaic of structure. Similarly, Kremsater and Bunnell (1990, 1992) and Nyberg and Janz (1990) reported that black-tailed deer preferred to be within 200 m either side of an edge, but noted that deer also occupied forest interior, particularly if the forest had small gaps permitting forage production.

We found little information on edge effects on large carnivores. A.N. Hamilton (pers. comm.) collected data from Vancouver Island showing that female black bears stayed close to edges, presumably to confer access to trees and safety.

Few studies document edge effects for non-game mammals. A. Chan-MacLeod's data (pers. comm.) from Alberta's boreal mixedwood forest show reduced abundance of deer mice, red-backed voles, and jumping mice in edges. Removal of downed wood from edges (to act as fire breaks) may have affected results. Preliminary data from dry Douglas-fir forests suggested that shrews prefer the clearcut part of south-facing edges in spring, show less attraction to edges in early summer, and are indifferent later in the year (Huggard and Klenner 1998). The authors speculated the spring pattern resulted from earlier snow melt (Huggard et al. 1998) and earlier plant growth likely causing higher invertebrate abundance. Godfryd and Hansell (1986) reported that abundance of some mammals (white-footed mice, meadow vole, masked shrews, northern short-tailed shrews) was positively associated more with edge length than edge width*. Rosenberg and Raphael (1986) reported two measures of edge (plot and 1 000-ha block). Sometimes species that are strongly associated with edges in plots (e.g., blue grouse) are negatively associated with high amounts of edge at the 1000 ha block scale. We summarize only results at the stand level. The three small mammals (wood mouse, deer mouse, and wood rat) that are positively associated with edge (Table 2) tend to omnivory. In Rosenberg and Raphael's study only one mammal spe-

---

* Length of edge may more clearly reflect total amount of edge habitat.

cies, the fisher, was negatively associated with edges at all scales of measurement, and could be considered a forest interior species. Mills (1995) reported that the southern red-backed vole (*Clethrionomys californicus*) has also been consistently documented as preferring forest interior.

Different responses to edge at different scales has also been documented in B.C. Huggard (pers. comm.) suggested that marten prefer edges at a local scale, but are negatively impacted by too many openings within larger areas.

Among mammals the primary effect of edges appears neutral or positive. Much of the positive response appears associated with foraging opportunities, provided by the edge itself or adjacent open areas, in close juxtaposition to taller vegetation that meets other needs. That appears true of members of the deer family responding positively to edges, omnivorous small mammals, predators, and bats (Table 2).

Among the mid-sized predators, the fisher is perhaps the most arboreal and most restricted to forested vegetation, but also ventures into openings to feed (reviews of Buskirk and Powell 1994 and Martin 1994). Smaller forest-dwelling species such as martens and weasels are reported as hunting along both permanent and successional edges (Buskirk and Powell 1994; Martin 1994; K. Lisgo pers. comm. Several studies have documented that bat activity is concentrated around edges (Krusic and Neefus 1996; Erickson and West 1996; Crampton and Barclay 1996). Although many bats may seek roost sites inside forest patches, their foraging is concentrated over clearcuts or near edges.

4.4 Birds

About one-third of commonly sampled bird species show no response to edge. Sisk and Margules (1993) report that 33% and 35% of species showed no response to edges of low and high contrast, respectively (some were habitat generalists; some, habitat specialists that avoid one of two juxtaposed habitats, but show no change in density near edges). Nonetheless, many studies report large numbers of bird species in edges (e.g., Lay 1938; Good and Dambach 1943; Edeburn 1947; Johnston 1947; Johnston and Odum 1956; Johnston 1970; Strelke and Dickson 1980). For example, most game birds seem to be edge-adapted (e.g., Dunn and Braun 1986), which is part of the reason they are accessible for hunting. Yahner (1983, 1984) reported high productivity of ruffed grouse and high diversity of birds in edges. Rosenberg and Raphael (1986) evaluated responses of 45 bird species (game and non-game) to edges in northwestern California. At the plot scale, 8 species were positively correlated with length of edge and 2 were negatively correlated ($p < 0.05$;Table 2). Data of Sisk and Margules (1993) also show more species responding positively to edges than avoiding them. Around edges of low contrast 46% of species peaked within about 150 m of edge and 21% of species avoided edges; around edges of high contrast 54% of species peaked near edges and 12% avoided edges. The greater disparity of response around edges of high contrast implies that structure elicits much of the response.

Increases in bird species richness at edges have been attributed to complex structure with shrubby vegetation (e.g., Anderson *et al.* 1977; Morgan and Gates (1982). Similarly, Strelke and Dickson (1980) attributed higher densities of great-crested flycatchers, eastern wood-pewees, and Carolina chickadees near edges to well-developed foliage layers. In British Columbia, Catt (1991) examined edges between old growth and pole/sapling stages or mature second growth and found increased bird density and bird species richness in edges. Although he found little evidence of distinct edge vegetation, his findings suggest that lower contrast edges (i.e., mature forest next to old growth) had a smaller increase in bird density and richness than did higher contrast edges (i.e., pole/sapling to old growth).

Edge effects: theory, evidence and implications

Table 2. Birds and mammals reported as more common at forest edges or forest interior

Prefer Edges	Prefer Interior
Predominantly eastern	*Predominantly eastern*
*bluejay [g/]	Acadian flycatcher [g/h/]
brown thrasher [ab/]	barred owl [o/]
Carolina chickadee [i/]	bay-breasted warbler [x/]
Carolina wren [q/]	black and white warbler [e/j/]
chestnut-sided warbler [a/s/x/]	blackburnian warbler [x/]
*eastern wood pewee [i/]	black-throated green warbler [f/]
field sparrow [m/v/]	*bluejay [j/s/x/]
*great-crested flycatcher [a/j/]	brown-headed nuthatch [o/]
indigo bunting [a/m/s/v/]	Canada warbler [a/]
magnolia warbler [a/]	cerulean warbler [f/]
northern bobwhite [h/]	eastern phoebe [h/]
northern cardinal [g/h/m/o/v/ab/]	*eastern wood pewee [a/h/]
orchard oriole [n/]	*great crested flycatcher [h/]
red-headed woodpecker [n/o/]	Kentucky warbler [h/m/]
summer tanager [g/h/]	northern parula [a/]
*tufted titmouse [g/q/]	ovenbird [e/g/h/s/z/]
white-eyed vireo [q/]	pine warbler [h/j/o/]
*woodthrush [q/x/]	*tufted titmouse [j/o/v/]
	*wood thrush [a/z/]
	worm-eating warbler [e/f/]
	yellow-throated vireo [f/z/]
Both eastern and western	*Both eastern and western*
American crow [m/]	Bewick's wren [ai/]
American goldfinch [m/y/]	*blue-gray gnatcatcher [f/ai]
American robin [n/x/]	brown creeper [b/f/]
blue grosbeak [n/]	downy woodpecker [h/m/]
*blue-gray gnatcatcher [l/]	golden-crowned kinglet [aa/]
brown-headed cowbird [p/s/]	*hairy woodpecker [h/]
chipping sparrow [l/n/w/]	hermit thrush [f/w/]
common yellowthroat [a/m/x/]	*orange-crowned warbler [ai/]
Eastern kingbird [n/]	*red-eyed vireo [m/o/s/]
gray catbird [s/ab/]	Swainson's thrush [y/]
*hairy woodpecker [ak/]	white-breasted nuthatch [o/]
house sparrow [n/]	winter wren [x/]
house wren [l/m/aa]	
loggerhead shrike [n/]	
northern flicker [l/h/]	
northern mockingbird [n/ai]	
*orange-crowned warbler [y/]	
purple finch [aa/ai]	
*red-eyed vireo [q/]	

Table 2. continuing

Prefer Edges	Prefer Interior
Both eastern and western	*Both eastern and western*
song sparrow [a/y/]	
sparrow hawk [o/]	
spotted towhee [a/g/h/s/w/ab/ai]	
Tennessee warbler [x/]	
warbling vireo [aa/]	
white-crowned sparrow [y/]	
Wilson's warbler [aa/]	
yellow-billed cuckoo [m/v/]	
Predominantly western	*Predominantly western*
brown towhee [ai/]	ash-throated flycatcher [ai/]
bushtit [ai/]	black-headed grosbeak [aa/]
California quail [ai/]	**bushtit** [ai/]
dark-eyed junco [ai/]	chestnut-backed chickadee [ai/]
dusky flycatcher [w/]	**dark-eyed junco** [ai/]
Hutton's vireo [ai/]	**Hutton's vireo** [ai/]
mountain bluebird [l/]	plain titmouse [ai/]
northern pygmy owl [aa/]	**rufous hummingbird (female)** [v/]
Nuttall's woodpecker [ai/]	Townsend's solitaire [aa/]
olive-sided flycatcher [aa/]	varied thrush [y/]
rufous hummingbird (male) [v/]	**western flycatcher** [aa/]
rufous hummingbird [y/]	**western wood pewee** [ai/]
scrub jay [ai/]	
violet-green swallow [ai/]	
western flycatcher [aa/]	
western wood pewee [aa/]	
wrentit [ai/]	
European	
chaffinch [i/]	brambling [l/]
great tit [i/]	capercallie [i/]
great spotted tit [i/]	chiffchaff [i/]
tree pipit [i/]	

Table 2. continuing

Prefer Edges	Prefer Interior
Mammals	*Mammals*
Columbian black-tailed deer[t/ab/ac/]	Sitka black-tailed deer[r]
moose[ad/]	
mule deer[u/]	
rocky mountain elk[v/]	
Roe deer[c/]	
Roosevelt elk[ab/]	
white-tailed deer[k/]	
woodmouse[c/]	California red-backed vole[d/]
deer mouse[aa/]	western gray squirrel[aa/]
wood rat[aa/]	
marten[af/]	striped skunk[aa/]
weasel[af/]	fisher[aa/]
black bears[ag/]	
bat spp.[ah/]	

[a] Small and Hunter 1989. Maine, forests surrounded by powerlines and rivers. [b] Keller and Anderson 1992. Wyoming, did not look at edges *per se* but at fragmented and unfragmented stands. [c] Hansson 1994. S.C. Sweden, boreal forest. [d] Mills 1995. S.W. Oregon, Klamath Mtn, only looked 90m into forest. [e] Hahn and Hatfield 1994. Eastern.U.S [f] Askins et al. 1987. Connecticut, oak-hickory-hemlock forest. [g] Kroodsma 1984. Eastern Tennessee, oak-hickory in agricultural matrix. [h] Kroodsma 1982. Eastern Tennessee. [i] Hansson 1983. Sweden, noted seasonal differences in edge/interior preference. [j] Strelke and Dickson 1980. East Texas, 30+ old pine in <30 yr old pine. [k] Clark and Gilbert 1982. Central Ontario. [l] O'Meara et al. 1981. Colorado, pinyon juniper. [m] Johnston 1947. Eastern U.S. [n] Johnston and Odum 1956 Georgia piedemont. [o] Lay 1938. Texas [p] Brittingham and Temple. 1983. Eastern deciduous forest in agricultural matrix. [q] Morgan and Gates 1982. Eastern U.S [r] Kirchhoff and Schoen 1983. Alaska. [s] Yahner 1987. Pennsylvania. [t] Hanley 1983. Alaska [u] Reynolds 1966. Arizona [v] Johnston 1947. Eastern U.S. [w] Sedgwick 1987. Pinyon-juniper. [x] Ferris 1979. Maine. [y] Hansen and Peterson. Unpublished data. Pacific Northwest forests. [z] Temple and Carey 1988. Eastern U.S. [aa] Rosenberg and Raphael 1986. Western U.S. [ab] Gates and Gysel 1978. Eastern U.S. [ac] Nyberg and Janz 1990. B.C. [ad] Kremsater and Bunnell 1992. B.C. [ae] Eastman 1976. B.C. [af] Buskirk and Powell 1994. [ag] Hamilton pers. comm. [ah] Krusic and Neefus 1996. [ai] Sisk and Margules 1993. [ak] Klenner and Huggard 1999.
[*] Indicates spp that have been found as 'edge' species in some studies and 'interior' in others.

The influence of structural differences around edges is evident in the work of Sisk and Margules (1993) who examined bird responses around edges of high contrast (grassland to oak woodland) and low contrast (chaparral to oak woodland). The dark-eyed junco and common bushtit were more abundant near edges of high contrast, but preferred oak interior when edges were of low or "soft" contrast. Figure 2 }summarizes broad patterns among species reported as preferring or avoiding edges (Table 2); species reported as both preferring and avoiding edges are excluded. Patterns consistent with findings reported for vegetation are apparent. Some species preferentially seek shrubby habitat; they represent 42% of species reported as preferring edges and 6% of species reported as avoiding edges (Figure 2b). Shrubs benefit both ground and shrub nesters. Ground and shrub nesters represent 56% of species reported as preferring edges (cowbirds excluded) and 29% of those avoiding edges (Figure 2a). Relations with "openness" were expressed as three classes: species preferring large openings (L), such as bobwhite or field sparrow; species preferring smaller openings in forested or wooded areas (S), such as Acadian flycatcher or indigo bunting; and species preferring closed canopies (C), such as Swainson's thrush or winter wren (Figure 2c). Species preferentially seeking large or small openings represent 89% of species reported as preferring edges; only one species (brown-headed nuthatch) avoiding edges appears to prefer larger openings to forage.

The above observations imply two points. First, that the richness around edges results from incursions by species preferring more open, often more shrubby, habitat. Second, that apparent avoidance of edges is more than an avoidance of openings (61% of species reported as avoiding edges in specific studies use at least small openings to forage). Understory foragers and ground gleaners represent 59% of species reported as preferring edges and 45% of species avoiding edges; bark gleaners and wood borers represent only 7% of species preferring edges, but 39% of species reported as avoiding edges (Figure 2d). Dietary habits also differ in ways not summarized in Fig. 2. A greater array of preferences is evident in species preferring edges (59% insectivorous, 24% omnivorous, 11% granivorous, 4% herbivorous, and 1 owl). Species avoiding edges were predominantly insectivorous (90%) plus 2 omnivores and an owl.

A wide range of distances of edge effects for birds has been documented. Most studies report edge effects of less than 50 m for birds, similar to that for vegetation. For example, in southcentral BC, Klenner and Huggard (1997) found that spruce grouse avoided only the first 5-10 m into the forest from an edge during winter and three-toed woodpeckers generally nested close to edges (see also Martin and Eadie 1998). Similarly, Gates and Gysel (1978) in Michigan found that over half of all bird nests were within 15 m of the edge of the woods. Gates and Mosher (1981) reported that depth of edge (indicated by nest distributions) ranged from 9 m to 64 m. Strelke and Dickson (1980) studied pine/hardwood stands greater than 30 years old adjacent to a pine plantations less than 3 years old. They found more birds in the first 50 m into the forest than at greater distances. Higher densities were associated with well-developed foliage layers. Hansson (1983) noted that more birds were observed within 50 m of edge inside the forest, particularly of tree gleaning species (implying a positive effect of edges on some insects). Several other Scandinavian studies report higher densities of birds within 50 m of an edge than in the forest interior (e.g., Haila *et al.* 1980; Hansson 1983; Helle 1983; Vickholm 1983 in Virkalla 1987; Helle 1986). Increased bird abundance in the edge appeared due to more shrubby cover (Helle 1983) or greater foliar volume at edges (Hansson 1983). Findings of a rather narrow influence of edges on abundance and richness is consistent with findings of workers in British woodlands (e.g., Fuller and Whittington 1987; Fuller 1988; Fuller and Warren 1991).

Edge effects extend into openings as well as into forests; just as openings increase bird species diversity in adjacent forests, forests influence bird species diversity in openings. Anderson *et al.* (1977) reported bird populations in a 12-m wide cleared swath were reduced relative to forest populations, but in a 30.5-m swath bird diversity and density were higher than in the forest. Swaths wider than 30.5 m had lower diversity of bird species (as compared to the forest) and contained many birds typically found in open habitats. Similarly, Niemuth (1993) found that species richness and abundance of grassland and shrubland birds were correlated positively with opening size (savannah patches) within Wisconsin pine barrens. Negative edge effects extending into openings have been documented with about the same frequency as negative effects into forests (review of Paton 1994; Burger et al. 1994). We found few researchers (e.g., Niemuth and Boyce 1997), however, that have suggested expanding opening sizes to reduce negative edge effects on bird species preferring open habitats.

There is little evidence that any species are found exclusively in edges; the higher richness near edges results from mixing of open and more closed communities. Laudenslayer and Balda (1976) could find no evidence for a distinct breeding bird population in narrow relatively permanent edges between stands of pinyon pine, ponderosa pine, and juniper. O'Meara et al. (1981) found one species, blue-gray gnatcatcher, occurred exclusively within narrow inherent edges in pinyon-juniper stands (but see Table 2). In their review of effects of forest practices on wildlife in Oregon, Hagar et al. (1995) reported that only the olive-sided flycatcher was strongly, and consistently, associated with edges; but this species is reported as preferring forest interior in other studies (e.g., Temple and Carey 1988; Sisk and Margules 1993).

Nor does there appear to be strong evidence of bird species requiring only forest interior habitat, especially in the Pacific Northwest (e.g., Table 2; Bunnell et al. 1997). In Sweden, Hansson (1983) noted that while edge effects disappeared rapidly into the forest, some rare bird species were observed only deep inside the forest during breeding. Reviewing vertebrate distributions relative to clearcut edges in the boreal forest, Hansson (1994) found no evidence of forest interior species. Although some species in the Pacific Northwest have been documented as more abundant when there is less edge or at greater distance from edge (Table 2), the review of Bunnell et al. (1997) suggested that there is no strong evidence that these species are area-sensitive when their abundance is evaluated over patches of different sizes (in a managed forest matrix).

Because a large proportion of literature about edge effects concerns effects on birds, we summarize the main points we derive from our review of edge effects and birds:

1) Whatever defines an "edge-related" or "forest-interior" bird species, it is not simply characteristics of that species.

The majority of species of Table 2 were "edge" or "interior" species in only single studies despite being evaluated in several. Moreover, 15 of 48 (31%) of "interior" bird species were reported as "edge" species in other studies. Variation in affinity reflects two issues: 1) there is no shared, repeatable definition of either "edge" or "interior" species, and 2) species are assigned to "edge" or "interior" categories empirically ("forest interior" species are absent from small forested tracts or are reported to avoid edges). Response to edge incorporates differences between the two adjacent habitats plus other species present in the area. Even if agreed-upon definitions of "edge" or "interior" were available, bird species' associations with "edge" or "interior" would remain variable across locations. Preference for edge or interior seems to depend on the suite of local species, the nature of the edges (forest patch size and matrix), and the scale of observations (i.e., stands versus landscapes).

2) Relations with "edge" or "interior" are more strongly expressed among eastern than among western studies.

Omitting 15 species reported as both "edge" and "interior" in different studies, Table 2 summarizes 48 North American bird species reported as positively associated with edges; 10 of those are western species. A total of 33 species were reported to avoid edges; 6 have predominantly western ranges. Among species occurring in both the east and west, of the 27 edge-related designations, 20 were documented in eastern populations; of the 12 interior designations, 8 were documented in eastern populations. The trend is sufficiently consistent and strongly expressed that it probably reflects more than a greater number of eastern studies. It is more difficult to detect "forest interior" species when dealing with a forested landscape than when dealing with wooded patches in an agricultural landscape (see

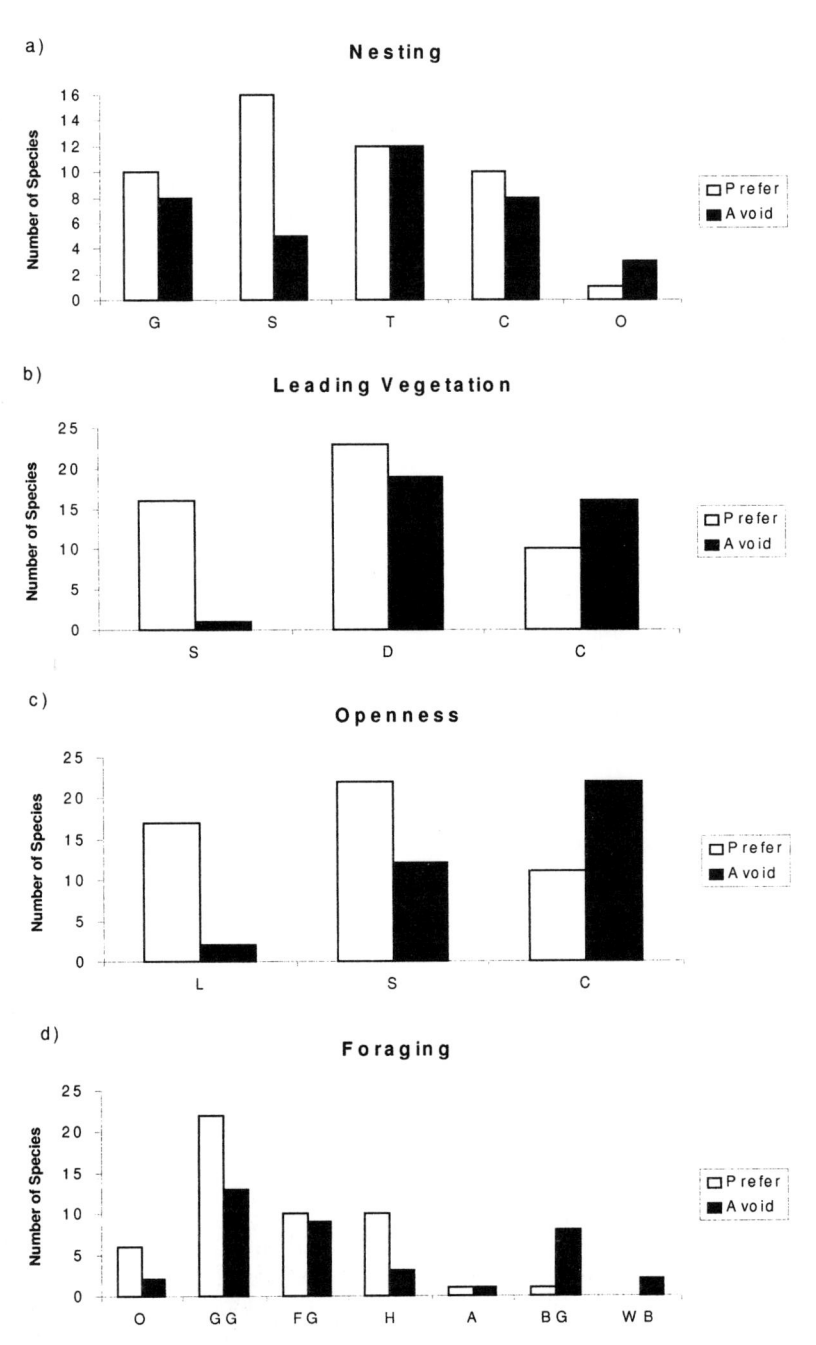

Figure 2. Number of birds reported to prefer or to avoid edges grouped by natural history features. a) preferred nesting sites: G = ground nesters; S = shrub nesters; T = tree nesters; C = cavity nesters; O = overhang (logs, bark slabs). b) preferred leading vegeatation: S = shrubby; D = deciduous trees; C = coniferous trees; c) preferred degrees of openness: L = large openings; S = small openings; C = closed canopy; d) foraging guilds. O = opportunistic (ground to canopy); GG = ground and understory understory gleaners; FG = tree foliage gleaners; H = hawk from perches; A = aerial; BG= bark gleaners; WB = wood borers.

Bunnell this volume). A large part of what determines an "interior" species appears to be its response to other species in adjacent habitat. The lack of consistent strong negative responses to edge in the Pacific Northwest forests is compatible with the weakly expressed response to stand age (Bunnell this volume) or patch size (Bunnell et al. 1997).

3) The variability in data suggests that it is unwise to transplant results of most eastern studies to managed western forests.

Unlike studies from areas where the matrix is more often agricultural or urban (see Marzluff and Restani this volume), the few studies from the managed forests of the Pacific Northwest (e.g., Schieck et al. 1995; McGarigal and McComb 1995) suggest little response to forest interior conditions. Where the matrix is agricultural or urban instead of early seral, or if the amount of forest area becomes more reduced than in those studies, more indications of negative edge effects will likely become apparent.

4) Structural features of edges promote increased stand-level richness of birds. It is not known whether the same or opposite may be true at regional levels. Abundant edge may affect regional diversity.

Blue grouse, for example, find their best spring and summer foraging in early seral stages or clearcuts (Campbell et al. 1990), but were negatively correlated with edge at landscape scales in the study of Rosenberg and Raphael (1986). (Perhaps winter habitat plays a role in response to measures of edge across landscapes – more information in necessary to interpret appropriate amounts of edge for several species.)

5) Changes in relationship to edge with scale of measurement and matrix habitat imply that more understanding is required before a monitoring approach can be recommended.

Length of edge, especially those of high contrast, is a potential interim measure, but even that can produce unreliable results, since results vary with scale and composition of matrix. The blue grouse example (noted in 4) above) also applies here.

6) Although there is no objective definition of "forest interior" conditions, for most bird species the major effect of edge occurs within the first 50 m from an opening.

Total forest area, the nature of the matrix, and mix of local species will influence assessments for both edge-related and apparent forest interior species. The bulk of data suggests that relations around edges potentially influencing forest interior habitat, are primarily a function of vegetative structure of adjacent habitat and the species inhabiting that habitat. Increased richness near edges appears to result from species (that prefer shrubby vegetation or more open areas) moving into forests. Many species reported as avoiding edges generally do not benefit from openness or shrubby vegetation and may simply be avoiding mortality agents near the edge. Mortality pressures, however, are not readily generalized. For example, species with open, cup-shaped nests are more vulnerable to both brood-parasitism and depredation. Such species represent 69% of those species reported as preferring edges but only 53% of those avoiding edges (the reverse of what would be expected if mortality agents were a dominating factor). Microclimate has not been documented as playing a role in preference for forest interior by some birds.

5. EDGES AND PROCESSES AROUND THEM

We considered abiotic processes that change the nature of edge habitat above (section 2). Processes of plant competition and invasion of alien species were discussed in section 3. Here we consider other processes acting around edges, including herbivory, nest parasitism and predation, and movements. We also treat human-induced effects such as roads, hunting, and pollution.

5.1 Herbivory

Deer, elk, and moose have all been reported as benefiting from openings and edges (Willms 1972; Eastman 1976: Hett et al. 1978; Alverson et al. 1988). Although edges have been documented as mostly beneficial, it is not clear whether preference for edges is because edges themselves have better forage abundance or quality than the opening or forest, or whether the apparent concentration of use around edges simply results from herbivores moving as little distance as possible to get from openings (preferred foraging areas) to forests (for shelter or concealment). Most likely the attraction of edges to herbivores is a function of both phenomena. A. N. Hamilton (pers. comm.) noted that berry-producing plants produced more fruit in the shade of edges; he also observed drying of understory shrubs on exposed edges after harvesting.

Because herbivory can be concentrated near edges, it may affect plant composition along edges, depress understories (Harmon and Franklin 1983), and even endanger rare plants (Bratton and White 1980). In Australia, differences in small mammal foraging patterns resulted in higher survival of rainforest seeds in pastures and edges than in forested habitats (Osunkoya 1994), perhaps because small mammals preferred forested habitats over pastures or edges. Also in Australia, Scougall et al. (1993) found that rates of removal of non-native seed decreased with distance into the forest; rates of removal of native seed were little affected.

5.2 Parasitism and predation

Literature documenting parasitism and predation around edges has focused on birds (Yahner 1988; Terborgh 1992; Paton 1994). Although generalist predators or competitors that benefit from edges can theoretically impact small mammals (Sievert and Keith 1985; Laurance and Yensen 1991), few studies address small mammals.

Parasitism of nests by cowbirds may have a significant impact in areas where wooded habitats are surrounded by agricultural areas or fields (Table 3). In eastern North America, where such conditions are relatively common, cowbird density has been inversely related to distance from open habitats (Brittingham and Temple 1983; Gates and Gysel 1978). Cowbirds typically feed in grasslands or agricultural areas and venture into forests to parasitize broods of other birds. Paton (1994) critically examined data around edges documenting brood parasitism. He found evidence of increased parasitism near edges in 3 of 6 samples.[†] Two of the 3 instances were studies documenting distance into openings from wooded elements (Best 1978; Johnson and Temple 1990). The third instance of elevated brood parasitism (by Brittingham and Temple 1983) documented rates of brood parasitism as a function of distance from 0.2 ha openings in the forest. Conclusions of Brittingham and Temple (1983) did much to stimulate concern about negative edge effects on neotropical migrants. Recently Bielefeldt and Rosenfield (1997) re-analyzed their data. The Acadian flycatcher provided the largest share (35%) of Brittingham and Temple's sample of 105 nests (13 species) and a majority (54%) of the 28 nests located >200 m from a non-forest opening. When the data were re-analyzed by treating the flycatcher separately from the other 12 species, the flycatcher experienced increased parasitism within the first 100 m of an opening and rates for other species were uniformly high over 300 m. Working in a forest type very similar to that studied by Brittingham and Temple (1983), Bielefeldt and Rosenfield (1997) found no evidence of increased parasitism near edges for the Acadian flycatcher.

In Table 3 we have summarized studies reported since Paton (1994) completed his review. Combining all studies there is evidence of increased brood parasitism near edges in 5 of 16 studies (31%). Even in agricultural settings, patterns of cowbird parasitism are not consistent. For example, Hahn and Hatfield (1994) observed cowbirds in interior forest habitats. They found birds nesting low or on the ground in forest interiors were more at risk from cowbirds than edge-using species. As well, cowbirds did not concentrate on species found to be hosts in most other areas. Similarly, Cooker and Capen (1995) found that presence of edges was less important than proximity to human development. They found remote forest openings in Vermont were unlikely to attract cowbirds. Trail and Baptista (1993) also found increasing incidences of

---

[†] Paton (1994) also re-analyzed data of Gates and Gysel (1978) which has been cited as documenting increased parasitism near edges. Rates of brood parasitism were 10%–12% within 25 m of the edge; and 3%–6% 50 m to 75 m from the edge. The difference was not significant (p = 0.395).

parasitism by cowbirds in areas close to human settlement (San Francisco). In the Pacific Northwest there is little indication of increased parasitism by cowbirds near forest edges. Schieck et al. (1995) reported cowbirds only rarely in coastal forests of British Columbia and noted 9 other studies of western forest birds with similar findings—brown-headed cowbirds were never or only rarely detected. We thus would not expect to find similar rates of parasitism in western, coastal forests as are documented in eastern and midwestern forests, and we do not (e.g., Schieck et al. 1995; K. MacKenzie pers. comm.).

As well as extending into forests, increased levels of brood parasitism also can extend from edges into openings (e.g., Best 1978; Johnson and Temple 1990; Niemuth and Boyce 1997). The pattern likely results because brown-headed cowbirds use trees as perches from which they search for nests to parasitize (Norman and Robertson 1975). Because rates of nest parasitism and depredation decline with distance into an opening, Niemuth and Boyce (1997) argued that forest practices in the pine barrens should be changed from numerous small, dispersed clearcuts to aggregations of clearcuts. Such a shift in practice should create fewer, larger openings, more forest interior, and less edge. Perches with a clear view of the surrounding habitat are probably very important to nest predators as well as nest parasites (e.g., Ratti and Reese 1988; Møller 1989). It is likely for this reason that shrub nesters in clearcuts with scattered retention of individual trees (or small groups of trees) seem to experience higher rates of depredation (e.g., Vega 1993).

There is increasing evidence and speculation in the eastern United States that generalist predators can negatively impact forest birds (Table 4; Bider 1968; Robbins 1979; Gates and Mosher 1981; Whitcomb et al. 1981; Chasko and Gates 1982; Ambuel and Temple 1983; Janzen 1986; Johnson and Temple 1986; Laudenslayer 1986; Temple and Carey 1988; Yahner and Scott 1988; Csuti 1991; O'Conner and Faaborg 1992). Corvids (jays, crows and ravens) are reportedly the primary avian predator of nests near edges (Møller 1989; Andren 1992), but many smaller mammals also prey on bird nests and may be under-represented in study results (Leimgruber et al. 1994; Hernandes et al. 1997; Marzluff and Restani this volume). Although some researchers have found greater activity of corvids near edges (e.g., Niemuth and Boyce 1997), many have found no relationships with distance to edge (Table 4; Angelstam 1986; Rudnicky and Hunter 1993). We should not expect consistent relations, even for the same prey species, simply because relationships likely are a local tradeoff between local risk of predation and forage abundance in the open and forested areas abutting the edge. Yahner and Wright (1985) found no difference in nest predation between 5 m and 50 m from edge, but that distance may be entirely within the "edge". Paton (1994) carefully reviewed studies documenting predation (and parasitism) of bird nests around edges, and concluded that the most conclusive studies indicate increased depredation rates within 50 m of an edge. Because of their sample design, studies showing greater depths of effects from predation or brood parasitism were less convincing (reasons vary among studies, but Paton's derivation of 50 m is rigorously derived). Moreover, many studies of predation used artificial nests, which may yield unreliable estimates of type of predator or rate of predation on natural nests (see Haskell 1995; King et al. 1996[‡], Marzluff and Restani this volume). Despite these shortcomings the evidence for increased rates of depredation near edges (in either openings or forests) is strong. Paton (1994) found evidence in 10 of 14 studies of artificial nests and 4 of 7 studies of natural nests. We have added 4 studies available since Paton's review (Table 4). The combined total of all studies (artificial and natural) yields evidence of increased depredation near edges in 15 of 25 studies (60%). Based on his review Paton (1994) concluded that habitat patches must be <10 ha to produce elevated depredation rates. The logistic response function of Niemuth and Boyce (1997) estimates the probability of parasitism 161 m into an opening as half (12%) of the probability

---

[‡] Inferences drawn from experiments with quail eggs may be spurious because small-mouthed predators often cannot break quail eggs which are larger than eggs of many passerines.

Table 3. Brood parasitism by cowbirds.

Trend into forest	Location	Source
Parasitism decreased from 0 m to 300 m into forest.	Eastern U.S.; variety of habitats in matrix; distance to non-forest opening.	Brittingham and Temple 1983.
No change in predation rate with distance from non-forest opening.	Similar to above; reanalysed Brittingham and Temple (1983) data and found an edge effect only for 1 of 13 species (predation elevated within 100 m of opening).	Bielefeldt and Rosenfield 1997.
Decreased 400 m into forest.	Eastern U.S.; field matrix.	O'Conner and Faaborg 1992.
Increased in forest interior (looked 0 m to 200 m, but categorical data).	Eastern U.S.; variety of habitats in matrix.	Hahn and Hatfield 1994.
No difference in cowbird predation between abrupt and gradual edges; predation did not differ between edges and 50 m inside forests.	Eastern U.S.; field and forest matrix.	Suarez et al. 1997.
None observed.	Eastern U.S.; young forest matrix.	King et al. 1996.
Few cowbirds in areas remote from chronic human disturbance (>7km away); more cowbirds in chronically disturbed areas (e.g., ski runs, hayfields) than in logged areas.	Eastern U.S.; logging and more permanent habitat disturbance in matrix.	Cooker and Capen 1995.
Parasitism 161 m into openings was half that at the forest edge.	Eastern U.S., savannah openings in jack pine matrix.	Niemuth and Boyce 1997.
Few or no cowbirds in forests near clearcut edges.	Western British Columbia.	Schieck et al. 1995; MacKenzie pers. comm., Bryant 1997.

0 m from the forest edge (24%). For parasitism rates on nests in the centre of a square opening to be reduced to half the rate at the forest edge, the patch would have to be >10.4 ha (close to Paton's estimate).

Because it is cited so frequently as suggesting edge effects extending 600 m into the forest, it is important to consider the work of Wilcove et al. (1986) carefully. They reported three samples. In one there was no increase in depredation rates near the edge (Paton 1994). The other two samples included artificial nests at the edge and 600 m into the forest. Significantly greater depredations rates were observed at the edge than 600 m into the forest. For no apparent reason, and without data, Wilcove et al. (1986) assumed a linear decline from the edge to 600 m. Actual depth of the edge effect was undocumented and could have occurred anywhere over the 600 m. We concur with Paton (1994): where it occurs (about 60% of studies), elevated depredation is concentrated within the first 50 m from an edge.

Generally, studies in European forests agree with studies in the eastern United States—nest predators can invade forests through fragmentation (Andren et al. 1985; Haila 1986; McLellan et al. 1986). The same phenomenon has not been found in western forests where harvesting is the primary anthropogenic disturbance (Tables 3 and 4; Rudnicky and Hunter 1993; Schieck et al. 1995; Bayne and Hobson 1997; Bunnell et al. 1997; K. MacKenzie pers. comm.). In western forests there is little evidence of negative edge effects (Table 3 and Marzluff and Restani this volume). This result reflects the fact that stands of different ages do not act as habitat islands, are not surrounded by agricultural habitat, and exist in a mosaic where edges are continually changing. It is possible that the more rugged nature and complex ecological mosaic of western forests creates more natural edge than is found in eastern forests, which may encourage more resilience to potential edge effects. More negative effects may become apparent in western forests as human activity and forest patterns continue to change. That notion is supported by findings of Hartley and Hunter (1998); predation was greater near edges in areas surrounded by less total forest than in areas surrounded by larger amounts of forest.

5.3 Dispersal, movement and isolation

Conceivably, edges could be unidirectional filters through which animals can pass and not return (speculated by Janzen 1986); they might also interfere with the continuity of some ecological processes, although this too has not been documented. It is more likely that habitats forming edges create dispersal barriers (Wiens et al. 1985). Whether edges themselves can limit movements presumably depends on what species are trying to cross the edge and on the structure of the edge habitat. We found no studies that documented edges themselves as barriers to movement.

A few studies speculate that isolation of forest patches is negatively impacting the ability of amphibians to disperse and recolonize (e.g., Welsh 1990), but studies have not examined the roles of edges themselves as dispersal barriers. Data are sparse, but those available indicate positive relations of amphibian diversity with edges (Rosenberg and Raphael 1986) and with small patches (e.g., Lehmkuhl and Ruggiero 1991). Given general amphibian intolerance to dessication, microclimate and microhabitat (e.g., downed wood) must play a role, but the role is more likely to be enacted within the open area forming an edge than in the edge itself.

Depending on the degree of contrast between edges and adjacent forests, edges together with one or the other adjacent habitats can act as barriers to dispersal of small mammals (e.g., Wegner and Merriam 1979; Chasko and Gates 1982; Bendell and Gates 1987). Yahner (1986) found that white-footed mice tended to avoid travelling through edges. Ferris (1979) noted that noise along highway edges may be a barrier to movement for a few species of forest birds (bay-breasted warbler, winter wren, blue jays, Blackburnian warblers), the same phenomenon may be true for mammals but is not documented.

Previous sections noted how edges can limit or allow abiotic movement of disseminules (e.g., seeds by wind transfer). Such abiotic transfers are strongly affected by the structure of the edge.

Table 4. Findings of studies evaluating depredation near forest edges.

Predator	Distance	Researcher	Location	Matrix
various predators on avian nests	increased predation near edge	Wilcove et al. 1986	Not recorded; looked 0 and 600m into forest	Field and suburbs
various predators on avian nests	large patches lowest predation	Wilcove 1985	Maryland & Tennessee; looked at patches ranging from 4 to 209,000 ha	Field and suburbs
various predators on avian nests	highest predation in 0 to 50 m from edge	Andren & Angelstam 1988	Southcentral Sweden (looked at edge (10 m outside to 10 m inside forest); 10-50; 50-200; and 200-500 m into forest).	Field
jays raven hooded crow badger	variable ranges of greatest predation depending on predator: 1500-2500m[a] 500-1500m 500-1000m 0 to 500m	Angelstam 1986	Central Sweden; based on locations of nests suffering predation	Field and mixed forest
no specific predators	no relation of predation rate to patch size	Rudnicky & Hunter 1993	Maine; patches 9 to 203 ha	Clearcut
raccoons, black rat snakes	abrupt edges had greater predation than gradual edge. Edges greater than 50 m from agricultural edge had no significant differences in predation from edges closer to agricultural edge	Suarez et al. 1997	S. Illinois; edges grouped by structure and by location: exterior (forest agricultural edges) and interior (edges more than 50 m away from agricultural edge.	Field
crows	no evidence of increase near edge	Yahner and Wright 1985	Pennsylvania (looked 50m into forest)	field
red squirrel least chipmunk	no evidence of increase near edge	King et al. 1996	New Hampshire (looked 400m into forest)	Clearcut
corvids, coyote, bear, chipmunks, jays, red squirrels	no relation with distance from edge	Ratti & Reese 1988	Idaho, looked 120m into forest	Field
raccoons, red squirrels, bluejay, black bear, N. blying squirrel	no difference in predation rate with distance from stream	Vander Haegan & DeGraaf 1996	Maine	Clearcut/ riparian buffer
rodents	no difference with distance from edge; nest predation greater in forest than in farmland.	Santos & Telleria 1992	Spain; study areas were 0.1 to 21 ha; plots ranged from 0 to 100 m from edge	field
eastern chipmunk, bluejay, owl	decrease predation into forest	Gates & Gysel 1978	Michigan, looked 123m into forest	Field

[a] These are distances to the closest field; all nests were within 50m of trails or rural roads which may be the edges to which predators were responding (Paton 1994).

5.4 Human activity

Many human activities, such as littering, drift of fertilizers, trampling from animals, grazing, escaping fires, poaching, and excessive human disturbance tend to be associated with edges (Bennett 1990a). Many mammals of suburban and urban and agricultural landscapes seem to be edge-adapted (Whitcomb et al. 1976; Butcher et al. 1981) and may adversely affect species typically found in forested environments. The primary edge effects resulting from harvesting, however, relate to the creation of roads and trails that allow access into forests. Documented effects of edges created by roads include increased rates of predation near roads, and changes in community composition close to roads. Other effects of roads such as creating barriers to movements of invertebrates and small mammals, and increasing traffic-caused mortality are not clearly edge effects and are not discussed here.

Roads are important avenues for human (McLellan 1990) and other animal predators and so prey near road edges may experience greater predation rates. For example, Rich et al. (1994) found that habitats adjacent to forest roads attracted more avian predators than those farther away, yet roads were not avoided by most prey species. Yahner and Mahan (1997) however, suggest roads edges are not as important predator movement corridors as are clearcut/forest edges. They found that artificial nests near clearcut/forest edges experienced greater predation than did those along road edges.

Edge effects around roads result from disturbance by traffic alone as well as from those direct effects of vehicular traffic and mortality. For deer and elk the effect extends beyond 100 m into the forest, depending on the vegetation cover available. Rost and Bailey (1979) found that deer used areas 300 to 400 m from the road three times more often than they used areas within 100 m of the road. Increased traffic tends to increase road avoidance. The type of vegetation around the road edge may affect how animals use nearby areas. For example, Nyberg and Janz (1990) reported that elk and deer used areas near roads when ample cover was available, but avoided roads where roadside cover was low. Deer may avoid roads more than elk, but elk are still affected by roads (Pedersen et al. 1979, Lyon 1979, Rost and Bailey 1979, Cole et al. 1997. Huggard (1993) reported that elk living greater than 1 km from the Trans Canada Highway in Banff, Alberta were on average 2.5 years older than those living closer than 1 km from the highway (as measured by wolf kills in the two distance bands) presumably because of highway mortality. Witmer and deCalesta (1983) reported that elk were located at half the expected frequency within a 500 m band surrounding paved roads. Significantly fewer observations were noted for a narrower band (125 m) around open spur roads, but no differences in observations were reported for spur roads closed to vehicles.

Bears generally avoid roads and human disturbance (Archibald et al. 1987; Garner and Vaughan 1987; Brody and Pelton 1989; McLellan 1990; Mattson 1990; Kazworm and Manley 1990). Archibald et al. (1987) found that for 14 hours a day bears avoided areas within 50 m to 300 m of roads. They used decibel levels to estimate the width of bands next to road affected by sounds from roads during hauling. A. N. Hamilton (pers. comm.) noted bears were affected by human disturbance associated with roads, but warned that the relation of bears to roads was complex. While adult male bears almost always avoid areas near roads, females may prefer those areas (perhaps as protection from males) and sub-adults may be displaced to those habitats (Mattson et al. 1987). As well, some bears habituate to humans, seek human presence, and do not avoid roads. McLellan and Shackelton (1988) reported similar habitat partitioning with dominant male grizzlies avoiding roads, and females and subadults being closer to roads. McLellen and Shackelton (1988) also reported that bears tend to avoid habitats within 100 m of roads during the day regardless of levels of vehicular traffic. In contrast, Mace et al. (1996) reported that bears used habitats within 500 m from roads when roads carried fewer than 10 vehicles per day, but avoided that habitat if more vehicles than that. Weilgus et al. (1998) reported that roads open to recreational use were avoided by bears whereas those restricted to use by forestry vehicles were not, suggesting that type of use (and likelihood of off-road activity) determined the likelihood of bear use near roads. Kazworm and Manley (1990) found black bears did not avoid roads by as great a distance as did grizzlies. Black bears demonstrated reduced use 0 to 122 m from roads, whereas grizzlies used habitats from 0 to 914 m less than expected. Like ungulates, bears may use night cover and vegetative screening to ameliorate effects of traffic and some may habituate to traffic. Schoen et al. 1990 (in Harrison 1992) studied bear use of habitats near roads and estimated that habitat suitability dropped by half if habitats were within 1.6 km of arterial or collector roads. The bulk of research has been on larger, generally hunted species; potential disturbance effects on smaller species is largely undocumented.

Some research describes how road edges are associated with undesirable human activities such as littering, igniting fires (Franklin and Forman 1987), poaching, removing snags for firewood, and excessive human disturbance (Bennett 1990a). Even footpaths can affect populations because of disturbance by frequent human intrusions (Yalden 1992 in Freedman et al. 1994). Invasion of exotic plant species have been documented by Tyser and Worley (1992) who suggested that 'alien' species were invading grasslands from roadside edges in Glacier National Park, Montana. They also found high levels of alien species 100 m from trail sides, suggesting that trails and especially roads were very vulnerable to colonization by exotic species.

Swaths cut for roads or transmission lines can change the species composition of faunal communities (Whitcomb et al. 1976, Noss 1983). Ferris (1979) noted that four bird species (bay-breasted warbler, blackburnian warbler, winter wren, and blue jay) seemed to avoid breeding near highways in Maine, although total breeding bird numbers were not different near and away from the road. These changes in composition were not attributed to disturbance, biological or physical edge effects. In some years the presence of edge species added to the abundance of birds near highways; they did not necessarily displace birds less associated with edges. The ready supply of grass on some roadsides can allow some microtine populations to flourish (Baker 1971). Adams and Geis (1983) noted abundant grassland species and habitat generalists in road right-of-ways. Roadside seeding can provide forage for ungulates (Bellis and Graves 1971) and bear and may alter movement patterns for those species. Although these species do not appear to be persistent, we found no work that examined seeding as a source of invasive species in forest habitats.

Microclimatic edge effects likely increase as road widths and permanence increase, although the literature does not report on those relations. Changes in microclimate may help explain the apparent increase of exotic plants along road edges. Vaillancourt (1995 in Reed et al. 1996) suggested 50 m edge effects around roads (but Reed et al. 1996 do not note the nature of the effects). Generally, roads create high contrast ecotones that do not decrease over time.

6. FOREST PRACTICES AND EDGES

Forest practices, especially in even-aged management, create edges. There obviously is tremendous variability in responses around edges, both for abiotic variables (Figure 1) and biotic responses (Tables 2 through 4). Findings from studies in a non-forested matrix (most of the current literature) do not transfer appropriately to conditions within an area of forest stands of different ages. We can offer little more than a summary of the major points and tentative recommendations derived from these.

Microclimatic edge effects are well documented, though highly variable in western, coastal forests

Microclimate changes near edges extend 1 to 3 tree heights into forests, depending on aspect and variable measured (Figure 1). Microclimatic effects appear to increase structural diversity of the vegetation near edges and increase vertebrate species richness, primarily in response to shrubby vegetation. Influences on invertebrates and insectivorous vertebrates are more direct. Although changes of some microclimate variables are detectable over 200 m, most detectable changes disappear within 100 to 150 m.

The largest effect on biological organisms occurs within 50 m of the edge.

Reported edge effects range from 0 to several hundred meters into forests. Although changes of microclimate variables are detectable over larger distances, edge influences on distributions of organisms or factors affecting organisms (such as predation and nest parasitism) are concentrated within 50 m of the edge. Effects have been reported as extending much farther, but these studies are usually either flawed or otherwise unconvincing (see Paton 1994 and Tables 3 and 4). Effects of disturbance from road traffic can extend farther (e.g., 400 m on deer and elk, particularly where roadside cover is low). Effects of hunting, poaching, and invasion of exotic species also likely extend farther but are poorly documented.

Few negative effects of edges have been documented in western coastal forests

Evidence for positive edge effects is similar in all studies that we looked at (species richness generally increases with edge and more species are positively associated with edges than are negatively associated with edges). Evidence of negative edge effects differs starkly between eastern or mid-western North America and coastal montane forests. In the east and midwest many studies document increased predation and parasitism near edges; in the Pacific Northwest where timber harvesting is the main disturbance researchers have found little effect of patch area or negative edge effects, and few species (Table 2) consistently associated with forest interior.

The kinds of edge effects important to forest management in the Pacific Northwest are largely undocumented, and must be inferred.

Most studies have focused on forest to field edges; fewer have examined forest to clearcut edges. All available data suggest relatively little effect across forest age classes as compared to forest to field contrasts. The difference occurs simply because there is less difference in vertebrate communities between a clearcut and adjacent forest than between a field and adjacent wooded area. Similarly, there are no applicable data on potential edge effects in the small openings induced by partial cutting in the Pacific Northwest, or on edges between old growth and older second-growth stands. Neither of these effects is likely to be large. Openings may be too small in the first instance; in both instances the structural contrasts are relatively small. It is these contrasts that appear most strongly associated with edge effects (Figure 2 for birds).

Data on edges induced by roads and most inherent edges such as streams and wetlands are sparse. Most inherent edges around streams and wetlands contain a large deciduous component that appears to contribute to their greater vertebrate richness. The available evidence suggests that as with successional forest edges, the greatest effects around streams occur within the first 50 m (e.g., Darveau 1995; Kinley and Newhouse 1997).

As an interim measure, use microclimate measures to define forest interior.

Although few negative edge effects are documented in Pacific Northwest Forests, more detrimental effects may become apparent if the total amount of forest area declines or if the matrix changes from successional forests to a more agricultural or urban character. To accommodate uncertainty some forest interior habitat should be provided either by aggregating harvest or otherwise expanding forest patch size to accommodate microclimate edge effects. Effects of microclimate may be particularly important for lichens, mosses, and invertebrates that are yet relatively unstudied.

7. LITERATURE CITED

Adams, L.W. and A.D. Geis. 1983. Effects of roads on small mammals. J. Appl. Ecol. 20: 403-415

Alexander, R.R. 1964. Minimizing windfall around clear cutting in spruce-fir forests. For. Sci. 10: 130-143.

Alverson, W.S., Walker, D.M. and S.L. Solheim. 1988. Forests too deer: edge effects in Northern Wisconsin. Cons. Biol. 2: 348-358

Ambuel, B. and S.A. Temple. 1983. Area-dependent changes in the bird communities and vegetation of southern Wisconsin forests. Ecology 64: 1057-1068

Anderson, S.H., Mann, K. and H.H. Shugart, Jr. 1977. The effect of tranmission-line corridors on bird populations. Amer. Midl. Nat. 97: 216-221

Andren, H. 1992. Corvid density and nest predation in relation to forest fragmentation: a landscape perspective. Ecology 73: 794-804

Andren, H. and P. Angelstam. 1988. Elevated predation rates as an edge effect in habitat islands: experimental evidence. Ecology 69: 544-547

Andren, J., Angelstam, P., Lindstrom, E. and P. Widen. 1985. Differences in predation pressure in relation to habitat fragmentation. Oikos 45: 273-277

Angelstam, P. 1986. Predation on ground nesting birds' nests in relation to predator densities and habitat edge. Oikos 47: 365-373

Angelstam, P. 1992. Conservation of communities — the importance of edges, surroundings and landscape mosaic structure. Pp. 9-70 in L. Hansson (ed.). Ecological Principles of Nature Conservation. Elsevier Applied Science, New York, NY

Archibald, W.R., R. Ellis, and A.N. Hamilton. 1987. Responses of grizzly bears to logging truck traffic in the Kimsquit River Valley, British Columbia. Int. Conf. Bear Res. and Manage. 7: 251-257. P. Zager ed. Port City Press Inc., Washington, D.C.

Askins, R.A., Philbrick, M.J. and D.S. Sugeno. 1987. Relationship between the regional abundance of forest and the composition of forest bird communities. Biol. Conserv. 39: 129-152

Baker, R.H. 1971. Nutritional strategies of myomorph rodents in North American grasslands. J. Mammal. 52: 800-805.

Bayne, E.M. and K.A. Hobson. 1997. Comparing the effects of landscape fragmentation by forestry and agriculture on predation of artificial nests. Cons. Biol. 11: 1418-1429

Bellis, E.D. and Graves H.B. 1971. Deer mortality on a Pennsylvannia interstate highway. J. Wildl. Manage. 35: 232-237

Bendell, P.R. and J.E. Gates. 1987. Home range and microhabitat partitioning of the southern flying squirrel (*Glaucomys volans*). J. Mammal. 68: 243-255

Bennett, A. 1990a. Habitat corridors and the conservation of small mammals in a fragmented forest environment. Landscape Ecol. 4: 109-122

Bennett, A.F. 1990b. Habitat corridors, their role in wildlife management and conservation. Department of Conservation and Environment, Victoria, Australia, Arthur Rylah Institute for Environmental Research.

Best , L.B. 1978. Field sparrow reproductive success and nesting ecology. Auk 95: 9-22

Bielefeldt, J., and R.N. Rosenfield. 1997. Reexamination of cowbird parasitism and edge effects in Wisconsin forests. J. Wildl. Manage. 61: 1222-1226

Bider, J.R. 1968. Animal activity in uncontrolled terrestrial communities as determined by sand transect techniques. Ecol. Monogr. 38: 269-308

Bratton, P. and P.S. White. 1980. Rare plant management — after preservation what? Rhodora 82: 49-75

Brett, D. 1989. Sea birds in the trees. Ecos 61: 4-8

Brittingham, M.C. and S.A. Temple. 1983. Have cowbirds caused forest songbirds to decline? Biosci. 33: 31-35

Brody, A.J. and M.R., Pelton. 1989. Effects of roads on black bear movements in western North Carolina. Wildl. Soc. Bull. 17: 5-10

Brown (in prep), cited in Lovejoy, T.E., Bierregaard, R.O. Jr., Rylands, A.B., Malcolm, J.R., Quintela, C.E., Harper, L.H., Brown, K.S. Jr., Powell, A.H., Powell, G.V.N., Schubart, H.O.R. and M.B. Hays. 1986. Edge and other effects of isolation on Amazon forest fragments. Pp. 257-285 in M.E. Soule (ed.). Conservation Biology: the Science of Scarcity and Diversity. Sinauer Associates, Inc., Sunderland, MA

Bryant, A.A. 1997. Effect of alternative silvicultural practices on breeding bird communities in montane forests. Report to MacMillan Bloedel Ltd, Nanaimo, B.C.

Buechner, M. 1987. Conservation in insular parks: simulation models of factors affecting the movement of animals across park boundaries. Biol Cons. 41: 57-76

Bunnell, F.L., McNay, R.S. and C.C. Shank. 1985. Snow and trees: deposition of snow on the ground — a review and synthesis. Ministries of Environment and Forests, IWIFR-17. Victoria, B.C

Bunnell, F. L., Kremsater,L.L. and R. W. Wells. 1997. Likely consequences of forest management on terrestrial, forest-dwelling vertebrates in Oregon. Oregon Forest Resources Institute, Portland, OR

Burger, L.D. Burger, L.W. and J. Faaborg. 1994. Effects of prairie fragmentation on predation on atrificial nests. J. Wildl. Manage. 58: 249-254

Burton, Phil. 1998. Pers. comm. Faculty of Forestry, University of British Columbia, 270-2357 Main Mall, Vancouver, B.C. V6T 1Z4.

Buskirk, S.W., and R.A. Powell. 1994. Habitat ecology of fishers and American martens. Pp. 283-296 in S.W. Buskrik, A.S. Harestad, M.G. Raphael, and R.A. Powell (eds.). Martens, sables, and fishers. Biology and conservation. Cornell University Press, Ithaca, NY.

Butcher, G.S., Niering, W.A., Barry, W.J. and R.H. Goodwin. 1981. Equilibrium biogeography and the size of nature preserves: an avian case study. Oecologia 49: 29-37

Cale, P. and R.J. Hobbs. 1991. Condition of roadside vegetation in relation to nutrient status. Pp. 353-362 in D.A. Saunders and R.J. Hobbs (eds.). The role of corridors. Surrey Beatty, Chipping Norton, New South Wales, Australia

Campbell, R.W., Dawe, N.K., Mctaggart-Cowan, I. [and others]. 1990. The birds of British Columbia. Vol.2, Diurnal birds of prey through woodpeckers. University of British Columbia Press, Vancouver, B.C.

Caruso, J.R. 1973. Regeneration within a middle-elevation Douglas-fir clearcut. Ph.D. dissertation, University of Washington, Seattle, WA

Catt, D. 1991. Bird communities and forest succession in the subalpine zone of Kootenay National Park, British Columbia. M.Sc. thesis, Simon Fraser University, B.C.

Chan McLeod, Ann Allaye. 1998. Pers. comm. Centre for Applied Conservation Biology, 270-2357 Main Mall, Vancouver, B.C., V6T 1Z4. allaye@unixg.ubc.ca

Chasko, G.G. and J.E. Gates. 1982. Avian habitat suitability along a transmission-line corridor in an oak-hickory forest region. Wildl. Monogr. 82

Chen, J., Franklin, J.F. and T.A. Spies. 1990. Microclimate pattern and basic biological responses at the clearcut edges of old-growth Douglas-fir stands. Northwest Env. 6: 424-425

Chen, J., Franklin, J.F. and T.A. Spies. 1992. Vegetation responses to edge environments in old-growth Douglas-fir forests. Ecol. Applic. 2: 387-396

Chen, J., Franklin, J.F. and T.A. Spies. 1993. Contrasting microclimates among clearcut, edge, and interior of old growth Douglas fir forest. Ag. For. Met. 63: 219-237

Chen, J., Franklin, J.F. and J.S. Lowe. 1996. Comparison of abiotic and structurally defined patch patterns in a hypothetical forest landscape. Cons. Biol. 10: 854-862

Clark, T.P. and F.F. Gilbert. 1982. Ecotones as measure of deer habitat quality in central Ontario. J. Appl. Ecol. 19: 751-758

Clements, F.E. 1907. Plant physiology and ecology. Henry Holt, New York, NY

Concannon, J.A. 1995. Characterizing structure, microclimate and decomposition of peatland, beachfront, and newly-logged forest edges in southeastern Alaska. Ph.D. dissertation. University of Washington, Seattle, W.A.

Cole, E.K., Pope, M.D. and R.G. Anthony. 1997. Effects of road management on movement and survival of Roosevelt elk. J. Wildl. Manage. 61: 1115-1126

Cooker, D.R. and D.E. Capen. 1995. Landscape-level habitat use by brown-headed cowbirds in Vermont. J. Wild. Manage 59: 631-637

Craig, V.J. 1995. Relationships between shrews (*Sorex* spp.) and downed wood in the Vancouver watersheds, B.C. M.Sc. thesis University of British Columbia, Vancouver, B.C.

Crampton, L.H. and R.M.R. Barclay. 1996. Habitat selection by bats in fragmented and unfragmented aspen mixedwood stands of different ages. Pp. 238-259 in R.M.R. Barclay and R.M. Brigham (eds.). Bats and forests symposium. Working Paper 23/1996, Research Branch. B.C. Ministry of Forests, Victoria, B.C.

Csuti, B. 1991. Introduction. Pp. 81-90 in W.E. Hudson (ed.). Landscape Linkages and Biodiversity. Defenders of Wildlife, Washington, D.C.

Darveau, C.H., Beauchesne, P., Belanger, L. (and others). 1985. Riparian forest strips as habitat for breeding birds in boreal forest. J. Wildl. Manage. 59: 67-78

deCasenave, J.L., Pelotto, J.P. and J. Protomastro. 1995. Edge-interior differences in vegetation structure and composition in a Chaco semi-arid forest, Argentina. For. Ecol. and Manage. 72: 61-69

DeMaynadier, P.G. and M.L. Hunter. 1997. Effects of silvicultural edges on the distribution and abundance of amphibians in Maine. Cons. Biol. 12: 340-352

Diamond, J.M. 1975. The island dilemma: lessons of modern biogeographic studies for the design of nature reserves. Biol. Cons. 7: 129-146

Dunn, P.O. and C.E. Braun. 1986. Summer habitat use by adult female and juvenile sage grouse. J. Wildl. Manage. 50: 228-235

Dunsworth, B.G. and J.T. Arnott. 1995. Growth limitations of regenerating montane conifers in field environments. In Proc. of the Montane Alternative Silviculture Systems Workshop. Edited by J.T. Arnott, W.J. Beese, A.K. Mitchell and J. Peterson. June 7-8, 1995. Courtenay, B.C., Can. For. Serv. and B.C. Min. For., Victoria, B.C. FRDA Rep. No. 238, pp.48-68

Eastman, D.S. 1976. Habitat selection and use in winter by moose in sub-boreal forests of north-central British Columbia and relationships to forestry. PhD thesis, University of British Columbia, 453 pp.

Edeburn, R.M. 1947. A study of the breeding distribution of birds in a typical upland area. Proceedings of the West Virginia Academy of Science, Volume 18, West Virginia University Bulletin Series 47, Number 9-I: 34-37

Edmonds, R.L. and C.M. Bigger. 1984. Decomposition and nitrogen mineralization rates in Douglas-fir needles in relation to whole tree harvesting practices. In Proceedings of the 1983 Society of American Foresters National Convention, Portland, OR

Erickson, J.L. and S.D. West. 1996. Managed forests in the western Cascades: the effects of seral stage on bat habitat use patterns. Pp. 215-227 in R.M.R. Barclay and R.M. Brigham (eds.). Bats and forests symposium. Working Paper 23/1996, Research Branch. B.C. Ministry of Forests, Victoria, B.C.

Ferris, C.R. 1979. Effects of Interstate 95 on breeding birds in northern Maine. J. Wildl. Manage. 43: 421-427

Forman, R.T.T. 1986. Emerging directions in landscape ecology and applications in natural resource management. Conference on science in the National Parks. pp 59-87 in: Conference on science in the National Parks, Proceedings of the 4th triennial conference on research in the National Parks and equivalent reserves. July 13-18, 1986. Fort Collins, CO. Colorado State University and the George Wright Society, Fort Collins, CO.

Forman, R.T.T. and J. Baudry. 1984. Hedgerows and hedgerow networks in landscape ecology. Environ. Manage. 8: 495-510

Forman, R.T.T. and M. Godron. 1986. Landscape Ecology. John Wiley and Sons, New York, NY

Franklin J.F., and R.T.T. Forman. 1987. Creating landscape patterns by forest cutting: ecological consequences and principles. Land. Ecol. 1: 5-18

Franklin J.F., Cromack, K. Jr., Denison, W., McKee, A., Maser, C., Sedell, J., Swanson, F. and G. Juday. 1981. Ecological characteristics of old-growth Douglas-fir forests. USDA Forest Service GTR PNW-118

Fraver, S. 1994. Vegetation responses along edge to interior gradients in the mixed hardwood forests of the Roanoke River Basin, North Carolina. Cons. Biol. 8: 822-832

Fritschen et al. 1971. cited in Harris, L.D. 1984. The Fragmented Forest. University of Chicago Press, Chicago.

Frost, E.J. 1992. The effects of forest-clearcut edges on the structure and composition of old growth mixed conifer stands in the western Klamath Mountains. M.Sc. thesis Humboldt State University, Arcata, CA

Fuller, R.J. 1988. A comparison of breeding bird assemblages in two Buckinghamshire clay vale woods with different histories of management. Pp. 53-65 in K.J. Kirby, and F.J. Wright (eds.). Woodland management and research in the caly vale of Oxfordshire and Buckinhamshire. Nature Conservancy Council, Peterborough, UK.

Fuller, R.J. and Warren, M.S. 1991. Conservation management in ancient and modern woodlands: response of fauna to edges and rotations. Pp. 445-471 in I.F. Spellerberg, F.B. Goldsmith, and M.G. Morris (eds.) The scientific management of temperate communities for conservation. Blackwell Scientific Pubs., Oxford, UK.

Fuller, R.J., and P.A. Whittington. 1987. Breeding bird distribution within Lincolnshire ash-line woodlands: the influence of rides and woodland edge. Acta Oecologica/Oecologica Generalis 8: 259-268

Garner, N.P. and M.R. Vaughan. 1987. Black bears' use of abandoned homesites in Shenandoah National Park. Int. Conf. Bear Res. And Manage. 7:151-157. P. Zager ed. Port City Press Inc., Washington, D.C.

Gates, J.E. 1991. Powerline corridors, edge effects, and wildlife in forested landscapes of the central Appalachians. Pp. 13-32 in J.E. Rodiek and E.G. Bolen (eds.) Wildlife and habitats in managed landscapes. Island Press. Covelo, California.

Gates, J.E. and L.W. Gysel. 1978. Avian nest dispersion and fledging success in field-forest ecotones. Ecology 59: 871-883

Gates, J.E. and J.A. Mosher. 1981. A functional approach to estimating habitat edge width for birds. Am. Midl. Nat. 105: 189-192

Geiger, R. 1965. The Climate Near the Ground. Harvard University Press, Cambridge, MA 611 Pp.

Gibbs, J.P. 1998. Amphibian movements in response to forest edges, roads, and streambeds in southern New England. J. Wildl. Manage. 62: 584-589

Giles, R.H. Jr. 1978. Wildlife management. W.H. Freeman, San Francisco, CA

Godfryd, A. and R.I.C. Hansell. 1986. Prediction of bird-community metrics in urban woodlots. Pp. 321-326 in J. Verner, M.L. Morrison, and C.J. Ralph (eds.). Wildlife 2000: Modelling habitat relationships of terrestrial vertebrates. University Wisconsin Press, Madison, WI

Good, E.E. and C.A. Dambach. 1943. Effect of land use practices on breeding bird populations in Ohio. J. Wildl. Manage. 7: 291-297

Gregory, P.T. 1978. Feeding habits and diet overlap of three species of garter snakes (*Thamnophis*) on Vancouver Island. Can. J. Zool. 56: 1967-1974

Gysel, L.W. 1951. Borders and openings of beech-maple woodlands in southern Michigan. J. For. 49: 13-19

Hagar, J.C., McComb, W.C. and C.C. Chambers. 1995. Effects of forest practices on wildlife, Chapter 9 in R.L. Beschta, J.R. Boyle, C.C. Chambers [and others]. Cumulative effects of forest practices in Oregon: literature and synthesis. Oregon State University, Corvallis, OR

Hahn, D.C. and J.S. Hatfield. 1994. Parasitism at the landscape scale: cowbirds prefer forests. Cons. Biol. 9: 1415-1424

Haila, Y. 1986. North European land birds in forest fragments: evidence for area effects? Pp 315-319 in J. Verner, M.L. Morrison, and C.J. Ralph (eds.). Wildlife 2000: Modeling habitat relationships of terrestrial vertebrates. University Wisconsin Press, Madison, WI

Haila, Y., Jarvinen, O. and R.A. Vaisanen. 1980. Effects of changing forest structure on long-term trends in bird populations in SW Finland. Ornis Scand. 11: 12-22

Hamilton, Tony N. 1998. Pers. comm. Ministry of Environment Lands and Parks, Wildlife Branch, PO Box 9374 Stn PROVGOV, Victoria, B.C. V8W 9M4.

Hanley, T.A. 1983. Black-tailed deer, elk, and forest edge in a western Cascades watershed. J. Wildl. Manage. 47: 237-242

Hansen, A. and J. Petersen. Unpublished data. Copies of overheads from a Coastal Oregon Productivity Enhancement Project conference, Newport, Oregon.

Hansson, L. 1983. Bird numbers across edges between mature conifer forests and clearcuts in central Sweden. Ornis Scand. 14: 92-103

Hansson, L. 1994. Vertebrate distributions relative to clear-cut edges in a boreal forest landscape. Landscape Ecology 9: 105-115

Harmon, M.E. and J.F. Franklin. 1983. Age distribution of western hemlock and its relation to Roosevelt elk populations in the south fork of the Hoh Valley, Washington. Northwest Science 57: 249-255

Harper, J.A. 1969. Relations of elk to reforestation in the Pacific Northwest. Pp. 67-71 in H.C. Black (ed.). Proc. symp. wildlife and reforestation in the Pacific Northwest. School Forestry, Oregon State Univ., Corvallis, OR

Harris, L.D. 1984. The fragmented forest. University of Chicago Press, Chicago, IL

Harris, L.D. 1989. A rationale for wildlife habitat superiority of southern bottomland hardwood communities. In J. Gosselink, L. Lee, and T. Muir. (eds.). 1990. Ecological Processes and Cummulative Impacts Illustrated by Bottomland Hardwood Ecosystems. Lewis publishers, Michigan

Hartley, M.J. and M.L. Hunter, Jr. 1998. A Meta-analysis of forest cover, edge effects, and artificial nest predation rates. Cons. Biol. 12: 465-469

Haskell, D.G. 1995. A reevaluation of the effects of forest fragmentation on rates of bird nest predation. Cons. Biol. 9: 1316-1318

Helle, P. 1983. Bird communities in open ground-climax forest edges in northeastern Finland. Oulanka Reports 3: 47-49

Helle, P. 1986. Effects of succession and fragmentation on bird communities and invertebrates in boreal forests. Acta Univ. Oul. A. 178 1986.

Hernandez, F. Rollins, D., and R. Cantu. 1997. Evaluating evidence to identify ground-nest predators in west Texas. Wildl. Soc. Bull. 25: 826-831.

Hett, J., Taber, R., Long, J. and J. Schoen. 1978. Forest management policies and elk summer carrying capacity in the *Abies amabilis* forest, western Washington. Environ. Manage. 2: 561-566

Huggard, D.J. 1993. Prey selectivity of wolves in Banff National Park. II. Age, sex, and condition of elk. Can. I. Zool. 71: 140-147.

Huggard, Dave. 1998. Pers. comm. Centre for Applied Conservation Biology, 270-2357 Main Mall, Vancouver, B.C, V6T 1Z4

Huggard, D. and W. Klenner. 1998. Effects of harvest type and edges on shrews at the Opax Mtn. silvicultural system site. in New information for the management of dry Douglas-fir forests: procedings of the dry Douglas-fir workshop. C. Hollstedt, A. Vyse, and D. Huggard (eds.) April 24-25 1996, Kamloops, B.C. B.C. Ministry of Foretst working paper *in press*.

Huggard, D. R. Walton, and W. Klenner. 1998. Depth and duration of snow at the Opax Mtn. silvicultural system site. in New information for the management of dry Douglas-fir forests: procedings of the dry Douglas-fir workshop. C. Hollstedt, A. Vyse, and D. Huggard (eds.) April 24-25 1996, Kamloops, B.C. B.C. Ministry of Foretst working paper *in press*.

Janzen, D.H. 1983. No park is an island: increase in interface from outside as park size decreases. Oikos 41: 402-410

Janzen, D.H. 1986. The eternal external threat. Pp. 286-303 in M.E. Soule (ed.). Conservation biology: the science of scarcity and diversity. Sinauer Associates, Inc., Sunderland, Massachusetts.

Johnson, R.G., and S.A. Temple. 1986. Assessing habitat quality for birds nesting in fragmented tallgrass prairies. Pp. 245-249 *in* J. Verner, M. Morrison, and C. Ralph, (eds.). Wildlife 2000: modelling habitat relationships of terrestrial vertebrates. University Wisconsin Press, Madison, WI

Johnson, R.G., and Temple, S.A. 1990. Nest predation and brood parasitism of tallgrass prairie birds. J. Wildl. Manage. 54: 106-111

Johnston, D.W. 1970. High density of birds breeding in a modified deciduous forest. Wilson Bull. 82: 79-82

Johnston, D.W. and E.P. Odum. 1956. Breeding bird populations in relation to plant succession on the Piedmont of Georgia. Ecology 37: 50-62

Johnston, V. 1947. Breeding birds of the forest edge in Illinois. Condor 49: 45-53

Julander, O. and D.E. Jeffery. 1964. Deer, elk, and cattle range relations on summer range in Utah. Trans. North Amer. Wildl. Nat. Res. Conf. 29: 404-414

Kazworm, W.F. and T.L. Manley. 1990. Road and trail influences on grizzly bears and black bears in northwest Montana. Int. Conf. Bear Res. And Manage. 8: 79-84. L. Darling and R. Archibald eds. Hemlock Printers, Victoria B.C.

Keller, M.E. and S.H Anderson, 1992. Avian use of habitat configuration created by forest cutting in southeastern Wyoming. Condor 94: 55-65

King, D.I., Griffin, C.R. and R.M. Degraaf. 1996. Effects of clearcutting on habitat use and reproductive success of the ovenbird in forested landscapes. Cons. Biol. 10: 1380-1386

Kinley, T.A. and N.J. Newhouse. 1997. Relationship of riparian reserve zone width to bird density and diversity in southeastern British Columbia. Northwest Sci. 71: 75-86

Kirchoff, M.D. and J.W. Schoen. 1983. Black-tailed deer use in relation to forest clear-cut edges in southeastern Alaska. J. Wildl. Manage. 47: 497-501

Klenner, W. and Huggard, D. 1997. Three-toed woodpecker nesting and foraging at Sicamous Creek. Pp. 224-233 in C. Hollstedt and A. Vyse (eds.), Sicamous Creek silvicultural systems project: workshop proceedings. April 24-25, 1996, Kamloops, B.C. Working Paper 24/1997, Research Branch, B.C. Ministry of Forests, Victoria, B.C. 283 pp.

Klenner, W. and Huggard, D. 1999. Nesting and foraging habitat requirements of woodpeckers in relation to experimental harvesting treatments at Opax Mountain. Pp. 252-266 in C. Hollstedt, A. Vyse and D. Huggard (eds.), New information for the management of dry Douglas-fir forests: proceedings of the dry Duoglas-fir workshop. B.C. Ministry of Forests, Victoria, B.C.

Koppenaal, R.S. and A. K. Mitchell. 1992. Regeneration of Montana forests in the coast for Western helmlock zone of British Columbia: a literature review. For. Can. and B.C.. Min. For. FRDA Rep. 192. 22 pp.

Kremsater, L.L. and F.L. Bunnell. 1990. Creating black-tailed deer winter range in second-growth forests. Northwest Environ. J. 6: 387-388

Kremsater, L.L. and F.L. Bunnell. 1992. Testing responses to forest edges: the example of black-taled deer. Can. J. Zool. 70: 2426-2435

Kroodsma, R.L. 1982. Edge effect on breeding forests birds along a power-line corridor. J. Appl. Ecol. 19: 361-370

Kroodsma, R.L. 1984. Effect of edge on breeding forest bird species. Wilson Bull 96: 426-436

Krusic, R.A., and C.D. Neefus. 1996. Habitat associations of bat species in the White Mountain National Forest. Pp. 185-198 in R.M.R. Barclay and R.M. Brigham (eds.). Bats and forests symposium. Working Paper 23/1996, Research Branch. B.C. Ministry of Forests, Victoria, B.C.

LaRue, P, Belanger, L. and J. Huot. 1995. Riparian edge effects on boreal balsam fir bird communities. Can. J. For. Res. 25: 555-566

Laudenslayer, W.F. Jr. 1986. Summary: predicting effects of habitat patchiness and fragmentation. Pp. 331-333 in J. Verner, M. Morrison, and C. Ralph, (eds.). Wildlife 2000: modelling habitat relationships of terrestrial vertebrates. University Wisconsin Press, Madison WI

Laudenslayer, W.F. Jr. and R.P. Balda. 1976. Breeding bird use of a pinyon-juniper-ponderosa pine ecotone. Auk 93: 571-586

Laurance, W.F. 1987. The rainforest fragmentation project. Laine 25: 9-12

Laurance, W.F. 1989. Ecological impacts of tropical forest fragmentation on non-flying mammals and their habitats. Ph.D. dissertation, Univ. California, Berkeley, CA

Laurance. W.F. and E. Yensen. 1991. Predicting the impacts of edge effects in fragmented habitats. Biol. Cons. 55: 77-92

Lay, D.W. 1938. How valuable are woodland clearings to birdlife? Wilson Bull. 50: 254-256

Lehmkuhl, J.F. and L. F. Ruggiero. 1991. Forest fragmentation in the Pacific Northwest and its potential effects on wildlife. Pp. 35-46 in L.F. Ruggiero, K.B. Aubrey, A.B. Carey, and M.H.Huff (eds.). Wildlife and vegetation of unmanaged Douglas-fir forests. USDA Forest Service PNW-GTR-285

Leimgruber, P, McShea, W.J., and J.H. Rappole. 1994. Predation on artificial nests in large forest blocks. J.Wildl. Manage. 58: 254-260

Leopold, A. 1933. Game Management. Charles Scribner's Sons, New York, NY

Levenson, J.B. 1981. Woodlots as biogeographic islands in southeastern Wisconsin. Pp. 13-39 in R.L. Burgess and D.M. Sharpe (eds.). Forest Island dynamics in Man-dominated Landscapes. Springer-Verlag, New York, NY

Lisgo, K. 1998. Personal communication. Centre for Applied Conservation Biology, University of British Columbia, Vancouver, B.C.

Lovejoy, T.E., Bierregaard, R.O. Jr., Rylands, A.B., Malcolm, J.R., Quintela, C.E., Harper, L.H., Brown, K.S. Jr., Powell, A.H., Powell, G.V.N., Schubart, H.O.R. and M.B. Hays. 1986. Edge and other effects of isolation on Amazon forest fragments. Pp. 257-285 in M.E. Soule (ed.). Conservation Biology: the Science of Scarcity and Diversity. Sinauer Associates, Inc., Sunderland, MA

Lyon, L.J. 1979. Habitat effectiveness as influences by roads and cover. J. Forestry 77: 658-660.

Lyon, L.J. and C.E. Jensen. 1980. Management implications of elk and deer use of clearcuts in Montana. J. Wildl. Manage. 44: 352-362

Mace R.D., Walker, J.S., Manley, T.D., Lyon L.J., and H. Zuuring. 1996. Relationships among grizzly bears, roads, and habitat in the Swan Mountains, Montana. J. Appl. Ecol 33: 1395-1404

MacKenzie, Ken. 1998. Pers. comm. Weldwood of Canada, P.O. Box 97, 100 Mile House, B.C. V0K 2E0

Martin, K., and J.M. Eadie. 1998. Nest webs: a community wide approach to the management and conservation of cavity-nesting forest birds. J. For. Ecol. & Manage. (in press).

Martin, S.K. 1994. Feeding ecology of American martens and fishers. Pp. 297-315 in S.W. Buskrik, A.S. Harestad, M.G. Raphael, and R.A. Powell (eds.). Martens, sables, and fishers. Biology and conservation. Cornell University Press, Ithaca, NY.

Martin, T.E. 1980. Diversity and abundance of spring migratory birds using habitat islands of the Great Plains. Condor 82: 430-439

Matlack, G.R. 1993. Microenvironment variation within and among forest edge sites in the eastern United States. Biol. Conserv. 66: 185-194

Mattson, D. 1990. Human impacts on bear habitat use. Int. Conf. Bear Res. And Manage. 8: 33-56. L. Darling and R. Archibald eds. Hemlock Printers, Victoria B.C.

Mattson, D.J., Knight, R.R. and D.M. Blanshard. 1987. The effects of development and primary roads on grizzly bear habitat use in Yellowstone National Park, Wyoming. Int. Conf. Bear Res. And Manage. 7:259-273 P. Zager ed. Port City Press Inc., Washington, D.C.

McCaffery, K.R. and W.A. Creed. 1969. Significance of forest openings to deer in northern Wisconsin. Tech. Bull. No. 44. Dept. Natural Resources, Madison, WI

McGarigal, K. and W.C. McComb. 1995. Relationships between landscape structure and breeding birds in the Oregon Coast Range. Ecol. Monographs 65: 235-260

McLellan, B. and D. Shackelton. 1988. Grizzly bears and resource extraction industries: effects of roads on behaviour, habitat use, and demography. J of Applied Ecology 25: 451-460

McLellan, B.N. 1990. Relationships between human industrial activity and grizzly bears. Int. Conf. Bear Res. And Manage. 8: 57-64. L. Darling and R. Archibald eds. Hemlock Printers, Victoria B.C

McLellan, C.H., Dobson, A.P., Wilcove, D.S. and J.F. Lynch. 1986. Effects of forest fragmentation on new- and old-World bird communities: empirical observations and theoretical implications. Pp. 305-313 in J. Verner, M. Morrison, and C. Ralph, (eds.). Wildlife 2000: modelling habitat relationships of terrestrial vertebrates. Univ. of Wisconsin Press, Madison, WI.

Mills, L.S. 1995. Edge effects and isolation: red-backed voles of forest remnants Cons. Biol. 9: 395-403

Moen, A.N. 1974. Turbulence and visualization of wind flow. Ecology 55: 1420-1424

Møller, A.P. 1989. Nest site selection across field woodland ecotones: the effect of nest predation. Oikos 562: 240-242

Morgan, K.A. and J.E. Gates. 1982. Bird population patterns in forest edge and strip vegetation at Remington Farms, Maryland. J. Wildl. Manage. 46: 933-944

Naiman, R.J., H. Decamps, and M. Pollack. 1993. The role of riparian corridors in maintaining regional diversity. Ecol. Applications 3: 209-212

Niemuth, N.D. 1993. Avian ecology in Wisonsin pine barrens. PhD thesis, University of Wyoming, Laramie, WY.

Niemuth, N.D., and M.S. Boyce. 1997. Edge-related nest losses in Wisconsin pine barrens. J. Wildlife Manage. 61: 1234-1239

Ng, F.S.P. 1983. Ecological principles of tropical lowland rain forest conservation. Pp. 359-376 in S.L. Sutton, T.C. Whitmore and A.C. Chadwick, (eds.).Tropical Rain Forest: ecology and management. Blackwell Scientific Publications, Oxford, England

Norman, R.F., and R.J. Robertson. 1975. Nest-searching behavior in the brown-headed cowbird. Auk 92: 610-611

Noss, R.F. 1983. A regional landscape approach to maintain diversity. Bioscience 33: 700-706

Nyberg, J.B. and D.W. Janz. 1990. Deer and elk habitats in coastal forests of southern British Columbia, B.C. Ministry of Forests, Special Report Series 5, co-published by B.C. Ministry of Environment in co-operation with Wildlife Habitat Canada. B.C. Ministry of Environment, Victoria, B.C

O'Conner, R.J. and J. Faaborg. 1992. The relative abundance of the brown-headed cowbird (*Molothrus ater*) in relation to exterior and interior edges in forests of Missouri. Trans. Miss. Acad. Sci. 26: 1-9

Odum, E.P. 1971. Fundamentals of Ecology. W.B. Saunders Co., Philadelphia, Pennsylvania

O'Meara, T.E, Haufler, J.B., Stelter, L.H. and J.G. Nagy. 1981. Nongame wildlife responses to chaining of pinyon-juniper woodlands. J. Wildl. Manage. 45: 381-389

Opdam, P. and A. Schotman. 1987. Small woods in rural landscape as habitat islands for woodland birds. Acta Oecologia 8: 269-274

Osunkoya, O.O. 1994. Postdispersal survivorship of north Queensland rainforest seeds and fruits: effects of forest, habitat and species. Aust. J. Ecol. 19: 52-64

Palik, B.J. and P.G. Murphy. 1990. Disturbance versus edge effects in sugar-maple/beech forest fragments. For. Ecol. Manage. 32: 187-202

Paton, P.W.C. 1994. The edge effect on avian nest success: how strong is the evidence? Cons. Biol. 8: 17-26

Pedersen, R.J., A.W. Adams, and J. Skovlin. 1979. Elk management in Blue Mountain habitats. Research Report, Oregon Dept. Fish and Wildlife, Portland, Oregon.

Pollard, E. 1992. Monitoring butterfly abundance in relation to the management of a nature reserve. Biol. Conserv. 24: 317-328

Planty-Tabacchi, A., E. Tabacchi, R.J. Naiman (and others) 1996. Invasibility of species-rich communities in riparian zones. Cons. Biol. 10: 598-607

Ranney, J.W., Bruner, M.C. and J.B. Levenson. 1981. The importance of edge in the structure and dynamics of forest islands. Pp. 67-95 in R.L. Burgess and D.M. Sharpe, (eds.). Forest Island Dynamics in Man-dominated Landscapes. Springer-Verlag, New York, NY.

Raphael, M.G. 1988. Long-term trends in abundance of amphibians, reptiles, and mammals in Douglas-fir forests of northwestern California. Pp. 23-31 in R.C. Szaro, K.E. Severson, and D.R. Patton (tech. eds.). Management of amphibians, reptiles, and small mammals in North America. Proceedings of a symposium. 19-21 July 1988, Flagstaff, AZ. USDA Forest Service, General Technical Report RM-166, Fort Collins, CO.

Ratti, J.T. and K.P. Reese. 1988. Preliminary test of the ecological trap hypothesis. J. Wildl. Manage. 52: 484-491

Reed, R.A., Johnson-Barnard J., and W.L. Baker. 1996. Contributions of roads to forest fragmentation in the Roacky Mountains. Cons. Biol. 10: 1098-1106

Reese, K. P. and J.T. Ratti. 1988. Edge effect: a concept under scrutiny. Pp. 127-136 in Trans. North American Wildl. and Nat. Res. Conference 53.

Reville, B.J., Tranter, J.D. and H.D. Yorkson. 1990. Impact of forest clearing on the endangered seabird, *Sula abbotti*. Biol. Cons. 51: 23-38

Reynolds, H.G. 1962. Use of natural openings in a ponderosa pine forest of Arizona by deer, elk, and cattle. USDA Forest Service Res. Note RM-78

Reynolds, H.G. 1966. Use of openings in spruce-fir forests of Arizona by elk, deer, and cattle. USDA Forest Service Res. Note RM-66

Rich, A.C., D.S. Dobkin and L. J. Niles. 1994. Defining forest fragmentation by corridor width: the influence of narrow forest-dividing corridors on forest-nesting birds in southern New Jersey. Cons. Biol. 8: 1109-1121

Robbins, C.S. 1979. Effect of forest fragmentation on bird populations. Pp. 198-212 in R. M. DeGraaf and K.E. Evans, (eds.). Management of north central and northeastern forests for non-game birds: workshop proceedings. USDA Forest Service GTR NC-51

Rosenberg, K.V. and M.G. Raphael. 1986. Effects of forest fragmentation on vertebrates in Douglas-fir forests. Pp. 263-272 in J. Verner, M. Morrison, and C. Ralph, (eds.). Wildlife 2000: modelling habitat relationships of terrestrial vertebrates. University Wisconsin Press. Madison, WI.

Rost, G.R., and J.A. Bailey. 1979. Distribution of mule deer and elk in relation to roads. J. Wildl. Manage. 43: 634-641

Rudnicky, T.C. and M.L. Hunter. 1993. Avian nest predation in clearcuts, forests and edges in a forest-dominated landscape. J. Wildl. Manage. 57: 358-364

Ruth, R.H. and R.A. Yoder. 1953. Reducing wind damage n the forests of the Oregon coast range. USDA Forest Service PNW RS-7

Sadoway, K.L. 1986. Effects of intensive forest management on amphibians and reptiles of Vancouver Island: problem analysis. B.C. Ministry of Forests, Victoria, B.C. IWIFR-23: 42 pp.

Santos. T., and J.L. Telleria. 1992. Edge effects on nest predation in Mediterranean fragmented forests. Biol. Cons. 60: 1-5.

Saunders, D.A., Hobbs, R.J. and C.R. Margules. 1991. Biological consequence of ecosystem fragmentation: a review. Cons. Biol. 5: 18-32

Schieck, J., Lertzman, K., Nyberg, B. and R. Page. 1995. Effects of patch size on birds in old-growth montane forests. Cons. Biol. 9: 1072-1084

Schoen, J.W., Flynn, R.W. Suring, L.H., and L.R. Beir. 1990. Brown bear habitat preferences and brown bear logging and mining relationships in southeast Alaska. Alaska Dept. Fish and Game Federal Aid Restoration Final Report, Project W-23-2 (cited in Harrison, R.L. 1992. Toward a theory of inter-refuge corridor design. Cons. Biol. 6: 293-295)

Scougall, S.A., Majer, J.D., and Hobbs, R.H. 1993. Edge effects in grazed and ungrazed Western Australian wheatbelt remnants in relation to ecosystem reconstruction. Pp. 163-178 in D.A. Saunders, R.H. Hobbs, and P.R. Ehrlich (eds.). Nature conservation 3: Reconstruction of fragmented ecosystems. Surrey Beatty & Sons, Chipping Norton, New South Wales, Australia. pp. 163-178.

Sedgewick, J.A. 1987. Avian habitat relationships in Pinyon-juniper woodlands. Wilson Bull. 93: 413-431

Shaloway, S.D. 1979. Breeding bird abundance and distribution in fencerow habitat. Ph.D. Diss. Mich. State University, East Lansing.

Sievert, P.R. and L.B. Keith. 1985. Survival of snowshoe hares at a geographical range boundary. J. Wildl. Manage. 49: 854-866

Sisk, T.D., and C.R. Margules. 1993. Habitat edges and restoration: methods for quantifying edge effects and predicting the results of restoration efforts. Pp 57-59 in D.A. Saunders, R.J. Hobbs and P.R. Ehrlich (eds.). Nature conservation 3: reconstruction 1987. Avian habitat relationships in Pinyon-juniper woodlands. Wilson Bull. 93: 413-431

Small, M.F. and M.L. Hunter. 1989. Response of passerines to abrupt forest-river and forest-powerline edges in Maine. Wilson Bull. 101: 77-83

Smith, D.M. 1962. The practice of Silviculture. John Wiley and Sons, New York, NY.

Soulé, M.E. 1986. The effects of fragmentation. Pp. 233-236 in ME Soule (ed.). Conservation Biology: the Science of Scarcity and Diversity. Sinauer Associates, Inc., Sunderland, Massachusetts

Spackman, S.C. and J.W. Hughes. 1995. Assessment of minimum stream corridor width for biological conservation: species richness and distribution along mid-order streams in Vermont, USA. Biol. Conserv. 71: 325-332

Start, A.N. 1991. How can edge effects be minimized? Pp. 417-418 in D.A. Saunders and R.J. Hobbs (eds.). The role of corridors. Surrey Beatty, Chipping Norton, New South Wales, Australia

Strelke, W.K., and J.G. Dickson. 1980. Effect of forest clear-cut edge on breeding birds in east Texas. J. Wildl. Manage. 44: 559-567

Suarez, A.V., Pfennig, K.S. and S.K. Robinson. 1997. Nesting success of a disturbance-dependent songbird on different kinds of edges. Conserv. Biol. 11: 928-935

Sutherland, Glen. 1998. Pers. comm. Centre for Applied Conservation Biology, 270-2357 Main Mall, Vancouver B.C. V6T 1Z4. (604) 822-0943. gsland@unixg.ubc.ca

Swanson, F.J., Kratz, T.K., Caine, N. and R.G. Woodmansee. 1988. Landform effects on ecosystem patterns and processes. Bioscience 38: 92-98

Temple, S.A. and J.R. Carey. 1988. Modelling dynamics of habitat-interior bird populations in fragmented landscapes. Cons. Biol. 2: 340-347

Terborgh, J.W. 1992. Why American songbirds are vanishing. Scientific American May 1992: 98-104

Thomas J.W., Maser, C. and J.E. Rodiek. 1979. Edges. Chapter 4 in J.W. Thomas (ed.) Wildlife Habitats in Managed Forests: The Blue Mountains of Oregon and Washington. Wildlife Management Institute, Washington, D.C. USDA For. Serv. Agric. Hand. 533

Trail, P.W. and L. F. Baptista. 1993. The impact of brown-headed cowbird parasitism on populations of the Nuttall's white-crowned sparrow. Cons. Biol. 7: 309-315

Tyser, R.W. and C.A. Worley. 1992. Alien flora in grassland adjacent to road and trail corridors in Glacier National Park, Montana (U.S.A.) Cons. Biol. 6: 253-262

Usher, M.B., Field, J.P. and S.E. Bedford. 1993. Biogeography and diversity of ground-dwelling arthropods in farm woodlands. Biodiversity Letters 1: 54-62

Vaillancourt, D.A. 1995. Structural and microclimatic edge effects associated with clearcutting in a Rocky Mountain forest. M.Sc. thesis. Dept. of Geography and Recreation, University of Wyoming, Laramie. In Reed, R.A., J. Johnson-Barnard, and W.L. Baker. 1996. Contribution of roads to forest fragmentation in the Rocky Mountains. Cons. Biol. 10: 1098-1106

Vander Haegen, W.M and R.M. DeGraaf. 1996. Predation on artificial nests in forested riparian buffer strips. J.Wildl. Manage. 60(3): 542-50

Vega, R.M.S. 1993. Bird communities in managed conifer stands in the Oregon Cascades: habitat associations and nest predation. MSc thesis, Oregon State University, Corvallis, OR.

Vickholm, M. 1983. Avoiten reunojen vaikutus metsalinnustoon. M.Sc. thesis, Dept. Zoology, Univ. Helsinki. Cited in Virkkala 1987. Effects of forest management on birds breeding in northern Finland. Ann. Zool. Fennici 24: 281-294

Virkkala, R. 1987. Effects of forest management on birds breeding in northern Finland. Ann. Zool. Fennici 24: 281-294

Wagner, R.G. 1980. Natural regeneration at the edges of Abies amabilis zone clearcuts on the west slope of the central Washington Cascades, M.Sc. thesis. University of Washington, Seattle, WA. Cited in Chen, J., J.F. Franklin, and T.A. Spies. 1992. Vegetation responses to edge environments in old-growth Douglas-fir forests. Ecol. Applic. 2: 387-396

Wales, B.A. 1972. Vegetation analysis of north and south edges in a mature oak-hickory forest. Ecological Monographs 42: 451-471

Wallmo, O.C. 1969. Response of deer to alternate-strip clearcutting of lodgepole pine and spruce-fir timber in Colorado. USDA For. Serv. Res. Note RM. 141.

Warren, M.S. 1985. The influence of shade on butterfly numbers in woodland rides, with special reference to the wood white Leptidae sinapsis. Biol. Conserv. 33: 147-164

Warren, M.S., Pollard, E., and Bibby, T.J. 1986. Annual and long-term changes in a population of the wood white butterfly Leptidae sinapsis. J. Anim. Ecol. 55: 707-719

Weger, E. 1977. Evaluation of winter use of second-growth stands by black-tailed deer. B.S.F. thesis, University British Columbia, Vancouver, B.C

Wegner, J.F. and H.G. Merriam. 1979. Movements by birds and small mammals between a wood and adjoining farmland habitats. J. Appl. Ecol. 16: 349-357

Weilgus, R, F.L. Bunnell, and P. Vernier. 1998. Effects of forestry on grizzly bear habitat use, population growth and population persistence. Report to Forest Renewal B.C. No. KB97489-ORE1.

Welsh, H.H. Jr and A.J. Lind. 1991. The structure of the herpetofaunal assemblage in the Douglas-fir/hardwood forests of northwestern California and southwestern Oregon. Pp. 395-413 in USDA Forest Service (ed.). Wildlife and vegetation of unmanaged Douglas-fir forests. USDA Forest Service, General Technical Report, PNW-GTR-285, Portland, OR.

Welsh, H.H. Jr. 1990. Relictual amphibians and old-growth forests. Cons. Biol. 3: 309-319

Whitcomb, R.F., J.F. Lynch, P.A. Opler, and C.S. Robbins. 1976. Island biogeography and conservation: strategy and limitations. Science 193: 1030-1032

Whitcomb, R.F., Robbins, J.J., Lynch, B.L., Whitcomb. M.K., Klimkiewits, K. and D. Bystrak. 1981. Chapter 8. Pp 125-204 in R.L. Burgess and D.M. Sharpe (eds.). Forest island dynamics in man-dominated landscapes. Springer-Verlag, New York, NY

Whitney, G.G. and J.R. Runkle. 1981. Edge versus area effects in the development of a beech maple forest. Oikos 37: 377-381

Wiens, J.A., Crawford, C.A. and J.R. Gosz. 1985. Boundary dynamics: a conceptual framework for studying landscape ecosystems. Oikos 45: 421-427

Wilcove, D.S. 1985. Nest predation in forest tracts and the decline of migratory songbirds. Ecology 66: 1211-1214

Wilcove, D.S., McLellan, C. and A. Dobson. 1986. Habitat fragmentation in the temperate zone. Pp. 237-256 in M.E. Soule (ed.). Conservation Biology: The Science of Scarcity and Diversity. Sinauer Associates, Sunderland, MA

Williams-Linera, G. 1990. Vegetation structure and environmental conditions on forest edges in Panama. J. Ecol. 78: 356-373

Willms, W.D. 1972. The influence of forest edge, elevation, aspect, site index, and roads on deer use of logged and mature forest, northern Vancouver Island. M.Sc. Thesis. University British Columbia, Vancouver, B.C.

Witmer, G.W. and D.S.deCalesta. 1983. Habitat use by female Roosevelt Elk in the Oregon Coast range J. Wildl. Manage. 47: 933-939

Wood, P.A. and M.J. Samways. 1991. Landscape element pattern and continuity of butterfly flight paths in an ecologically landscaped botanic garden, Natal, South Africa. Biol. Cons. 58: 149-166

Yahner, R.H. 1983. Seasonal dynamics, habitat relationships, and management of avifauna in farmstead shelterbelts. J. Wildl. Manage. 47: 85-104

Yahner, R.H. 1984. Effects of habitat patchiness created by a ruffed grouse management plan on avian communities. Am. Midl. Nat. 111: 409-413

Yahner, R.H. 1986. Spatial distribution of white-footed mice (*Peromyscus leucopus*) in fragmented forest stands. Proceedings of the Pennsylvania Academy of Science 60: 165-166

Yahner, R.H. 1987. Use of even-aged stands by winter and spring bird communites. Wilson Bull. 99: 218-232

Yahner, R.H. 1988 Changes in wildlife communities near edges. Cons. Biol. 2: 333-339

Yahner, R.H. and C.G. Mahan. 1997. Effects of logging roads on depredation of artificial ground nests in a forested landscape. Wildl. Soc. Bull. 25(1): 158-162

Yahner, R.H. and D.P. Scott. 1988. Effects of forest fragmentation on depredation of artificial avian nests. J. Wildl. Manage. 52: 158-161

Yahner, R.H. and A.L. Wright. 1985. Depredation on artificial nests: effects of edge and plot age. J. Wildl. Manage. 49: 508-513

Yalden, D.W. 1992. The influence of recreational disturbance on common sandpipers Actitus hypoleucos breeding by an upland reservoir, in England. Biol. Conser. 61: 41-49. in Freedman, B., S. Woodley, and J. Loo. 1994. Forestry practices and biodiversity, with particular reference to Maritime Provinces of eastern Canada. Environ. Rev. 2: 33-77

Young, A.G. 1988. The ecological significance of the edge effect in a fragmented forest landscape. M.Sc. thesis, University Auckland, New Zealand. Cited in A.F. Bennett. 1990. Habitat corridors, their role in wildlife management and conservation. Department of Conservation and Environment, Arthur Rylah Institute for Environmental Research, Victoria, Australia.

Young, A. and N. Mitchell. 1994. Microclimate and vegetation edge effects in a fragmented podocarp-broadleaf forest in New Zealand. Biol. Conserv. 67: 63-72

Zen, M.C. 1995. Growth and morphological development of Douglas-fir and western hemlock on forest edges in the Alberni Valley, B.C. M.Sc. thesis, University Washington, Washington, D.C.

The Effects of Forest Fragmentation on Avian Nest Predation

John M. Marzluff, Marco Restani

One important mechanism that links forest fragmentation to avian population viability is nest predation. We synthesized the literature (1984-1998) to evaluate the extent of this process in forested fragments. Edge effects were more common in forest fragments within urban or agricultural matrices than in fragments within a regenerating forest matrix. However, even within urban and agricultural landscapes, edge effects were not observed in half of the studies reviewed. Increases in nest predation were associated with reduced patch size and increased landscape fragmentation more often than with distance from edge, but this pattern was not universal. We suspect that edge and fragmentation effects were uncommon because the community of nest predators was diverse, including both birds and mammals. Predator communities varied from edges to interior forests such that the overall rate of predation was similar along the habitat continuum. We propose that managers (1) focus their concern on the detrimental effects of urbanization and agricultural intensification, (2) conserve what can effectively be conserved in urban environments, (3) rely on commercial forests to maintain biodiversity, (4) reduce the conversion of commercial forest land into urban developments, and (5) remain vigilant for effects other than predation in fragmented forest settings.

Key words: agriculture, artificial nests, edge effects, fragmentation, managed forest, nest predation, urbanization.

1. INTRODUCTION

Forest fragmentation reduces the amount of habitat, isolates remaining forest tracts, and increases edge (Harris, 1984; Laurance and Bierregaard, 1997). To understand the impact of fragmentation on wildlife, we must link forest fracturing to wildlife population viability. A mechanistic understanding of habitat selection in fragmented environments is necessary to understand why animals may or may not occupy forest fragments (Wiens, 1994). Here, we extend this argument by focusing on the mechanisms that affect animal reproduction and survivorship in fragmented landscapes. Such an approach can identify causal relationships thereby increasing the effectiveness and flexibility of management (Marzluff et al., in press).

Although forest fragmentation has many direct and indirect effects on the abiotic and biotic components of remaining forests (Laurance and Yensen, 1991), two biotic effects have received the most attention regarding impacts on birds: nest predation and nest parasitism. Nest predation is possibly the primary mechanism that limits reproduction and population viability of songbirds in fragmented landscapes (Robinson and Wilcove, 1994).

Ecosystem Sciences Division, College of Forest Resources, University of Washington, Seattle, WA 98195-2100

Predators commonly destroy 40-60% of nests (Nice, 1957; Ricklefs, 1969; Martin, 1995), but this figure rises to 80% in fragmented habitats (Robinson and Wilcove, 1994). Clearly, a thorough understanding of this pervasive process is needed for effective management of songbirds.

Conservation biologists have recently conducted many empirical and theoretical studies of the effects of forest fragmentation on rates of nest predation (reviewed by Paton, 1994; Robinson and Wilcove, 1994; Andren, 1995; Murcia, 1995). The impact of fragmentation appears to be determined by the extent of fragmentation, the density and diversity of nest predators, the matrix surrounding fragments, and the techniques used by researchers. Below, we briefly summarize these propositions and then report on the results of a new synthesis that includes recent studies from managed forest landscapes. A recent increase in sampling the effects of diverse predator communities and the increasing investigation of forest fragments in forested matrices, as opposed to urban or agricultural matrices, prompted this review, which also serves to update previous reviews.

The primary objective of this paper is to review the literature to determine what types of studies have been conducted on the effects of forest fragmentation. We then synthesize results and reappraise the effects that predators have in fragmented landscapes with special reference to forests of western North America. Finally, we present several new hypotheses concerning the effects of fragmentation on nest predation.

2. PREVIOUS IDEAS ABOUT HOW AND WHY FRAGMENTATION AFFECTS AVIAN NEST PREDATION

Researchers typically use artificial nests to measure the influence of fragmentation on nest predation by (1) testing for a decrease in predation rates from the forest edge to the interior (i.e., edge effects), (2) testing for an increase in predation as forest fragment size decreases (i.e., patch size effects), and (3) comparing rates of predation in forest stands surrounded by fragmented versus contiguous forest landscapes (i.e., landscape effects).

Fundamentally, the behavior, density and diversity of nest predators determines the effect of fragmentation. In order for fragmentation to elevate rates of nest predation, predators must either (1) reach highest density or forage preferentially along forest edges, (2) reach highest density or forage preferentially in small forest fragments or fragmented landscapes, or (3) exhibit highest community diversity at forest edges, in small fragments, or in fragmented landscapes. The first two circumstances are commonly reported and occur when predators specialize on edge habitats or when generalist predators common to non-forested habitat penetrate into forested habitat (Andren, 1995).

Early studies suggested that the composition of matrix habitat is an important determinant of nest predation rates (Gates and Gysel, 1978; Whitcomb et al., 1981; Wilcove, 1985). Fragments surrounded by agriculture or urban habitat typically showed elevated rates and edge-specific occurrence of predation. Corvid predators common to urban and agricultural habitats were able to penetrate into adjacent forest fragments and lower songbird productivity (Andren et al., 1985; Angelstam, 1986; Andren, 1992). In contrast, forest fragments surrounded by young forest, which typically occur in landscapes managed for timber production, did not show elevated or edge-specific rates of predation (DeGraaf and Angelstam, 1993; DeGraaf, 1995). This occurred because few predators were thought to use young forests and clearcuts. Overall predator densities were also reduced because their preferred habitat (interior forest) was reduced.

Research techniques used to measure predation rates can profoundly influence the determination of fragmentation effects. The most common approach is to place artificial nests at varying distances from forest edges, in fragments of various size, or in landscapes characterized by varying degrees of fragmentation. Artificial nests are typically filled with relatively large, thick-shelled eggs (chicken or quail eggs) which cannot be opened by small mammalian predators (e.g., chipmunks and mice; Roper, 1992; Haskell, 1995). When all potential predators are sampled with wax-coated or clay eggs (teeth impressions indicate predation even if the egg is not opened or removed), fragmentation effects are reduced or negated (Haskell, 1995).

Recent studies that sampled the entire predator community often failed to find edge effects because different predators were important at different distances from the stand edge (Nour et al., 1993). When researchers used more sophisticated eggs to identify the assemblage of nest predators, the simple relationship between surrounding matrix and predation in forest fragments became less clear (Haskell, 1995).

3. ANOTHER LOOK AT THE PUBLISHED LITERATURE

We searched the published literature (e.g., BIOSIS, Current Contents, AGRICOLA) for articles that directly tested the effects of fragmentation on nest predation in forest fragments. We identified 47 articles published from 1984–1998 (Table 1). We excluded studies of grassland fragmentation and studies that simply reported nest predation without comparing rates among areas or locations varying in their level of fragmentation. Our search was not confined geographically.

3.1 What Types of Studies Have Been Conducted?

Researchers usually investigated fragmentation by determining if nest predation rates varied with distance from the forest fragment's edge (65 %, N = 40 of 62 total tests). Landscape fragmentation (i.e., patch and landscape effects) was investigated by comparing predation among fragments of varying size (19 %, N = 12), or between landscapes varying in overall degree of fragmentation (16%, N = 10).

The typical assessment of fragmentation's effects on nest predation was a single season's artificial nest experiment. Over half of the studies (58%, N = 27) lasted a single year and 83% (N = 39) lasted 1 or 2 years. Only three studies lasted five or more years (Kuitunen and Helle, 1988; Robinson, 1992; Robinson et al., 1995). Real nests were monitored only 10 times (23%) compared to 38 experiments (77%) using artificial nests. Regardless of nest type, the vast majority of studies (96%, N = 42) were conducted at shrub or ground level. Two studies (4%) used cavity nests and no studies investigated the forest canopy. Although some biological reality was sacrificed by using artificial nests, statistical power was obtained; twenty-eight studies (76%) placed artificial nests in a replicated experimental design. However, even these experiments rarely determined causal relationships; none compared predation before and after fragmentation and none randomly allocated treatments to stands.

Most studies (53%) were conducted in forest fragments embedded in landscapes dominated by agriculture and urbanization (N = 32% agriculture, 11% urban, 11% agriculture and urban). Although recent interest in fragmentation of managed forest landscapes was evident, only 34% (N = 16) of the studies we reviewed assessed nest predation in forest fragments surrounded by managed forested landscapes. The remaining studies examined nest predation in a mixture of agriculture and managed forest.

The most glaring omission from studies of nest predation was a lack of study of the predators themselves. Thirteen studies (28%) did not identify predators (Table 1). Only eight studies (17%) surveyed predators and only two (Angelstam, 1986; Tewksbury et al., 1998) related predator occurrence to habitat type and edge. In fact, behavioral studies of species that are common nest predators were rare overall; we only found six additional studies that investigated the behavior of nest predators (Bowman and Harris, 1980; Reitsma et al., 1990; Vickery et al., 1992; DeGraaf and Maier, 1996; Lariviere and Messier, 1997; King et al., 1998b).

Table 1. Studies of fragmentation effects on remaining forest fragments used in meta-analysis. Matrix is the dominant landscape remaining around the studied forest fragment. Outcomes of tests within studies are reported as NS if results were not significant (P> 0.05), EDGE if predation or parasitism was significantly higher at forest fragment edges relative to interior, FRAG-S if predation or parasitism was significantly higher in small compared to large forest fragments, or FRAG-L if predation or parasitism was significantly higher in fragmented relative to contiguous landscapes. Properties of experimental design are reported as CORR if natural variation in fragmentation was related to predation or parasitism, REPLICATE if spatial replicates of stands varying in level of fragmentation were used, or MANIPULATE if the degree of fragmentation was randomly assigned to stands. Responses were obtained with dummy nests or real nests and consisted of predation or parasitism alone unless otherwise indicated.

Matrix	Outcome	Predators	Design	Response	Duration	Location	Reference
Ag	FRAG-S	Corvids	Repeat	Artificial	2	Sweden	Andren 1992
Ag	EDGE FRAG-S	No ID	Repeat	Artificial	3	Sweden	Andren & Angelstam 1988
Ag	FRAG-S	Corvids	Repeat	Artificial	2	Sweden	Andren et al. 1985
Ag	NS Edge	Corvids, Meso Mam	Corr	Artificial	1	Sweden	Angelstam 1986
Managed Forest	FRAG-S NS Edge	Meso Mam, Small Mam	Repeat	Artificial,	1	Colombia	Arango-Velez & Kattan 1997
Managed Forest Ag	NS Edge NS Edge FRAG	Corvids, Small Mam, Meso Mam	Repeat	Artificial	2	Saskatchewan	Bayne et al. 1997
Ag	EDGE	No ID	Corr	Real Nests	4	Sweden	Bjorklund 1990
Managed Forest	NS Edge	Corvids, Small Mam	Repeat	Artificial and Real Nests	1	Alberta	Boag et al. 1984
Managed Forest	EDGE	Meso Mam	Corr	Artificial	1	Belize, Mexico	Burkey 1993
Urban	EDGE	Corvids, Small Mam, Meso Mam	Repeat	Artificial	2	NH, MA	Danielson et al. 1997
Managed Forest	NS Edge	Corvids, Small Mam, Meso Mam	Manipulate	Artificial	3	Quebec	Darveu et al. 1997
Managed Forest	NS Frag-L	Small Mam, Meso Mam, Large Mam	Repeat	Artificial	1	NH	DeGraaf 1995
Managed Forest	NS Frag-L	Meso Mam	Corr	Artificial	1	NH	DeGraaf & Anglestam 1993
Managed Forest	EDGE	Small Mam, Meso Mam	Repeat	Artificial	1	MN	Fenske-Crawford & Niemi 1997

Table 1. Continued.

Matrix	Outcome	Predators	Design	Response	Duration	Location	Reference
Managed Forest Ag	EDGE NS Edge	Meso Mam	Repeat	Artificial	1	Costa Rica	Gibbs 1991
Ag	NS Frag-L	Corvids, Small Mam	Corr	Artificial	2	Alberta	Hannon & Cotterill 1998
Managed Forest	NS Edge	Corvids, Small Mam, Meso Mam	Corr	Real Nests	1	MN	Hanski et al. 1996
Ag	NS Edge	Corvids, Small Mam, Meso Mam	Repeat	Artificial	2	NY	Haskell 1995
Urban/Ag	FRAG-S	No ID	Corr	Real Nests	2	PA	Hoover et al. 1995
Managed Forest Ag	NS Edge NS Frag - SEDGE	Corvids, Small Mam, Meso Mam	Repeat	Artificial	1	Finland	Huhta et al. 1996
Urban	NS Edge FRAG-L	Small Mam, Meso Mam	Corr	Artificial	1	AL	Keyser et al. 1998
Managed Forest	EDGE	No ID	Corr	Artificial Nests	1	NH	King et al. 1998a
Ag	EDGE FRAG-S	No ID	Corr	Real Cavity Nests	10	N. and W. Europe	Kuitunen & Helle 1988
Managed Forest Ag	NS Edge NS Edge	Small Mam	Repeat	Artificial	1	Australia	Laurance et al. 1993
Ag	EDGE	Corvids	Repeat	Artificial	3	Denmark	Møller 1988
Ag	EDGE	Corvids	Repeat	Artificial	1	Denmark	Møller 1989
Ag	NS Edge NS Frag-S	Corvids, Small Mam	Repeat	Artificial	1	Belgium	Nour et al. 1993
Managed Forest	NS Edge	Corvids, Small Mam, Meso Mam	Repeat	Artificial	1	ID	Ratti & Reese 1988
Ag	FRAG-L	No ID	Corr	Real Nests	6	IL	Robinson 1992
Urban Ag	EDGE	Corvids, Small Mam, Meso Mam	Corr	Real Nests	1	Midwestern US	Robinson & Wilcove 1994
Ag/Urban	FRAG-L	No ID	Corr	Real Nests	5	Midwestern US	Robinson et al. 1995
Managed Forest	NS Edge	No ID	Repeat	Artificial	2	ME	Rudnicky & Hunter 1993
Urban	NS Edge	Corvids, Small Mam	Repeat	Artificial	1	NY	Russo & Young 1997
Ag	EDGE	No ID	Repeat	Artificial Cavity Nests	1	Sweden	Sandstrom 1991

Table 1. Continued.

Matrix	Outcome	Predators	Design	Response	Duration	Location	Reference
Ag	NS Edge	Small Mam	Repeat	Artificial	1	Spain	Santos & Telleria 1992
Managed Forest	NS Edge	Birds, Mammals	Corr	Artificial	1	SC	Sargent et al. 1998
Ag	NS Edge						
Urban/Ag	NS Frag-L	No ID	Corr	Artificial	1	PA	Seitz & Zeglers 1993
Managed Forest	NS Edge	Corvids, Small Mam, Meso Mam	Repeat	Artificial	2	AK and British Columbia	Sieving & Willson 1998
Managed Forest	FRAG-S NS Edge	Corvids, Meso Mam	Corr	Artificial	1	ME	Small & Hunter 1988
Managed Forest	NS Edge	Meso Mam	Repeat	Artificial	3	Germany	Storch 1991
Ag	NS Frag-S	No ID	Repeat	Artificial	2	Spain	Telleria & Santos 1992
Urban/Ag	EDGE	Corvids, Grackles	Corr	Real Nests	2	WI	Temple & Cary 1988
Managed Forest	NS Frag-L	No ID	Corr	Real Nests	2	MT	Tewksbury et al. 1998
Ag	NS Edge						
Urban	FRAG-S	No ID	Corr	Artificial	1	MD, TN	Wilcove 1985
Urban	NS Edge NS Frag-S	Reptiles, Small Mam, Meso Mam	Repeat	Artificial	1	Singapore, Pulau Ubin	Wong et al. 1998
Managed Forest	FRAG-L	Corvids	Repeat	Artificial	1	PA	Yahner & Scott 1988
Managed Forest	NS Edge	Corvids	Repeat	Artificial	1	PA	Yahner et al. 1989

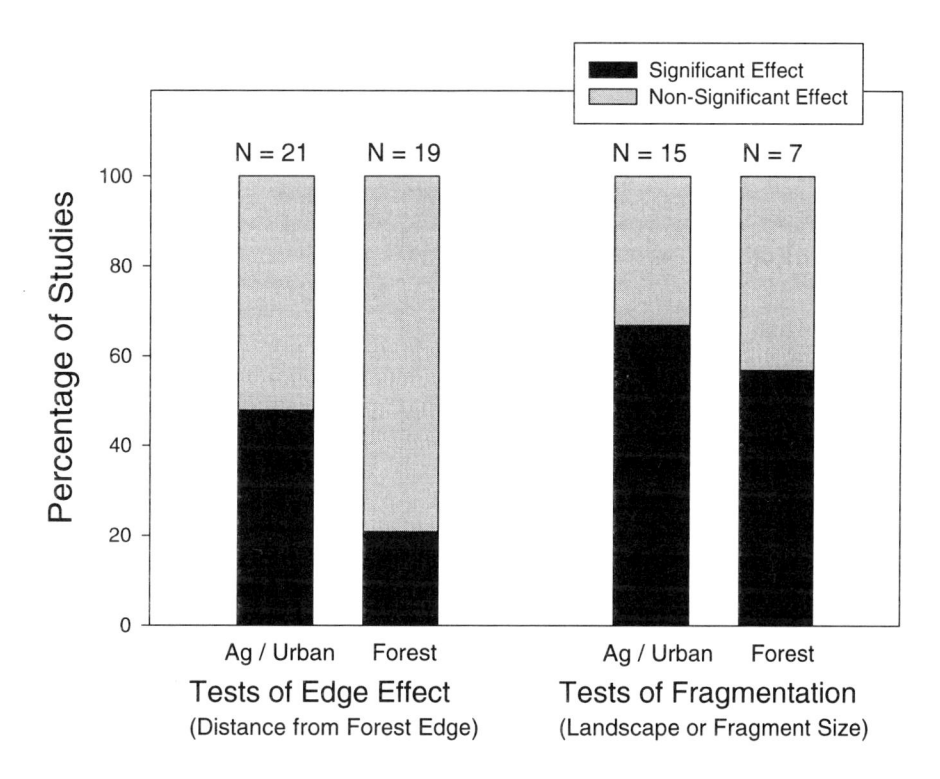

Figure 1. Documentation of edge and fragmentation (patch size and landscape fragmentation) effects in forest fragments imbedded in two types of landscapes (agricultural or urban versus managed forest). Black portion of bars show the percentage of studies finding significant effects (P < 0.05). Sample size (number of studies) is given above bars. Studies contributing to this figure are listed in Table 1.

3.2 Does Fragmentation Affect Nest Predation in Forest Fragments?

Nearly two-thirds of the studies (65%) we reviewed failed to find a significant decrease in nest predation from the forest edge to the forest interior. Our review suggests that, contrary to popular belief, edge effects may be an uncommon occurrence in forest fragments. Likewise, studies comparing habitat patches of various size or landscapes varying in degree of fragmentation often produce inconsistent results. Most studies (59%) indicate an increase in nest predation, but over a third (41%) found no such effect.

Two factors discussed earlier - composition of the matrix and diversity of the predator community - explained some of the variation in whether edge effects were or were not observed. Forest plots in an urban or agricultural matrix were more likely to experience edge effects than were plots surrounded by young forest (Fig. 1; $X^2(1) = 3.10$, $P = 0.079$). However, half of forest fragments in agricultural or urban landscapes did not have significant edge effects (Fig. 1), and this appeared due in part to predator diversity (Fig. 2A). Studies reporting non-significant edge effects typically (60%) reported diverse predator assemblages of birds and mammals or a variety of mammals (28%). In contrast, studies that found edge effects typically (33%) did not identify predators or implicated only mammals or only corvids as nest predators (Fig. 2A). Only one study that failed to find a significant edge effect in agricultural or urban settings implicated

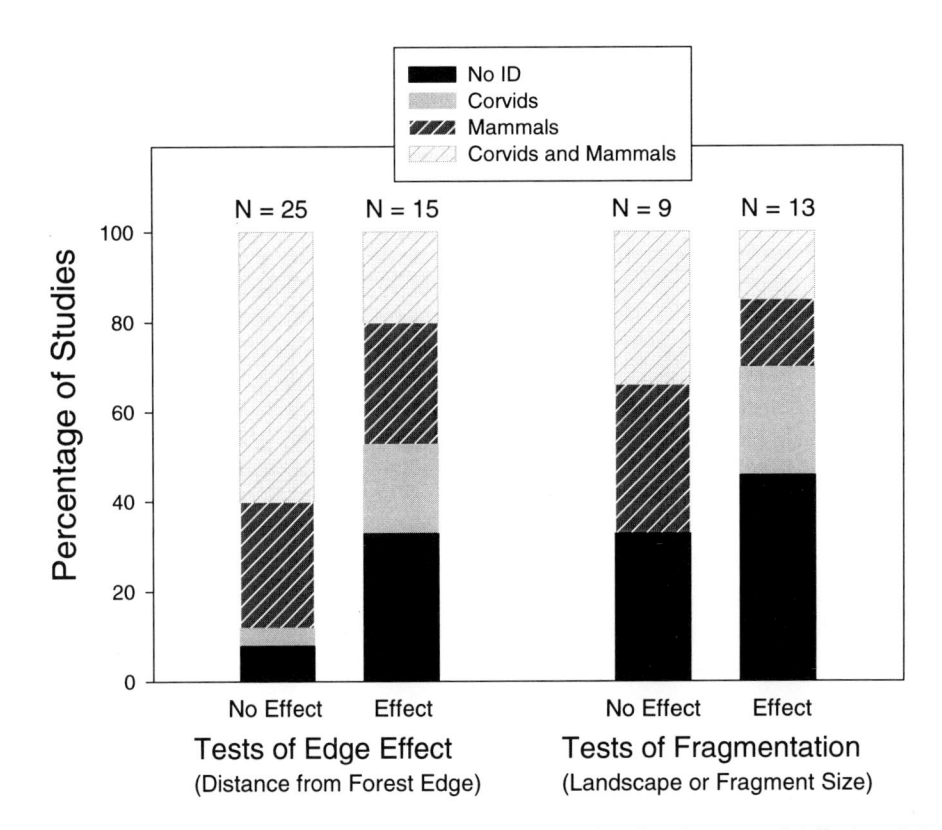

Figure 2. Predator communities identified as important in studies that documented (effect) or failed to document (no Effect) edge and fragmentation (patch size and landscape fragmentation) effects in two types of landscapes (agricultural and urban versus managed forest). Sample size (number of studies) is given above bars. Studies contributing to this figure are listed in Table 1.

corvids alone as nest predators (Nour et al., 1993).

A wide range of mammalian and avian predators have been implicated in studies of fragmentation. Avian predators are primarily members of the Family Corvidae, notably the common raven (*Corvus corax*), hooded crow (*C. corone*), American crow (*C. brachyrhynchos*), black-billed magpie (*Pica pica*), jay (*Garrulus glandarius*), Steller's jay (*Cyanocitta stelleri*), blue jay (*C. cristata*), and gray jay (*Periosorius canadensis*). Mammalian predators can be subdivided by size: "large" (bears [*Ursus spp.*], monkeys [*Cebidae spp.*]); "meso" (foxes [*Vulpes spp.*], badgers [*Taxidea taxus*], skunks [*Mephitis spp.*], weasels [*Mustela spp.*], and raccoons [*Procyon spp.*]); and "small" (squirrels [e.g., *Sciurus* and *Tamiasciurus spp.*], mice [e.g., *Peromyscus maniculatus*], shrews [*Sorex spp.*], and voles [*Microtus spp.*]). Corvids may be the primary drivers of edge effects in urban and agricultural settings, but it seems unlikely that study areas contain only corvid nest predators. Rather, most study designs preclude complete detection of the predator community, especially small mammalian predators. When entire predator communities are assayed, significant edge effects become unlikely.

Patch size and overall degree of landscape fragmentation appeared to elevate nest predation more frequently than did proximity to edge, but the result was inconsistent and not clearly related to the landscape matrix or predator diversity (Figs. 1 and 2). Over one-third of studies (40%) failed to detect a significant effect regardless of surrounding landscape or patch size (Fig. 1; Fisher's Exact Test P = 0.376). Predator communities were also similar in studies detecting versus not detecting patch size or landscape fragmentation effects (Fig. 2).

Why have some studies failed to detect patch size and landscape fragmentation effects? Some studies may have considered too small a range in fragment sizes to detect changes in the level of nest predation (e.g., Santos and Telleria, 1992; Nour et al., 1993; Huhta et al., 1996 did not consider fragments any larger than 50 ha). Removing these studies from the analysis, two remaining studies showed convincingly that fragmentation from timber harvest in the eastern US did not lead to increased predation (DeGraaf and Angelstam, 1993; DeGraaf, 1995). In urban and agricultural landscapes, the effects were more consistent; only two failed to show significant effects of fragmentation (Telleria and Santos, 1992; Seitz and Zeigler, 1993), one of which reported a nearly significant trend.

In summary, patch size and degree of landscape fragmentation effects were related to elevated rates of predation more frequently than proximity to forest fragment edge. The rigorous studies conducted in the midwestern USA (Robinson et al., 1995) clearly demonstrated a significant increase in nest predation with increased landscape fragmentation by agriculture and urbanization. In contrast, fragmentation resulting from timber harvest was less closely associated with elevated rates of predation (DeGraaf and Angelstam, 1993; Rudnicky and Hunter, 1993; DeGraaf, 1995; Hanski et al., 1996). Nest predation associated with forest patches created by timber harvest did not increase and this pattern was also supported by less frequent edge effects in forested landscapes. These results suggested that juxtaposition of forest patches varying in age and structure was less likely to increase and focus predation at mature forest edges than was juxtaposition of forest patches with agricultural or urban parcels.

4. WHEN WILL FOREST FRAGMENTATION AFFECT AVIAN PRODUCTIVITY, WITH SPECIAL REFERENCE TO WESTERN USA LANDSCAPES?

Edge effects, which decrease overall productivity of a fragmented region by reducing the amount of interior forest (Laurance and Yensen, 1991), appear to be predictable, at least theoretically. To produce edge effects, nest predation must be low or moderate in intensity because high density and mobility of predators will increase predation even within the interior of forest fragments, which would yield uniformly high predation along the matrix - edge - interior continuum (Santos and Telleria, 1992; Russo and Young, 1997; Wong et al., 1998). Predators with generalized habitat requirements are most likely to drive edge effects because they make forays from urban and agricultural landscapes deep into forests (Andren, 1995). Predator communities must also be relatively simple to produce accentuated edge effects. Complex predator communities will produce high rates of predation throughout matrix, edge, and forested habitats as different predator species peak in abundance at varying points along the habitat continuum from matrix to forest interior.

We predict that forest fragments in the western U.S. will rarely exhibit strong edge effects. This region is currently experiencing rapid population growth and urbanization, thereby isolating forest fragments in a suburban and urban matrix. However, rather than introducing a single, generalized, human commensal that preys on eggs or nestling birds, a suite of generalist predators commonly accompanies urbanization (e.g., American crows, cats, dogs, raccoons, rats, and non-native squirrels). This community mixes with native forest predators (e.g., native squirrels, Steller's jay, gray jay, common raven, deer mice) to produce a vibrant predator community that spans the spectrum of habitat from matrix to edge to forest interior (Fig. 3A). The effect of edges in western urban and agricultural landscapes is

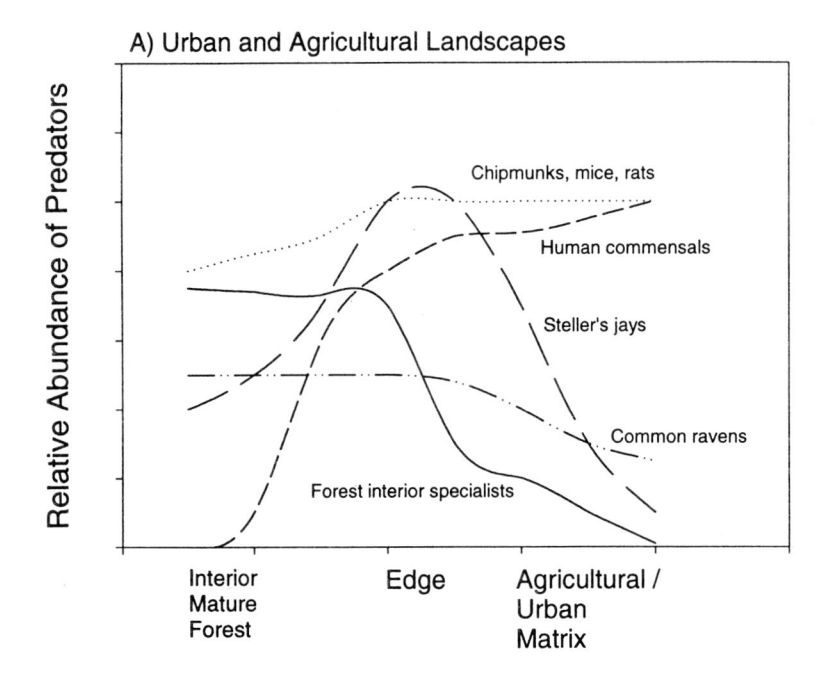

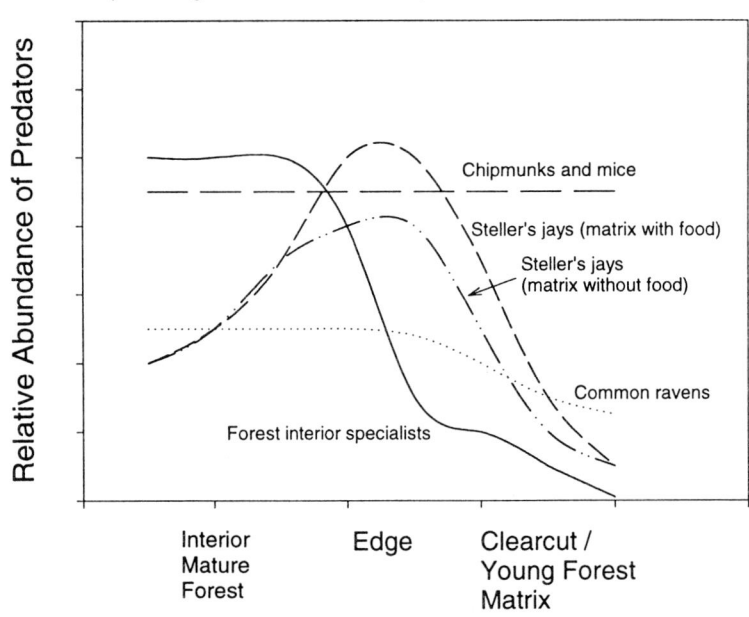

Figure 3. Hypothesized distributions of avian nest predators with respect to forest fragment edges in western U.S. urban / agricultural landscapes (A) versus managed forest landscapes (B). Interior forest predators include gray jays, red squirrels, flying squirrels, mustelids, northern goshawk, and barred owl. Human commensal predators include American crows, cats, dogs, eastern gray squirrels. Steller's jays in forested landscapes are plotted separately for clearcuts with berry crops (food) and without berry crops (no food). Distributions are based on Rosenberg and Raphael (1983) and Marzluff et al. (1998b).

likely to hinge on the importance of American crows as nest predators because they are strongly associated with human activity and occur only a short distance into forests abutting human-dominated areas.

Forest fragments in urbanized landscapes should, however, suffer greatly reduced productivity. Overall rates of predation are expected to be high regardless of proximity to edge so that significant patch size and landscape fragmentation effects should result when forest fragments become isolated in urban landscapes. Small to medium-sized predators should be especially common and exert significant effects on productivity because some "meso" predators (e.g., northern goshawks, barred owls, badgers) are unlikely to tolerate human disturbance. However, others, such as raccoons, skunks, and crows may thrive in urban/suburban landscapes.

Edge effects in forested landscapes managed for timber harvest are infrequent (Fig. 1). This is because (1) few new predators accompany the creation of clearcuts and young forests, (2) loss of large predators may not be absolute so "meso-predator release" (Soule et al., 1988) may be tempered compared to that in urban settings, (3) forest nest-predator communities are diverse, being comprised of interior and edge forest corvids, abundant small mammals, moderate-sized mammals, and birds of prey, and (4) edge habitats within managed forests represent early seral stages.

Therefore, we predict that forest fragments in western U.S. landscapes dominated by regenerating, managed forests will rarely show edge or fragmentation effects related to nest predation (Fig. 3B). Nest predator communities are diverse and typically comprised of forest specialists that rarely venture into clearcuts and regenerating forests (flying squirrels [*Glaucomys sabrinus*], red squirrel [*T. hudsonicus*], accipiter hawks [*Accipiter spp.*], gray jay, and common raven; Rosenberg and Raphael, 1986; Marzluff et al., 1998b). Where clearcuts are dominated by berry-producing plants, some nest predators may increase travel across edges to forage on berries and incidentally elevate nest predation on edges as well (e.g., Vickery et al., 1992). The most important western forest predator within this category is the Steller's jay, a species that is abundant in edge habitats (Marzluff et al., 1998b). We hypothesize that this corvid will be an important determinant of edge and fragmentation effects in western forest landscapes. Steller's jays should concentrate their activity and cause edge effects in settings where understory plants provide abundant food in clearcuts and in openings along forest edges (Fig. 3B).

Finally, we suggest that avian population viability in fragmented, western, forested landscapes should be determined more by absolute habitat loss and habitat isolation, whose effects will be determined by species-specific dispersal abilities, than by inflated rates of nest predation. Nest predation rates are high regardless of fragmentation, and edge effects are unlikely to elevate these rates. Isolation of breeding songbird populations may reduce population size thereby increasing the probability of local extinction. The ability of birds to recolonize habitat patches that have experienced local extinction will be an important key to avian diversity. The relative amount of habitat remaining and the distance between patches will likely affect the process of extinction and recolonization more than will nest predation.

5. MANAGEMENT IMPLICATIONS

5.1 Be Concerned About Urbanization and Agricultural Intensification

The primary implication of this review is that resource managers should be more concerned with forest fragmentation that results from urbanization and agricultural intensification than with fragmentation resulting from commercial timber harvest. This is especially true with respect to edge effects because they are most pronounced in urban and agricultural landscapes (Fig. 1). Moreover, the spatial arrangement of urban and managed forest land is an important determinant of biodiversity. Interspersing land uses increases the amount of detrimental interface between agricultural or urban parcels and forests.

Forest managers should develop partnerships with urban planners to minimize the overall amount of urban interface with forests. Edge could be reduced by designing urban developments that are compact and of high density. Unfortunately, exactly the opposite type of development is most common where urbanization contacts forests (Marzluff et al., 1998a). Remnant habitat patches within urban developments that are designed to provide "green space" may actually be sink habitats or ecological traps (Gates and Gysel, 1978). Forest managers and agricultural interests should forge similar partnerships to reduce edge between forest fragments and cropland.

Policy makers, with guidance from resource managers, can make important contributions to conservation by reducing the conversion of forest land to urban and agricultural holdings. Maintaining economically viable commercial forestry is important in this respect because low land prices make conversion to urban developments more likely. Partitioning the landscape into distinct land uses at a gross scale would provide for large forested areas such that other impacts of fragmentation (e.g., isolation) may be lessened.

5.2 Manage What You Can in Urban and Agricultural Landscapes

We know very little about the conservation value of urban or agricultural "green space." Given this poor knowledge, we should (1) determine what species can exist in the face of such land use, (2) determine critical needs for species of special concern in such landscapes, and (3) identify and monitor sink habitats. Some species (e.g., corvids, "meso" predators) may tolerate or even thrive on urban and agricultural intensification (Marzluff et al., 1998a). Managing for such species can make a contribution to regional biodiversity (i.e., beta diversity) and managers should understand what features of the habitat are needed by species that flourish in such landscapes. However, we also need to determine how to provide for rare species (e.g., some neotropical migrants and many cavity nesters). Minimizing edge effects by clustering forested areas into larger, contiguous areas will likely be necessary to conserve rare species. Moreover, maintaining fewer, but larger, forested tracts within urban and agricultural settings may be the only way to maintain viable populations of many bird species. Small stands are likely to be ecological traps (Gates and Gysel, 1978) or sinks (Pulliam, 1988) because abundant predators limit reproduction below levels necessary for long-term viability.

5.3 Rely on Commercial Forests to Contribute to Biodiversity

Our review suggests that forest fragments within managed forests may contain healthier bird populations than populations in fragments surrounded by urban development or agriculture because a major factor limiting productivity - nest predation - is not universally increased by fragmentation caused by timber harvest. Forests managed for timber production may make substantial contributions to biodiversity, even in tropical areas where degradation is most feared (Cannon et al., 1998). However, we recognize that alpha diversity (i.e., within habitat diversity) may not automatically fare well in managed forests because other aspects of fragmentation limit biodiversity: isolation, the rate of habitat shifting, and overall habitat loss.

The potential impact of isolation and habitat shifts in managed forests requires a thorough understanding of dispersal. Managers must monitor spatial requirements of animals and determine their willingness to cross forest breaks of various size and composition. A majority of songbird species preferred to move within continuous forest rather than cross gaps (Desrochers and Hannon 1997), which indicated that habitat connectivity has important implications for large scale management prescriptions (see also Beier and Noss, 1998). Habitat types (e.g., young versus older seral stages) shift location in managed forest landscapes much more quickly than in forests affected by natural processes like fire. Moreover, the habitat juxtaposition resulting from management differs from that created through natural processes. How does wildlife respond to these relatively rapid, unnatural changes? Managers may be able to minimize the potential negative impacts of these types of habitat change by placing clearcuts near stands of mature forest. How near will depend on the dispersal ability of target species.

Managing large forest landscapes to minimize change in adjacent forest stands would be enhanced if cutting proceeded in a coordinated fashion from one landscape edge through the landscape to the far edge. As the cut progressed through the landscape over time, contiguous mature forest would remain in front of the cut and regrowing, contiguous forest would develop behind the cut. Such a pattern would minimize overall fragmentation (see also King et al., 1998) and minimize dispersal distances for individuals displaced by cutting. Although theoretically appealing, implementing such a forest harvest, whether conducted in a linear or circular fashion, faces considerable challenges. For example, managers would need to determine suitable rates of cutting and, more importantly, current landscape conditions would require extensive coordination among state, federal, and private timber interests.

5.4 Remain Vigilant in Forested Landscapes

Managers in forested landscapes should not become complacent of edge effects. Relatively benign human activities at forest edges may quickly transform a forested landscape into a functional urban landscape. Camps and other recreation centers may increase edge effects in managed forests. Managers should monitor human commensals such as Amercian crows and generalist predators like Steller's jays that may be indicative of increasingly likely edge effects.

6. FUTURE RESEARCH NEEDS

The most obvious research need is a detailed understanding of nest predators. We are aware of only one study that quantified the foraging behavior of predators in fragmented environments (Bowman and Harris, 1980). Without understanding predator foraging behavior, habitat selection, and demography in relation to forest fragmentation, we will not be able to anticipate when fragmentation will elevate predation or how to manage predators in fragmented landscapes. We need to determine what affects predator population density, the scale at which predator's perceive edges and fragments, and how predators search for nests. For example, in some fragmented landscapes the predator's population may be managed by reducing supplemental food (e.g., garbage in urban areas, waste grain in agriculture), yet such prescriptions remain untested until we demonstrate that increased predator populations result from food supplementation.

We need to pay greater attention to the effects of fragmentation on real nests and nests in locations other than the forest understory and ground. Real nests are difficult to find and monitor, but the extensive work on artificial nests warrants validation by well designed, carefully executed studies on real nests. Although trends in nest predation by fragment size were consistent between artificial and real nests, predators destroyed artificial nests more than real nests (Wilson et al. 1998). The effects of fragmentation on nest predation of canopy nesters is virtually unknown. Real canopy nests have not been monitored and artificial nest experiments have not been conducted in the canopy zone. Our ongoing research (Marzluff et al., 1998b) is simulating canopy nests of Marbled Murrelets (*Brachyramphus marmoratus*) and determining that a diversity of mammalian and avian nest predators also attack canopy nests. The relationship of canopy nest success to forest fragment edge and landscape fragmentation is complex. Like ground and shrub nests, predation of canopy nest is probably affected by the matrix surrounding the fragment, the age of the forest fragment, and the proximity of the fragment to other forest fragments and to areas of human habitation and recreation fragments.

Acknowledgements. J.M. Scott, E. Arnett, R. Sallabanks, M. Raphael, and L. Kremsater *provided helpful comments on this manuscript. JMM's research on fragmentation has been funded by the U.S. Fish and Wildlife Service, U.S. Forest Service, National (U.S.) Park Service, Rayonier, Washington Department of Natural Resources, NCASI, Boise Cascade Corporation, Willamette Industries, and the Oregon Department of Forestry.*

LITERATURE CITED

Angelstam, P., 1986, Oikos 47, 365-373.
Andren, H., 1992, Ecology 73, 794-804.
Andren, H., 1995, Mosaic Landscapes and Ecological Processes, Chapman and Hall, London, pp. 225-255.
Andren, H. and Angelstam, P., 1988, Ecology 69, 544-547.
Andren, H., Angelstam, P., Lindstrom, E. and Widen, P., 1985, Oikos 45, 273-277.
Angelstam, P., 1988, Oikos 47, 365-373.
Arango-Velez, N. and Kattan G. H., 1997, Biological Conservation 81, 137-143.
Bayne, E. M., Hobson, K. A. and Fargey, P., 1997, Ecography 20, 233-239.
Beier, P. and Noss, R. F. 1998. Conservation Biology 12, 1241-1252.
Bjorklund, M., 1990, Ibis 132, 613-617.
Boag, D. A., Reebs, S. G. and Schroeder, M. A., 1984, Canadian Journal of Zoology 62, 1034-1037.
Bowman, G. B. and Harris, L. D., 1980, Journal of Wildlife Management 44, 806-813.
Burkey, T. V., 1993, Biological Conservation 66, 139-143.
Cannon, C. H., Peart, D. R. and Leighton, M., 1998, Science 281, 1366-1368.
Danielson, W. R., DeGraaf, R. M. and Fuller, T. K., 1997, Landscape and Urban Planning 38, 25-36.
Darveau, M., Belanger, L., Huot, J., Melancon, E. and DeBellefeuille, S., 1997, Ecological Applications 7, 572-580.
DeGraaf, R. M., 1995, Forest Ecology and Management 79, 227-234.
DeGraaf, R. M. and Angelstam, P., 1993, Forest Ecology and Management 61, 127-136.
DeGraaf, R. M. and Maier, T. J., 1996, Wilson Bulletin 108, 535-539.
Desrochers, A. and Hannon, S. J., 1997, Conservation Biology 11, 1204-1210.
Fenske-Crawford, T. J. and Niemi, G. J., 1997, Condor 99, 14-24.
Gates, J. E. and Gysel, L. W., 1978, Ecology 59, 871-883.
Gibbs, J. P., 1991, Oikos 60, 155-161.
Hannon, S. J. and Cotterill, S. E., 1998, Auk 115, 16-25.
Hanski, I. K., Fenske, T. J. and Niemi, G. J., 1996, Auk 113, 578-585.
Harris, L. D., 1984, The Fragmented Forest, University of Chicago Press, Chicago.
Haskell, D. G., 1995, Conservation Biology 9, 1316-1318.
Hoover, J. P., Brittingham, M. C. and Goodrich, L. J., 1995, Auk 112, 146-155.
Huhta, E., Mappes, T. and Jokimaki, J., 1996, Ecography 19, 85-91.
Keyser, A. J., Hill, G. E. and Soehren, E. C., 1998, Conservation Biology 12, 986-994.
King, D. I., DeGraaf, R. M. and Griffin, C. R., 1998a, Conservation Biology 12, 1412-1415.
King, D. I., Griffin, C. R. and DeGraaf, R. M., 1998b, Forest Ecology and Management 104, 151-156.
Kuitunen, M. and Helle, P., 1988, Ornis Fennica 65, 150-155.
Lariviere, S. and Messier, F., 1997, American Midland Naturalist 137, 393-396.
Laurance, W. F. and Bierregaard, B. O., Jr., 1997, Tropical Forest Remnants, University of Chicago Press, Chicago.
Laurance, W. F., Garesche, J. and Payne, C. W., 1993, Wildlife Research 20, 711-723.
Laurance, W. F. and Yensen, E., 1991, Biological Conservation, 55, 77-92.
Martin, T. E., 1995, Ecological Monographs 65, 101-127.
Marzluff, J. M., Gehlbach, F. R. and Manuwal, D. A., 1998a, Avian Conservation: Research and Management, Island Press, Washington, DC., pp. 283-306.
Marzluff, J. M., Luginbuhl, J. M., Bradley, J. E., Raphael, M. G., Evans, D. E., Varland, D. E., Young, L. S., Horton, S. P. and Courtney, S. P., 1998b. Annual Report, Sustainable Ecosystems Institute, Portland, OR.

Marzluff, J. M., Raphael, M. G. and Sallabanks, R., In Press, Wildlife Society Bulletin.
Møller, A. P., 1988, Oikos 53, 215-221.
Møller, A. P., 1989, Oikos 56, 240-246.
Murcia, C., 1995, Trends in Ecology and Evolution 10, 58-62.
Nice, M. M., 1957, Auk 74, 305-321.
Nour, N., Matthysen, E. and Dhondt, A. A., 1993, Ecography 16, 111-116.
Paton, P. W. C., 1994, Conservation Biology 8, 17-26.
Pulliam, H. R., 1988, American Naturalist 132, 652-661.
Ratti, J. T. and Reese, K. P., 1988, Journal of Wildlife Management 52, 484-491.
Reitsma, L. R., Holmes, R. T. and Sherry, T. W., 1990, Oikos 57, 375-380.
Ricklefs, R.E., 1969, Smithsonian Contributions in Zoology 9, 1-48
Robinson, S. K., 1992, Ecology and Conservation of Neotropical Migrant Landbirds, Smithsonian Institution Press, Washington D.C., pp. 408-418.
Robinson, S. K. and Wilcove, D. S., 1994, Bird Conservation International 4, 233-249.
Robinson, S. K., Thompson, F. R. III, Donovan, T. M., Whitehead, D. R. and Faaborg, J., 1995, Science 267, 1987-1990.
Roper, J. J., 1992, Oikos 65, 528-530.
Rosenberg, K. V. and Raphael, M. G., 1986, Wildlife 2000: Modeling Habitat Relationships of Terrestrial Vertebrates, University of Wisconsin Press, Madison, pp. 263-272.
Rudnicky,T. C. and Hunter, M. L. Jr., 1993, Journal of Wildlife Management 57, 358-364.
Russo, C. and Young, T. P., 1997, Urban Ecosystems 1, 171-178.
Sandstrom, U., 1991, Ornis Fennica 68, 93-98.
Santos, T. and Telleria, J. L., 1992, Biological Conservation 60, 1-5.
Sargent, R. A. , Kilco, J. C., Chapman, B. R. and Miller, K. V., 1998, Journal of Wildlife Management 62, 1438-1442.
Seitz, L. C. and Zegers, D. A., 1993, Condor 95, 297-304.
Sieving, K. E. and Willson, M. F., 1998, Ecology 79, 2391-2402.
Small, M. F. and Hunter, M. L., 1988, Oecologia 76, 62-64.
Storch, I., 1991, Ornis Scandinavica 22, 213-217.
Telleria, J. L. and Santos, T., 1992, Biological Conservation 62, 29-33.
Temple, S. A. and Cary, J. R., 1988, Conservation Biology 2, 340-347.
Tewksbury, J. J., Hejl, S. J. and Martin, T. E., 1998, Ecology 79, 2890-2903.
Vickery, P. D., Hunter, M. L. Jr. and Wells, J. V., 1992, Oikos 63, 281-282.
Whitcomb, R. F., Lynch, J. F., Klimkiewicz, M. K., Robbins, C. S., Whitcomb, B. L. and Bystrak, D., 1981, Forest Island Dynamics in man-dominated landscapes, Springer-Verlag, New York, pp.125-205.
Wiens, J. A., 1995, Ibis 137 (Supplement), 97-104.
Wilcove, D. S., 1985, Ecology 66, 1211-1214.
Wilson, G. R., Brittingham, M. C. and Goodrich, L. J., 1998, Condor 100,357-364.
Wong, T. C. M., Sodhi, N. S. and Turner, I. M., 1998, Biological Conservation 85, 97-104.
Yahner, R. H. and Scott, D. P., 1988, Journal of Wildlife Management 52, 158-161.
Yahner, R. H., Morrell, T. E. and Rachael, J. S., 1989, Journal of Wildlife Management 53, 1135-1138.

The Role of Genetics in Understanding Forest Fragmentation

L. Scott Mills[1] and David A. Tallmon[2]

Recent advances in genetic analysis have demonstrated clearly that population fragmentation can lead to a loss of genetic variation. Whether such loss will occur on any particular forest fragment is less certain, and will depend in part on whether the population on the fragment is completely isolated, the size of the fragment, the time since the isolation occurred, and the underlying historical genetic structure that exists due to natural barriers to movement. A loss in genetic variation can cause inbreeding depression and decreased ability to adapt to long-term changes. Inbreeding depression, a decrease in birth and/or survival rates due to a decrease in heterozygosity, has been found in a wide variety of wild plant and animal populations, and has been shown to decrease population growth rate and population persistence. Genetic tools can assay the extent to which fragmentation has occurred in a landscape by facilitating comparison between past and present levels of connectivity and population sizes. Genetic aspects of forest fragmentation cannot be ignored, because the population consequences are real. However, genetic analysis of fragmented populations is of limited utility unless placed in an ecological context based on field research.

The definition of fragmentation proposed for this conference – the process of reducing the size and connectivity of stands that compose a forest – begins to make the concept of fragmentation operational, and objectively defined. But two of the most important terms in this definition – size and connectivity -- are vague. The critical question is, how much of a reduction in the size and connectivity of wildlife populations can occur before a forest fragmentation "alarm bell" should go off? Of course, the answer will depend on the species, the place, and a host of other factors, but remarkable technological and analytical advances of the last decade have revolutionized the role that genetics can play in considering this crucial question. We will consider how changes in forest fragment size and connectivity can alter genetic structure of wildlife populations, and how those changes can alter population dynamics. We also explore how genetic tools can help us to decide how small is too small for a forest fragment, or how unconnected is too unconnected. These issues are not strictly genetic issues. A theme we hope to weave throughout is that genetic aspects are an important consideration when evaluating the consequences of forest harvest, both because genetic changes can negatively impact wildlife populations, and because they can tell us something about when a forest fragment has become small and isolated. We first define some terms that are relevant to evaluating genetic changes following fragmentation, then describe some known consequences of forest fragmentation. Finally, we discuss genetic tools that can help evaluate changes in popu-

---

[1] Wildlife Biology Program, School of Forestry, University of Montana, Missoula, MT
[2] Division of Biological Sciences, University of Montana, Missoula, MT

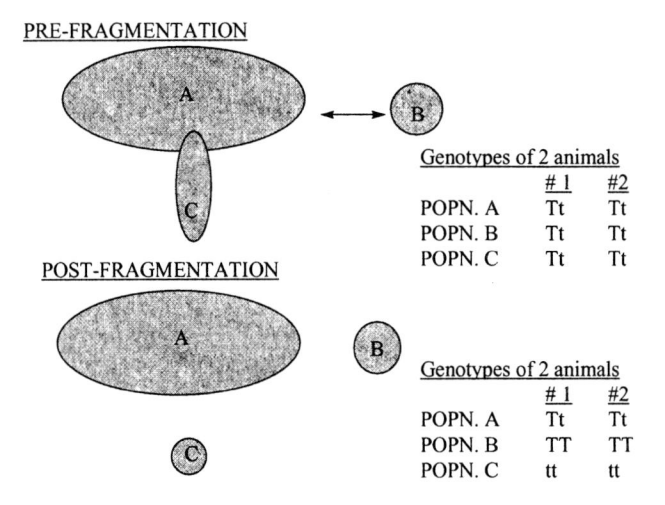

Figure 1. Cartoon demonstration of potential genetic consequences following forest fragmentation events that decrease population size and connectivity among populations. Encircled areas represent suitable habitat and contain wildlife populations. Arrows represent gene flow across a semi-hospitable habitat. The genotypes of only two individuals and one gene are shown per population, although in real studies many genes and many individuals would be sampled.

lation size and connectivity, thereby making "fragmentation" a more operational term.

1. RELATIONSHIP BETWEEN GENETIC VARIATION, POPULATION SIZE, AND CONNECTIVITY

1.1 Background

At the outset, it is important to emphasize that although we cannot see genes (or the DNA building blocks that comprise genes) with the human eye, their dynamics and impacts on populations are not merely theoretical. Both morphological characteristics and behavioral traits have some genetic component, making genetic composition as real and important to organisms as the air they breathe or the food they eat.

Genetic variation in a wildlife population can be described by its heterozygosity and allelic diversity (see Chambers, 1983 for concise description; Hartl and Clark, 1997 for very readable details). Heterozygosity is the proportion of genotypes in the population that have different alleles, or forms of a gene (e.g. genotype Tt vs. TT or tt). Allelic diversity can be described by the number of different alleles in a population.

Fragmentation can affect genetic variation in a variety of ways. We will consider two cases of special relevance to forest fragmentation, realizing that these changes would be complicated by the strength and mode of natural selection, as well as factors such as mutation rate and interactions among genes. The first case is for single populations that go from being large to being small, as might be expected from habitat loss via forest fragmentation. In a large population, the frequency of different alleles, and therefore heterozygosity and allelic diversity, will change slowly in response to natural selection under local conditions. In contrast, in a small isolated population, natural selection becomes overwhelmed by sampling error. That is, because only a small number of individuals reproduce and they produce only a limited number of offspring, only certain subsets of alleles are

Allele Frequencies in Fragmented Populations

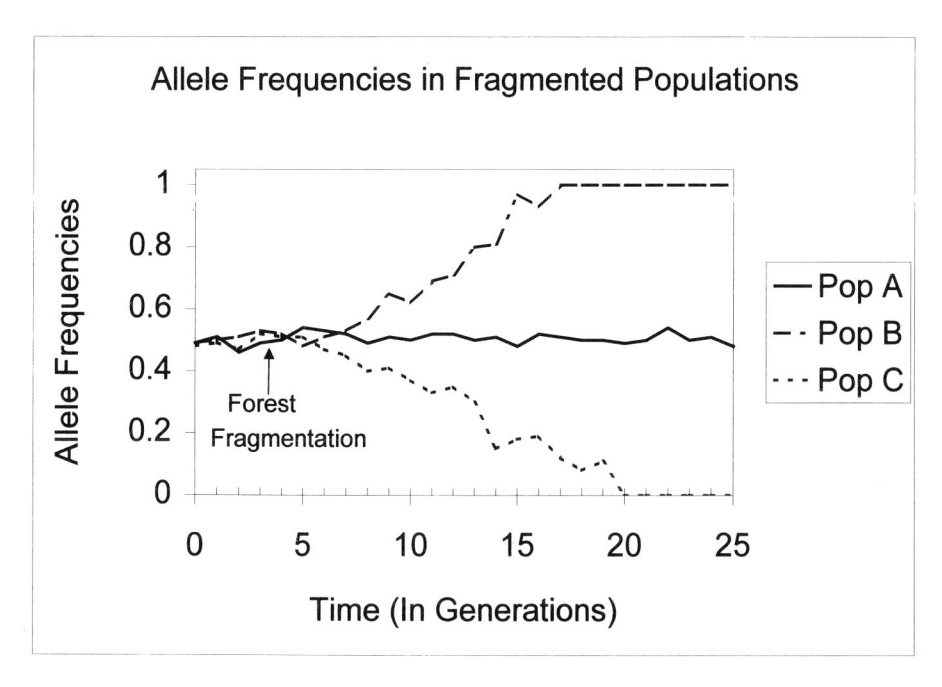

Figure 2. Changes in the frequencies of two alleles (T and t) of a single gene found in three populations following a hypothetical forest fragmentation event, such as that in Fig. 1. Here, the fragmentation event occurs in the third generation. Thereafter, allele frequencies in the reduced size (Population C) and reduced connectivity (Population B) populations diverge from each other and from population A; the time for divergence could be hundreds of generations (see text). Allele frequencies in population A remain relatively stable due to its large size.

passed on to the offspring of the next generation. This process of genetic drift, or sampling error, can lead to rapid and random changes in allele frequencies over a few generations. Consequently, heterozygosity and allelic diversity are lost, as some alleles reach a frequency of one while others vanish (Allendorf, 1983).

For example, consider a simple hypothetical situation of a fragmented landscape (Fig. 1). Fragmentation has caused population C to become small after being severed from the large population A, with changes in gene frequencies and loss of heterozygosity and allelic diversity over subsequent generations (Fig. 2). In contrast, there is little genetic drift (sampling error) each generation in population A because a relatively much larger number of offspring is created each generation from the pool of parents (Fig. 1, Fig. 2). Therefore, we see genetic drift playing a large role in eliminating genetic variation in the small population compared to the large population.

The second, related way fragmentation affects genetic variation can be seen by looking across populations transformed from being connected to isolated. Although population B in Fig. 1 is small, it is connected by movement of animals between B and A prior to fragmentation. This frequent movement of individuals between the two populations causes them to have similar allele frequencies and levels of genetic variation, because gene flow from population A counteracts genetic drift in population B. After fragmentation, population B has become isolated, and as genetic drift takes effect on these now independent populations, allele frequencies in population B will diverge over time from those of population A (Fig. 2). In addition, within the isolated populations, heterozygosity and allelic diversity decrease (Fig. 1, Fig. 2).

With these cartoon characterizations, a couple of important points emerge. First, allelic diversity can be high across suites of isolated populations, as different alleles are lost within different populations. Secondly, even though allelic diversity can be retained in sets of small, isolated populations, heterozygosity and allelic diversity within any one population will be lost eventually due to genetic drift, even if there are fairly strong selective forces acting on the population. As will be discussed shortly, genetic changes are important because they can lead to inbreeding depression in the short-term, and decreased long-term ability of a population to respond to selective pressures (Briscoe et al., 1992).

1.2 Results From Field Studies

Genetic changes in populations are real and measurable, and not merely theoretical. At the extreme of isolation, we know that genetic variation tends to be lower in populations on islands than for those on mainlands (Frankham, 1997). As just a few of many examples, genetic variation is lower on islands than the mainland for wolves (Wayne et al., 1991), Channel island foxes (Gilbert et al., 1990), and silvereye birds (Degnan, 1993). Genetic variation also tends to decrease in general as population size decreases for a wide range of wildlife species (see Frankham, 1996).

Thus, there is no doubt that genetic variation can be lost when isolation is complete and/or population sizes are small. But do these expected genetic changes occur with habitat fragmentation in general, and forest fragmentation in particular? As expected from the previous discussion, genetic consequences of habitat fragmentation will vary widely, depending on the size of the population, how much populations have become truly isolated, how long ago the isolation occurred, and other factors. But reductions in genetic variation have certainly been documented. For example, heterozygosity and allelic diversity declined in greater prairie chickens (Bouzat et al., 1998a, Bouzat et al. 1998b), hairy-nosed wombats (Taylor et al., 1994), and koalas (Houlden et al., 1996), all of whose ranges and population sizes have been reduced by habitat fragmentation, disease, and/or overharvest over the last century. Similarly, population differentiation or loss of genetic variation has accompanied habitat fragmentation and harvesting of wild turkeys (Leberg, 1991) and brown bears (Paetkau et al., 1998), as well as frogs whose dispersal has become limited by urban sprawl (Hitchings and Beebee, 1997).

Forest fragmentation plays a role in many of these instances, but studies focusing on forest fragmentation per se are limited. One example is that of Sarre (1995) on geckos in forest remnants in Australia. By examining genetic variation in nine remnant populations and three large nearby nature reserves across a landscape, Sarre (1995) found lower variation in remnant than in nature reserve populations and attributed the changes to recent forest fragmentation. Similarly, Wauters et al. (1994) found that populations of the Eurasian red squirrel on five woodland fragments had lower genetic variation than did two large control populations. On the other hand, Leung et al. (1993) found that although genetic variation in an Australian rodent was lower on a true island than in a control, there was no loss of variation for 3 forest fragments created about 60-70 years ago. Similarly, in our own work we have found strong ecological effects of forest fragmentation on California red-backed voles in Oregon, resulting in both negative edge effects and apparent isolation (Mills, 1995; Mills 1996); however, our preliminary genetic analyses of highly sensitive microsatellite DNA markers show no differences in levels of genetic variation between 5 large control and 13 small remnant populations (Tallmon et al., In Prep).

The studies that demonstrated a reduction in genetic diversity on remnants support our expectations about the effects of fragmentation. But the studies that fail to demonstrate a reduction in genetic diversity do not necessarily reject the possibility that fragmentation affects genetic variation. There are at least three reasons, probably acting together, that explain why a loss of genetic variation may not be detected in many studies of forest fragmentation. The first explanation is that isolation may not exist, despite fragmentation. Even a small amount of connectivity can minimize the loss of genetic variation. One to ten migrating individuals per generation are sufficient to minimize the loss of genetic variation in any one population (Spielman and Frankham, 1992; Mills and Allendorf, 1996; Newman and Tallmon, In Prep), although more may be necessary if populations fluctuate greatly over time (Vucetich and Waite, In Review).

A second reason why decreased genetic variation may not be detected in a study of an apparently isolated species is that not enough time has passed to initiate a change in genetic composition. After a drastic change in population size or connectivity (say due to forest fragmentation), anywhere from several to hundreds of generations may pass before measures of variation reflect those changes (Steinberg and Jordan, 1997; Lindenmayer and Lacy, 1995; Slatkin, 1994). This "time to equilibrium" will be faster for truly isolated populations if they are relatively small (Allendorf and Phelps, 1981; Varvio et al., 1986). It will also depend on the genetic tool used to assess genetic variation; protein electrophoresis is the least likely to reveal recent changes since proteins have low levels of variation (see Leung et al., 1993), whereas DNA markers are more sensitive (Avise, 1994). The time to equilibrium explains why islands that have been isolated for hundreds or thousands of years so consistently show low levels of variation compared to relatively recently-formed habitat remnants.

The third major reason why forest fragmentation may not lead to detectable changes in genetic variation is that population differentiation can occur "naturally" across a landscape. That is, historical barriers to dispersal such as rivers and mountain ranges can contribute statistical noise that can obscure the detection of subtle, on-going genetic changes due to recent fragmentation (Avise, 1994). For example, Cunningham and Moritz (1998) found that the effects of recent clearing had less effect on genetic diversity in a rainforest lizard than long-term climatic and geological processes.

It is worth noting that the last two of these factors – time lags and statistical noise – will similarly affect ecological studies of edge effects and forest fragmentation described elsewhere in this book. Before concluding that fragmentation has "no effects," one should consider whether the time scale of the study is appropriate, and whether the study has sufficient statistical power to detect effects that do in fact exist in a "noisy" landscape.

2. POPULATION-LEVEL CONSEQUENCES OF CHANGES IN GENETIC VARIATION

2.1 Background

Although the evidence is incontrovertible that fragmentation can have consequences on genetic variation at varying time scales, genetic concerns only become relevant if they compromise population persistence. Genetic changes due to forest fragmentation can have such effects, both in the short term, primarily through inbreeding depression, and in the long term through loss of adaptive genetic variation and through the accumulation of mutations that decrease adaptive potential. Inbreeding depression is the reduction in fitness of individuals produced from mating among relatives, which becomes unavoidable in a small population. Adaptive potential is the ability to evolve to future conditions, which is constrained by underlying levels of genetic variation in a population. These genetic consequences of fragmentation can contribute significantly to population decline.

2.2 Short-Term Effects: Inbreeding Depression

Like the loss of variation in small, isolated populations, the phenomenon of inbreeding depression is real and has been observed. Awareness of the demographic consequences of inbreeding depression probably traces back to the dawn of domestication (Wright, 1977). More recently, the cost of inbreeding has been quantified in an array of wild mammal species in captivity, with varying levels of reduced juvenile survival (Laikre and Ryman, 1991; Ralls et al., 1988; Lacy, 1993), reduced reproduction (Brewer et al., 1990; Ballou, 1997; Lacy and Horner, 1997; Lacy and Ballou, 1998; Margulis, 1998), and reduced adult survival (Jiménez et al., 1994).

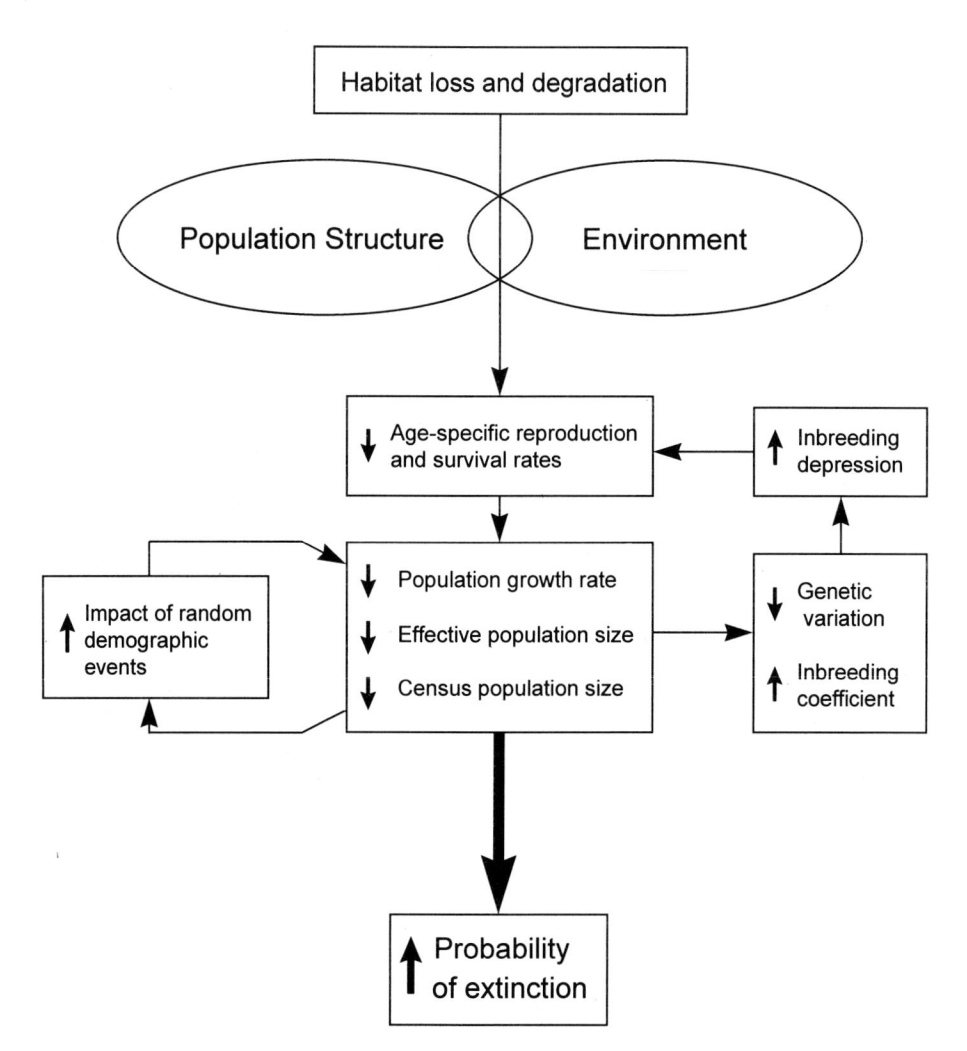

Figure 3. Simplified representation of a positive feedback extinction vortex. "Habitat loss and degradation" may include invasion of exotics, overharvest, etc. "Population structure" includes the age-structure, sex ratio, behavioral interactions, distribution, physiological status, and intrinsic birth and death rates. "Environment" includes habitat as well as extrinsic factors that vary, such as weather, competition, predators, and food abundance. Each turn of the feedback cycle increases extinction probability. The extinction vortex model predicts that some small populations are more likely to become smaller, and go extinct, each passing generation from the interaction of genetic and non-genetic factors. Adapted from Soulé and Mills (1998).

Importantly, inbreeding depression arising from small population size has also been demonstrated in free-living populations of plants, snails, fish, frogs, birds, and mammals (for reviews see Frankham, 1995; Lacy, 1997). Inbreeding depression is probably underestimated in captivity because negative fitness effects might not show up in the relatively tranquil captive environment. Field studies also tend to underestimate overall inbreeding depression because any one study typically measures only one or a few of the many aspects of fitness likely to be simultaneously affected by inbreeding depression (Keller et al., 1994; Jiménez et al., 1994; Lacy, 1997). In two well-studied species where several measures of fitness have been measured, African cheetahs and Florida panthers, inbreeding depression affects physiological traits ranging from sperm production to immune response to heart function (O'Brien, 1994; Roelke et al., 1993). These effects translate into potential impacts on a wide range of life history characteristics. Ultimately, inbreeding depression arising from reduced genetic variation can decrease not only individual birth and death rates, but also population growth rate (Leberg, 1990) and probability of extinction (Newman and Pilson, 1997; Saccheri et al., 1998). This means inbreeding depression can have important consequences for population persistence.

Fortunately, the discussion of the consequences of inbreeding depression on the persistence of fragmented populations has matured beyond a point where genetic issues are considered apart from other factors. Following a decade where genetic factors were considered preeminent, by the late 1980's the pendulum had swung to the opposite extreme, with a prevailing view that populations would go extinct due to other factors before inbreeding translated into a substantial loss in fitness (for review see Mills and Smouse, 1994; Hedrick et al., 1996). Subsequently, a number of studies have shown that such a radical dichotomy – splitting apart "genetic" vs. "nongenetic" factors in demographic analyses – was false (Mills and Smouse, 1994; Lynch et al., 1995; Newman and Pilson, 1997; Saccheri et al. 1998). It is now widely accepted that inbreeding depression is just one of several demographic impacts that interact to increase extinction probability via an "extinction vortex" (Gilpin and Soulé, 1986) in a fragmented population (Figure 3; Soulé and Mills, 1998).

Unfortunately, not many studies are comprehensive enough to demonstrate how genetic and non-genetic factors interact to lead to population decline or to an extinction vortex following fragmentation (Figure 3). However, in a recent example, Westemeier et al. (1998) tracked demographic and genetic changes in greater prairie chickens in Illinois for 35 years. They noted a steep population decline due to habitat loss as the population became isolated and declined to a low of <50 birds. During the same time period, populations in neighboring states remained large and widespread. The Illinois population declined despite intense and somewhat successful efforts to control predators and increase the quality and quantity of habitat. The conclusion that genetics played a role in the continuing downward spiral of this population came from two observations: first, the decline in prairie chicken numbers was accompanied by a decrease in genetic variation, both for the Illinois birds compared to the larger, nearby populations (Bouzat et al., 1998b), and for the present population in Illinois compared to historical samples collected before the demographic contraction (Bouzat et al., 1998a). Second, translocations of prairie chickens from one of the neighboring states to Illinois increased egg fertility and hatching success (Westemeier et al., 1998), the life stage that surpasses all others in its importance to population growth rate in greater prairie chickens (Wisdom and Mills, 1997). This increase in egg hatching success, without any obvious changes in environmental conditions, implies that inbreeding depression on hatch rate was "broken" by the new breeders. A similar story of inbreeding depression causing subtle changes in demographic rates and exacerbating population decline has been documented for adders isolated by agricultural expansion in Sweden (Madsen et al., 1996).

2.3 Long-Term Effects

Although short-term effects of inbreeding generally receive the most attention, it is important to remember that conservation efforts and goals must be placed in an evolutionary time scale (Meffe, 1996). Long-term adaptation to climate, disease, or other changes will ultimately be limited by available genetic variation (Soulé, 1980). This is especially true if movement of individuals to track optimal habitat conditions is limited by habitat fragmentation (Lande, 1996). Populations that have been reduced in size will have decreased genetic variation as demonstrated above, and so will have decreased ability to adapt to changing conditions.

In addition to a loss of adaptive variation, isolated and small populations are threatened with a "mutational meltdown" (Lande, 1995; Lynch et al., 1995). Although mutations are the ultimate source of the genetic variation that allows populations to evolve and to adapt to changing conditions, most mutations are deleterious. As stated earlier, large populations can respond to natural selection, allowing many deleterious mutations to be removed effectively by natural selection. In contrast, small populations are dominated by random genetic drift and are therefore less able to shed deleterious mutations. As a result, over many generations small populations become less able to persist as mutations that reduce fitness become fixed by random chance (Lynch et al., 1995). This will further decrease population size, leading to increased accumulation of deleterious mutations, a decrease in fitness, again and again – a "mutational meltdown". Conversely, because large populations can respond to natural selection, those few mutations that are beneficial are more likely to be maintained in these populations than in small ones.

3. INSIGHTS INTO DEGREE OF FRAGMENTATION USING GENETIC TECHNIQUES

3.1 Background

If reductions in population size and connectivity, the twin components of forest fragmentation, can lead to genetic changes that negatively impact wildlife populations, how might genetic tools help us make operational the concepts of "fragmentation"? One approach would be to compare genetic variation in a population to a "standard" value: if the target population has lower genetic variation than the standard, it is too small; if the population has genetic structure indicating lower connectivity than the standard, it is too isolated. Unfortunately, no such standard exists for an "appropriate" level of genetic variation; in fact, as a general rule over the short term the absolute amount of heterozygosity is less important than the rate of loss of heterozygosity (Frankel and Soulé, 1981; Hedrick and Miller, 1992). As for connectivity, there is a standard, or rule of thumb, mentioned earlier: that one [to ten] breeding immigrants per generation (generation time will be species specific) will minimize the loss of heterozygosity in a population while allowing for local adaptation (Mills and Allendorf, 1996). Although this rule may be a useful starting point in the absence of other information, it is not entirely satisfactory because it is based purely on genetic considerations, and ignores many nuances of behavior and population and genetic structure.

Instead of searching for a universal rule, perhaps genetic tools can help us derive from the animals themselves, at a particular place and particular time, some operational definitions of when fragmentation has occurred. For example, just as we might evaluate population density or rates of movement using radiotelemetry or capture-recapture studies in an unfragmented landscape, and use those as goals for desired population density or rate of connectivity in a fragmented area, we may be able to do something similar using genetic techniques. A problem with this approach is that large, unfragmented "control" areas do not exist for many species and many areas; also, distinct processes may be at work in different locations (Steinberg and Jordan, 1997).

3.2 Historical Levels of Connectivity and Population Size

Fortunately, genetic techniques can provide a window into the past that other field techniques cannot. Samples collected prior to fragmentation can be compared to those collected in the same location after fragmentation to evaluate whether genetic variation has been lost. Thanks to recent advances in analysis of DNA, the potential exists to analyze very old samples from wall mounts or museum specimens (Ellegren, 1991; Mundy et al., 1997), as well as contemporary samples collected non-invasively (without having to capture the animal) from hair, feces, urine, and so on (Morin and Woodruff, 1996; Kohn and Wayne, 1997; Schwartz et al., 1998). If variation is lower in contemporary samples, increased isolation of existing populations is implied

(Bouzat et al., 1998a; Bouzat et al., 1998b; Taylor et al., 1994; Houlden et al., 1996).

It is even possible to learn about past levels of connectivity by looking at current genetic population structure across a fragmented landscape. Because isolated populations evolve independently (Fig. 1, Fig. 2), more isolated populations become more genetically differentiated compared to populations with high connectivity. This level of differentiation can be quantified with genetic analyses, and the average level of movement leading to that differentiation can be estimated (Avise, 1994; Slatkin, 1994; Slatkin, 1995; Goudet, 1995; Templeton and Georgiadis, 1995, Paetkau et al., 1998).

The most commonly used way to quantify connectivity indirectly via population differentiation is to use Wright's Fst statistic (but see Slatkin, 1985; Slatkin and Maddison, 1990; and Neigel, 1996 for other approaches). Fst is a measure of population subdivision calculated from allele frequencies in samples collected from populations across a landscape. It is based upon the distribution of genetic variation within versus between populations. If variation is distributed between populations (see Figure 1 Post-Fragmentation) then Fst is large. Conversely, if most genetic variation is found within populations (Fig. 1 Pre-Fragmentation) then Fst is small. The Fst derived from genetic samples can be used to calculate Nm, the number of individuals exchanged between populations per generation (Slatkin and Barton, 1989):

$Nm \approx (¼)(1/Fst - 1)$.

A low Nm value, or low connectivity, would be indicated by a large Fst value, whereas a high Nm would be implied by a small Fst value. A Nm of one would correspond to one immigrant per generation. Estimates of population subdivision (Fst) or connectivity (Nm) should always be accompanied by estimates of error, which can be quite large (Steinberg and Jordan, 1997).

There are many analogs to Fst that can be used to estimate migration depending upon the type of genetic marker used (Slatkin, 1995; Rannala and Hartigan, 1996; Weir and Cockerham, 1984). However, all Fst and analogous measures are based upon the same general principles and are subject to the same basic assumptions. For example, these methods assume an infinite number of islands all equally accessible to each other. Although this will obviously be violated in all real-world cases, the model is relatively robust to these violations (Mills and Allendorf, 1996). More importantly, estimates of migration based on population subdivision assume equilibrium levels of divergence among populations. This is a very important assumption, because it can take hundreds of generations for equilbrium levels of divergence to be reached (Steinberg and Jordan, 1997). Therefore, levels of divergence measured for recently fragmented populations will reflect past population structure and not recent changes, unless population sizes are extremely small.

Consequently, because patterns seen in genetic markers reflect previous equilibrium levels of connectivity, genetic markers have the potential to tell us about historic movement rates. As mentioned earlier, the number of generations required to reach equilibrium will vary widely according to mutation rates and selection pressures, among other factors, but will often be on the order of 100 generations (Crow and Aoki, 1984; Slatkin, 1994; Slatkin and Barton, 1989). This is widely viewed as a criticism of these techniques for estimating current levels of gene flow (Neigel, 1996; Ims and Yoccoz, 1997; Bossart and Prowell, 1998).

However, long times to equilibrium may actually be a benefit of these approaches for operationalizing connectivity because human-caused fragmentation is typically on the order of just a few generations for most vertebrate species. Thus, historic levels of connectivity can help direct current management to maintain long term processes by facilitating comparison against current gene flow determined from direct ecological approaches or from molecular tools that reach equilibrium Nm more quickly (e.g. Avise, 1994; Slatkin, 1994; Slatkin, 1995). For example, Forbes and Boyd (1996) relate historical to current rates of gene flow for recolonizing wolves in Montana.

3.3 Current Levels of Connectivity and Population Size

The use of genetic tools is not limited to insights into past levels of fragmentation. Genetic tools can also be used to estimate important demographic components such as current connectivity among fragmented populations, recent population bottlenecks, and current population sizes. Waser and Strobeck (1998) outline an "assignment test" that can assign individuals in a population to the population from which they originated. In short, a disperser or its offspring will look genetically distinct relative to a background of individuals with very different genetic make-ups (consider how a Post-Fragmentation disperser from population B to C in Figs. 1 or 2 would "stand-out" if its genes were examined in a sample of individuals from C). The assignment test takes advantage of the genetic differences among semi-isolated populations and provides a method to detect putative dispersal events without detailed monitoring via telemetry or mark-recapture (Davies et al., 1999). For example, using this technique Favre et al. (1997) inferred sex-biased dispersal of shrews, Paekau et al. (1995) described immigration in polar bears, and Haig et al. (1997) identified the breeding population of origin for shorebirds on fall migration. A limitation of this technique is that it can only distinguish dispersers in cases where the genetic profiles of populations are differentiated enough that individuals originating from a different population "stand out" genetically or many genetic markers are used (Rannala and Mountain, 1997). Importantly, statistical power to detect immigrants can be assessed for any data set using the approach outlined by Rannala and Mountain (1997).

DNA markers can also be used to investigate whether a population has passed through a small-population bottleneck and is therefore susceptible to inbreeding depression. When a population passes through a bottleneck, allelic diversity is reduced faster than heterozygosity, because rare alleles are lost quickly and these alleles have little effect on heterozygosity (Nei et al., 1975; Allendorf, 1986; Leberg, 1992; Hartl and Pucek, 1994). Building on this, Luikart and Cornuet (1998) developed a test for detecting bottlenecks by tracking allele changes. With analysis of at least 5-10 highly polymorphic microsatellite markers in 20-30 individuals, the test successfully detected bottlenecks that occurred in the last several to tens of generations in a variety of animals (Luikart and Cornuet, 1998).

A final way genetic techniques can be used to evaluate fragmentation is in the direct estimation of current population size. On the heels of molecular tools that allow individual identification of animals using creatively-collected bits of tissue (Haig, 1998; Schwartz et al., 1998), there is now the potential to use unobtrusive or non-invasive microsatellite techniques to "mark" and then "recapture" animals. These data can be used in traditional mark-recapture estimators of population size (e.g. Woods et al., 1996; Palsboll et al., 1997). These non-invasive approaches offer advantages over conventional mark-recapture approaches, including the potential to minimize or even eliminate some of the problems that plague demographic estimates, such as tag loss, the need to handle animals, or even trap response or individual heterogeneity in capture probability (Foran et al., 1998). The technique can be limited by underlying low levels of genetic variation, however. We have evaluated the use of non-invasive DNA sampling to estimate population size for rare carnivores (Mills et al., In Press), and found that this technique will lead to a negative bias in population size estimates unless greater than about five to seven microsatellite markers are analyzed. Thus, DNA sampling holds great promise for estimating current population size of hard-to-sample forest animals, but it is not a silver bullet.

3.4 Genetic Markers

Since the development of protein markers in the late 1960's, an array of genetic markers have become available for evaluating the aspects of forest fragmentation that we have discussed (Avise, 1994; Hillis et al., 1996). The workhorse of traditional population genetics, protein electrophoresis, is relatively insensitive to fragmentation effects because the genetic markers are usually not highly variable, although the approach is relatively inexpensive and can be used effectively to understand long-term evolutionary history. Unfortunately protein electophoresis often requires sacrificing individuals for genetic study, because tissues such as muscle, heart, liver, or brain are usually required.

In the past decade, the widespread adoption of PCR based techniques -- a method of making many copies of target DNA from small amounts of DNA sampled in the field -- has increased the availability of genetic approaches to wildlife studies and the number of questions that can be addressed (Parker et al., 1998). Only small amounts of DNA, as might be obtained from a few hair follicles, are sufficient for many genetic analyses. Several textbooks address the application of genetic markers to conservation issues in detail (Loeschcke et al., 1994; Avise and Hamrick, 1996; Smith and Wayne, 1996). For example, microsatellite DNA markers typically show extremely high allelic diversity, and so are very sensitive to fragmentation events that might otherwise go undetected with other markers (Davies et al., 1999). In addition, sex-biased dispersal can be identified by comparing patterns in uniparentally inherited markers, such as mitochondrial DNA or sex chromosome markers, to patterns from protein and nuclear DNA that are passed to offspring from both parents (e.g. Bowen et al., 1992). Overall, the genetic tools to address most of the questions surrounding forest fragmentation are available, but the most important step is to choose the correct tool to best answer a question (Milligan et al., 1994). Often, interpretation will be most robust when different types of molecular markers are used simultaneously, because each marker has strengths and weaknesses (Haig, 1998).

4. GENETIC ISSUES MEET OTHER FACTORS

By the time genetic changes are detected in forest fragments, populations may already be in trouble from ecological and/or demographic impacts. This does not mean that genetic factors can be dismissed, because there are plenty of instances demonstrating genetic changes over the short time scale of human-induced forest fragmentation, and those changes can further affect population persistence. Gaines et al. (1997) review several studies of the effects of habitat fragmentation on small mammals at different scales and note that extreme fragmentation can lead to loss of genetic variation and increased genetic differentiation. However, they also note that levels of fragmentation that can be demographically important do not necessarily result in genetic changes. In other words, detectable genetic changes imply severe isolation and/or population size reduction has occurred, and the genetic changes will interact with other factors to affect population dynamics and persistence (Soulé and Mills, 1998).

There is now no doubt that movement between populations that have been isolated by human activities can increase population persistence for both genetic and non-genetic reasons. In experiments, movement between populations has been shown to reduce probability of extinction due to demographic and environmental variation, and increase probability of recolonization (e.g. Forney and Gilpin, 1989; Fahrig and Merriam, 1994; Sjogren Gulve, 1994). Similarly, experiments have shown that movement between inbred small populations has decreased the chance of extinction due to inbreeding depression (Spielman and Frankham, 1992).

In sum, fragmentation becomes a threat to populations from a genetic perspective when populations become small and isolated, causing them to become inbred and to lose adequate levels of genetic variation for future adaptation to an ever-changing environment. These threats are real, well-documented, and not mere theoretical conjecture. However, genetic factors cannot be addressed adequately in the absence of knowledge of demographic and environmental changes. Genetic techniques are powerful supplements to ecological approaches to monitoring populations and can be a powerful and effective tool to direct management activities across fragmented landscapes.

Acknowledgments. We thank Len Broberg, Kathy Ralls, Erran Seaman, and Paul Spruell for comments on the manuscript. The paper was completed with research support from USDA (NRI

Competitive Grant Program USDA 97-35101-4355, and McIntire-Stennis), and National Science Foundation (MONTS 291835). David Tallmon was supported by the University of Montana Training WEB Program (NSF #DGE9553611).

LITERATURE CITED

Allendorf, F. W. 1983. Isolation, gene flow, and genetic differentiation among populations. Pp 51-65 in *Genetics and conservation: a reference for managing wild animal and plant populations.* Benjamin/Cummings, Menlo Park, Ca.

Allendorf, F. W. 1986. Genetic drfit and loss of alleles versus heterozygosity. *Zoo Biology* 5:181-190.

Allendorf, F.W, and S. A. Phelps. 1981. Use of allelic frequencies to describe population structure. *Canadian Journal of Fisheries and Aquatic Sciences* 38:1507-1514.

Avise, J. C. 1994. *Molecular markers, natural history, and evolution.* Chapman and Hall, New York.

Avise, J. C., and J. Hamrick, eds. 1996. *Conservation Genetics: Case Histories from Nature.* Chapman and Hall, New York.

Ballou, J. D. 1997. Ancestral inbreeding only minimally affects inbreeding depression in mammalian populations. *Journal of Heredity* 88:169-178.

Bossart, J. L., and D. P. Prowell. 1998. Genetic estimates of population structure and gene flow: limitations, lessons and new directions. *Trends in Ecology and Evolution* 13:202-206.

Bouzat, J. L., H. A. Lewin, and K. N. Paige. 1998a. The ghost of genetic diversity past: historical DNA analysis of the greater prairie chicken. *American Naturalist* 152:1-6.

Bouzat, J. L., H. H. Cheng, H. A. Lewin, R. L. Westemeier, J. D. Brawn, and K. N. Paige. 1998b. Genetic evaluation of a demographic bottleneck in the greater prairie chicken. *Conservation Biology* 12:836-843.

Bowen, B.W., A. B. Meylan, J. P. Ross, C. J. Limpus, G. H. Balazs, and J. C. Avise. 1992. Global population structure and natural history of the green turtle (*Chelonia mydas*) in terms of matriarchal phylogeny. *Evolution* 46:865-881.

Brewer, B. A., R. C. Lacy, M. L. Foster, and G. Alaks. 1990. Inbreeding depression in insular and central populations of *Peromyscus* mice. *Journal of Heredity* 81:257-266.

Briscoe, D. A., J. M. Malpica, A. Robertson, G. J. Smith, R. Frankham, R.G. Banks, and J.S.F. Barker. 1992. Rapid loss of genetic variation in large captive populations of *Drosophila* flies: Implications for the genetic management of captive populations. *Conservation Biology* 6:416-425.

Chambers, S. M. 1983. Genetic principles for managers. Pp 15-46 . In Schonewald-Cox, C. M., S. M. Chambers, B. F. Macbryde, and W. L. Thomas. *Genetics and conservation: a reference for managing wild animal and plant populations.* Benjamin/Cummings, Meno Park, Ca.

Crow, J. F., and K. Aoki. 1984. Group selection for a polygenic behavioral trait: estimating the degree of population subdivision. *Proceedings of National Academy of Sciences* 81:6073-6077.

Cunningham, M., and C. Moritz. 1998. Genetic effects of forest fragmentation on a rainforest restricted lizard (Scincidae: *Gnypetoscincus queenslandiae*). *Biological Conservation* 83:19-30.

Davies, N., F. X. Villablanca, and G. K. Roderick. 1999. Determining the source of individuals: multilocus genotyping in nonequilibrium population genetics. *Trends in Ecology and Evolution* 14:17-21.

Degnan, S. M. 1993. Genetic variability and population differentiation inferred from DNA fingerprinting in Silvereyes (Aves: Zosteropidae). *Evolution* 47:1105-1117.

Ellegren, H. 1991. DNA typing of museum birds. *Nature* 354:113.

Fahrig, L., and G. Meriam. 1994. Conservation of fragmented populations. *Conservation Biology* 8:50-59.

Favre, L., F. Balloux, J. Goudet, and N. Perrin. 1997. Female biased dispersal in the monogamous mammal *Crocidura russula*: evidence from field data and microsatellite patterns. *Proceedings of the Royal Society of London, Series B* 264:127-132.

Foran, D. R., S. C. Minta, and K. S. Heinemeyer. 1998. DNA-based analysis of hair to identify species and individuals for population research and monitoring. *Wildlife Society Bulletin* 25:840-847.

Forbes, S. H., and D. K. Boyd. 1996. Genetic variation of naturally colonizing wolves in the central rocky mountains. *Conservation Biology* 10:1082-1090.

Forney, K., and M. E. Gilpin. 1989. Spatial structure and population extinction: a study with *Drosophila* flies. *Conservation Biology* 3:45-51.

Frankel, O. H., and M. E. Soule. 1981. *Conservation and Evolution*. Cambridge University Press. Cambridge, Massachusetts.

Frankham, R. 1995. Conservation genetics. *Annual Review of Genetics* 29:305-327.

Frankham, R. 1996. Relationship of genetic variation to population size in wildlife. *Conservation Biology* 10:1500-1508.

Frankham, R. 1997. Do island populations have less genetic variation than mainland populations? *Heredity* 78:311-327.

Gaines, M. S., J. E. Diffendorfer, R. H. Tamarin, and T. S. Whittam. 1997. The effects of habitat fragmentation on the genetic structure of small mammal populations. *Journal of Heredity* 88:294-304.

Gilbert, D. A., N. Lehman, S. J. O'Brien, and R. K. Wayne. 1990. Genetic fingerprinting reflects population differentiation in the California Channel Island fox. *Nature* 344:764-766.

Gilpin, M. E., and M. E. Soulé. 1986. Minimum viable populations: processes of species extinction. Pages 18-34 in M. E. Soulé, ed. *Conservation biology: the science of scarcity and diversity*. Sinauer, Sunderland, Mass.

Goudet, J. 1995. FSTAT (Version 1.2): a computer program to calculate F-statistics. *The Journal of Heredity* 86(6):485-486.

Haig, S.M., C. L. Gratto-Trevor, T.D. Mullins, and M. A. Colwell. 1997. Population identification of western hemisphere shorebirds throughout the annual cycle. *Molecular Ecology* 6:413-427.

Haig, S.M. 1998. Molecular contributions to conservation. *Ecology* 79:413-427.

Hartl, D. L., and A. G. Clark. 1997. *Principles of population genetics*, Third Ed. Sinauer Assoc. Inc., Sunderland, Mass.

Hartl, G.B., and Z. Pucek. 1994. Genetic depletion in European bison (*Bison bonasus*) and the significance of eletrophoretic heterozygosity for conservation. *Conservation Biology* 8:167-174.

Hedrick, P. W., and P. S. Miller. 1992. Conservation genetics: techniques and fundamentals. *Ecological Applications* 2:30-46.

Hedrick, P. W., R. C. Lacy, F. W. Allendorf, and M. E. Soule. 1996. Directions in conservation biology: comments on Caughley. *Conservation Biology* 10:1312-1320.

Hillis, D. M., C. Moritz, and B. K. Mable. 1996. *Molecular Systematics*. Sinauer Associates, Sunderland, MA.

Hitchings, S. P., and T. J. C. Beebee. 1997. Genetic substructuring as a result of barriers to gene flow in urban *Rana temporaria* (common frog) populations: implications for biodiversity conservation. *Heredity* 79:117-127.

Houlden, B. A., P. R. England, A. Taylor, W. D. Greville, and W. B. Sherwin. 1996. Low genetic variability of the koala (*Phascolarctos cinereus*) in south eastern Australia following a severe population bottleneck. *Molecular Ecology* 5:269-281.

Ims, R. A., and N. G. Yoccoz. 1997. Studying transfer processes in metapopulations. Emigration, migration, and colonization. Pages 247-265 in *Metapopulation Biology*, Academic Press.

Jiménez, J. A., K. A. Hughes, G. Alaks, L. Graham, and R. C. Lacy. 1994. An experimental study of inbreeding depression in a natural habitat. *Science* 266:271-273.

Keller, L. F., P. Arcese, J. N. M. Smith, W. M. Hochachka, and S. C. Stearns. 1994. Selection against inbred song sparrows during a natural population bottleneck. *Nature* 372:356-357.

Kohn, M. H., and R. K. Wayne. 1997. Facts from feces revisited. *Trends in Ecology and Evolution* 12:223-227.

Lacy, R. C. 1993. Impacts of inbreeding in natural and captive populations of vertebrates: implications for conservation. *Perspectives in Biology and Medicine* 36:480-496.

Lacy, R. C. 1997. Importance of genetic variation to the viability of mammalian populations. *Journal of Mammalogy* 78:320-335.

Lacy, R. C., and B. E. Horner. 1997. Effects of inbreeding on reproduction and sex ratio of

Rattus villosissimus. Journal of Mammalogy 78:877-887.

Lacy, R. C., and J. D. Ballou. 1998. Effectiveness of selection in reducing the genetic load in populations of *Peromyscus polionotus* during generations of inbreeding. *Evolution* 52:900-909.

Laikre, L. and N. Ryman. 1991. Inbreeding depression in a captive wolf (*Canis lupus*) population. *Conservation Biology* 5:33-40.

Lande, R. 1995. Mutation and conservation. *Conservation Biology* 9:782-791.

Lande, R. 1996. The meaning of quantitative genetic variation in evolution and conservation. Pages 27-40 in R. C. Szaro and D. W. Johnston, eds. *Biodiversity in Managed Landscapes*. Oxford University Press.

Leberg, P. L. 1990. Influence of genetic variability on population growth: implication for conservation. *Journal of Fisheries Biology* 37:(A)193-195.

Leberg, P. L. 1991. Influence of fragmentation and bottlenecks on genetic divergence of wild turkey populations. *Conservation Biology* 5:522-530.

Leberg, P. L. 1992. Effects of population bottlenecks on genetic diversity as measured by allozyme electrophoresis. *Evolution* 46:477-494.

Leung, K.P., C. R. Dickman, and L. A. Moore. 1993. Genetic variation in fragmented populations of an Australian rainforest rodent, *Melomys cervinipes*. *Pacific Conservation Biology* 1:58-65.

Lindenmayer, D. B., and R. C. Lacy. 1995. Metapopulation viability of leadbeater's possum, *Gymnobelideus leadbeateri*, in fragmented old-growth forests. *Ecological Applications* 5:164-182.

Loeschcke, V., J. Tomiuk, and S. K. Jain. 1994. *Conservation genetics*. Birkhauser Verlag. Basel, Switzerland.

Luikart, G., and J. M. Cornuet. 1998. Empirical evaluation of a test for identifying recently bottlenecked populations from allele frequency data. *Conservation Biology* 12:228-237.

Lynch, M., J. Conery, and R. Burger. 1995. Mutation accumulation and the extinction of small populations. *American Naturalist* 146:489-518.

Madsen, T., B. Stille, and R. Shine. 1996. Inbreeding depression in an isolated population of adders *Vipera berus*. *Biological Conservation* 75:113-118.

Margulis, S.W. 1998. Differential effects of inbreeding at juvenile and adult life-history stages in *Peromyscus polionotus*. *Journal of Mammalogy* 79:326-336.

Meffe, G. K. 1996. Conserving genetic diversity in natural systems. Pages 41-57 in R. C. Szaro and D. W. Johnston, eds. *Biodiversity in Managed Landscapes*. Oxford University Press.

Milligan, B. G., J. Leebens-Mack, and A. E. Strand. 1994. Conservation genetics: beyond the maintenance of marker diversity. *Molecular Ecology* 3:423-435.

Mills, L. S. 1995. Edge effects and isolation: red-backed voles on forest remnants. *Conservation Biology* 9:395-403.

Mills, L. S. 1996. Fragmentation of a natural area: Dynamics of isolation for small mammals on forest remnants. Pages 199-219 In Wright, G., editor. *National Parks and Protected Areas: Their Role in Environmental Protection*. Blackwell Press.

Mills, L. S., and F. W. Allendorf. 1996. The one-migrant-per-generation rule in conservation and management. *Conservation Biology* 10:1509-1518.

Mills, L.S., and P.E. Smouse. 1994. Demographic consequences of inbreeding in remnant populations. *American Naturalist* 144:412-431.

Mills, L. S., J. J. Citta, K. P. Lair, M. K. Schwartz, and D. A. Tallmon. In Press. Estimating animal abundance using non-invasive DNA sampling: promise and pitfalls. *Ecological Applications*.

Morin, P. A., and D. S. Woodruff 1996. Noninvasive genotyping for vertebrate conservation. Pages 298-313 in Smith, T. B., and R. K. Wayne, eds. *Molecular Genetic Approaches in Conservation*. Oxford University Press, New York.

Mundy, N. I., C. S. Winchell, T. Burr, and D. S. Woodroff. 1997. Microsatellite variation and microevolution in the critically endangered San Clemente Island loggerhead shrike. *Proceedings of the National Academy of Sciences* 264:869-875.

Nei, M., T. Maruyama, and R. Chakraborty. 1975. The bottleneck effect and genetic variability in populations. *Evolution* 29:1-10.

Neigel, J. E. 1996. Estimation of effective population size and migration parameters from genetic data. In Smith, T. B., R. K. Wayne, eds. *Molecular Genetic Approaches in Conservation*. Pp. 329-346. Oxford Univ Press, New York.

Newman, D., and D. Pilson. 1997. Increased probability of extinction due to decreased genetic effective population size: experimental populations of *Clarkia pulchella*. *Evolution* 51:354-362.

Newman, D., and D. A. Tallmon. In Prep. Increased fitness and decreased population divergence due to gene flow.

O'Brien, S. J. 1994. A role for molecular genetics in biological conservation. *Proceedings of the National Academy of Sciences* 91:5748-5755.

Paetkau, D., W. Calvert, I. Stirling, and C. Strobeck. 1995. Microsatellite analysis of population structure in Canadian polar bears. *Molecular Ecology* 4:347-354.

Paetkau, D., L. P. Waits, P. L. Clarkson, L. Craighead, E. Vyse, R. Ward, and C. Strobeck. 1998. Variation in genetic diversity across the range of North American brown bears. *Conservation Biology* 12:418-429.

Palsboll, P. J., J. Allen, M. Berube, P. J. Clapham, T. P Fedderson, P. S. Hammond, R. R. Hudson, H. Jorgensen, S. Katona, A. H. Larsen, F. Larsen, J. Lien, D. K. Mattila, J. Sigurjonsson, R. Sears, T. Smith, R. Sponer, P Stevick, and N. Olen. 1997. Genetic tagging of humpback whales. *Nature* 388:767-769.

Parker, P. G., A. A. Snow, M. D. Schug, G. C. Booton, and P. A. Fuerst. 1998. What molecules can tell us about populations: choosing and using a molecular marker. *Ecology* 79:361-382.

Ralls, K., J. D. Ballou, and A. Templeton. 1988. Estimates of lethal equivalents and the cost of inbreeding in mammals. *Conservation Biology* 2:185-193.

Rannala, B., and J. A. Hartigan. 1996. Estimating gene flow in island populations. *Genetical Research* 67:147-158.

Rannala, B., and J. L. Mountain. 1997. Detecting immigration by using multilocus genotypes. *Proceedings of the National Academy of Sciences* 94:9197-9201.

Roelke, M. E., J. S. Martenson, S. J. O'Brien. 1993. The consequences of demographic reduction and genetic depletion in the endangered Florida panther. *Current Biology* 3:340-350.

Saccheri, I. Kuussaari, M. Kankare, M., Vikman, P., Fortelius, W., and Hanski, I. 1998. Inbreeding and extinction in a butterfly metapopulation. *Nature* 392:491-494.

Sarre, S. 1995. Mitochondrial DNA variation among populations of *Oedura reticulata* (Gekkonidae) in remnant vegetation: implications for metapopulation structure and population decline. *Molecular Ecology* 4:395-405.

Schwartz, M. K., D.A. Tallmon, G.H. Luikart. 1998. Review of DNA-based census and effective population size estimators. *Animal Conservation* 1:293-299.

Sjogren Gulve, P. 1994. Distribution and extinction patterns within a northern metapopulation of the pool frog, *Rana lessonae*. *Ecology* 75:1357-1367.

Slatkin, M. 1985. Gene flow in natural populations. *Annual Review of Ecology and Systematics* 16:393-430.

Slatkin, M. 1994. Gene flow and population structure. Pages 3-17 in Real, L. A. ed. *Ecological Genetics*. Princeton University Press. Princeton, N.J.

Slatkin, M. 1995. A measure of population subdivision based on microsatellite allele frequencies. *Genetics* 139:457-462.

Slatkin, M., and N. H. Barton. 1989. A comparison of three indirect methods for estimating average levels of gene flow. *Evolution* 43:1349-1369.

Slatkin, M., and W. P. Maddison. 1990. A cladistic measure of gene flow inferred from the phylogenies of alleles. *Genetics* 123:603-613.

Smith, T. B., and R. K. Wayne. 1996. *Molecular genetic approaches in conservation*. Oxford University Press. Oxford.

Soulé, M. E. 1980. Thresholds for survival: maintaining fitness and evolutionary potential. Pages 151-170 in Soulé, M. E., and B. A. Wilcox, eds. *Conservation biology: an evolutionary-ecological perspective*. Sinauer, Sunderland, Mass.

Soulé, M. E., and L. S. Mills. 1998. No need to isolate genetics. *Science* 282:1658-1659.

Spielman, D., and R. Frankham. 1992. Modeling problems in conservation genetics using captive *Drosophila* populations: improvement of reproductive fitness due to immigration of one individual into small partially inbred populations. *Zoo Biology* 11:343-351.

Steinberg, E. K., and C. E. Jordan. 1997. Using molecular genetics to learn about the ecology of threatened species: the allure and the illusion of measuring genetic structure in natural populations. Pp. 440-460 in *Conservation Biology for the Coming Decade* (P. Fiedler and P. Kareiva, eds.). Chapman and Hall, New York.

Tallmon, D. A., L. S. Mills, and F. W. Allendorf. In Prep. Genetic effects of ecological isolation on California red-backed voles.

Taylor, A. C., W. B. Sherwin, and R. K. Wayne. 1994. Genetic variation of simple sequence loci in a bottlenecked species: The decline of the northern hairy-nosed wombat (*Lasiorhinus krefftii*). *Molecular Ecology* 3:277-290.

Templeton, A. R., and N. J. Georgiadis. 1995. A landscape approach to conservation genetics: conserving evolutionary processes in the African *Bovidae*. Pages 398-430 in Avise, J., and J. Hamrick, eds. *Conservation Genetics: Case Histories from Nature*. Chapman and Hall, New York.

Varvio, S., R. Chakraborty, and M. Nei. 1986. Genetic variation in subdivided populations and conservation genetics. *Heredity* 57:189-198.

Vucetich, J. A., and T. A. Waite. In Review. n-Migrants-per-generation for the genetic management of fluctuating populations. *Proceedings of the Royal Society*.

Waser, P.M., and C. Strobeck. 1998. Genetic signatures of interpopulation dispersal. *Trends in Ecology and Evolution* 13:43-44.

Wauters, L. A., Y. Hutchinson, D. T. Parkin, and A. A. Dhondt. 1994. The effects of habitat fragmentation on demography and on the loss of genetic variation in the red squirrel. *Proceedings of the Royal Society of London, Series B*. 255:107-111.

Wayne, R. K., and 10 co-authors. 1991. Conservation genetics of the endangered Isle Royale gray wolf. *Conservation Biology* 5:41-51.

Weir, B.S., and C.C. Cockerham. 1984. Estimating F-statistics for the analysis of populations structure. *Evolution* 38:1358-1370.

Westemeier, R. L., J. D. Brawn, S. A. Simpson, T. L. Esker, R. W. Jansen, J. W. Walk, E. L. Kershner, J. L. Bouzat, and K. N. Paige. 1998. Tracking the long-term decline and recovery of an isolated population. *Science* 282:1695-1698.

Wisdom, M. J., and L. S. Mills. 1997. Sensitivity analysis to guide population recovery: prairie-chickens as an example. *Journal of Wildlife Management* 61:302-312.

Woods, J.G., B. McLellan, D. Paetkau, C. Strobeck, M. Proctor. 1996. DNA fingerprinting applied to mark-recapture bear studies. *International Bear News* 5:9-10.

Wright, S. 1977. Evolution and the genetics of populations, Vol. 3: Experimental results and evolutionary deductions. University of Chicago.

Forest Fragmentation of the Inland West

Issues, Definitions, and Potential Study Approaches for Forest Birds

Rex Sallabanks[1], Patricia J. Heglund[2], Jonathan B. Haufler[3], Brian A. Gilbert[4], William Wall[5]

There is considerable evidence from eastern landscapes of the U.S. that fragmentation of forests has been a significant factor in the decline of some bird populations. Although primarily unstudied, the perceived drivers of fragmentation effects (e.g., agricultural intensification, timber harvest, and urban development) are often assumed to be operating similarly in western forest landscapes. In this paper, with special emphasis on forests of the inland west, we provide a brief review of studies of fragmentation, and discuss the prevalent land-use practices that influence landscape structure. In addition, we describe how forested landscapes differ in the inland west compared with other regions of the U.S. and articulate how and why these differences may alleviate some of the typical effects of fragmentation (e.g., patch size and edge effects that result in increased rates of nest predation and brood parasitism).

One of the central tenets of our paper is how to define fragmentation. We argue that any definition must include a reference to historical landscape conditions, and propose the following: fragmentation is *"an anthropogenic disruption (including disruption of natural disturbance regimes) in continuity of habitat in relation to historical condition."* By "historical," we refer to the period 100-400 years prior to major European-settlement. Our definition distinguishes habitat fragmentation as a land management issue from the natural patchiness of a landscape, which others have also referred to as fragmentation. We suggest that habitat fragmentation refers to anthropogenic changes, and that naturally patchy environments should be described as such. Finally, we conclude our paper by offering some study approaches and research objectives and, by way of example, present an experimental design of a study that is the first large-scale regional assessment of the effects of landscape composition on avian productivity in conifer forests of the inland west.

Key words: silviculture, agriculture, avian productivity, edge effects, forest fragmentation, historical conditions, natural disturbance regimes, inland west, landscape composition, patch size effects.

1. FRAGMENTATION ISSUES

Fragmentation has been defined as the conversion of large areas of contiguous native forest to other types of vegetation and/or land use (i.e., food crops, pasture, residential areas) leaving remnant patches of forest that vary in size and isolation (Saunders et al., 1991; Murcia, 1995). Fragmentation, as defined, reduces the total area covered by forest and creates "permanent" sharp boundaries between the remaining timber and the surrounding lands. This process potentially exposes organisms that remain in the fragment, particularly along

---
[1]*Sustainable Ecosystems Institute, Meridian, ID;* [2]*Turnstone Ecological Research Associates, Moscow, ID;* [3]*Boise Cascade Corporation, Boise, ID;* [4]*Plum Creek Timber Company, Veradale, WA;* [5]*Potlatch Corporation, Lewiston, ID.*

its boundary, to the conditions of a very different surrounding microclimate (sensu "edge effects"; Giles, 1978; Saunders et al., 1991; Bolger et al., 1997). Fragmentation of this nature often creates an environment favorable to interspecific competitors, nest predators, and avian brood parasites (Askins, 1995; Murcia, 1995; Johannesen and Ims, 1996; Donovan et al., 1997) and may change the types and availability of both food and nesting resources for birds. Both interspecific and intraspecific mechanisms, including resource utilization, social interaction, and dispersal rate, may influence survival and recruitment rates of breeding populations in forest remnants (Johannesen and Ims, 1996). There is much evidence that fragmentation of forested habitat is a significant factor in the apparent decline of some bird populations in midwestern (Ambuel and Temple, 1983; Robinson, 1992; Donovan et al., 1995; Van Horn et al., 1995) and eastern (Whitcomb et al., 1981; Wilcove, 1985; Askins et al., 1990; Hoover et al., 1995) North America. In addition, there is some evidence (Böhning-Gaese et al., 1993) that increased forest fragmentation in these regions has played a larger role in the decline of forest-nesting migratory birds, by way of leading to increased nest predation on the breeding grounds, than deforestation on tropical wintering grounds.

The term fragmentation has often been applied to large expanses of managed and unmanaged forests where harvest practices and variation in climate and physical site condition lead to a mosaic of forest successional conditions and composition (Harris, 1984; Murcia, 1995; McGarigal and McComb, 1995; McGarigal and McComb, this volume). Thus, Wiens (1995) suggested fragmentation can refer either to the spatial pattern of patchiness of a habitat, or to the process that produces such a pattern, including fire or forest management. This process has been defined as a "disruption of habitat continuity" (Lord and Norton, 1990; Harris and Silva-Lopez, 1992; Green, 1995) and, as such, may be natural or human-induced.

When considering the meaning of fragmentation, it is important to remember that although both temperate and boreal forests cover huge regions of the North American continent, they have historically displayed great differences in patch dynamics (Nilsson and Ericson, 1992). In general, temperate deciduous forests may be characterized by small-scale patch dynamics with a mean gap size of 0.01 ha (Runkle, 1985). In contrast, many western and boreal forests have large-scale patch dynamics of up to 10,000 ha or more due to fire and disease pathogens (Pickett and Thompson, 1978). These factors produce a heterogeneous landscape mosaic of forest and grassland vegetation types that vary naturally with respect to forest cover. Based upon this natural variation in landscape composition, if we define "fragmentation" as "disruption of habitat continuity" or "disruption of percent forest cover" (e.g., Donovan et al., 1995; Robinson et al., 1995a), then western forests have always been extremely dynamic and "fragmented," at least to some degree. Plant and animal species of dynamically fragmented forests are thought to have evolved a variety of mechanisms to cope with such fragmentation (Nilsson and Ericson, 1992). Thus, species that evolved in large expanses of forest characterized by a variety of large-scale disturbances should be resistant to forest fragmentation as long as that fragmentation is an historically dynamic process. This is especially true of forest landscapes in the inland west, where, relative to forests in the eastern U.S. and along the Pacific coast, contiguous forest cover has always been disrupted due to disturbance processes such as wildfire and pathogens in combination with a highly variable landscape topography. For example, the moist forest conditions of the coastal Cascades and Olympic mountain ranges resulted in a disturbance cycle characterized by high intensity, stand-replacing fires at long, albeit highly variable, intervals (Fahnestock and Agee, 1993). This disturbance regime resulted in relatively homogeneous landscapes with a stand development pattern that was relatively consistent over wide areas. In contrast, forests of the Northern Rocky Mountain region contained a diversity of disturbance regimes ranging from low intensity under-burns, to variable intensity mosaic burns, and high intensity, stand-replacing fires (Agee, 1993; Agee, this volume). Especially today, fire and insect outbreaks are common occurrences in forests of the northern Rockies and, when combined with changes in macro-

and micro-climate, soil fertility, and soil drainage, have resulted in a mosaic of disturbance-induced plant communities (Daubenmire and Daubenmire, 1968). However, in a review of forest fragmentation in Minnesota, Green (1995) posed the question, "... is a forest landscape composed of a patchwork of forest types and ages a fragmented landscape?" Clearly, nomenclature and context are important components in any discussion of forest fragmentation and we agree with Fahrig (this volume) that the definition of fragmentation is not simply an issue of semantics.

Fragmentation in managed forests often leads to major increases in edge. Studies providing empirical evidence demonstrating negative effects on forest-nesting bird populations have mostly been conducted in the eastern and midwestern U.S. where much of the forested area has been cleared and permanently converted to agricultural cropland or human habitations (Kremsater and Bunnell, this volume). The result of such land management has been "islands" of forests far removed from other such forests and a concomitant reduction in the total amount of forest remaining in an area (Askins and Philbrick, 1987; Hunter, 1990). Paton (1994) reviewed studies that addressed the influence of edge effects on avian nest success and concluded that forest fragmentation seems to result in increased nest predation and parasitism. However, this review included just two studies from the western U.S. (Paton 1994). The first study, conducted by Boag et al. (1984), examined egg depredation among artificial nests and natural spruce grouse (*Dendragapus canadensis*) nests in lodgepole pine (*Pinus contorta*) forests of southwestern Alberta in relation to trails and stand structural characteristics. Boag et al. (1984) observed highest predation in areas >15m from trails that lacked shrub cover under the forest canopy. The second study, conducted by Ratti and Reese (1988), investigated depredation of artificial nests placed on the ground and in shrubs in plots along a grassland-conifer forest ecotone in northern Idaho. Ratti and Reese (1988) reported that nests placed adjacent to high contrast (grassland-forest) edges (sensu Thomas (1979)) had higher predation rates than nests placed adjacent to forest stands in early successional stages. They suggested birds may be poorly adapted to nesting along high-contrast edges and that such habitats may have a barrier effect on nesting birds and create travel lanes for predators. For an updated review of edge effects, with special reference to western forests, see Marzluff and Restani (this volume).

Whereas the studies reviewed above address edge effects, they provide little guidance on the effects of fragmentation and patch size on forest-nesting bird populations in the west. Donovan et al. (1997) made a plea for scientists to forego documenting the occurrence of edge effects in favor of examining the conditions that lead to edge effects such as landscape composition. Although a few studies from the "west" have examined avian occurrence in relation to patch size effects (Schieck et al., 1995), forest edges (Rosenburg and Raphael, 1986; Ratti and Reese 1988), landscape structure (McGarigal and McComb, 1995; McGarigal and McComb, this volume), and fragmentation (Lemkuhl et al., 1991; Schmiegelow et al., 1997; Schmiegelow and Hannon, this volume), we know of only one study of avian productivity in relation to fragmentation in the inland west (Tewksbury et al., 1998). Tewksbury et al. (1998) conducted a comparative study of the effects of anthropogenically caused forest fragmentation in relation to naturally fragmented forests on the breeding productivity of songbirds in riparian areas of the northern Rocky Mountain region. They provide strong evidence that predation rates in naturally fragmented forests were greater than in anthropogenically influenced landscapes. In addition, they showed that although brood parasitism decreased with increasing forest cover, the strongest predictor of predation and parasitism was not forest cover, but the abundance of human development on the landscape and the density of Brown-headed Cowbird (*Molothrus ater*) hosts.

In reviewing the literature, we found limited evidence to suggest that populations of forest birds are declining in the west. Indeed evidence to the contrary comes from analyses of North American Breeding Bird Survey (BBS) data which reveal that more species are increasing in the western states than are decreasing (Sauer and Droege, 1992; Peterjohn et al., 1995). It is clear that when dealing with large expanses of managed forest, rather than an agricultural landscape containing isolated patches of remnant forest, we first need to assess the applicability of island biogeography theory to regions where forests still dominate the land (Hunter, 1990; Martin, 1992). If western forests are more naturally fragmented or have larger average gap dynamics than eastern forests, then the birds that breed in them have had a longer evolutionary history with edge-related factors such as predators and cowbirds. We hypothesize that the typical effects of forest fragmentation (as seen in eastern landscapes) might not be so heavily expressed in western systems. Studies by Rosenburg and Raphael (1986), Lemkuhl et al. (1991), McGarigal and McComb (1995), Schieck et al. (1995), Schmiegelow et al. (1997), and Tewksbury et al. (1998) support this hypothesis. Studies that have examined natural forest fragments elsewhere have found mixed effects (Dobkin and Wilcox, 1986; Sallabanks et al., *submitted manuscript a*). One goal of the conference on which these proceedings are based was to address such a hypothesis and explore evidence for or against its acceptability.

Two land-use practices that are prevalent in western forest landscapes are agriculture and silviculture, and both can significantly reduce the amount of contiguous mature and old-growth forest cover. Agriculture occurs in the form of pasture development for cattle grazing, cattle feed lots, arable crop fields, hay meadows, ranches, and sheep farms; silviculture includes clearcutting, seed tree retention, shelterwood cutting, thinning, and group selection. Agriculture (especially the development of farmland) is detrimental to the reproductive success of many breeding birds (e.g., Galbraith, 1988; Andrén, 1992; Berg, 1992; Gates et al., 1994; Warner, 1994; Fuller et al., 1995; Newton, 1998) and has had a greater impact on populations of Neotropical migratory birds than any other human activity (Rodenhouse et al., 1995). The role of silviculture in naturally fragmented forests is less clear and early evidence suggests that impacts vary with bird species (Thompson et al., 1992; Hejl et al., 1995; Tewksbury et al., 1998; Sallabanks et al., *in press*). Silvicultural practices may be less influential compared with agricultural practices because the forest edges that are created by silviculture are of lower contrast (less "abrupt") over time (DeGraaf, 1992; Suarez et al., 1997), thereby reducing edge-related phenomena such as nest predation and brood parasitism (Rudnicky and Hunter, 1993; Hanski et al., 1996; King et al., 1996). In addition, the effects of fragmentation by silviculture in a managed forest landscape may tend to "dissolve" at any one location over time as clearcuts or heavily thinned stands return to later seral stages (Hagan et al., 1996). Indeed, studies of fragmentation in industrial forest landscapes of northern Europe have generally not found forest patch size effects (Haila et al., 1989; Huhta, 1998).

If forest fragmentation in the west leads to increased rates of nest predation and brood parasitism as it does in the east (but see Tewksbury et al., 1998), then both agriculture and silviculture have the potential to threaten the viability of bird populations in western forest ecosystems. In other regions of the U.S., nest predation is the most important source of reproductive failure for forest-nesting songbirds (Martin, 1993a; 1993b; Howlett and Stutchbury, 1996); brood parasitism by cowbirds is also significant (Brittingham and Temple, 1983; Robinson et al., 1995b; Thompson et al., *in press*). What is unknown is whether these mechanisms of reproductive failure are important processes in western systems; predators and cowbirds are certainly locally abundant depending on habitat condition, but how they are influenced by agriculture and silviculture at a landscape scale has not yet been studied.

2. THE CURRENT STATE OF FRAGMENTATION IN WESTERN FORESTS

Studies of fragmentation in the eastern and midwestern U.S. have identified several important mechanisms that negatively affect forest birds and these same concerns have been raised about forest fragmentation for the inland forests of the Pacific Northwest. But does fragmentation in inland forests operate in the same manner as in eastern and midwestern forests? Can we use the same definitions of fragmentation as have been used in other regions? There are a number of significant differences in western forests that require that fragmentation be approached differently compared with past work in other regions.

Unlike many eastern or midwestern forests, inland forests of the west, prior to European settlement, were strongly influenced by the role of fire as a disturbance agent (Agee, this volume). Low elevation forests (e.g., Douglas-fir [*Pseudotsuga menziesii*] and grand fir [*Abies grandis*] habitat types; Steele et al., 1981) in much of the inland west were influenced by frequent, low intensity ground fires which reduced densities of trees and surface vegetation. Crane and Fischer (1986) reported average fire return intervals for these forests of 10 – 30 years. This disturbance regime produced low elevation forests characterized by open, park-like stands of all-ages but predominantly large ponderosa pine (*Pinus ponderosa*), with a grass dominated understory. These forests were regularly interrupted by shrublands which occurred on the dryer areas such as many south-facing slopes. These conditions can be readily seen in old landscape photographs (Boise National Forest, 1993). High elevation forests in the inland west were historically influenced by a mixed fire regime that included considerable high intensity stand-replacing fires. The result was a mosaic of successional stages across the landscape with higher tree densities than low elevation forests. Shrubland openings, though more dispersed than at lower elevations, still were a regular feature of the landscape (Boise National Forest, 1993). Thus, historically, forests of the inland west occurred in patches of varying size and shape, and were regularly affected by either frequent understory burns or longer interval stand-replacing burns, depending on the elevation and resulting moisture regimes.

These same forests support much different conditions today. Low elevation forests have been strongly influenced by the combined effects of fire suppression, livestock grazing, and timber harvest (Agee, 1993). Currently, these forests support significantly higher tree densities than occurred historically, and many of the historically-occurring openings have filled with trees (Boise National Forest, 1993). In addition to higher tree densities, the composition of the tree species has also shifted significantly (Haufler et al., 1996). At higher elevations, the effects of fire suppression have not been as dramatic, as the longer historical fire return intervals of these forests has not caused them to be as altered as the low elevation forests.

Concerns over fragmentation in forests of the inland west are often directed toward proposed logging activity in lower elevation forests. But what is fragmentation in these forests? As discussed, these forests today support much higher densities of trees over more continuous forest cover than occurred under historical disturbance regimes. This has led to many changes in animal species distributions. For example, the northern Idaho ground squirrel (*Spermophilus brunneus brunneus*) has been proposed for listing under the Endangered Species Act by the U.S. Fish and Wildlife Service. Primary threats to its existence appear to be the changes in habitat conditions caused by fire exclusion in low elevation forest ecosystems (Yensen and Sherman, 1997). The lack of fire has isolated openings containing pockets of preferred habitat in certain soil conditions, and has encouraged the development of inhospitable habitat for dispersal among the existing forest stands. Historically, fire maintained more openings and more favorable dispersal habitat in park-like ponderosa pine forests. Thus, the ground squirrel is imperiled by fragmentation of its habitat that is caused by the higher densities and greater continuity of forest conditions than occurred historically. Similarly, habitat conditions for white-headed woodpeckers (*Picoides albolarvatus*), white-breasted nuthatches (*Sitta carolinensis*), and pygmy nuthatches (*Sitta pygmaea*), all species that were reported to occur in open ponderosa pine forests as occurred historically, are absent from much of the present low elevation forests (Sallabanks et al., *submitted manuscript b*). How much of the historical habitat requires restoration, in terms of patch sizes and distributions, to allow for reestablishment of populations of such bird species is uncertain. However, it is clear that their preference is for a more open and patchy landscape that more closely resembles the historical conditions that supported earlier populations. To the detriment of these species, lack of appropriate types and amounts of disturbance has created habitat changes resulting in greater continuity of forest conditions. Under these conditions, what then constitutes fragmentation of forest habitat?

For the reasons stated above, we feel that it is imperative that fragmentation be defined relative to historical landscape conditions. We propose that an appropriate definition of habitat fragmentation is: "*an anthropogenic disruption (including disruption of natural disturbance regimes) in continuity of habitat in relation to historical condition.*" Under this definition, fragmentation can only be properly described when placed in the context of historical disturbance regimes, and the type, size, and distribution of ecological communities produced by these disturbance regimes. This complicates its measurement (quantification of edges of different types is only relevant to differences from historical amounts and types of edges), but has far more ecological significance. We would further propose that an appropriate historical time-frame for comparative purposes would be 100-400 years prior to major European-settlement of a landscape (Steele, 1994), and as further discussed by Morgan et al. (1994). Our definition distinguishes habitat fragmentation as a land management issue from the natural patchiness of a landscape, which many have also referred to as fragmentation. We suggest that habitat fragmentation refers to anthropogenic changes, and that naturally patchy environments should be described as such.

3. APPROACHES TO STUDYING FRAGMENTATION IN THE WEST

Given the definition of habitat fragmentation proposed above, how should concerns over habitat fragmentation be addressed in forests of the inland west? It is clear that we cannot compare the effects of current landscape conditions with those of historical times because in the inland west there are no landscapes in historical conditions left to study today. Examining the central issues of forest fragmentation in the inland west (as we perceive them) is an impossible task. Known drivers of fragmentation effects from other regions of the U.S. include agriculture, silviculture, and urban development (reviewed above). Alternatively, therefore, one approach to studying fragmentation in the inland west might be to assess the influence of these drivers on western bird populations. Using historical data and known relationships between birds and habitat, our understanding of fragmentation effects could then be enhanced by comparing the current condition of the landscapes studied to what they were probably like historically. Our knowledge could then be further developed by reconstructing historical landscapes using forestry to restore habitat to its historical range of variability (i.e., creating more open, park-like stands of ponderosa pine). Potential changes in bird populations between historical and current conditions could be extrapolated from known bird-habitat relationships, and actual changes in restored forest stands could be monitored.

With these information needs in mind, we propose a series of regional assessments of songbird nesting success in relation to landscape composition. Such assessments should contain studies from regions with different historical patch size dynamics and moisture regimes. Specific study objectives might be to: 1) Determine whether there are avian productivity problems in western forest landscapes; 2) Assess the influence of forest fragmentation (by agriculture and active silviculture) on avian population viability in western forest ecosystems with varying patch dynamics; and 3) Identify the mechanisms responsible for low nesting success if avian productivity problems are found to exist (Haufler 1998). To exemplify one approach to studying fragmentation in western landscapes, we present details of a study design from our own research in Idaho and Montana. We consider the degree to which a primarily forested landscape is comprised of nonforested "agricultural lands" (e.g., cultivated fields, residential housing, farms, ranches, and meadows), managed forest components (e.g., recent clearcuts, early seral forest, and selectively logged forest) and/or unmanaged forest (old-growth, recent or historical burns, insect damaged or diseased patches, etc.). In our study, "landscapes" are circles with 3.22 km (two mile) radii (Fig. 1). To be considered human-induced fragmentation, closed-canopy forest cover must have been disrupted by nonforest (i.e., agricultural habitat) and/or open-canopy forest created by active silviculture.

To evaluate the influence of landscape condition on the nesting success of forest-breeding

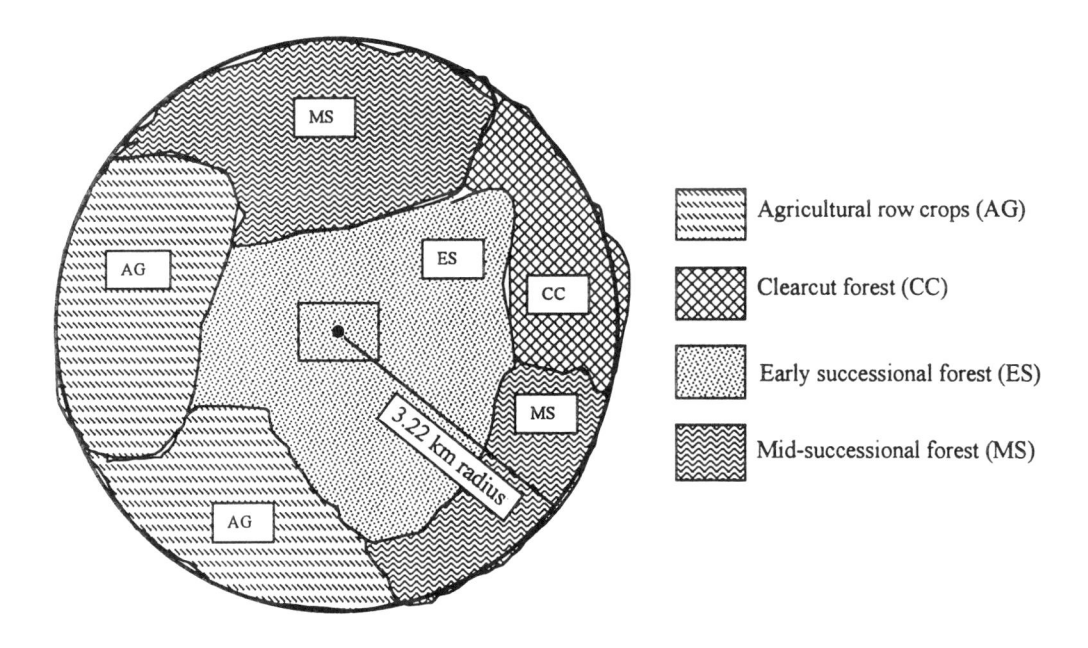

Figure 1. GIS image of a landscape being studied as part of a regional assessment of the effects of landscape composition on avian productivity in the inland west. In this particular case, the landscape shown represents an "agriculturally influenced managed forest landscape." One of three landscape "treatments" being studied (see text for details), this type of landscape has been replicated four times in each of three geographical regions of the inland west (west-central Idaho, northern Idaho, and western Montana). Note that the landscape has a radius of 3.22 km (2 miles) and that nest monitoring occurs in the centrally located 50-ha study plot.

songbirds, we have suggested a trio of treatments and controls be examined. The first "treatment" (agriculturally influenced managed forests) is a managed forest landscape containing agricultural inclusions (5%–50% of the landscape; illustrated in Fig. 1). The second "treatment" (silviculturally managed forests) is a landscape that does not contain any agricultural component (note that much of the western U.S. is "open range"; grazing under forest canopy is an activity that is therefore allowed in this treatment). The third "treatment" (unmanaged forest) consists of a landscape containing no human-induced forest changes (although again, some grazing may occur). These treatments have been replicated in west-central Idaho and northern Idaho; treatments one and two have also been replicated in western Montana. Within each treatment, we selected four replicate landscapes per region (Table 1). Each study plot where data are collected is approximately 50 ha in size and centered within the 3.22-km radius (3,253 ha) landscape unit (Fig. 1). All landscapes are mutually exclusive (i.e., they do not overlap) and were selected using land-use classification data in existing Geographical Information System (GIS) databases. Elevation, cover type, habitat type (plant association), and successional stage have been held as constant as possible within and among study regions.

In this example, we focused on the influence of landscape composition on avian productivity, rather than the size of isolated forest fragments as has typically been the case in other regions of the U.S. (e.g., Galli et al., 1976; Freemark and Merriam, 1986; Blake and Karr, 1987; Blake, 1991). We argue that traditional definitions of fragmentation that include references to factors such as patch size and degree of iso-

Table 1. Experimental design for our study of avian nest success in forested landscapes of the inland west

Landscape treatments:	Agriculturally influenced managed forest	Silviculturally managed forest	Unmanaged forest
Landscape composition:	Agriculture and active silviculture	Active silviculture only	Neither agriculture or active silviculture
Study regions:			
West-central Idaho	$n = 4$ landscapes	$n = 4$ landscapes	$n = 4$ landscapes
Northern Idaho	$n = 4$ landscapes	$n = 4$ landscapes	$n = 4$ landscapes
Western Montana	$n = 4$ landscapes	$n = 4$ landscapes	not sampled

lation may not be applicable to western forests. As discussed earlier, such habitats have not been reduced to island woodlots of variable size. Previous studies of fragmentation have also primarily focused on how avian nesting success can change when landscapes become fragmented by agriculture, rather than by forestry. Few studies have attempted to examine both as we have here. One recent study from central Canada (Bayne and Hobson, 1997) is similar to the one we describe above in that nest success is addressed in landscapes fragmented by both agriculture and silviculture. Bayne and Hobson (1997) monitored artificial nests and found that ground nests were depredated more frequently in landscapes fragmented by agriculture compared with those fragmented by logging.

Our experimental design is also similar to two studies from: 1) the Midwest, where avian demography was compared between forest fragments in an agricultural landscape and contiguous forests (Donovan et al., 1995); and 2) the northeast, where abundance and productivity indices of bird species were compared between industrial forest landscapes that were fragmented by silviculture and those that were not (Hagan et al., 1996). Both studies found deleterious effects of fragmentation, either in the form of reduced nest success (Donovan et al., 1995) or reduced pairing success (Hagan et al., 1996). As reviewed earlier, there have been no previous studies of fragmentation on a regional scale in forests of the inland west.

Are such studies as the one we illustrate feasible? What are the limitations, constraints, and problems associated with conducting such work? Based upon our experience, such studies are certainly feasible. Pooling data among all three regions of our study, we found and monitored 1,818 nests of 62 bird species in 1997 and 1998 combined. Those species for which we found the most nests include the dusky flycatcher (*Empidonax oberholseri*; $n = 243$), dark-eyed junco (*Junco hyemalis oreganus*; $n = 226$), American robin (*Turdus migratorius*; $n = 225$), chipping sparrow (*Spizella passerina*; $n = 181$), Swainson's thrush (*Catharus ustulatus*; $n = 148$), and warbling vireo (*Vireo gilvus*; $n = 122$). The number of nests found varied among regions and treatments. Most nests were found in west-central Idaho ($n = 905$), followed by northern Idaho ($n = 530$), and western Montana ($n = 383$). This variation was due to a combination of differences among forest habitat conditions, nest searcher experience, and time spent searching for nests. With respect to treatment type, more nests were found in agriculturally influenced managed forest landscapes ($n = 957$) and silviculturally managed forest landscapes ($n = 671$), compared with unmanaged forest landscapes ($n = 190$). The large discrepancy here was due to the lack of sampling in unmanaged landscapes in western Montana during both years (Table 1) and in west-central Idaho in 1997. In 1999, unmanaged forest landscapes will again be sampled in both northern Idaho and west-central Idaho and nest sample sizes should become significantly higher for this important treatment. Unmanaged forest landscapes have not been sampled in western Montana because all potential options for study sites were too steep or too high in elevation to serve as valid comparisons in this region of our study. Finding representative landscapes is therefore one example of potential logistical constraints. Another potential problem with a regional assessment such as ours is maintaining similar forest habitat types (and bird communities) among regions for comparative analyses at the end of the study. Finally, the costs associated with large-scale studies such as the one we highlight here are high and partitioning large projects into smaller, connected research efforts as we have done helps offset these costs. Building funding and research cooperatives that involve multiple partners is one way to overcome this constraint.

4. CONCLUSIONS

The first question we have posed is: "Are there avian productivity problems in forests of the inland west?" If we find that bird populations, at least for those species for which we have good data, appear to be healthy, then we can assume that current conditions are adequate to support viable source populations for many birds. Using our data, management programs for the region would then have the potential to ensure that habitat remains available in sufficient quantity and quality to meet species' needs. If, on the other hand, we discover that, at least for some species, productivity problems exist, we can use our data to better understand the cause. The second question we have asked is therefore: "If productivity problems occur, are they influenced by landscape composition and what's the underlying mechanism causing nest failure?" One aspect of our work that we have not discussed is an effort to identify predators using clay eggs placed in artificial nests, as well as a camera component designed to photograph visitors to such nests. In addition to fragmentation effects, we will therefore be able to assess the role of nest predation as a factor contributing to low nesting success, and even identify the predators in some cases. For this scenario, management recommendations will therefore hinge upon our results, and such recommendations could be at the stand-level, landscape-level, or both.

In collaboration with other land-management agencies (e.g., the U.S. Forest Service), management action might also come in the form of habitat restoration for key species that we identify as being at risk (e.g., those species identified earlier as needing open, park-like stands of ponderosa pine). Our research might find that, while most species are doing fine, there are a few that are not. The habitat requirements of these "at-risk species" could then be better evaluated and future monitoring efforts be targeted toward these important birds. Management recommendations would then be focused on the needs of these species in particular. We might also discover that there are some habitat conditions that do not correspond well with their historical range of variability (e.g., they are far less abundant now than they were). Through management workshops and planning sessions, the trade-offs associated with restoring such conditions to the landscape could then be evaluated, especially with the needs of bird species dependent on those conditions in mind.

In summary, we urge that researchers studying forest fragmentation in the western U.S. carefully consider traditional definitions of the term and whether it is appropriate to apply them to their study systems. Whether the patch size and edge effects found to be so common in the eastern and midwestern U.S. also occur in the west remains to be seen, although there is some evidence that they may not (e.g., McGarigal and McComb, 1995; Schieck et al., 1995; Schmiegelow et al., 1997; Tewksbury et al., 1998; McGarigal and McComb, this volume; Schmiegelow and Hannon, this volume). Finally, we emphasize the need to relate current landscape conditions to those of historical times, and argue that forests of the inland west have always been fragmented to some degree. A long evolutionary history with natural openings and their associated edge effects may have decreased the sensitivity of some bird species to the "typical" effects of fragmentation that have been witnessed elsewhere. To help us better understand the effects of forest fragmentation in the west, we have attempted to outline the important issues in need of consideration and have suggested potential study approaches.

Acknowledgments. Our research on fragmentation in the interior Pacific Northwest has been supported by the National Fish and Wildlife Foundation, Boise Cascade Corporation, Potlatch Corporation, Plum Creek Timber Company, Idaho Department of Fish and Game, the U.S. Forest Service, the U.S. Fish and Wildlife Service, the University of Idaho, Boise State University, Turnstone Ecological Research Associates, and the Sustainable Ecosystems Institute. C. K. Friers, J. M. Marzluff, and M. Restani provided constructive comments on a previous version of this paper.

LITERATURE CITED

Agee, J.K., 1993, *Fire ecology of Pacific northwest forests*. Island Press, Washington, D.C. 493pp.

Ambuel, B. and Temple S.A., 1983, Area-dependent changes in the bird communities and vegetation of southern Wisconsin forests. *Ecology* 64, 1057-1068.

Andrén, H., 1992, Corvid density and nest predation in relation to forest fragmentation: A landscape perspective. *Ecology* 73, 794-804.

Askins, R.A., 1995, Hostile landscapes and the decline of migratory birds. *Science* 67, 1956-1957.

Askins, R.A., Lynch J.F. and Greenberg R., 1990, Population declines in migratory birds in eastern North America. *Current Ornithology* 7, 1-57.

Askins, R.A. and Philbrick M.J., 1987, Effect of changes in regional forest abundance on the decline and recovery of a forest bird community. *Wilson Bulletin* 99, 7-21.

Bayne, E.M. and Hobson K.A., 1997, Comparing the effects of landscape fragmentation by forestry and agriculture on predation of artificial nests. *Conservation Biology* 11, 1418-1429.

Berg, A., 1992, Factors affecting nest-site choice and reproductive success of Curlews *Numenius arquata* on farmland. *Ibis* 134, 44-51.

Blake, J.G., 1991, Nested subsets and the distribution of birds on isolated woodlots. *Conservation Biology* 5, 58-66.

Blake, J.G. and Karr J.R., 1987, Breeding birds of isolated woodlots: area and habitat relationships. *Ecology* 68, 1724-1734.

Boag, D.A., Reebs S.G. and Schroeder M.A., 1984, Egg loss among spruce grouse inhabiting lodgepole pine forests. *Canadian Journal of Zoology* 62, 1034-1037.

Böhning-Gaese, K., Taper M.L. and Brown J.H., 1993, Are declines in North American insectivorous songbirds due to causes on the breeding range? *Conservation Biology* 7, 76-86.

Boise National Forest, 1993, *Snapshot in time: repeat photography on the Boise National Forest 1870 - 1992*. July 1993, U.S. Dept. Agriculture, Forest Service, Intermountain Region, Ogden, UT. 239pp.

Bolger, D.T., Alberts A.C., Sauvajot R.M., Potenza P., McCalvin C., Tran D., Mazzoni S. and Soulé M.E., 1997, Response of rodents to habitat fragmentation in coastal southern California. *Ecological Applications* 7, 552-563.

Brittingham, M.C. and Temple S.A., 1983, Have cowbirds caused forest songbirds to decline? *BioScience* 33, 31-35.

Crane, M.F. and Fischer W.C., 1986, *Fire ecology of the forest habitat types of central Idaho*. U.S. For. Serv. Gen. Tech. Rep. INT-218. 86pp.

Daubenmire, R. and Daubenmire J.B., 1968, *Forest vegetation of Eastern Washington and Northern Idaho*. Washington Agricultural Experiment Station Technical Bull 60. College of Agriculture, Washington State University, Pullman, WA. 104pp.

DeGraaf, R.M., 1992, Effects of even-aged management on forest birds at northern hardwood stand interfaces. *Forest Ecology and Management* 46, 95-110.

Dobkin, D.S., and Wilcox, B.A., 1986, Analysis of natural forest fragments: riparian birds in the Toiyabe Mountains, Nevada. *Wildlife 2000: Modeling habitat relationships of terrestrial vertebrates* (J. Verner, M.L. Morrison and C.J. Ralph, eds.), University of Wisconsin Press, Madison, Wisconsin, pp. 293-299.

Donovan, T.M., Jones P.W., Annand E.M. and Thompson III F.R., 1997, Variation in local-scale edge effects: mechanisms and landscape context. *Ecology* 78, 2064-2075.

Donovan, T.M., Thompson III F.R., Faaborg J. and Probst J.R., 1995, Reproductive success of migratory birds in habitat sources and sinks. *Conservation Biology* 9, 1380-1395.

Fahnestock, G.R. and Agee J.K., 1983, Biomass consumption and smoke production by prehistoric and modern fires in western Washington. *Journal of Forestry* 62, 799-805.

Freemark, K.E. and Merriam H.G., 1986, Importance of area and habitat heterogeneity to bird assemblages in temperate forest fragments. *Biological Conservation* 36, 115-141.

Fuller, R.J., Gregory R.D., Gibbons D.W., Marchant J.H., Wilson J.D., Baillie S.R. and Carter N., 1995, Population declines and range contractions among lowland farmland birds in Britain. *Conservation Biology* 9, 1425-1441.

Galbraith, H., 1988, Effects of agriculture on the breeding ecology of Lapwings *Vanellus vanellus*. *Journal of Applied Ecology* 25, 487-503.

Galli, A.E., Leck C.F. and Forman R.T.T., 1976, Avian distribution patterns in forest islands of different sizes in central New Jersey. *The Auk* 93, 356-364.

Gates, S., Gibbons D.W., Lack P.C. and Fuller R.J., 1994, Declining farmland bird species: modeling geographical patterns of abundance in Britain. *Large-scale ecology and conservation biology* (P.J. Edwards, R.M. May and N.R. Webb, eds.), Blackwell Science, Oxford, pp. 153-177.

Giles, Jr. R.H., 1978, *Wildlife Management*. Freeman, San Francisco. 416pp.

Green, J., 1995, *Birds and forests: a management guide*. Minnesota Dept. of Nat. Res., St. Paul, MN.

Hagan, J.M., Vander Haegen W.M. and McKinley P.S., 1996, The early development of forest fragmentation effects on birds. *Conservation Biology* 10, 188-202.

Haila, Y., Hanski I.K. and Raivio S., 1989, Methodology for studying the minimum habitat requirements of forest birds. *Annales Zoologici Fennici* 26, 173-180.

Hanski, I.K., Fenske T.J. and Niemi G.J., 1996, Lack of edge effect in nesting success of breeding birds in managed forest landscapes. *The Auk* 113, 578-585.

Harris, L.D., 1984, *The fragmented forest*. University of Chicago Press, Chicago, IL. 211pp.

Harris, L.D. and Silva-Lopez G., 1992, Forest fragmentation and the conservation of biological diversity. *Conservation biology, the theory and practice of nature conservation, preservation, and management* (P. Fiedler and S. Jain, eds.), Chapman and Hall, New York, pp. 197-237.

Haufler, J.B., 1998, A strategy for bird research in forested ecosystems of the western U.S.. *Avian Conservation: Research and Management* (J.M. Marzluff and R. Sallabanks, eds.), Island Press, Covelo, CA, pp. 220-229.

Haufler, J.B., Mehl C.A. and Roloff G.J., 1996, Using a coarse-filter approach with species assessment for ecosystem management. *Wildlife Society Bulletin* 24, 200-208.

Hejl, S.J., Hutto R.L., Preston C.R. and Finch D.M., 1995, Effects of silvicultural treatments in the Rocky Mountains. *Ecology and management of Neotropical migratory birds: A synthesis and review of critical issues* (T.E. Martin and D.M. Finch, eds.). Oxford University Press, New York, pp. 220-244.

Hoover, J.P., Brittingham M.C. and Goodrich L.J., 1995, Effects of forest patch size on nesting success of Wood Thrushes. *The Auk* 112, 146-155.

Howlett, J.S. and Stutchbury B.J., 1996, Nest concealment and predation in Hooded Warblers: Experimental removal of nest cover. *The Auk* 113, 1-9.

Huhta, E., Jokimaki J. and Helle P., 1998, Predation on artificial nests in a forest dominated landscape - the effects of nest type, patch size, and edge structure. *Ecography* 21, 464-471.

Hunter, M.L., 1990, *Wildlife, forests, and forestry: principles of managing forests for biological diversity*. Prentice Hall Career and Technology, Englewood Cliffs, NJ. 370pp.

Johannesen, E. and Ims R.A., 1996, Modeling survival rates: habitat fragmentation and destruction in root voles experimental populations. *Ecology* 77, 1196-1209.

King, D.I., Griffin C.R. and DeGraaf R.M., 1996, Effects of clearcutting on habitat use and reproductive success of the Ovenbird in forested landscapes. *Conservation Biology* 10, 1380-1386.

Lemkuhl, J.F., Ruggiero L.E. and Hall P.A., 1991, Landscape patterns of forest fragmentation and wildlife richness and abundance in the southern Washington Cascade range. *Wildlife and vegetation of unmanaged Douglas-fir forests* (L.F. Ruggiero, K.B. Aubry, A.B. Carey and M.H. Huff, tech. coords.), U.S. For. Serv. Gen. Tech. Rep. PNW-285, pp. 425-442.

Lord, J.M. and Norton D.A., 1990, Scale and the spatial concept of fragmentation. *Conservation Biology* 4, 197-202.

Martin, T.E., 1992, Landscape considerations for viable populations and biological diversity. *Transactions of the North American Wildlife and Natural Resources Conference* 57, 283-291.

Martin, T.E., 1993a, Nest predation among vegetation layers and habitat types: Revising the dogmas. *American Naturalist* 141, 897-913.

Martin, T.E., 1993b, Nest predation and nest sites: New perspectives on old patterns. *BioScience* 43, 523-532.

McGarigal, K. and W.C. McComb, 1995, Relationships between landscape structure and breeding birds in the Oregon Coast Range. *Ecological Monographs* 65, 235-260.

Morgan, P., Aplet G.H., Haufler J.B., Humphries H.C., Moore, M.M. and Wilson W.D., 1994, Historical range of variability: a useful tool for evaluating ecosystem change. *Journal of Sustainable Forestry* 2, 87-112.

Murcia, C., 1995, Edge effects in fragmented forest: implications for conservation. *Trends in Ecology and Evolution* 10, 58-62.

Newton, I., 1998, Bird conservation problems resulting from agricultural intensification in Europe. *Avian Conservation: Research and Management* (J.M. Marzluff and R. Sallabanks, eds.). Island Press, Covelo, CA, pp. 307-322.

Nilsson, S.G. and Ericson L., 1992, Conservation of plant and animal populations in theory and practice. *Ecological Principles of Nature Conservation* (L. Hansson, ed.). Elsevier Applied Science, London, pp. 71-112.

Paton, W.C., 1994, The effect of edge on avian nesting success: How strong is the evidence? *Conservation Biology* 8, 17-26.

Peterjohn, B.G., Sauer J.R. and Robbins C.S., 1995, Population trends from the North American Breeding Bird Survey. *Ecology and management of Neotropical migratory birds: A synthesis and review of critical issues* (T.E. Martin and D.M. Finch, eds.). Oxford University Press, New York, pp. 3-39.

Pickett, S.T.A. and J.N. Thompson, 1978, Patch dynamics and the design of nature reserves. *Biological Conservation* 32, 335-353.

Ratti, J.T. and Reese K.P., 1988, Preliminary test of the ecological trap hypothesis. *Journal of Wildlife Management* 52, 484-491.

Robinson, S.K., 1992, Population dynamics of breeding Neotropical migrants in a fragmented Illinois landscape. *Ecology and conservation of neotropical migratory landbirds* (J.M. Hagan III and D.W. Johnston, eds.). Smithsonian Institution Press, Washington, D.C., pp. 408-418.

Robinson, S.K., Thompson III F.R., Donovan T.M., Whitehead D.R. and Faaborg J., 1995a, Regional forest fragmentation and the nesting success of migratory birds. *Science* 267, 1987-1990.

Robinson, S.K., Rothstein S.I., Brittingham M.C., Petit L.J. and Gryzbowski J.A., 1995b, Ecology and behavior of cowbirds and their impact on host populations. *Ecology and management of Neotropical migratory birds: A synthesis and review of critical issues* (T.E. Martin and D.M. Finch, eds.). Oxford University Press, New York, pp. 428-460.

Rodenhouse, N.L., Best L.B., O'Connor R.J. and Bollinger E.K., 1995, Effects of agricultural practices and farmland structures. *Ecology and management of Neotropical migratory birds: A synthesis and review of critical issues* (T.E. Martin and D.M. Finch, eds.). Oxford University Press, New York, pp. 269-293.

Rosenburg, K.V. and Raphael, M.G., 1986, Effects of forest fragmentation on wildlife communities of Douglas-fir. *Wildlife 2000: Modeling habitat relationships of terrestrial vertebrates* (J. Verner, M.L. Morrison and C.J. Ralph, eds.), University of Wisconsin Press, Madison, Wisconsin, pp. 263-272.

Rudnicky, T.C. and Hunter M.L., 1993, Avian nest predation in clearcuts, forests, and edges in a forest-dominated landscape. *Journal of Wildlife Management* 57, 358-364.

Runkle, J.R., 1985, Disturbance regimes in temperate forests. *The ecology of natural disturbance and patch dynamics* (S.T.A. Pickett and P.S. White, eds.). Academic Press, New York, pp. 17-33.

Sallabanks, R., Arnett E.B., Wigley T.B. and Irwin L.L., *In press*, Accommodating birds in managed forests of North America: A review of bird-forestry relationships. *NCASI Technical Bulletin*, No. 000.

Sallabanks, R., Walters J.R. and Collazo J.A., *Submitted manuscript a*, Breeding bird abundance in bottomland hardwood forests: habitat, edge, and patch size effects. *Wilson Bulletin*.

Sallabanks, R., Haufler J.B. and Mehl C.A., *Submitted manuscript b*, Forest birds of the Southern Idaho Batholith landscape: A demonstration of an ecosystem diversity matrix. *Ecological Applications*.

Sauer, J.R. and Droege S., 1992, Geographical patterns in population trends of Neotropical migrants in North America. *Ecology and conservation of neotropical migratory landbirds* (J.M. Hagan III and D.W. Johnston, eds.). Smithsonian Institution Press, Washington, D.C., pp. 26-42.

Saunders, D.A., Hobbs R.J. and Margules C.R., 1991, Biological consequences of ecosystem fragmentation: a review. *Conservation Biology* 5, 18-32.

Schieck, J., Lertzman K., Nyberg B. and Page R., 1995, Effects of patch size on birds in old-growth montane forests. *Conservation Biology* 9, 1072-1084.

Schmiegelow, F.K., Machtans C.S. and Hannon S.J., 1997, Are boreal birds resilient to forest fragmentation? An experimental study of short-term community responses. *Ecology* 78, 1914-1932.

Steele, R., 1994, The role of succession in forest health. *Journal of Sustainable Forestry* 2, 183-190.

Steele, R., Pfister R.D., Ryker R.A. and Kittams J.A., 1981, *Forest habitat types of central Idaho*. U.S. For. Serv. Gen. Tech. Rep. INT-114.

Suarez, A., Pfenning K. and Robinson S.K., 1997, Nesting success of a disturbance-dependent songbird on different kinds of edges. *Conservation Biology* 11, 928-935.

Tewksbury, J.J., Hejl S.J. and Martin T.E., 1998, Breeding productivity does not decline with increasing fragmentation in a western landscape. *Ecology*, 79, 2890-2903.

Thomas, J.W., Maser C. and Rodiek J.E., 1979, Edges. *Wildlife habitats in managed forests: the Blue Mountains of Oregon and Washington* (J.W. Thomas, ed.). USDA, Forest Service, Agricultural Handbook 553, U.S. Government Printing Office, Washington, D.C., pp. 48-59.

Thompson III F.R., Dijak W.D., Kulowiec T.G. and Hamilton D.A., 1992, Breeding bird populations in Missouri Ozark forests with and without clearcutting. *Journal of Wildlife Management* 56, 23-30.

Thompson III F.R., Donovan T.M., Robinson S.K., Faaborg J. and Whitehead, D.R. *In press*, Biogeographic, landscape and local factors affecting cowbird abundance and host parasitism levels. *The ecology and management of cowbirds* (T.L. Cook, S.K. Robinson, S.I. Rothstein and S.G. Sealy, eds.). University of Texas Press, Austin.

Van Horn, M.A., Gentry R.M. and Faaborg J., 1995, Patterns of ovenbird (*Seiurus aurocapillus*) pairing success in Missouri forest tracts. *The Auk* 112, 98-106.

Warner, R.E., 1994, Agricultural land use and grassland habitat in Illinois: Future shock for midwestern birds? *Conservation Biology* 8, 147-156.

Whitcomb, R.F., Robbins C.S., Lynch J.F., Whitcomb B.L., Klimkiewicz M.K. and Bystrak D., 1981, Effects of forest fragmentation on avifauna of the eastern deciduous forest. *Forest island dynamics in man-dominated landscapes* (R.L. Burgess and D.M. Sharpe, eds.). Springer-Verlag, New York, pp. 125-206.

Wiens, J.A., 1995, Habitat fragmentation: Island v landscape perspectives on bird conservation. *Ibis* 137, S97-S104.

Wilcove, D.S., 1985, Nest predation in forest tracts and the decline of migratory songbirds. *Ecology* 66, 1211-1214.

Yensen, E. and Sherman, P.W., 1997, *Spermophilus brunneus*. *Mammalian Species* 560, 1-5.

Forest-level Effects of Management on Boreal Songbirds: the Calling Lake Fragmentation Studies

Fiona K.A. Schmiegelow[1] and Susan J. Hannon[2]

The advent of industrial logging in Canada's boreal mixedwood forest has raised concerns about losses of wildlife habitat. Although habitat fragmentation has been implicated in the decline of many North American songbird populations, most studies to date have been constrained by limitations of scale, and lack of an experimental approach. Over the past six years (1993-98), we have studied the effect of fragmentation on avian communities in older, mixedwood forest using both experimental and descriptive approaches. Our study area encompassed ca. 400 km^2 of forest in north-central Alberta, Canada. A fragmentation experiment created isolated and connected forest fragments of 1, 10, 40 and 100 ha, each replicated three times, matched with controls. We sampled the bird community in these areas before, and in each of 5 years after, forest harvesting. Movement of birds through the riparian buffer strips connecting reserves was monitored to assess the efficacy of these potential corridors. Artificial nest experiments were conducted along clearcut edges, in forest fragments and in forest interiors. A study of naturally-isolated patches of similar habitat was used to assess longer-term vulnerability of certain species to fragmentation by forest harvesting. Finally, a study of two, adjacent, harvested landscapes compared patch vs. landscape-scale interpretations of fragmentation effects. Our results highlight the role that context plays in influencing fragmentation effects, the value of adopting an experimental approach to assess impacts, and the dangers of basing management recommendations on short-term studies.

Key words: boreal mixedwood forest, songbirds, experimental fragmentation, nest predation, connectivity, landscape-scale

1. INTRODUCTION

Due to the demands of burgeoning human populations, and our increasing efficiency at exploiting resources, habitat fragmentation has become one of the most ubiquitous, and visible, ecological problems, contributing to concerns about a global collapse of biological diversity. In agricultural and urban landscapes, permanent conversion of the matrix surrounding habitat fragments severely constrains options for species occupying remnant patches. There are more opportunities in forested landscapes to incorporate scientific knowledge into land-management decisions. For example, the potentially negative effects of harvesting-induced fragmentation on the diversity and resilience of forest ecosystems might be reduced if harvesting occurs at scales similar to the dominant natural disturbances in the area (e.g. Hunter 1992, Haila et al. 1994, Bunnell 1995). Concerns about the long-term sustainability of managed, forested ecosystems have resulted in demands for forest

---

[1] Department of Renewable Resources, University of Alberta, Edmonton, AB Canada T6G 2H1
[2] Department of Biological Sciences, University of Alberta, Edmonton, AB Canada T6G 2E9

companies to develop management plans that incorporate maintenance of non-timber values. Nevertheless, because uncertainty is often viewed as justification for inaction (e.g. Campbell 1996), researchers must demonstrate a need and options for alternative harvesting practices, if changes in policy are to take place.

Studies of habitat fragmentation to date have been constrained by limitations of scale, and by the lack of an experimental approach. Communities of species in fragmented habitats have seldom been studied before fragmentation. The conclusions of such studies therefore rest on several untested assumptions. For example, habitat fragmentation has been implicated as a major factor in the decline of many North American songbird populations (e.g. Wilcove and Robinson 1990, papers in Hagan and Johnston 1992). For Neotropical migrant species, however, there has been considerable debate over whether habitat changes in breeding or wintering areas have played a greater role (e.g. Terborgh 1989;1992, Böhning-Gaese et al. 1993, Sherry and Holmes 1995). Some researchers have even questioned whether populations have actually declined (e.g. James and McCulloch 1995). A common problem has been an inability to identify causal factors through controlled experimentation at appropriate scales. Small-scale experimental model systems, though of heuristic value, have yielded little information that can be applied to the challenge of maintaining sensitive species in real landscapes.

Forest management planning takes place over very large spatial and over moderately long temporal scales. Harvesting of trees over extensive areas can radically change landscape patterns. Given favorable socio-economic and political climates, research through management is possible at scales commensurate with the level at which land-management decisions are made. In 1993, we described the opportunities for such research in the western boreal forest, much of which had just been allocated for industrial forestry (Schmiegelow and Hannon 1993). We outlined the development of a large, experimental fragmentation study in the boreal mixedwood forest of Northern Alberta. The Calling Lake Fragmentation Project was designed to assess the outcome of the current harvesting guidelines on fragmentation of older mixedwood forest, and the related effects on avifauna in the region, and to explore alternative management options. Short-term results from this research, i.e. 1-2 years after harvesting of our study area, have been reported elsewhere (Machtans et al. 1996, Schmiegelow 1997, Schmiegelow et al. 1997). Here, we summarise these findings, and present new results and conclusions from data collected 5 years after isolation of our experimental sites. We also present overviews of complementary, contemporary studies conducted in adjacent areas. We conclude with a discussion of the implications of our work to management of the boreal mixedwood forest, and recommendations for future research directions in this area.

2. BACKGROUND

2.1 The Boreal Mixedwood Forest

The Boreal Mixedwood Ecoregion forms a broad band north of the aspen parkland and south of the northern coniferous forest, primarily in the provinces of Alberta and Saskatchewan (Rowe, 1972; Fig. 1). In Alberta, it covers an area of 290,000 km2 ; about 40% of the province. The boreal mixedwood forest is a natural mosaic of stands of different ages and species composition interspersed with peatlands, other wetlands, and lakes and streams. In upland mesic habitats, the dominant tree species are trembling aspen (Populus tremuloides) and white spruce (Picea glauca), occurring most commonly as mixed stands, but also as pure stands. Black spruce (Picea mariana), balsam poplar (Populus balsamifera), paper birch (Betula papyrifera) and tamarack (Larix laricina) dominate wetter sites. Jack pine (Pinus banksiana) is found primarily in xeric sites; balsam fir (Abies balsamea) is relatively less common.

2.2 Natural and Anthropogenic Disturbances

Disturbances such as insect outbreaks and fire have been major historical forces promoting the mosaic found in the boreal Mixedwood forest. Forest tent caterpillar (Malacosoma disstria) is the main herbivore of deciduous trees in this region, but rarely destroys entire stands (Peterson and Peterson 1992). Blow downs caused by windstorms are important on a small scale. The major natural disturbance agent in the boreal mixedwood forest is fire (Johnson 1992). Fires are episodic, with large fires occurring periodically. Estimates of fire cycles in the mixedwood vary from approximately 60 (Murphy 1985) to 240 years (Cumming 1997), and there is considerably controversy over actual rates (Armstrong et al. in press). Fire suppression and changes in land-use practices have altered the natural frequency and intensity of insect outbreaks and fire (e.g. Murphy 1985, Roland 1993, Cumming 1997).

The boreal forest in Alberta has been increasingly impacted by anthropogenic disturbance. Clearing for agriculture is prevalent along the southern fringe and in the Peace River area. Exploration for, and development of, oil and gas deposits has resulted in over 9,000km2 of boreal forest being cleared for well sites, cut lines for seismic activities, pipelines and access roads; planned subsurface extraction of oil from tar sands may deforest another 3,450km2 (Alberta Environmental Protection 1998). Small to medium-scale harvesting of white spruce for sawlogs has a long history in some areas, but broad-scale harvest of aspen dates back to only 1992. Previously, aspen was considered a "weed" tree by foresters and considerable effort and expense was allocated to reducing its abundance locally. Now, however, aspen is economically important in the production of pulp and paper. Thus, the pure aspen and aspen-dominated mixedwood forests are coming under increasing pressure from logging companies. As of November 1997, 40% of the boreal forest (and >75% of the mixedwood) had been allocated to logging companies under Forest Management Agreements with the provincial government.

Mature (50-80 yr) and old (>80 yr) aspen forests are slated to be cut first. The rotation period will be 40-70 years (Alberta Energy/Forestry, Lands and Wildlife 1992), so that few stands of aspen will reach the "old" stage. The current operating ground rules call for an average cutblock size of 40 ha (maximum 60 ha) with a two- or three-pass clear-cutting system. The second pass will occur when trees on the first pass are about 3-m tall. This will result in high fragmentation of the forest, high edge/area ratios in the remaining uncut portions of the forest, and a lack of large continuous stands of older aspen. The ground rules also specify leave strips 100-m wide around permanent lakes and 30-m wide along streams.

Potential impacts of this harvesting on ecosystem processes and biodiversity are unknown, but since the harvest is expected to proceed on a 70-year rotation, old aspen-dominated mixedwood forests will become rare and increasingly fragmented. For this reason, our studies focussed on that forest type. Further, old aspen forests are structurally unique compared to younger stands. They have more canopy gaps, more developed understory, have accumulated deadwood (snags and downed woody debris), and have a mix of tree species and ages (Stelfox 1995). Old stands have higher species richness than young or mature forests, and several bird species occur at their highest abundances in old forests (Schieck et al.1995).

3. STUDY AREA AND DESIGN

We conducted this research near Calling Lake, in north-central Alberta, Canada (55° N, 113° W). Our study area encompassed ca. 400 km2 of boreal mixedwood forest and was comprised of 4 contiguous townships (each ~10x10km) (Fig. 1). The northwest township remained unharvested for the duration of the study, the southwest and northeast townships experienced the first pass of a two-pass clear-cut harvest in the winters of 1993/1994 and 1994/95, respectively. Merchantable forest within these townships ("leave blocks") will be harvested in a second pass in approximately 10 years time. Cutblocks varied in size from 9 to 59 ha, were a maximum of 400-m wide and were long and narrow with irregular boundaries. Up to 10% of trees were left standing on the cutblocks, either singly or in clumps. Due to the presence of leave blocks and unmerchantable areas (water bodies,

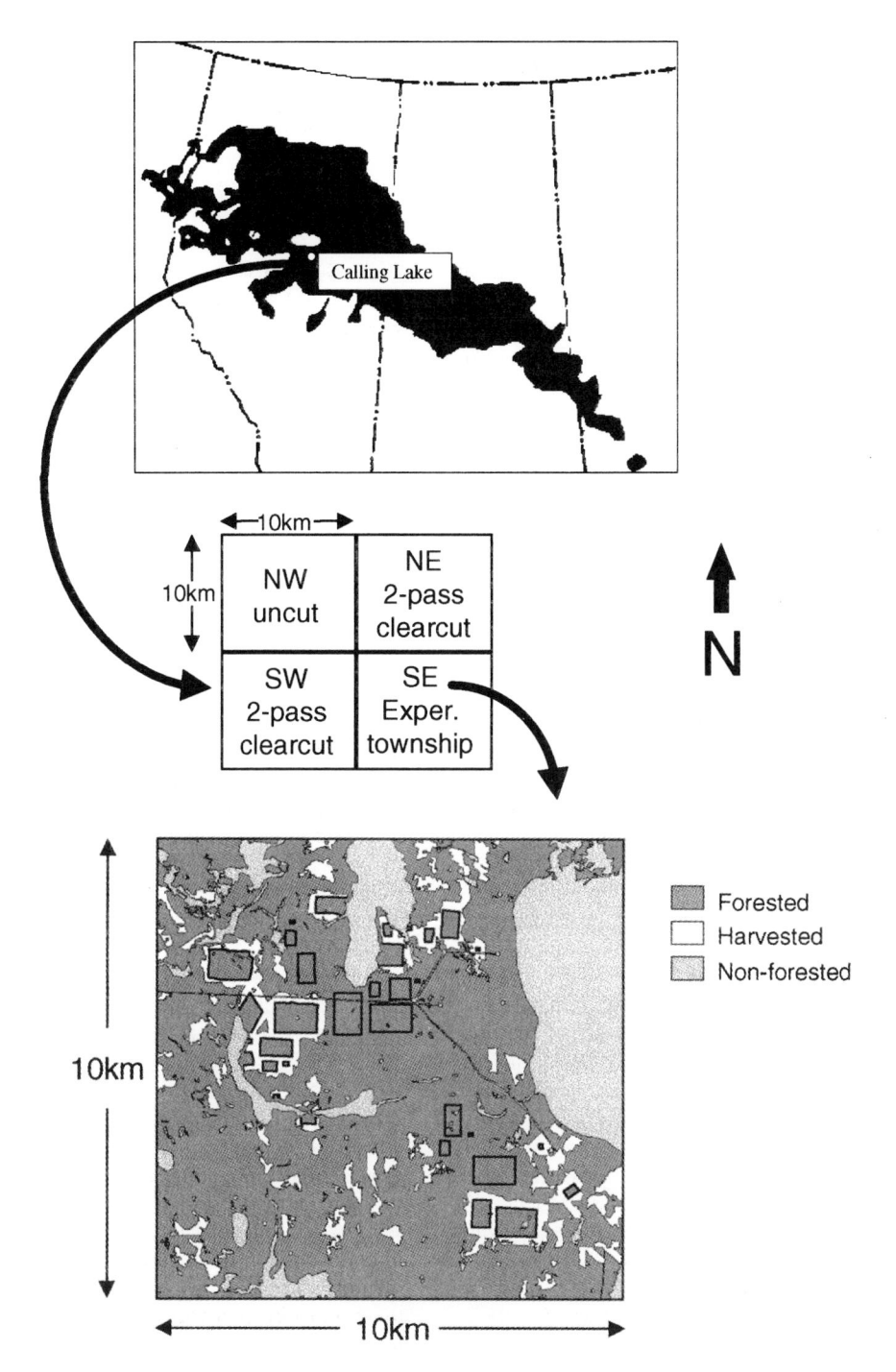

Figure 1. Study area near Calling Lake, Alberta showing the extent of the boreal mixedwood (black) in Alberta, Saskatchewan and Manitoba, the 4 townships which comprise the study area, and an enlargement of the experimental township (dark gray=forest, light gray=water, white=clearcut). Control sites within the continuous forest are outlined in black.

wetlands, black spruce bogs), approximately 13.4 and 10.9% of the total area of the southwest and northeast townships was logged.

The southeast township (hereafter called the experimental township) was experimentally harvested in the winter of 1993-94, between November and March.

The experimental design involved two treatments: isolated and connected forest fragments, with common controls. Isolates were created by clear-cut logging a 200 m wide strip around forest patches. A minimum isolation distance of 200-m was chosen because previous studies have demonstrated that 100-m or less can act as a barrier to bird species sensitive to fragmentation (Soule et al. 1988, Bierregard and Lovejoy 1989). Connected patches of forest were isolated by 200 m of clear-cutting on three sides, with the fourth side connected to 100-m wide riparian buffer strips (Fig. 1). Isolated forest fragments were 1, 10, 40 and 100 ha in size; connected fragments were 1, 10 and 40 ha in size. We did not include any 100 ha connected fragments because sufficient, suitable forest adjacent to riparian areas was not available. Paired controls of 1, 10, 40 and 100 ha were placed within ca. 4000 ha of continuous, adjacent forest. Each size class was replicated three times, within each treatment and control, as suggested by a priori power analyses (Schmiegelow and Hannon 1993). All sampling sites were in old (80-130 years), aspen-dominated forest, similar in canopy height, canopy closure, tree species composition and understory features. For the experiment, variation in these features was stratified across replicate groups and size classes, within treatments and controls. For example, replicates in group 1 occurred in the youngest forest (80-100 yrs); replicates in group 2 were in older, relatively pure aspen forest (100-130 yrs), and replicates in group 3 were in older aspen forest (100-130 yrs) with some white spruce in the canopy. Each replicate group was represented by one site in each size class, in each treatment or control.

The experimental design we adopted was based largely on planned forestry operations, which dictated the spatial scale at which our research was conducted. For the experimental fragmentation study, sites of 1, 10, 40 and 100 ha were chosen to bracket the mean cutblock size (40 ha) required by provincial regulations. Whether this scale corresponds to the ecological processes underlying patterns in the bird communities we studied is not known, but, given that forest harvesting is now the major disturbance in this region, it seems reasonable to study the system at a scale commensurate with these activities. Replication level was determined statistically (see Schmiegelow and Hannon 1993). Although the sites selected were as similar as possible with respect to habitat characteristics, some differences did exist, and these were most apparent in the smaller study sites. Results presented in this paper span the period from 1993 (pre-harvest in the experimental and southwest township) through 1998.

4. RESULTS FROM THE FRAGMENTATION EXPERIMENT

Point counts were used to document breeding bird community patterns in the experimental township. Permanent sampling stations were located on fixed grids, at 200-m intervals, resulting in a total of 93 stations in control sites and in isolated fragments, and 33 stations in connected fragments. Our use of a 100-m sampling radius in all but the 1-ha sites resulted in the area sampled representing approximately 63% of the actual area of each study site. In each year, each point count station was sampled 5 times during the breeding season, from mid-May to late June, following a standardised protocol, with annual means representing relative abundance (see Schmiegelow et al. 1997). Our earlier reporting used data from all 5 sampling rounds. For the more recent analyses presented here, we used data from only the last 4 rounds of sampling, when most migratory species had arrived, and individuals had settled on territories. Results are discussed for passerines only. Tests for treatment effects incorporate before-after and treatment-control comparisons.

4.1 Short-term Effects

Short-term results from the fragmentation experiment, i.e. 1-2 years immediately after harvesting, have been reported elsewhere (e.g., Machtans et al. 1996, Schmiegelow 1997, Schmiegelow et al. 1997). Here, we summarise the main findings.

4.1.1 Summer

Two years after harvest, there were some significant changes in the breeding bird community in response to fragmentation (Schmiegelow et al. 1997). Species richness did not decline in either the isolated or connected fragments. However, species turnover in isolated fragments, a measure of change in community composition, was higher than in control sites and connected fragments, indicating that species replacement had occurred. Diversity in the 1 and 10 ha isolates was reduced, resulting in a significant diversity-area relationship for the isolated fragments. We documented some crowding effects (sensu Whitcomb et al. 1981) of Neotropical and short-distance migrants in the isolated fragments 1 year after harvest, but by the second year, Neotropical migrants were 6-18% less abundant in isolated and connected fragments, with some species in this group experiencing absolute population declines of 10-60%. Differences in numbers of short-distance migrants were no longer significant. The strongest response to fragmentation was exhibited by resident birds, which were 32-38% less abundant in isolated and connected patches of forest in the second year following harvest. This effect was most pronounced in small, isolated fragments (≤ 10 ha), from which resident species became locally extinct. We related variation in short-term response between birds with different migratory strategies to breeding phenology, and degree of habitat specialization (Schmiegelow et al. 1997).

The presence of 100-m wide strips of forest joining adjacent patches of forest helped to mitigate the negative effects of fragmentation on breeding birds observed in the completely isolated fragments, although differences were small. Higher variation in community metrics in the connected fragments, due to a smaller range of fragment sizes (1-40 ha only), reduced our ability to detect effects, so conclusions should be drawn with caution.

4.1.2 Winter

Birds were censused 3 times each winter, in 1993, 1994 and 1995, using fixed-width transects that overlapped the point count stations used for summer surveys. Transect sampling is more appropriate than point counts in winter, as birds are not singing and are more mobile. The wintering bird communities in fragments of older, deciduous-dominated forest also changed after fragmentation (Schmiegelow 1997). The response of resident passerines varied with fragment size. There were too few observations from 1 and 10 ha sites for meaningful comparisons, but numbers in the 40 ha isolated fragments declined by approximately 45% following fragmentation, while numbers in the 100 ha fragments remained stable, and numbers in the controls increased. Resident species used the 100-m wide riparian buffer strips in winter, perhaps explaining why numbers were maintained in the connected fragments. The presence of mature coniferous trees (primarily white spruce) was also an important factor in winter habitat selection among all resident species. This relationship probably accounts for the seasonally increased area requirements of resident passerines we observed, as these birds depend on mature conifers for both food and shelter in winter, and the total amount of white spruce in older, deciduous-dominated forest generally increases with area (Schmiegelow, unpub. data).

4.1.3 Conclusions

In a similar study in tropical forest fragments (Stouffer and Bierregaard 1995), isolation by as little as 100 m led to immediate losses of some species from 1 and 10 ha sites. However, as these authors point out, there is no seasonal migration of breeding birds into these areas, so species persistence is largely dependent on delaying extinction by maintaining larger populations. In contrast, we did not document widespread species loss immediately following fragmentation. Whether this was due to (1) insensitivity to fragmentation, (2) delayed responses, or (3) local extinctions being offset by recolonizations is not clear. For resident passerines, some species loss and substantial abundance declines were apparent two years after fragmentation, consistent with the results for tropical forest residents. For Neotropical migrants, we did not detect species loss, and abundance declines were less severe. This pattern is consistent with (2) and (3), above. However, delayed responses to forest fragmentation by birds breeding in temperate areas, followed by gradual, long-term population declines of habitat specialists have been predicted by Hagan et al. (1996).

4.2 Mid-term Effects

Constraints on the generality of the short-term results from our experimental work are that the patterns evident in patches of habitat shortly after the surrounding forest was cut may not be representative of the longer-term effects of habitat fragmentation on these bird communities, and that large amounts of suitable habitat were still present regionally. Over time, due both to lags in ecological response (Wiens 1989a), and changes to the surrounding landscape (e.g. McGarigal and McComb 1995, Robinson et al. 1995), other fragmentation effects may become apparent. A complementary study of the bird communities in patches of older, aspen-dominated forest naturally isolated from one another by a surrounding matrix of black spruce (Schmiegelow 1997) suggests that both Neotropical migrant and resident species may experience further population declines and local extinctions in the fragments created by forest harvesting, consistent with predictions by Hagan et al. (1996). Significantly fewer species and lower numbers of individuals were recorded in the natural isolates than in similar sized control sites within continuous forest, even though isolation distances were small (20-180 m), and there was a high abundance of aspen-dominated forest in the surrounding area.

There are several reasons why the response of some songbird species to habitat fragmentation by forest harvesting might be delayed. First, return of individuals present prior to fragmentation, due to site fidelity, could mask abundance effects until mortality is sufficient that reduced recruitment is apparent. Second, the development of secondary fragmentation effects such as competition, predation, parasitism and other edge effects will lag behind the fragmentation event to varying degrees. Five years after harvesting, the intervals period for which we now have response data, confounding effects such as crowding should have dissipated, and negative effects may be more discernible. We hereafter refer to these 5-year responses as mid-term, recognising that our choice of time is somewhat arbitrary.

4.2.1 Assessment of Risk

Using an objective set of criteria, we identified 6 species for which we expected different mid-term responses to experimental fragmentation at Calling Lake. Our assessment of risk was based on area-sensitivity, reflected by species distribution patterns in naturally-isolated mixedwood forest (Schmiegelow 1997), early evidence of population decline from the experimental fragmentation study (Schmiegelow et al . 1997), and reliance of the species on older forest (Schieck and Nietfeld 1995, Kirk et al. 1996, Schmiegelow unpubl. data). Species meeting all 3 criteria were assigned to a high-risk category, while those fulfilling 2 of 3 criteria were placed in a moderate-risk category. When 0 or only 1 criteria were satisfied, species were considered at low risk. We predicted that:

1) species considered at high risk, the Black-throated Green Warbler (*Dendroica virens*) and the Red-breasted Nuthatch (*Sitta canadensis*) would exhibit continued declines in the fragments, relative to the controls;

2) moderate-risk species, the Black-capped Chickadee (*Parus atricapillus*) and the Yellow Warbler (*Dendroica petechia*), should either remain at the population levels observed 2 years after harvesting, or decline further; and

3) trends in numbers of species considered at low risk, the White-throated Sparrow (*Zonotrichia albicolis*) and the Yellow-rumped Warbler (*Dendroica coronata*) should not differ among the fragments and controls.

We tested our predictions by comparing abundance data for these species from our fragments and controls, before harvesting, and 2 and 5 years after harvesting. We calculated contrast values for the abundance of our 6 target species at all point count stations at which they were recorded, for the 2 time intervals of interest: 1993 to 1995 (2 years after harvest), and 1993 to 1998 (5 years after har-

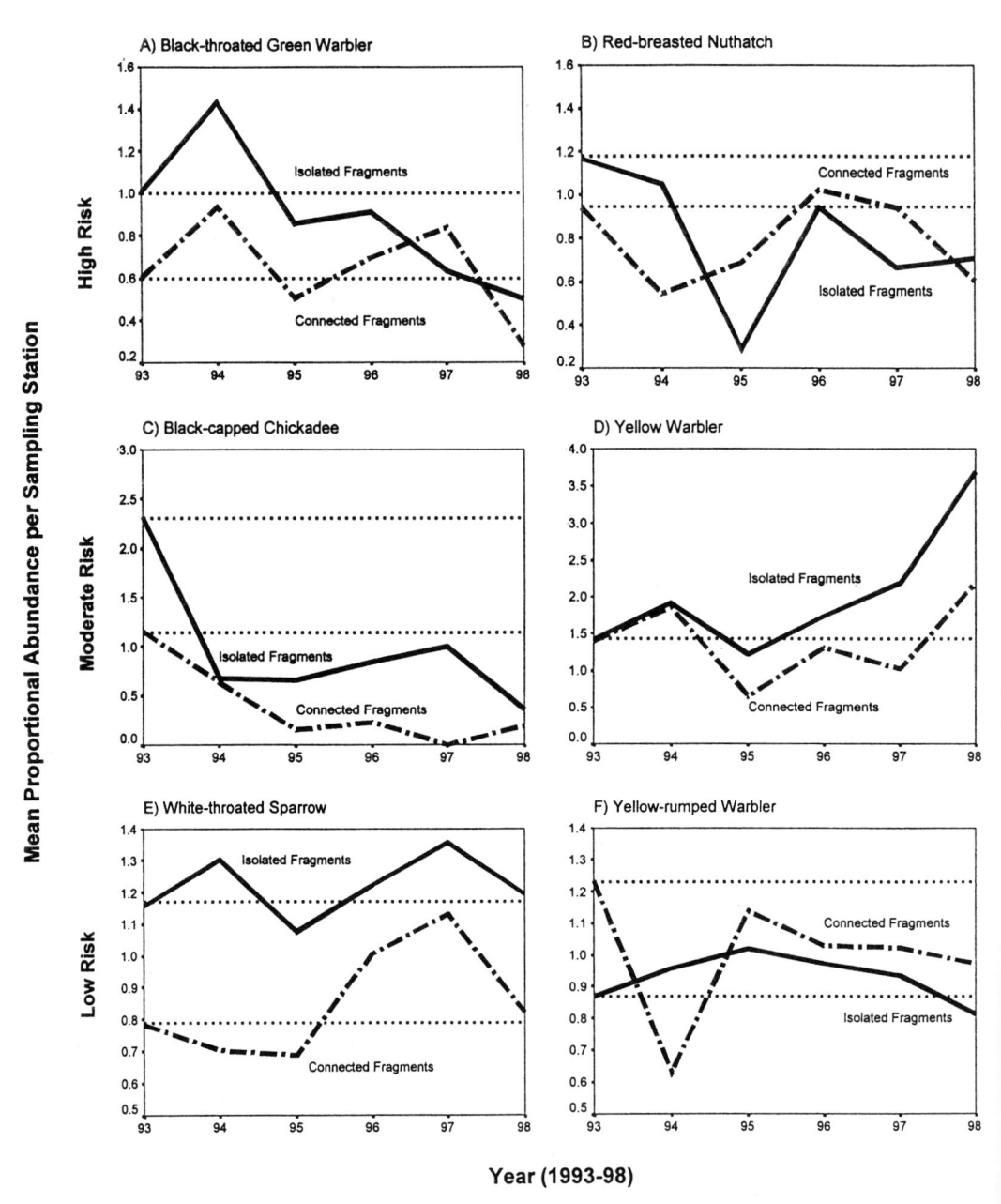

Figure 2. Abundance response of 6 bird species to experimental forest fragmentation 1-5 years after isolation by clearcut harvesting at Calling Lake, Alberta. Connected fragments (n=9; # sampling stations=33) were harvested on 3 sides but remained connected to riparian buffer strips. Isolated fragments (n=12; # sampling stations=93) were harvested on all 4 sides. Response was measured by the mean change in abundance of each species per station, relative to abundance changes within control sites (n=12; # sampling stations=93), and is represented as a proportional change that accounts for natural, annual variation.

vest). In this case, contrasts simply represent the absolute difference in abundance values over the specified time intervals. We then compared the mean values between isolated and connected fragments and controls, using pairwise t-tests.

4.2.2 Results

Species considered at high risk, the Black-throated Green Warbler (BGNW) and the Red-breasted Nuthatch (RBNU), showed significant, continued declines relative to the controls, consistent with our first prediction (BGNW - $p<0.0001$, $p=0.023$; RBNU - $p=0.030$, $p=0.023$, isolated and connected fragments, respectively) (Fig. 2A&B). Numbers of Black-throated Green Warblers were reduced by over 50% in both the isolated and connected fragments 5 years after harvesting. Absolute numbers of Red-breasted Nuthatches increased in all areas between 1993 and 1998, but the relative gains in the isolated and connected fragments were 40% lower than in the controls.

Results for species considered moderately at risk, the Black-capped Chickadee and the Yellow Warbler, varied (Fig. 2C&D). Black-capped Chickadee numbers remained depressed in the isolated ($p=0.003$) and connected ($p=0.009$) fragments, at roughly 45% below pre-fragmentation levels, consistent with our second prediction. Over the same time period, numbers of this species in the controls increased 4-fold. In contrast, numbers of Yellow Warblers declined in the controls relative to the isolated fragments ($p=0.030$), but were not different from the connected fragments ($p=0.634$). Yellow Warblers have colonised the regenerating cutblocks adjacent to the isolated fragments (Schmiegelow, unpub. data).

The response of species we considered low risk generally agreed with our third prediction. Abundance patterns of White-throated Sparrows (WTSP) and Yellow-rumped Warblers (YRWA) in the forest fragments either mirrored those in the controls, or were highly variable (Fig. 2E&F), with no significant abundance response to fragmentation apparent in either species 5 years after harvest (WTSP - $p=0.441$, $p=0.484$; YRWA - $p=0.584$, $p=0.261$, isolated and connected fragments, respectively. There is, however, a non-significant downward trend in Yellow-rumped Warblers in the connected fragments.

4.2.3 Conclusions

These mid-term results suggest that local populations of some species are still adjusting 5 years after harvesting of the experimental area and isolation of the fragments. A more thorough evaluation of all species must be conducted, as well as detailed analyses of community metrics, such as species richness, diversity and turnover, to assess whether certain species have been lost from fragmented areas, or community structure significantly altered. As well, consideration of factors that could affect reproductive success is necessary before further conclusions can be drawn.

5. PREDATION, PARASITISM AND AVIAN RESPONSES TO EDGE

Clutch predation and cowbird parasitism can greatly reduce reproductive output of birds breeding in fragmented forests (Robinson et al. 1995), particularly those nesting at forest edges (Andren 1995, Paton 1995). This "negative edge effect" may be more common in landscapes fragmented by agriculture or urbanization than in areas fragmented by forestry activities (Andren 1995, Donovan et al. 1997). In the Calling Lake studies, we used a two-pronged approach to examine the potential impacts of predation and nest parasitism. The first approach was to conduct artificial nest experiments (Cotterill, Hannon and Song). The second approach was to monitor nest success of a "target" species, the American Redstart (*Setophaga ruticilla*) in the experimental reserve township one year prior to logging and three years post-logging in control and isolated reserves (Villard and Hannon).

Artificial nest experiments were conducted in the southwest and northeast townships, away from the experimental reserves, where the landscapes had experienced the first pass of a 2-pass harvest. The experiments took place at three types of locations: 1) on transects that spanned leave blocks from edge to

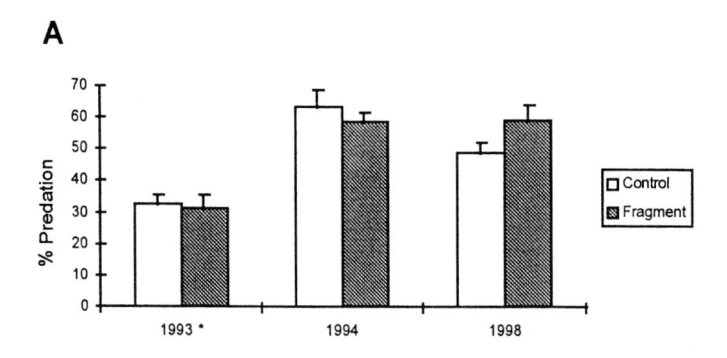

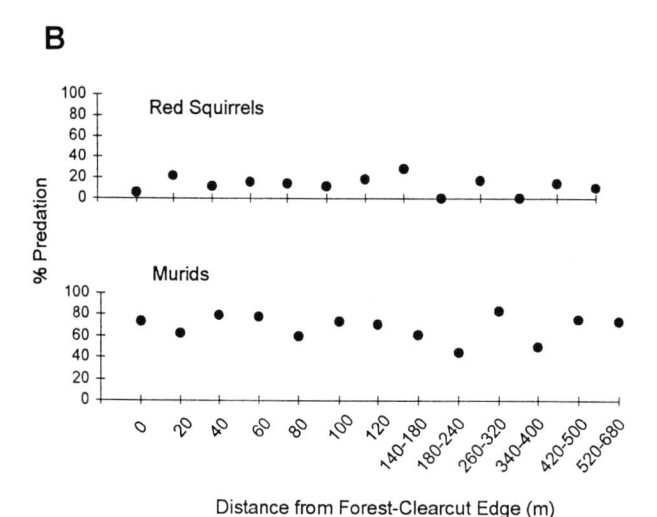

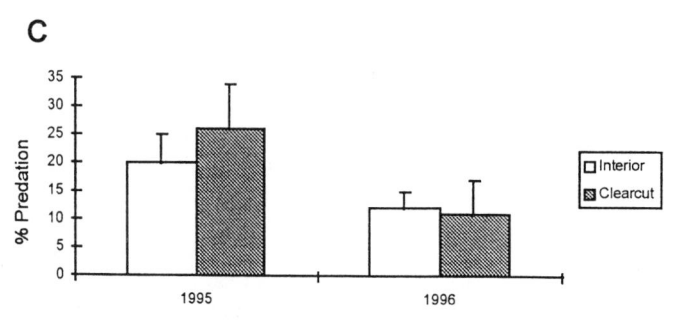

Figure 3. Percentage of artificial ground and shrub nests taken by predators at Calling Lake. A) % predation of nests in forest fragments and unharvested control blocks in 1993, 1994 and 1998 (*1993 data are from pre-harvest period). Repeated measures ANOVA indicated that year was significant (1993 lower than 1994, 1998, $F=26.88$, $P<0.0001$), but year ($F=0.17$, $P=0.68$) and year X treatment interaction were not significant ($F=1.76$, $P=0.18$). B) % predation by red squirrels and murid rodents of nests placed in forest at different distances from a cutblock edge in 1994 (from Cotterill 1996). G-test: red squirrels $G=8.86$, $P=0.72$; murids: $G=9.89$, $P=0.81$).
C) % predation of nests placed in forest along a clearcut edge and in the forest interior in 1995 and 1996. 2-way ANOVA indicated that year was significant ($F=4.25$, $P=0.048$), but not treatment ($F=0.16$, $P=0.69$) nor year X treatment interaction ($F=0.35$, $P=0.56$).

edge (i.e. adjacent to cutblocks), compared with similarly situated transects in unharvested forest, conducted 1 year prior to harvest and 1 and 5 years after harvest (Cotterill 1996, Hannon unpub. data); 2) on transects starting at the forest/cutblock edge and penetrating up to 680m into uncut forest (Cotterill 1996); and 3) on transects in forest along the forest/cutblock edge, compared with transects in interior forest (Song 1998, Song and Hannon, in press). In experiments 1) and 2), shrub and ground nests each contained a quail egg and a plasticine egg. In 3), shrub and ground nests contained a single quail egg.

Results from the three nest predation experiments were consistent. Harvesting did not increase nest predation rates (pre-harvest, post-harvest comparison Fig. 3A) and there did not appear to be an increase in predation at the forest/clearcut edge (Fig. 3B&C). Experiments 1 and 3 were multi-year experiments and indicated large year effects in predation rates, but no treatment effects. Predators (identified by marks in plasticine eggs) of ground nests were primarily murid rodents (Peromyscus or Clethrionomys) and red squirrels (Tamiasciurus hudsonicus) were the main identified predators of shrub nests (Cotterill 1996). However, many eggs from shrub nests were missing and hence the predators not identified.

The results of the redstart study mirror those of the nest predation studies: we found annual variation in predation rates (in the same direction as the artificial nest study of 1993/94) but no effects of treatment (i.e. nests in fragments were just as successful as those in continuous forest) (Villard and Hannon unpub. data). In addition, we found that cowbird parasitism was extremely low (only 4 of 245 nests over 4 years). The Calling Lake redstart population exhibits very few behavioural defense mechanisms against cowbirds (Hobson and Villard 1998), suggesting that cowbird parasitism has not been an important cause of nest failures in the region historically.

Numerous authors have criticised artificial nest experiments (reviewed in Major and Kendal 1996). We argue that they are useful indicators of spatial and temporal patterns of predation, particularly given the concordance in the patterns of predation we found using artificial nests and real nests. However, we recognize that the use of plasticine eggs could bias predator identification towards murid rodents (Bayne et al. 1997) and hence use of nest cameras or other methods may be more reliable for determining predator identity.

How do we explain the absence of elevated predation rates at edges and in fragments in the Calling Lake Study? We examine 3 hypotheses.

1) Landscape-level changes in abundance of nest predators swamp treatment effects. Numbers of red-backed voles, deer mice and red squirrels vary considerably over years. From 1993-1994, for example, red-backed voles increased 4-fold and deer mice 2-fold (Chan-McLeod unpub. data) and red squirrels increased by about 3 times (Cotterill and Hannon unpub. data.).

2) Predators found at Calling Lake do not congregate at edges or are not increased by forest fragmentation. Red squirrel and murid rodent predation was constant from edge to interior (Fig. 3B) and both groups are distributed throughout the forest and in the case of murids, in the cutblocks as well (Chan-McLeod unpublished data). Hannon and Cotterill (1996) performed similar experiments in an agricultural landscape 100 km to the south and found there that corvids were more important as predators. Rodents did not show an edge-related increase in predation rates, but corvids did (Hannon and Cotterill 1998). There are more species of corvids and they are over 5 times as abundant in the agricultural landscape than at Calling Lake (Hannon unpub. data). These results are consistent with Bayne and Hobson's (1997) work in mixedwood forests in Saskatchewan: they found higher predation and a greater diversity of predators along edges in agricultural landscapes than in logged landscapes.

3) Songbirds do not increase in abundance, and concentrate predators, at forest edges in the naturally patchy boreal forest. At Calling Lake, Song (1998) found no difference in the total abundance of songbirds at 2-3yr old forest/cutblock edges and forest interior, nor was there a difference when species were lumped into migratory, nesting or foraging guilds. Vegetation structure at the clearcut edges was not different from the interior. At

sites in mature forest, insufficient time may have elapsed for changes in understory development to occur. At sites in old forest, canopy gaps are common and understory is already well developed. Hence, increases in insolation at clearcut/old forest edges may not radically change structure of the understory. Given the rapid regeneration in aspen cutblocks, large future changes in vegetation structure at edges might not be expected (Song 1998).

6. CONNECTIVITY AND USE OF CORRIDORS

A contentious issue in Conservation Biology is the efficacy of corridors to enhance animal movements through fragmented landscapes (reviewed by Beier & Noss 1998). Intuitively, one might question the utility of corridors for birds, given their vagility, however a number of studies have found that isolated forest patches have low occupancy probabilities for some species of birds (reviewed in Desrochers et al. in press). At Calling Lake, Machtans et al. (1996) measured "flows" of birds through 100-m wide forested buffer strips between the lakeshore and cutblocks, through unharvested lakeshore forest and across clearcuts between connected reserves at two sites in the experimental township. In the first two, flow was measured by capturing birds in 100m wide mists nets perpendicular to the lakeshore, and in the latter by counting birds flying across the cutblocks over the same sampling area as that sampled by the nets. Captures of juveniles (presumably during dispersal) increased in the buffer strip from one year pre-harvest to one and two-years post-harvest. However, comparisons between the buffer strip and spatial control gave conflicting results. At one site, captures were >50% higher in the buffer strip than in the control, while at the other site, captures were similar between the buffer and control. The two sites differed in their configurations, making interpretation difficult (Machtans et al. 1996). This experiment requires more replication at sites with different configurations and at both lakeshore and upland corridors.

In terms of relative observation rates between buffer strips and cutblocks, forest passerines were rarely seen crossing cutblocks but were captured in the corridor, shrub-associated species were equally likely to be seen in the cutblocks and captured in the corridor, and the Yellow-bellied Sapsucker, the only woodpecker studied, was seen more frequently crossing cutblocks than caught in the corridor (Fig. 4). Although there are obviously differences in detectability between the two methods, the results in Figure 4 are conservative in that detectability is higher in the clearcuts. Hence, at least when a corridor is present, forest birds are more likely to use the corridor than cross the cutblock. These results were confirmed by "gap-crossing" experiments conducted in Quebec (Desrochers and Hannon 1998), where birds were enticed to cross gaps of different widths, some with the option of using forested corridors. Here results were species-specific, but in general birds were twice as likely to travel through 50m of forest than 50m of open area. In addition, birds preferred to detour through the woods even if the distance was three times longer than taking a shortcut in the open.

We can use the data from the corridor and gap-crossing studies to rank bird species as to their relative probabilities to cross-gaps.

We supplemented these data with results from bird responses to openings in the forest canopy caused by partial cutting (Tittler 1998). For the six species that we previously ranked for sensitivity to fragmentation (see Section 4.2.1), we ranked relative probabilities to cross-gaps as low (Black-throated Green Warbler, Black-capped Chickadee), moderate (Redbreasted Nuthatch, Yellow-rumped Warbler), and high (White-throated Sparrow and Yellow Warbler). We then calculated differences in proportional abundances of these species between controls and isolated fragments 5 years after harvest (from Fig. 2) to determine which species were most affected by isolation in the fragments over the longer-term. Species were affected by fragmentation (from most to least) in the following order: Black-throated Green Warbler, Black-capped Chickadee, Redbreasted Nuthatch, Yellow-rumped Warbler, Yellow Warbler, and White-throated Sparrow, congruent with our rankings of gap-sensitivity.

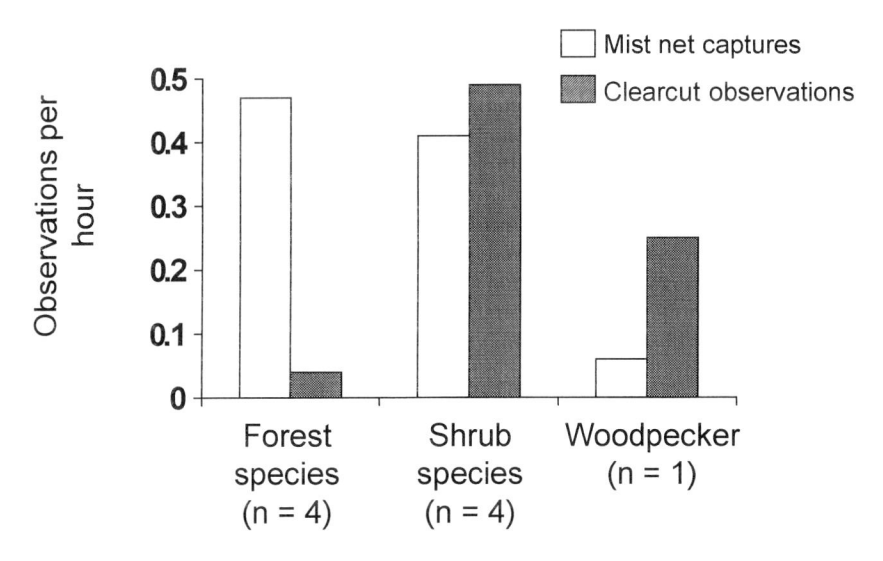

Figure 4. Number of mist-nest captures/hour in forest and number of clearcut observations/hour for 4 forest species (Least Flycatcher, Ovenbird, Swainson's Thrush, Yellow-rumped Warbler), 4 shrub species (White-throated Sparrow, Mourning Warbler, Song Sparrow, Yellow Warbler) and the Yellow-bellied Sapsucker. Based on 396 hours of mistnetting and 137 hours of observation in clearcuts; area of observation standardized to area netted by mistnets. Data reworked from Table 2 of Machtans et al. (1996).

Hence, our behavioural measures appear to be good predictors of species vulnerability to isolation effects.

7. IMPLICATIONS FOR FOREST MANAGEMENT

We chose to focus on older stands of aspen forest because they are most at risk from harvesting of the boreal mixedwood forest. Such older stands provide critical habitat for many (25% or more) boreal birds found in deciduous-dominated forest (Schieck and Nietfeld 1995, Kirk et al. 1996, Timoney and Robinson 1996). Furthermore, older, mixed stands have a higher diversity and abundance of breeding birds than either pure deciduous or pure coniferous stands (Schmiegelow, unpub. data).

Most older aspen stands contain small amounts of white spruce (Lee et al. 1995, Schmiegelow unpub. data); however, partitioning of the mixedwood landbase in Alberta between deciduous pulpwood production and coniferous sawlog exploitation may virtually eliminate the region's characteristic mixed stands (Cumming et al. 1994, Lieffers and Beck 1994).

When forested areas are fragmented by urban, agricultural or industrial expansion, surrounding habitats are often permanently transformed into unsuitable environments for many species. However, some might characterise fragmentation by forest harvesting as a form of habitat transition, whereby the landscape remains primarily forested, but logging activities dictate the initiation and culmination of succession.

Natural regeneration of harvested sites in our experimental area has been rapid, and recolonisation of these areas by songbirds has been dramatic (Fig. 5). Nevertheless, regenerating areas are unlikely to provide productive habitat for species dependent on older forest between subsequent passes (Schmiegelow, unpub. data), or within the short (40-70 year)

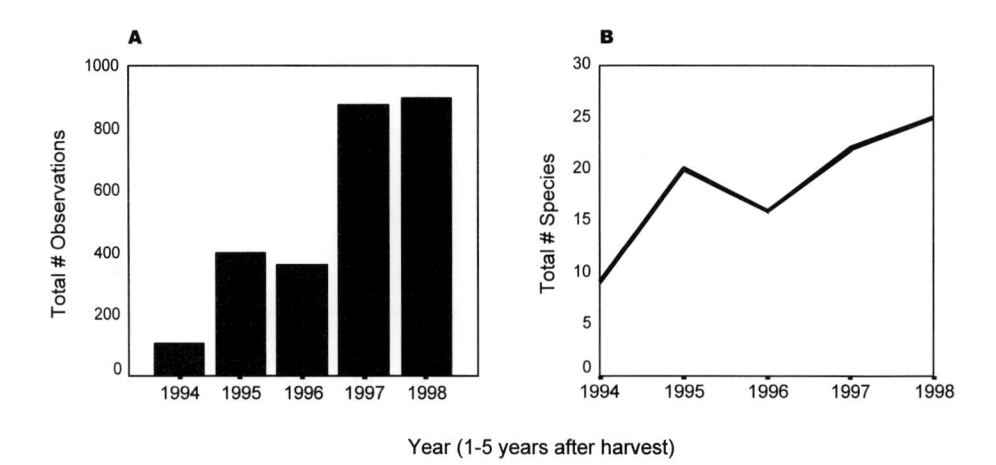

Figure 5. Trends in abundance (A) and richness (B) of the bird community colonising the regenerating cutblocks surrounding experimentally-created forest fragments near Calling Lake, Alberta. Data are from 24 point count stations sampled from 1 to 5 years after harvest (1994-98).

rotation periods planned for pulpwood production (Schieck and Nietfeld 1995). Hence, planned harvesting operations will result in functional conversion of older forest, once harvested, into unsuitable habitats for these species. This conclusion, in conjunction with the results presented here, leads to several immediate management recommendations for forest songbirds in the boreal mixedwood.

First, greater allowances must be made for maintenance of older deciduous-dominated forest than are currently in place in Alberta. Second, to reduce the additive effects of fragmentation, and provide for year-round requirements of resident species, patches of older forest should be left in several large (> 100 ha) blocks within logged townships (ca. 100 km^2). Much larger reserves may be necessary at regional levels, depending on the ability of the smaller, localised patches to maintain species over time (e.g. Robinson et al. 1995). Third, management strategies aimed at reducing cutblock sizes are misguided because they will result in greater fragmentation of available habitat, both in the short and long-term (see Wells et al., this volume). Fourth, the value of older forest maintained in riparian buffer strips for vulnerable bird species should be carefully evaluated. These linear habitats presently account for almost all merchantable leave areas, and occupy ca. 5% of available forest. Finally, the present practice of leaving small residual patches of old forest in cutblocks should be re-evaluated, since these areas are unlikely to provide productive habitat for vulnerable species. In fact, leaving up to 40% of the trees behind on cutblocks does not retain all forest species, and abundances of forest species are lower on retention blocks as compared to continuous forest (Tittler 1998). While these patches may serve as stepping-stones for movement through regenerating forest by species of older forest (a testable hypothesis), trade-offs in reserve allocation, as with the riparian buffer strips, must be considered.

8. ECOLOGICAL EXPERIMENTATION

Most ecological experiments happen at small spatial scales, are short in duration, and study only one or two species (Kareiva 1994). There is good reason for this. Economic and logistical constraints hinder replication of treatments at larger scales, and it is difficult to identify appropriate replicate units (Scheiner 1993). Funding cycles make it difficult to plan studies for longer than two or three years, and the complexity of interpreting communities

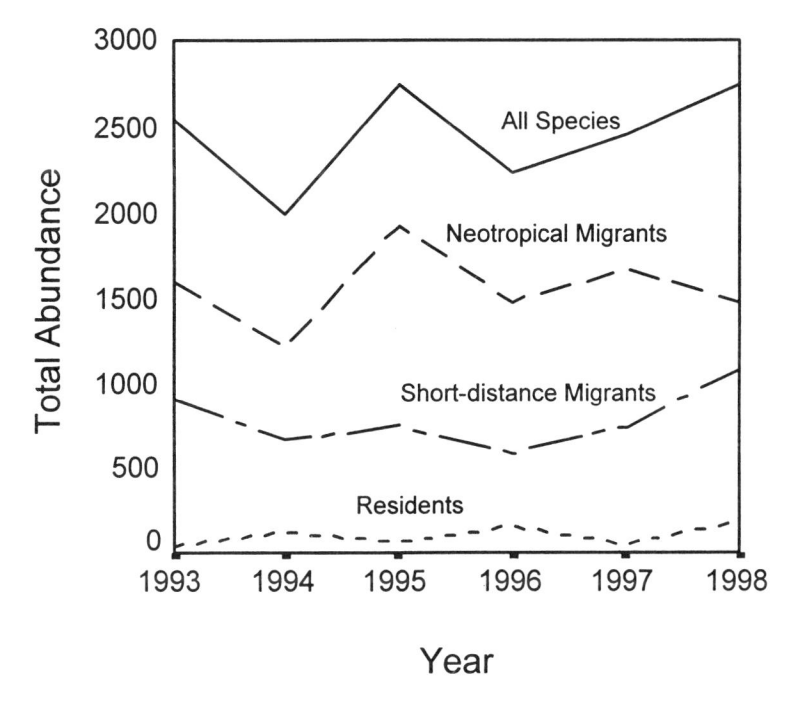

Figure 6. Annual trends in abundance of birds in older, deciduous-dominated forest near Calling Lake, Alberta. Data are from 93 point count stations located in continuous forest (i.e. the control sites for the fragmentation experiment), sampled over a 6 year period (1993-98).

encourages scientific controversy (Wiens 1989b). Unfortunately, the answer to certain questions may depend on whether we have chosen the appropriate scale (May 1994), and in choosing certain scales (spatial, temporal or taxonomic), we impose a bias on our study of ecological systems (Levin 1992).

Despite such limitations, adopting an experimental approach to assess environmental impacts avoids many of the pitfalls associated with alternative study designs (see Wiens and Parker 1995). Two factors from this experiment illustrate this point. First, despite our best efforts to control for variations in factors other than site area, there were baseline differences in the bird communities between our control and treatment sites. Second, there was high annual variability in both species composition and abundance in our control sites (Fig. 6). Without both before/after and control/treatment comparisons, it would have been impossible to (1) detect real effects of the experimental fragmentation, and (2) assess the magnitude of these effects.

Our results also highlight the dangers of basing management recommendations on short-term, non-experimental studies (see also Wiens and Parker 1995). Had we not analysed the data using both before/after and control/treatment comparisons, we would have been guilty of both Type I and Type II errors. The ecological and economic costs associated with such errors, if they result either in action or inaction, can be significant (e.g. Peterman 1990, Schmiegelow 1992). Even given the rigour of our experimental design, rare species might have declined substantially, or even disappeared, without detection. Although 58 passerine species were recorded during the 1993-95 breeding seasons, 21 of these were so rare that conventional statistical analyses of abundance data were not possible. However, rare species may be among those most at risk. We are currently exploring alternative approaches to addressing concerns about these species using more extensive data sets. Finally, the differences between our short- and mid-term results reinforce the need for longer-term studies. Whether the statistically significant effects we report represent biologically meaningful effect sizes is not known. The assessment of meaningful effect size is highly subjective. Longer-term monitoring of the experimental fragmentation sites is necessary to determine whether additional population declines and local extinctions will occur. However, we argue that the fragmentation and overall loss of older habitat occurring over very large scales warrants changes in forest management if future options are to be maintained.

9. A CAUTIONARY NOTE

Theoretical approaches to studying habitat fragmentation have tended to be patch-centred (e.g. Theory of Island Biogeography, Metapopulation Theory), and researchers have generally viewed fragments apart from the landscape in which they are embedded, but the surrounding matrix may not be wholly inhospitable (Wiens 1994). Some species may be able to shift to other habitat types, when their preferred habitat has been harvested, a process we term "habitat compensation" (Norton et al. submitted). Hence, a patch-centred view may overestimate the effects of habitat loss and fragmentation for some species. We examined the effects of harvesting deciduous forest on bird abundance in two adjacent landscapes: one in the experimental township and the second in north- and southwest townships. We compared our interpretations of the results when the study was done at the patch-scale (i.e. abundances in the aspen patches left after harvest) and at the landscape scale (abundances in the patches plus the surrounding landscape (clearcuts and coniferous forest) (Norton et al. submitted)). The shift from a patch to a landscape view altered our interpretation of half of the most common species' responses to logging. In addition these responses were not always consistent between the two landscapes. These results lead us to propose two cautionary notes: 1) studies of habitat fragmentation must be interpreted in the context of the entire landscape and 2) applying the results of localised studies to broader regions requires landscape-scale replication.

10. FUTURE DIRECTIONS

Large, continuous and relatively undisturbed expanses of forest may be important as population sources to ensure occupancy of habitat patches in more fragmented landscapes (e.g. Robinson et al. 1995). The boreal mixedwood has high species richness, and for species for which we have reproductive data, appears to be producing a "surplus" of young (Hannon unpub. data). At a site 100km south in an agricultural landscape, reproductive success, as measured by artificial nest predation and by monitoring redstart nests, is lower than at Calling Lake and the area may be a population sink (Hannon and Cotterill 1998, Hannon unpub. data). Clearly, information on juvenile dispersal is critical to determine the dynamics between potential sources and sinks and to determine at what level of fragmentation or habitat loss dispersal is thwarted. Direct measurement of juvenile movements and trajectories is problematic and awaits the development of smaller satellite transmitters that can be used for passerines. However, testing the responses of birds to potential movement barriers and measuring rates of movements through different landscape elements may help us to predict the impacts of landscape change (reviewed in Desrochers et al. in press). Schmiegelow is presently exploring the use of spatial and temporal occupancy patterns of selected species in continuous and fragmented sites within our experimental township as a surrogate measure of some of these processes. In addition, use of genetic markers may allow us to infer population exchanges and differentiation (see Prior et al. 1997 for an approach) and to monitor loss of heterozygosity in isolated populations. With our limited knowledge of dispersal distances, it is difficult to predict at what scale this will be useful for neotropical migrants, but this might be a good approach for resident species.

Another fruitful research direction will be to determine the relative importance of habitat loss, habitat fragmentation and the spatial configuration of landscape elements in explaining the occupancy patterns of birds in real landscapes (e.g. Kareiva and Wennergren 1995, Fahrig 1997). Currently, Hannon is examining the relative importance of local vegetation characteristics vs landscape structure in determining species abundances in 3 landscape types: logged, burned and continuous (Hannon unpub. data). We plan to verify some of the ensuing relationships on the Calling Lake landscape. In a complementary study, Schmiegelow and colleagues are analysing bird and vegetation data sets of differing spatial and temporal resolution to test for statistical signals of local vs. regional effects. Theoretical work has indicated that forest cover of 20-30% is an important threshold in explaining the relative impacts of habitat loss and fragmentation on population persistence and connectivity (Andren 1994, Fahrig 1997). However, more empirical work remains to be done to determine if this threshold is important in nature (but see Freemark and Collins 1992).

Our work has indicated that the current North American emphasis on the sensitivity of neotropical migrants to fragmentation may have been misplaced. We also documented declines in abundance of resident species in isolates, which mirrors results found in European forests (e.g. van Dorp and Opdam 1987, Verboom et al. 1991). More work is required on responses of resident species to fragmentation, particularly in winter when increases in energetic costs of movement, roosting or foraging could be fatal. For example, St. Clair et al. 1998 found that Black-capped Chickadees in winter avoided crossing gaps greater than 50 m and used treed fencerows to move around the landscape, and abundances of chickadees were lower in fragments greater than 50 m away from other forest (Desrochers et al. in press).

11. CONCLUSIONS

In 1997, the World Resources Institute identified Canada's boreal forests as containing significant amounts of the last 20% of frontier forest in the world. The WRI defined frontier forests as those "large enough to provide havens for indigenous species and to survive indefinitely without human intervention, if protections are put in place now" (WRI press release, March 4th, 1997). All other forests were considered too small or fragmented. However, systematic habitat fragmentation and loss of older forest, due to both forestry and energy sector activities, is happening throughout Canada's boreal mixedwood, in those very areas identified as globally significant for their "intact" forests.

The importance of these areas for North American songbird populations is unknown. These forests may serve as continental source pools for migrant populations in more southern, degraded forests. The persistence of these species in boreal forests will depend on their resilience to the major, widespread disturbances now underway. While relatively intact forests still exist, there is an opportunity for researchers to collect information that incorporates foresight, rather than hindsight, into future land management decisions, and for industrial and institutional stewards of the land to be proactive, rather than reactive.

Acknowledgements. Much of the work presented here has been collaborative, and we thank S. Cotterill, C. Machtans, M. Norton, S. Song, R. Tittler and M-A. Villard for their contributions. C. McCallum and T. Morcos assisted with data preparation. Many field assistants were instrumental in the collection of this data, and we are grateful for their dedication. K. With and F. Bunnell made helpful comments on an earlier version of this manuscript. Funding for this research was provided by the Alberta Fish and Wildlife Trust Fund, Alberta Forest Development Trust Fund, Alberta-Pacific Forest Industries, Canada/Alberta Partnership Agreement in Forestry, Canadian Circumpolar Institute, Canadian Forest Service, Canadian Wildlife Service, Cooper Foundation, Daishowa-Marubeni International, Tri-council Eco-research Program, Natural Sciences and Engineering Research Council, Sustainable Forest Management Network, Weyerhaeuser Canada, and Wildlife Habitat Canada.

LITERATURE CITED

Alberta Energy/Forestry, Lands and Wildlife. 1992. Timber harvesting planning and operating ground rules. Alberta Energy/Forestry, Lands and Wildlife Pub. 67. 75 pp.

Alberta Environmental Protection, Natural Resources Service, Recreation and Protected Areas Division, Natural Heritage Planning and Evaluation Branch. 1998. The Final Frontier: protecting landscape and biological diversity within Alberta's boreal forest natural region. Protected Areas Report No. 13.

Andren, H. 1994. Effects of habitat fragmentation on birds and mammals in landscapes with different proportions of suitable habitat: a review. Oikos 71: 355-366.

Andren, H. 1995. Effects of landscape composition on predation rates at habitat edges. Pp. 225-255 in Hansson, L., L. Fahrig, and G. Merriam (Eds.), Mosaic Landscapes and Ecological Processes. Chapman & Hall, London.

Armstrong, G.W., S.G. Cumming, and W.L. Adamowicz. In press. Timber supply implications of natural disturbance management. The Forestry Chronicle.

Bayne, E.M., and K.A. Hobson. 1997. Comparing the effects of landscape fragmentation by forestry and agriculture on predation of artificial nests. Conservation Biology 11: 1418-1429.

Bayne, E.M., K.A. Hobson, and P. Fargey. 1997. Predation on artificial nests in relation to forest type: contrasting the use of quail and plasticine eggs. Ecography 20: 233-239.

Beier, P. and R.F. Noss. Do habitat corridors provide connectivity? Conservation Biology 12:1241-1252.

Bierregaard. R.O., and T.E. Lovejoy. 1989. Effects of forest fragmentation on Amazonian understory bird communities. Acta Amazonica 19:215-241.

Böhning-Gaese, K., M.L. Taper, and J.H. Brown. 1993. Are declines in North American insectivorous songbirds due to causes on the breeding range? Conservation Biology 7:76-86.

Bunnell, F.L. 1995. Forest-dwelling vertebrate faunas and natural fire regimes in British Columbia: patterns and implications for conservation. Conservation Biology 9:636-644.

Campbell, P. 1996. Playing the trump card of uncertainty. Nature 380:9.

Cotterill, S.E. 1996. Effect of clearcutting on artificial egg predation in boreal mixedwood forests in north-central Alberta. MSc. Thesis, University of Alberta, Edmonton.

Cumming, S.G. 1997. Landscape Dynamics of the Boreal Mixedwood Forest. Ph.D. Thesis, University of British Columbia, Vancouver, BC.

Cumming, S.G., P.J. Burton, S. Prahacs, and M.R. Garland. 1994. Potential conflicts between timber supply and habitat protection in the boreal mixedwood of Alberta, Canada: a simulation study. Forest Ecology and Management 68: 281-302.

Desrochers, A., and S.J. Hannon. 1998. Gap crossing decisions by forest songbirds during the post-fledging period. Conservation Biology 11: 1204-1210.

Desrochers, A., S.J. Hannon, and C.C. St. Clair. In press. Movement of songbirds in fragmented forests: Can we "scale up" from behaviour to explain occupancy patterns in the landscape? Ostrich.

Donovan, T.M., P.W. Jones, E.M. Annand, and F.R. Thompson III. 1997. Variation in local-scale edge effects: mechanisms and landscape context. Ecology 78: 2064-2075.

Fahrig, L. 1997. Relative effects of habitat loss and fragmentation on population extinction. Journal of Wildlife Management 61: 603-610.

Freemark, K., and B. Collins. 1992. Landscape ecology of birds breeding in temperate forest fragments. Pp. 443-454 in J.M. Hagan III and D.W. Johnston (Eds.), Ecology and conservation of neotropical migrant landbirds. Smithsonian Institution Press, Washington and London.

Hagan, J.M., and D.W. Johnston, eds. 1992. Ecology and conservation of Neotropical migrant landbirds. Smithsonian Press, Washington, DC, USA.

Hagan, J.M., W.M.Vander Haegen, and P.S. McKinley. 1996. The early development of forest fragmentation effects on birds. Conservation Biology 10: 188-202.

Haila, Y., I.K. Hanski, J. Niemeiä, P. Punttila, S. Raivio, and H. Tukia. 1994. Forestry and the boreal fauna: matching management with natural forest dynamics. Annales Zoologici Fennici 31: 187-202.

Hannon, S.J., and S.E. Cotterill. 1996. Nest predation in aspen woodlots in an agricultural area in Alberta: the enemy from within. Auk 115: 16-25.

Hobson, K.A. and M-A. Villard. 1998. Forest fragmentation affects the behavioral response of American Redstarts to the threat of cowbird parasitism. The Condor 100:389-394.

Hunter, M.L. 1992. Paleoecology, landscape ecology, and the conservation of Neotropical migrant passerines in boreal forests. Pp. 511-523 in J.M. Hagan and D.W. Johnston (Eds.), Ecology and conservation of Neotropical migrant landbirds. Smithsonian Institution Press.

James, F.C., and C.E. McCulloch. 1995. The strength of inferences about causes of trends in populations. Pp. 40-54 in T.E. Martin and D.M. Finch (Eds.). Ecology and management of Neotropical migratory birds. Oxford University Press, USA.

Johnson, E.A. 1992. Fire and vegetation dynamics: studies from the North American boreal forest. Cambridge University Press, Cambridge.

Kareiva, P. 1994. Higher order interactions as a foil to reductionist ecology. Ecology 75: 1527-1528.

Kareiva, P., and U. Wennergren. 1995. Connecting landscape patterns to ecosystem and population processes. Nature 373: 299-302.

Kirk, D.A., A.W. Diamond, K.A. Hobson, and A.R. Smith. 1996. Breeding bird communities of the western and northern Canadian boreal forest: relationship to forest type. Canadian Journal of Zoology 74: 1749-1770.

Lee, P.C., S. Crites, and J.B. Stelfox. 1995. Changes in forest structure and floral composition in a chronosequence of aspen mixedwood stands in Alberta. Pp. 29-48 in J.B. Stelfox (Ed.), Relationships between stand age, stand structure, and biodiversity in aspen mixedwood forest in Alberta. Jointly published by Alberta Environmental Centre, Vegreville, AB, and Canadian Forest Service, Edmonton, AB, Canada.

Levin, S.A. 1992. The problem of pattern and scale in ecology. Ecology 73: 1943-1967.

Lieffers, V.J., and J.A. Beck. 1994. A semi-natural approach to mixedwood management in the prairie provinces. Forestry Chronicle 70:260-264.

Machtans, C.S., M-A. Villard, and S.J. Hannon. 1996. Use of riparian buffer strips as movement corridors by forest birds. Conservation Biology 10: 1366-1379.

Major, R.E., and C.E. Kendal. 1996. The contribution of artificial nest experiments to understanding avian reproductive success: a review of methods and conclusions. Ibis 138: 298-307.

McGarigal, K., and W.C. McComb. 1995. Relationships between landscape structure and breeding birds in the Oregon Coast Range. Ecological Monographs 65: 235-260.

Murphy, P.J. 1985. History of forest and prairie fire control policy in Alberta. Report T/77, Alberta Energy and Natural Resources, Edmonton, Alberta.

Norton, M.R., S.J. Hannon, and F.K.A. Schmiegelow. Fragments are not islands: patch vs landscape perspectives on songbird presence and abundance in a boreal forest fragmented by logging. Submitted to Ecography.

Paton, P.W.C. 1995. The effect of edge on avian nest success: how strong is the evidence? Conservation Biology 8: 17-26.

Peterman, R.M. 1990. Statistical power analysis can improve fisheries research and management. Canadian Journal of Fisheries and Aquatic Sciences 47: 2-15.

Peterson, E.B., and N.M. Peterson. 1992. Ecology, management, and use of aspen and balsam poplar in the prairie provinces, Canada. Forestry Canada, Northwest Region, Northern Forest Centre, Edmonton, Alberta. Special Report 1.

Prior, K.A., H.L. Gibbs, and P.J. Weatherhead. 1997. Population genetic structure in the Black Rat Snake: implications for management. Conservation Biology 11: 1147-1158.

Robinson, S.K., F.R. Thompson III, T.M. Donovan, D.R. Whitehead, and J. Faaborg. 1995. Regional forest fragmentation and the nesting success of migratory birds. Science 267: 1987-1990.

Roland, J. 1993. Large-scale forest fragmentation increases the duration of tent caterpillar outbreak. Oecologia 93: 25-30.

Rowe, J.S. 1972. Forest regions of Canada. Canadian Forestry Service Publication 1300, Ottawa.

Scheiner, S.M. 1993. Theories, hypotheses, and statistics. Pages 1-13 in S.M. Scheiner and J. Gurevitch (editors), Design and analysis of ecological experiments. Chapman and Hall, NY, USA.

Schieck, J., and M. Nietfeld. 1995. Bird species richness and abundance in relation to stand age and structure in aspen mixedwood forests in Alberta. Pp. 115-158 in J.B. Stelfox (Ed.), Relationships between stand age, stand structure, and biodiversity in aspen mixedwood forest in Alberta. Jointly published by Alberta Environmental Centre, Vegreville, AB, and Canadian Forest Service, Edmonton, AB, Canada.

Schieck, J., M. Nietfeld, and J.B. Stelfox. 1995. Differences in bird species richness and abundance among three successional stages of aspen-dominated boreal forests. Canadian Journal of Zoology 73: 1417-1431.

Schmiegelow, F.K.A. 1992. The use of atlas data to test appropriate hypotheses about faunal collapse. Pp. 67-74 in G.B. Ingram and M.R. Moss (Eds.), Landscape approaches to wildlife and ecosystem management in Canada. Polyscience, Montreal.

Schmiegelow, F.K.A. 1997. The effect of experimental fragmentation on bird community dynamics in the boreal mixedwood forest. Ph.D. thesis, University of British Columbia, Vancouver, Canada.

Schmiegelow, F.K.A., and S.J. Hannon. 1993. Adaptive management, adaptive science and the effects of forest fragmentation on boreal birds in northern Alberta. Transactions 58th North American Wildlife and Natural Resources Conference: 584-598.

Schmiegelow, F.K.A., C.S. Machtans and S.J. Hannon, 1997. Are boreal birds resilient to forest fragmentation? An experimental study of short-term community responses. Ecology 76:1914-1932.

Sherry, T.W., and R.T. Holmes. 1995. Summer versus winter limitation of populations: what are the issues and what is the evidence? Pp. 85-120 in T.E. Martin and D.M. Finch (Eds.), Ecology and management of Neotropical migratory birds. Oxford University Press, USA.

Song, S.J. 1998. Effect of natural and anthropogenic forest edge on songbirds breeding in the boreal mixed-wood forest of northern Alberta. PhD. Thesis, University of Alberta, Edmonton.

Song, S.J., and S.J. Hannon. Unpublished manuscript. Predation in heterogeneous forests: a comparison at natural and anthropogenic edges. Eoscience (in press).

Soulé, M.E., D.T. Bolger, and A.C. Alberts. 1988. Reconstructed dynamics of extinctions of chapparal-requiring birds in urban habitat islands. Conservation Biology 2:75-92.

St. Clair, C.C., M. Bélisle, A. Desrochers, and S.J. Hannon. 1998. Winter responses of forest birds to habitat corridors and gaps. Conservation Ecology [online] 2(2): 13.

Stelfox, J.B., 1995. Relationships between stand age, stand structure and biodiversity in aspen mixedwood forests in Alberta. Alberta Environment Centre (AECV95-R1), Vegreville, AB and Canadian Forest Service (Proj. No. 0001A), Edmonton, AB.

Stouffer, P.C., and R.O. Bierregaard, Jr. 1995. Use of Amazonian forest fragments by understory insectivorous birds. Ecology 76: 2429-2445.

Terborgh, J. 1989. Where have all the birds gone? Princeton University Press, USA.

Terborgh, J. 1992. Perspectives on the conservation of Neotropical migrant landbirds. Pp. 7-12 in J.M. Hagan and D.W. Johnston (Eds.), Ecology and conservation of Neotropical migrant landbirds. Smithsonian Institution Press, USA.

Timoney, K. P., and A.L. Robinson. 1996. Old-growth white spruce and balsam poplar forests of the Peace River Lowlands, Wood Buffalo National Park, Canada: development, structure and diversity. Forest Ecology and Management 81: 179-196.

Tittler, R. 1998. Effects of residual tree retention on breeding songbirds in Alberta's boreal mixed-wood forest. MSc. Thesis, University of Alberta, Edmonton.

van Dorp D., and P. Opdam. 1987. Effects of patch size, isolation and regional abundance on forest bird communities. Landscape Ecology 1: 59-73.

Verboom, J., A. Schotman, P. Opdam, and J.A.J. Metz. 1991. European Nuthatch metapopulations in a fragmented agricultural landscape. Oikos 61: 149-156.

Whitcomb, R.F., C.S. Robbins, J.F. Lynch, B.L. Whitcomb, M.K. Klimliewcz, and D. Bystrak. 1981. Effects of forest fragmentation on the avifauna of the eastern deciduous forest. Pages 125-205 in R.C. Burgess and D.M. Sharpe, editors. Forest island dynamics in man-dominated landscapes. Springer-Verlag, N.Y., USA.

Wiens, J.A. 1989a. The ecology of bird communities. Vol. 2. Processes and variations. Cambridge University Press.

Wiens, J.A. 1989b. The ecology of bird communities. Vol. 1. Foundations and patterns. Cambridge University Press.

Wiens, J.A., 1994. Habitat fragmentation: island vs. landscape perspectives on bird conservation. Ibis 137: s97-s104.

Wiens, J.A., and K.R. Parker. 1995. Analyzing the effects of accidental environmental impacts: approaches and assumptions. Ecological Applications 5: 1069-1083.

Wilcove, D.S., and S.K. Robinson. 1990. The impact of forest fragmentation on bird communities in Eastern North America. Pp. 319-331 in A. Keast (Ed.), Biogeography and ecology of forest bird communities. SPB Publishing, The Hague.

Forest Fragmentation Effects on Breeding Bird Communities in the Oregon Coast Range

Kevin McGarigal and William C. McComb

We investigated the relationship between landscape structure and breeding bird abundance in the central Oregon Coast Range. We sampled vegetation and birds in 30 landscapes (250-300 ha) distributed equally among 3 basins. Landscapes represented a range in structure based on the proportion of the landscape in a late-seral forest condition and the spatial configuration of that forest within the landscape. We computed several landscape metrics from digital vegetation cover maps. Using analysis of variance and regression procedures, we quantified the independent effects of late-seral forest area and configuration on 55 bird species.

Species varied in the strength and nature of the relationship between abundance and gradients in late-seral forest area and configuration. Landscape structure typically explained less than 50% of the variation in each species' abundance among the landscapes. An equal number of species were positively and negatively associated with late-seral forest area, and species' abundances were generally greater in the more fragmented landscapes. Given the scope and limitations of our study, the strength of our findings provides strong empirical evidence that landscape structure at the scale of 300-ha watersheds is probably important to several bird species in the central Oregon Coast Range. Our study also provides empirical evidence that late-seral forest area is probably more important than late-seral forest configuration. Land managers should focus first on maintaining sufficient late-seral forest area and secondarily consider the details of how late-seral forest is arranged.

Key words: Birds, Oregon Coast Range, late-seral forest relationships, fragmentation, late-seral forest area, late-seral forest configuration, landscape pattern.

1. INTRODUCTION

Habitat fragmentation is a landscape-level process in which a specific habitat is progressively sub-divided into smaller, geometrically more complex, and more isolated fragments as a result of both natural and human activities, and it involves changes in landscape composition, structure, and function at many scales and occurs on a backdrop of a natural patch mosaic created by changing landforms and natural disturbances. This definition has several key points. First, fragmentation is a landscape-level process, not a patch-level process (i.e., fragmentation alters the area and distribution of habitat patches within a broader habitat mosaic or landscape), yet nearly all of the field studies on fragmentation have employed a patch-centered sampling scheme in which independent habitat patches, not landscapes, were sampled (e.g., Rosenberg and Raphael 1986, Lehmkuhl et al. 1991). Based on the relationship derived between species richness or abundance and a variety of 'patch' characteristics, such as patch size and isolation, inferences often have been made about how fragmenta-

---

Forestry and Wildlife Management Department, University of Massachusetts, Amherst, MA

tion at the landscape level affects wildlife populations. Yet, it is unclear whether relationships derived at the patch level can be extrapolated to the landscape level (Wiens et al. 1987 and 1993; Wiens 1989a,b).

Second, fragmentation is a habitat-specific process (i.e., fragmentation occurs only in reference to a specific habitat). Since habitat is an organism-specific concept, the target habitat must be clearly specified and at a resolution relevant to the organism(s) under consideration. For example, the target habitat may be forest, late-seral forest, or old-growth Douglas-fir (*Pseudotsuga menziesii*) forest. The choice of resolution would depend on the habitat specificity of the organism(s) under consideration. The attention to resolution is crucial because the fragmentation trajectory within the same landscape can differ markedly among habitats in relation to habitat specificity. For example, in some Pacific Northwest (PNW) landscapes, old-growth Douglas-fir forest has become highly fragmented due to logging, yet late-seral forest (a broader, more inclusive definition) remains abundant and highly connected (McGarigal and McComb 1995).

Third, fragmentation is a scale-dependent process, both in terms of how we (humans) perceive and measure fragmentation and in how organisms perceive and respond to fragmentation. The extent of the landscape considered, in particular, can have an important influence on the measured fragmentation level; a highly fragmented habitat at one scale may be comparatively unfragmented at a much coarser (or finer) scale (e.g., when fragmented woodlots occur within a forested region). In addition, for habitat fragmentation to be consequential, it must occur at a scale (both extent and grain) that is functionally relevant to the organism(s) under consideration; yet more often than not, we do not know how each organism scales the environment. The lack of correspondence between the scale of measured fragmentation and the scale of functional fragmentation represents a great challenge to future studies on fragmentation.

Fourth, habitat fragmentation results from both natural and anthropogenic causes. Most studies have focused on anthropogenic causes and ignored the backdrop of the natural patch mosaic created by changing landforms and natural disturbances. Much of the fragmentation dogma, for example, stems from field studies conducted in the eastern deciduous forest of North America on forest fragmentation caused by agricultural development and urbanization (e.g., Whitcomb et al. 1981, Robbins et al. 1989, Terborgh 1989). More recently, the anthropogenic causes of forest fragmentation have been expanded to include commercial timber management (e.g., Lehmkuhl and Ruggiero 1991, Hunter et al. 1995, McGarigal and McComb 1995). Unfortunately, similar attention has not been given to forest fragmentation caused by natural disturbances, yet it is increasingly clear that coniferous forest landscapes in western North America are naturally highly dynamic, and that many forested habitats periodically experience rapid fragmentation and recovery following large-scale catastrophic fire disturbances (Wallin et al. 1996). It seems likely, therefore, that anthropogenic disturbances, such as logging, primarily affect the scale, rate, and pattern of fragmentation. It is still unclear how native fauna are affected by anthropogenic disturbances in landscapes also subject to natural disturbance-induced fragmentation.

Finally, in real landscapes, fragmentation involves both (1) habitat loss and (2) fragmentation per se (i.e., the subdivision of habitat into isolated fragments); that is, these components are almost always confounded in the real world. The latter effect is more generally referred to as habitat configuration (i.e., the spatial pattern of habitat), and most of the recent scientific attention has focused on this component of fragmentation (e.g., Fahrig and Merriam 1994). The emphasis on spatial pattern stems, in part, from the presumed importance of habitat configuration on ecosystem and population processes, especially population extinction (Kareiva and Wennergren 1995). Unfortunately, although several theoretical and empirical studies have examined the effects of either habitat area or configuration on population persistence, there have been few theoretical studies and even fewer empirical studies that have examined the relative effects of both habitat area and configuration (e.g., see Fahrig 1997). Distinguishing between the effects of habitat area and configuration has important implications for conservation biology. As

Fahrig (1997) notes, "if habitat configuration is important, then within some limits it should be possible to mitigate effects of habitat loss by ensuring that remaining habitat is not fragmented. On the other hand, if the effects of fragmentation are trivial in comparison to effects of habitat loss, then the assumption that loss can be mitigated by reduced fragmentation has potentially dangerous consequences for conservation."

Commercial timber management is the major anthropogenic cause of late-seral forest loss and fragmentation in the PNW (Lehmkuhl and Ruggiero 1991). Commercial timber management alters landscape structure by changing the areal extent and configuration of plant communities and seral stages across the landscape, and this occurs on a backdrop of a natural patch mosaic created by changing landforms and natural disturbances (Swanson et al. 1988). In this scenario, the natural landscape is a spatially and temporally dynamic mosaic of forest patches (i.e., shifting mosaic) due to natural disturbances (e.g., fire and windthrow) and forest regrowth, and it is the scale and structure of this mosaic that is dramatically altered by timber management activities (Swanson et al. 1990). Vertebrate population responses in forested landscapes altered by timber management activities are likely to differ from those in landscapes experiencing forest loss and fragmentation due to urbanization and agricultural development. Sharp forest/nonforest edges are transient in managed forest landscapes because of forest regrowth, and regenerating plantations do not function as nonhabitat for many species, but rather as habitat of variable quality and permeability to animal movements. Late-seral forest patches may never be truly isolated or may be isolated for only a brief period of years. Moreover, vertebrate populations in the PNW have evolved in a naturally heterogeneous landscape and may have evolved strategies for coping with the interspecific interactions (e.g., competition and predation) associated with fragmented habitats. For these reasons, it is questionable whether the empirical findings on forest fragmentation from urban or agricultural landscapes extend to the dynamic forest landscapes of the PNW and elsewhere (Hejl 1992). Indeed, it seems likely that vertebrate response to habitat fragmentation would be less pronounced.

Over the past several years there has been a great deal of attention on the fragmentation of late-seral forests in the PNW. Initially, the focus of attention was on spotted owls (*Strix occidentalis*) and old-growth Douglas-fir ecosystems (Thomas et al. 1990). However, the debate has since broadened to include all species and all late-seral forests (FEMAT 1993). We conducted a study to determine how changes in late-seral forest area and configuration affect breeding bird populations in the central Oregon Coast Range. Our specific objective was to quantify the relationship between the extent and configuration of late-seral forest habitats and the abundance of breeding bird species.

2. STUDY AREA

The study was conducted in the central Oregon Coast Range in three river basins (Fig. 1), as described in detail by McGarigal and McComb (1995). Briefly, elevation ranges from sea level to 968 m; climate is maritime, characterized by mild, wet winters and cool, dry summers; annual precipitation ranges from 150-300 cm and occurs primarily during the winter months (Franklin and Dyrness 1973:71-72). The area is characterized by steep slopes and deeply-cut drainages. The area is almost entirely forested and lies almost exclusively within the western hemlock (*Tsuga heterophylla*) vegetation zone (Franklin and Dyrness 1973:70-108). Coniferous forest communities dominated by seral Douglas-fir cover most of the landscape, although broadleaf forests, primarily red alder (*Alnus rubra*), are locally common and often comprise mixed-species stands with Douglas-fir. Franklin and Dyrness (1973:70-108) provide a complete description of the vegetation.

The current landscape structure reflects a complex history of natural and anthropogenic disturbances, and reflects differences in forest management practices on public and private lands. Representative of the historic natural disturbance regime within the Coast Range physiographic province, the entire study area experienced a stand-replacement fire in the

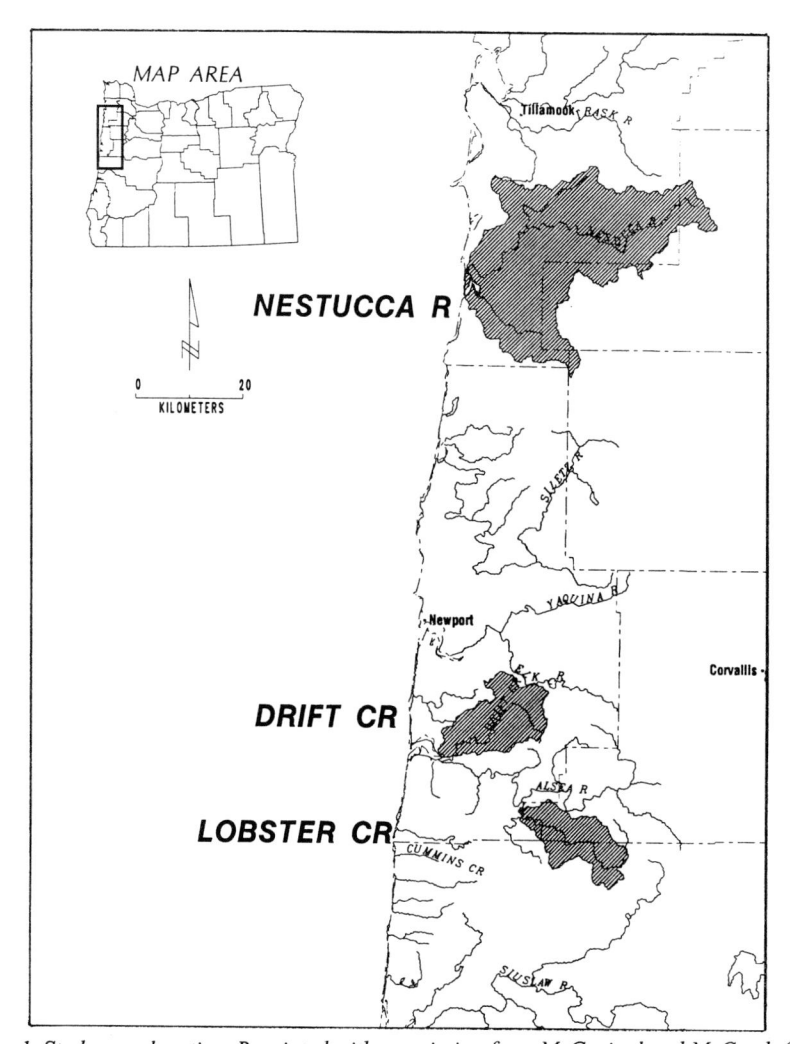

Figure 1. Study area location. Reprinted with permission from McGarigal and McComb (1995).

mid-1800's and regenerated naturally (Spies and Cline 1988).

As a result, virtually the entire study area developed as a relatively even-aged coniferous forest. This forest matrix contained a legacy of scattered old-growth forest patches--remnants of the pre-disturbance forest. Small, scattered disturbance patches caused by mass soil movements (e.g., debris slides, debris flows), wind storms, root disease, and a variety of other small-scale disturbance agents, including human settlement, perforated the forest matrix during its successional development.

Until the mid 1900's, the maturing coniferous forest matrix and old-growth remnant patches represented a relatively unfragmented habitat for organisms associated with late-seral coniferous forest. For our purposes, late-seral forest included large sawtimber (> 20% overstory cover comprised of trees with a mean dbh > 53.3 cm) and remnant old-growth (> 20% overstory cover comprised of trees with a mean dbh > 81.3 cm and multistory canopy) patches scattered irregularly throughout the large sawtimber.

Following World War II, private and public land managers began harvesting the late-seral forest at an accelerated rate to meet the growing demands for wood products. Extensive road networks were developed to access the

Late-Seral Forest Area (% of subbasin)

Spatial Configuration

Simple / Complex — 100%, 80%, 60%, 40%, 20%, 0%

300-ha subbasin

Late-Seral Forest | Early-seral forest

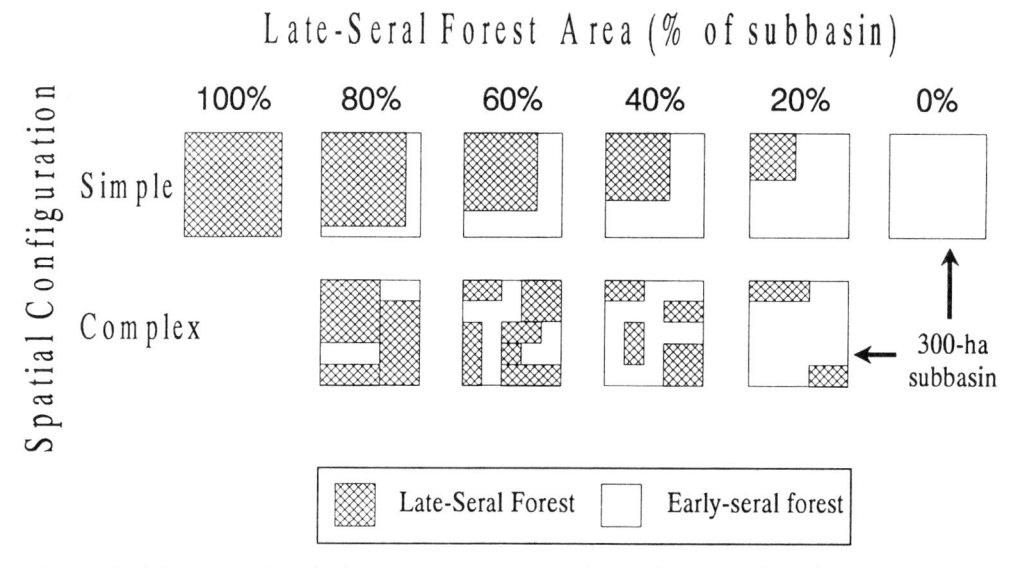

Figure 2. Schematic of study design, representing 1 of 3 replicates, with each replicate in a separate basin. Late-seral forest includes all conifer-, mixed-, and hardwood-large sawtimber; early-seral forest includes all other patch types. Reprinted with permission from McGarigal and McComb (1995).

timber, resulting in a rapid dissection of the late-seral forest matrix. In general, private industrial forest lands were clearcut quickly and extensively, and often progressively in large swaths. Consequently, the process of late-seral forest removal progressed rapidly; in many cases, resulting in attrition and eventual loss of late-seral forest. As a result, landscapes dominated by private industrial ownership currently consist largely of extensive, young (0-40 years), even-aged, Douglas-fir plantations, although isolated, small patches of late-seral forest exist in some areas. In contrast, federal land managers used the dispersed-patch or staggered-setting system of clearcutting, which maximizes the dispersion of the residual late-seral forest (Franklin and Forman 1987); 10- to 20-ha patch cuts are interspersed with uncut forest areas of at least equal size (Smith 1986). Due in part to limits on the allowable cut and the pattern of cutting, the process of late-seral forest removal progressed slowly on federal lands. Due to recent events associated with the management of late-successional forests in the Pacific Northwest (FEMAT 1993), the loss and fragmentation process has been held largely in check. As a result, landscapes managed by federal agencies currently consist of a matrix of late-seral, unmanaged forest dominated by Douglas-fir and red alder embedded with numerous small (8-25 ha), young, even-aged, Douglas-fir plantations.

This complex disturbance history has created a wide range of landscape structural conditions within the study area, particularly with respect to the extent and configuration of late-seral forest, and these patterns vary as a function of scale. At the scale of a 300-ha subbasin, all landscape structural conditions are represented on federal lands, whereas attrition and complete loss of late-seral forest characterize most private lands. At the scale of a 10,000-ha basin, in an area dominated by federal ownership, late-seral forest still represents the matrix within which young forest and other disturbance patches are embedded. At the scale of the entire Coast Range physiographic province, late-seral forest represents fragments (i.e., remnant patches) within a disturbance matrix dominated by young forest plantations. Thus, the extent and configuration (i.e., fragmentation) of late-seral forest within our study area is a function of the scale under consideration.

3. METHODS

3.1 Study Design

We selected 10 landscapes in each of 3 basins (i.e., 3 replicates, $n = 30$ landscapes) based on the proportion of each landscape in a late-seral forest condition and the spatial configuration (i.e., relative fragmentation defined on the basis of late-seral forest edge density) of late-seral forest within the landscape (Fig. 2). This 2-dimensional design ensured that we sampled a wide range of landscape structural conditions with respect to late-seral forest. Moreover, this design allowed us to separate the potentially confounding effects of late-seral forest area and configuration on the bird community. We defined landscapes as 250- to 300-ha areas, generally conforming to second- or third-order watersheds. This landscape extent represents the reference scale for this investigation, and although a wide variety of a late-seral forest fragmentation levels are represented at this scale, these experimental units are nested within a larger landscape context in which late-seral forest is the matrix.

3.2 Bird Sampling

We sampled breeding birds at systematic sample points distributed in a 200- x 400-m grid in each landscape. Based on calculated effective detection distances, this resulted in a 10-65% sampling intensity among bird species (Appendix A). We sampled the landscapes in a different basin each year between 1990-1992. The consequences of confounding year and basin were generally insignificant (McGarigal and McComb 1995). Each year, we sampled birds 4 times in each of the 10 landscapes at nearly regular intervals during the breeding season using standard variable circular plot techniques (Reynolds et al. 1980, McGarigal and McComb 1995).

3.3 Vegetation Mapping

To map vegetation, we defined 27 patch types, including 5 nonforested patch types and 22 forested patch types; the latter varied on the basis of plant community, seral condition, and canopy closure (see McCarigal and McComb for details). These patch types correspond to broad habitat types that are widely recognized to be meaningful to a wide variety of wildlife in western Oregon and Washington (Bruce et al. 1985).

Late-seral forest, as we defined it, included hardwood-, mixed-, and conifer-large sawtimber. In some landscapes, late-seral forest also included scattered remnant old-growth trees and patches. We defined minimum patch size as 0.785 ha and ≥ 50 m wide in the narrowest dimension. This minimum area corresponds roughly to the smallest estimated home range size of any bird species found in the study area (Brown 1985). We mapped vegetation in each landscape on aerial photos (1988-1989 color infrared, 1:20000) using a stereoscope, ground-truthed 100 percent of each landscape, and then transformed vegetation cover maps into planimetrically-corrected digital coverages.

3.4 Data Analysis

3.4.1 Bird Abundance

For each species, we calculated an index of abundance for each landscape to serve as dependent variables in the analyses of bird-habitat relationships. We calculated the average number of bird detections per visit per station (i.e., sampling point within landscape) for each species, including detections at any distance from the station. Because we did not make any statistical comparisons among species, it was not necessary to compute density estimates for each species based on each species' effective detection distance, or to consider differences in detectability among species. Dependent variables were log-transformed to improve the distribution of the residuals in the analysis of variance (ANOVA) and regression analyses described below; subsequent analyses of the residuals indicated that normality and variance assumptions were adequately met in most cases.

3.4.2 Vegetation Patterns

We imported the digital vegetation maps of each landscape into the Arc/Info Geographic

Table 1. *Indices used to quantify the spatial extent and configuration of late-seral forest in 30 300-ha landscapes in Drift Creek, Lobster Creek, and Nestucca River basins in Benton, Lincoln, and Tillamook Counties, Oregon, 1990-92*

Index Name (units)	Description[a]
Percent of landscape (%)	Percentage of the landscape comprised of late-seral forest (see text for definition of late-seral forest).
Patch density (#/100 ha)	Density of late-seral forest patches.
Mean patch size (ha)	Average size of late-seral forest patches.
Contrast-weighted edge density (m/ha)	Density of edge involving late-seral forest patches weighted by the degree of structural and floristic contrast between adjacent patches; equals unweighted edge density when all edge is maximum contrast and approaches 0 when all edge is minimum contrast.
Mean shape index	Mean patch shape complexity; equals 1 when all patches are circular and increases as patches become noncircular
Total core area index (%)	Total percentage of late-seral forest area that is greater than 100-m from the nearest edge (patch perimeter) of each patch.

[a] See McGarigal and Marks (1995) for a complete description and definition of each index.

Information System and used the program FRAGSTATS (McGarigal and Marks 1995) to calculate the area (defined here as the proportional abundance of late-seral forest in each landscape) and spatial configuration of late-seral forest (i.e., large sawtimber patch types pooled together). The other patch types present in each landscape (pooled together as early-seral forest in Figure 2) affected the configuration of late-seral forest only by affecting the contrast between each late-seral forest patch and its neighborhood.

We used a combination of procedures to statistically remove any relationship between late-seral forest area and each configuration index (see McGarigal and McComb 1995 for details) and then selected 5 somewhat independent indices representing different components of configuration (Li and Reynolds 1995; Table 1). It is important to note that these adjusted indices quantify late-seral forest configuration independent of late-seral forest area.

3.4.3 Bird-Habitat Relationships

We assessed the effects of late-seral forest area and configuration on each species' abundance using 3-way ANOVA (3 basins by 6 levels of area by 1 or 2 levels of configuration depending on area, $n = 30$) with basin as a block and configuration nested within area. We excluded uncommon (<35 total detections) and poorly distributed (<33% occurrence among subbasins) species in these analyses. However, we qualitatively examined the pattern of occurrence in relation to late-seral forest area and configuration for each of these species.

Although we categorized both area and configuration variables for purposes of this ANOVA study design (Fig. 2), we did this in part to ensure representation of the full range of these inherently continuous gradients in landscape structure. Moreover, the ANOVA design treated configuration as a simple dichotomous variable defined on the basis of edge density (i.e., high or low density), yet we recognized that configuration consists of many aspects. Therefore, we also assessed the relationship between late-seral forest area and configuration and each dependent variable using regression procedures, as follows.

To determine the strength and nature of the relationship between *late-seral forest area* and each dependent variable, we regressed late-seral forest area on each species' abundance index separately using general linear and non-linear (quadratic polynomials) models. We treated all 30 landscapes as independent observations. Similarly, to assess the relationship

Table 2. Relationship between late-seral forest area and the abundance of breeding bird species in 30 300-ha subbasins in Drift Creek, Lobster Creek, and Nestucca River basins in Benton, Lincoln, and Tillamook Counties, Oregon, 1990-92. Data represent birds detected at any distance during 4 visits to 32-38 sampling points within each subbasin, including only detections of new individuals from separate sampling points within a subbasin during a visit. Species are grouped according to the nature of the relationship and are listed in order of decreasing significance of the area effect.

Species[a]	ANOVA[b] F	P	Regression[c] b_1	R^2	P	Relationship[d]
GROUP A:	Abundance increases linearly with increasing late-seral forest area						
Gray jay	14.24	<0.001	+	63%	<0.001	
Brown creeper	13.40	<0.001	+	59%	<0.001	
Winter wren	8.57	<0.001	+	53%	<0.001	
Varied thrush	7.76	<0.001	+	52%	<0.001	
Chestnut-backed chickadee	4.63	0.007	+	24%	0.007	
Evening grosbeak	4.14	0.011	+	31%	0.002	
Pacific-slope flycatcher	3.49	0.022	+	15%	0.037	
Hammond's flycatcher	3.24	0.029	+	28%	0.003	
Pileated woodpecker	2.51	0.068	+	18%	0.019	
Red-breasted nuthatch	2.26	0.092	+	19%	0.015	
Hermit warbler	1.40	0.271	+	12%	0.063	
GROUP B	Abundance highest at intermediate levels of late-seral forest area (i.e., curvilinear relationship between late-seral forest area and abundance)						
Hairy Woodpecker	5.50	0.003	+-	46%	<0.001	
Steller's jay	2.94	0.041	+-	23%	0.030	
American robin	1.65	0.199	+-	23%	0.031	
Song sparrow	2.75	0.051	+-	35%	0.003	
GROUP C:	Abundance lowest at intermediate levels of late-seral forest area (i.e., curvilinear relationship between late-seral forest area and abundance)						
Common nighthawk	2.22	0.097	−+	29%	0.009	
Black-throated gray warbler	3.57	0.020	−+	34%	0.004	
Bewick's wren	0.76	0.590	−+	14%	0.041	
Rufous hummingbird	1.61	0.208	−+	21%	0.012	
GROUP D:	Abundance decreases linearly with increasing late-seral forest area						
Yellow-rumped warbler	0.68	0.642	-	12%	0.063	
Mountain quail	2.10	0.113	-	15%	0.033	
Purple finch	2.11	0.111	-	13%	0.054	
Northern flicker	2.16	0.104	-	15%	0.034	
Warbling vireo	2.22	0.097	-	13%	0.055	
Western wood-pewee	2.42	0.076	-	1%	0.588	
Wrentit	2.56	0.064	-	24%	0.006	
MacGillivray's warbler	3.21	0.030	-	23%	0.007	
Rufous-sided towhee	3.35	0.026	-	34%	<0.001	
Band-tailed pigeon	4.35	0.009	-	10%	0.094	
Orange-crowned warbler	5.39	0.003	-	53%	<0.001	
Hutton's vireo	7.11	<0.001	-	50%	<0.001	

[a] Scientific names in Appendix A.

[b] 3-way ANOVA [model: $\log_{10}$(abundance) = basin + area + pattern(area)] with basin as a block; $n = 30$ subbasins; area effect 5,18 degrees of freedom.

[c] Simple linear regression [Group A and D model: $\log_{10}$(abundance) = $b_0 + b_1$(area); Group B and C model: $\log_{10}$(abundance) = $b_0 + b_1$(area) + b_2(area)^2]; $n = 30$ subbasins.

[d] Idealized relationship between late-seral forest area and bird abundance.

between *late-seral forest configuration* and each dependent variable, we regressed each configuration index on each species' abundance index separately using general linear and nonlinear (quadratic polynomials) models. To assess the additional explanatory power of late-seral forest configuration after accounting for late-seral forest area, we used partial regression. Specifically, we created a parsimonious general linear or nonlinear model using stepwise regression that included late-seral forest area and any partially significant configuration indices as independent variables. The general partial F-statistic and associated partial R^2 for the configuration indices measure the additional explanatory contribution of configuration after area has been taken into account. We conducted partial regression analyses only for species with significant area relationships.

4. RESULTS

We recorded 82,639 detections representing 85 breeding bird species during 4 visits to 1046 sampling stations distributed among the 30 landscapes. Number of detections varied greatly among species, in part because of differences in abundance, but also because of differences in effective detection distances (Appendix A). Fifty-five of 85 species were sufficiently abundant and widespread to include in the species-specific analyses.

4.1 Late-seral Forest Area

Fifteen species exhibited a strong statistical association with late-seral forest at the patch level (see McGarigal and McComb 1995 for details). Twelve of these species exhibited strong and exclusive statistical selection for late-seral forest based on an analysis of use versus availability at the patch level, although all of these species were detected in younger seral conditions as well, and some (e.g., Winter Wren) were even relatively common in younger seral conditions. The remaining 3 species were less common but widely distributed (Olive-sided Flycatcher, Red-tailed Hawk and Western Wood-pewee) and associated with the juxtaposition of late-seral forest and early-seral, open-canopied habitats. Two species (Golden-crowned Kinglet and Hairy Woodpecker) exhibited disproportionate use of large sawtimber, but showed disproportionate use of other seral conditions as well. A few additional species (e.g., Spotted Owl, Vaux's Swift, Bald Eagle) seemed to be strongly associated with late-seral forest, but were detected too infrequently (≤ 18) or were absent from too many (≥ 20) landscapes to warrant conclusions regarding habitat associations and were not included in the analysis.

The relationship between late-seral forest area and species' abundances at the landscape level varied dramatically among species, even for those strongly associated with late-seral forest at the patch level. Thirty-one of 55 (56%) species exhibited significant area relationships based on either the ANOVA or the regression analyses (Table 2). Species logically fell into four groups based on the nature of the relationship (Table 2). The abundances of 11 species increased linearly with increasing late-seral forest area (Group A), although the strength of the relationship varied considerably (R^2, 12-63%). Four species exhibited curvilinear relationships in which abundances peaked at intermediate levels of late-seral forest area (Group B; R^2, 23-46%). Four species exhibited curvilinear relationships in which abundances decreased at intermediate levels of late-seral forest area (Group C; R^2, 14-34%). Finally, the abundances of 12 species decreased linearly with increasing late-seral forest area (Group D; R^2, 1-50%). Several species (Red Crossbill, Olive-sided Flycatcher, Red-breasted Sapsucker, and Red-tailed Hawk) showed virtually no ($R^2 \leq 3\%$) relationship to late-seral forest area at the subbasin scale, despite the fact that these species demonstrated a strong association with late-seral forest at the patch level (McGarigal and McComb 1995).

4.2 Late-seral Forest Configuration

Based on the ANOVA, only 5 species (mountain quail, olive-sided flycatcher, western wood-pewee, western bluebird, and pine siskin) were affected by late-seral forest configuration when evaluated as a simple dichotomous variable nested within 4 levels of late-seral forest area (20%, 40%, 60%, and 80% of landscape area in late-seral forest); all spe-

Table 3. Effects of late-seral forest configuration (high vs. low fragmentation) on the abundance of breeding bird species in 30 300-ha subbasins in Drift Creek, Lobster Creek, and Nestucca River basins in Benton, Lincoln, and Tillamook Counties, Oregon, 1990-92. Data represent birds detected at any distance during 4 visits to 32-38 sampling points within each subbasin, including only detections of new individuals from separate sampling points within a subbasin during a visit. Only species exhibiting a significant effect based on ANOVA are shown. Species are listed in order of decreasing significance of the configuration effect.

Species[a]	ANOVA[b] F	P	Configuration Components[c] PD/MPS	TCAI	CWED
Mountain quail	4.83	0.008	27% (0.003)	-----	----
Olive-sided flycatcher	3.31	0.034	12% (0.059)	14% (0.046)	48% (<0.001)
Western woodpewee	3.07	0.043	11% (0.068)	18% (0.019)	40% (<0.001)
Western bluebird	2.56	0.074	----	17% (0.023)	26% (0.004)
Pine siskin	2.52	0.078	16% (0.092)	----	----

[a] Scientific names in Appendix A.
[b] 3-way ANOVA [model: $\log_{10}$(abundance) = basin + area + configuration(area)] with basin as a block; n = 30 subbasins; configuration effect 4,18 degrees of freedom.
[c] R^2 and associated P-value for simple linear or nonlinear (quadratic polynomial) relationship between species abundance and 3 major configuration components; n = 30 subbasins; PD = patch density; MPS = mean patch size; TCAI = total core area index; and CWED = contrast-weighted edge density.

cies were generally more abundant in the more fragmented landscapes (Table 3). However, each of these species was associated with a different combination of configuration components. Olive-sided flycatcher and western woodpewee, for example, were strongly associated with abundant, high-contrast edge; whereas, mountain quail was associated with high patch densities.

Several other species were associated with one or more of the individual configuration gradients. Figure 3 represents plots of species in two-dimensional landscape structure space. In each plot, the vertical axis represents a gradient in late-seral forest area; species located at the top end of the gradient have positive associations with late-seral forest area, whereas species located at the bottom end of the gradient have negative associations with late-seral forest area (i.e., greater abundance in early-seral landscapes). The horizontal axis represents a gradient in one component of late-seral forest configuration. The distance away from the origin along either dimension represents the strength of the relationship; the farther away, the stronger the relationship. Comparing the distance away from the origin along the x- and y-axis provides a simple way to assess the relative importance of late-seral forest area versus configuration.

Fifteen species demonstrated a significant relationship with patch density, although the percent of variation in species' abundance accounted for by this variable was relatively small (R_1^2 9-27%; Fig. 3a). Roughly the same number of species' had positive and negative relationships, and almost half of the species had stronger relationships with late-seral forest area than patch density. The results were similar for mean patch size, although for a different set of species (Fig. 3b). Thirteen species exhibited a significant relationship with mean shape index, although the R^2 values were again relatively small (10-23%; Fig. 3c). Most species were positively associated with this gradient, indicating an association with landscapes having more complex patch shapes than the average condition. Twenty of 21 species had significant relationships with total core area index and contrast-weighted edge density, respectively, with R^2 values as high as 50% (Fig. 3d-e). Most of these species were associated with the more fragmented landscape conditions (i.e., less core area and higher edge contrast) and did not have significant area relationships.

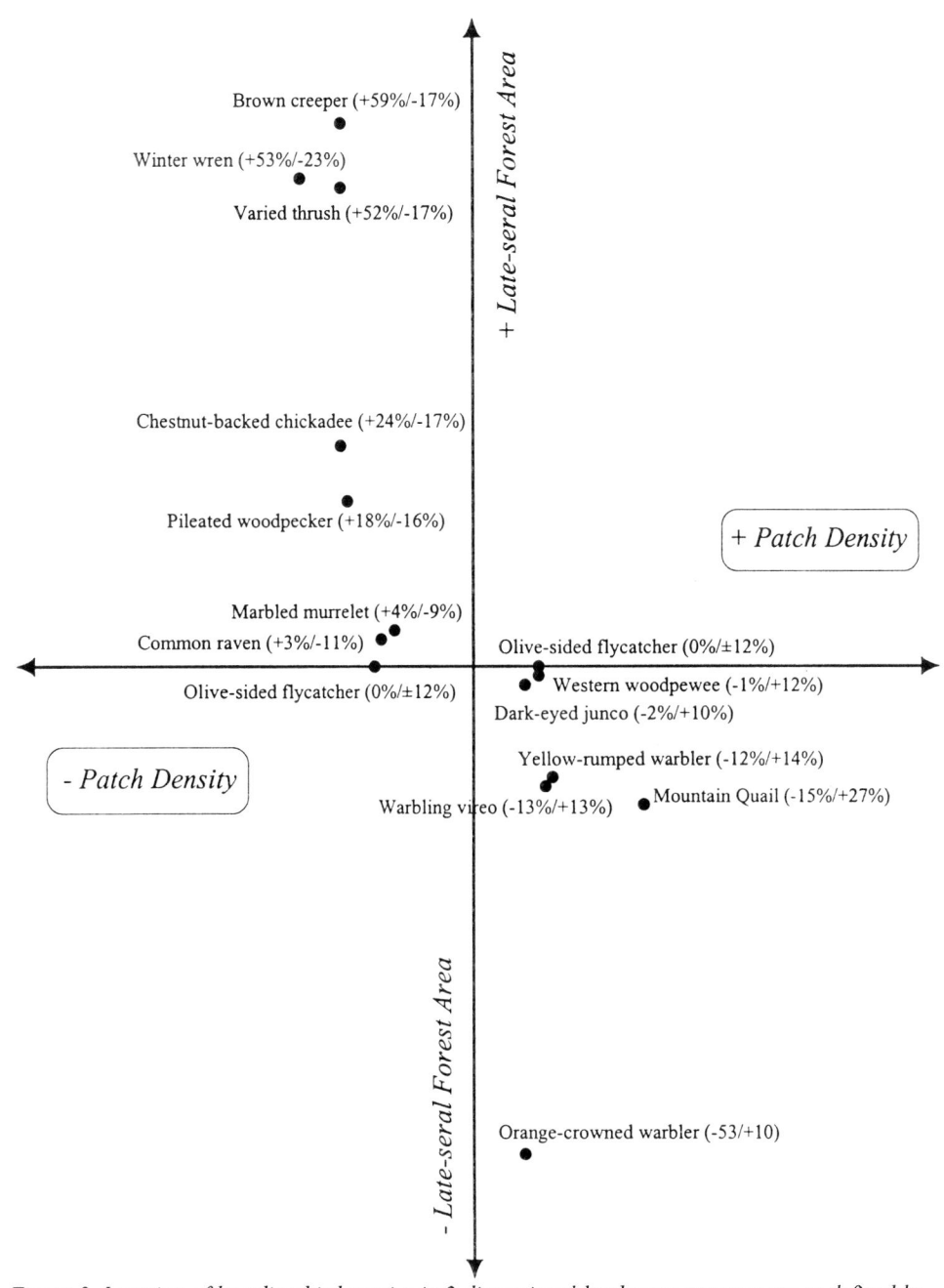

Forest Fragmentation Effects on Breeding Bird Communities 233

Figure 3. Location of breeding bird species in 2-dimensional landscape structure space defined by the magnitude (R^2) and direction (sign of coefficient) of the species' late-seral forest area relationship (y-axis) and configuration relationship (x-axis) for gradients in (a) patch density, (b) mean patch size, (c) mean shape index, (d) total core area index, and (e) contrast-weighted edge density in 30 300-ha landscapes in Drift Creek, Lobster Creek, and Nestucca River basins in Benton, Lincoln, and Tillamook Counties, Oregon, 1990-92. The numbers in parentheses are R^2 values associated with the late-seral forest relationship (Table 2) and the corresponding configuration gradient. See Appendix A for scientific names.

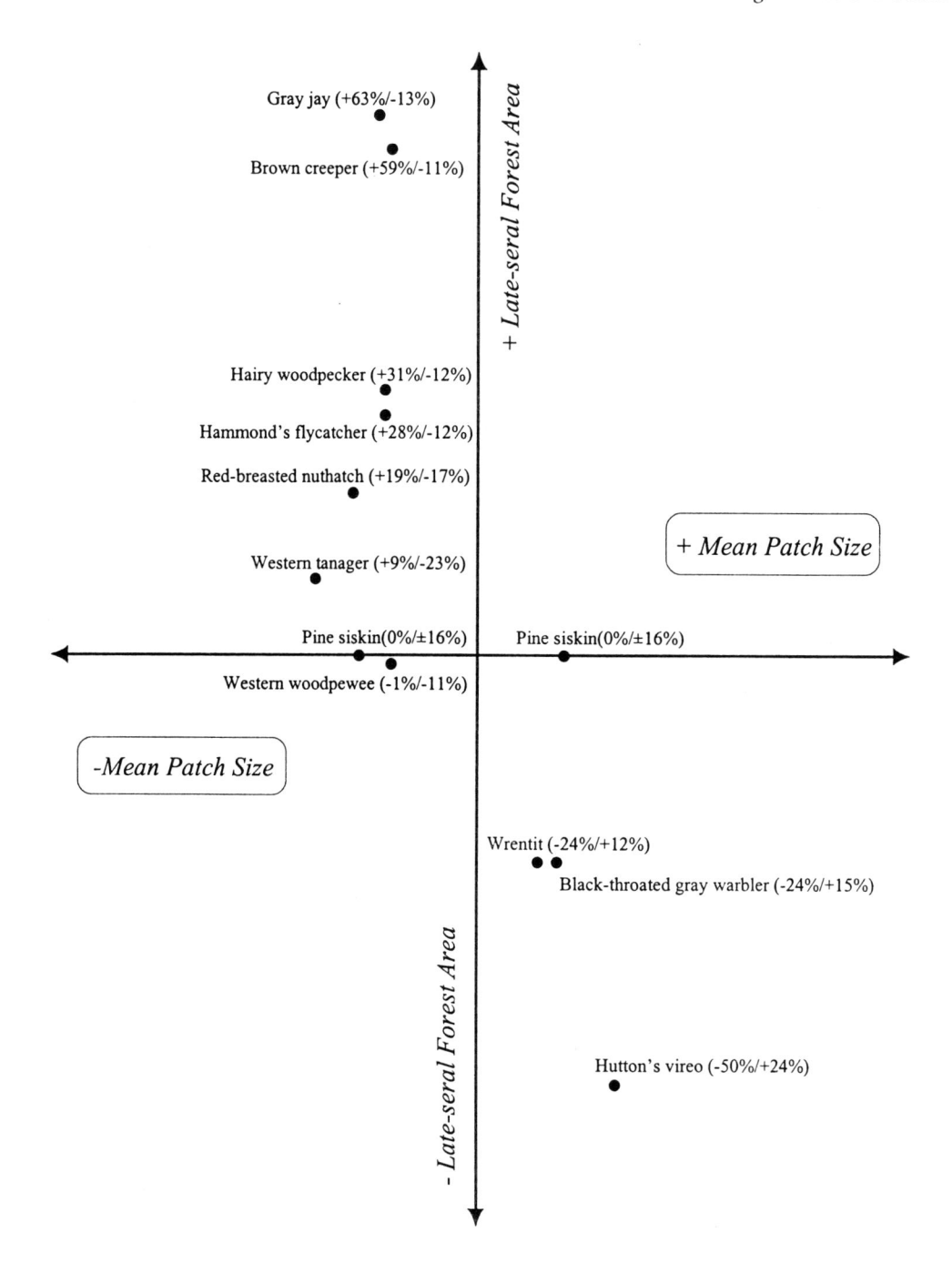

Figure 3b.

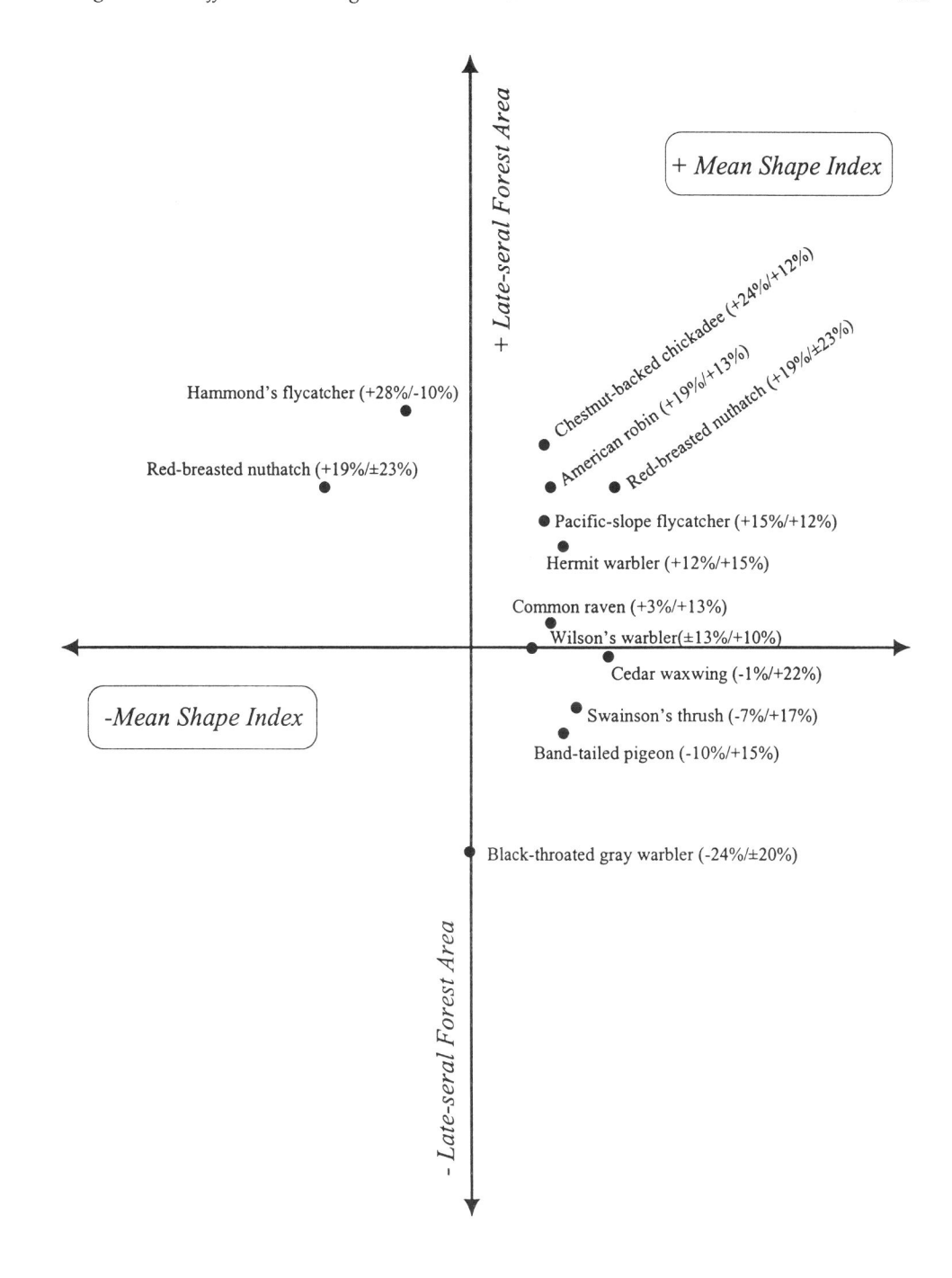

Forest Fragmentation Effects on Breeding Bird Communities 235

Figure 3c.

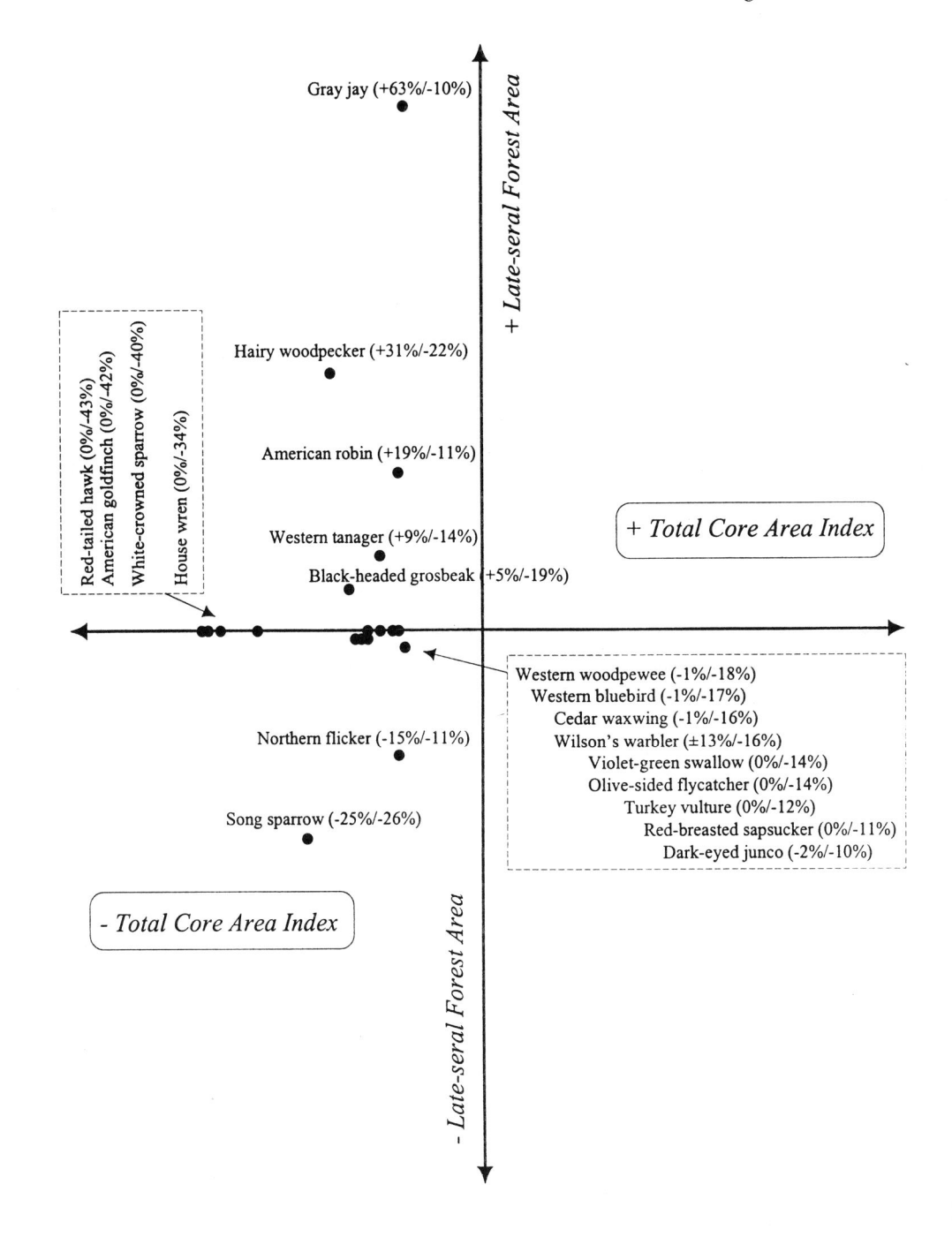

Figure 3d.

Forest Fragmentation Effects on Breeding Bird Communities 237

+ *Late-seral Forest Area*

+*Contrast-Weighted Edge Density*

Hairy woodpecker (+31%/±42%)

Western tanager (+9%/+18%)

Hairy woodpecker (+31%/± 42%)

Dark-eyed junco (-2%/+13%)

Red crossbill (+3%/± 20%)
Red-tailed hawk (0%/+22%)
Violet-green swallow (0%/+29%)
American goldfinch (0%/-31%)
Olive-sided flycatcher (0%/+41%)
White-crowned sparrow (0%/+50%)

Red crossbill (+3%/± 20%)

Ruffed grouse (-2%/-17%)

Red-breasted sapsucker (0%/+19%)
Turkey vulture (0%/+19%)
Western bluebird (-1%/+26%)
Western woodpewee (-1%/+34%)
House wren (0%/+48%)

−*Contrast-Weighted Edge Density*

Northern flicker (-15%/+14%)

Black-throated gray warbler (-24%/-13%)

Macgillivray's warbler (-23%/+13%)
Song sparrow (-25%/+17%)

− *Late-seral Forest Area*

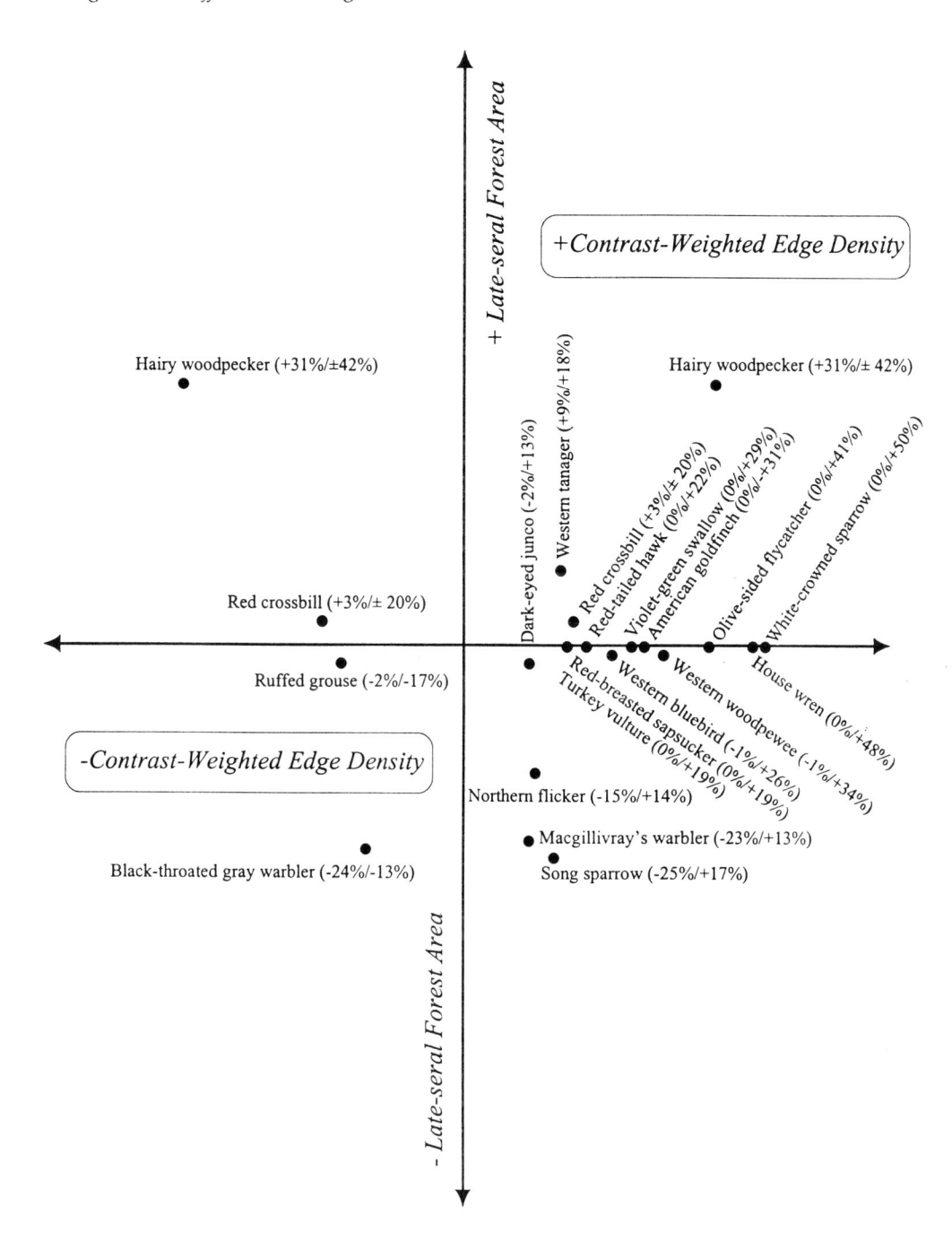

Figure 3e.

4.3 Late-seral Forest Area versus Configuration

Of the 31 species with a significant late-seral forest area relationship at the landscape level, a parsimonious combination of configuration metrics explained an additional 0% to 53% of the variation in abundance among landscapes.

Twelve of these species did not have a significant late-seral forest configuration relationship after area was taken into account. For most species (81%), late-seral forest area was the most significant explanatory variable, although the relative importance of configuration increased as the strength of the area relationship decreased (Fig. 4). Specifically, late-seral forest configuration was generally unimportant for species with strong area relationships, and was of equal or greater importance for species with weak area relationships.

5. DISCUSSION

Our results demonstrate that abundances of many breeding bird species are related to the extent and configuration of late-seral forest at the scale of 300-ha landscapes, although the magnitude and nature of these relationships vary dramatically among species. Based on a conservative ANOVA approach, 42% (23/55) of the species demonstrated a significant effect of late-seral forest area on abundance (Table 2) and 9% (5/55) of the species showed a significant configuration effect (Table 3). Based on the more liberal regression approach, 56% (31/55) and 82% (45/55) of the species had significant late-seral forest area and configuration relationships, respectively. Consequently, we conclude that forest management activities that alter the extent and configuration of late-seral forest will likely affect the abundances of many breeding bird species. However, roughly an equal number of species demonstrated positive and negative area relationships. Therefore, any change in late-seral forest area will likely have a compensatory effect on the community; that is, an equal number of species will benefit as will suffer by changes in late-seral forest area. In addition, several species exhibited complex, nonlinear relationships between abundance and late-seral forest area. In particular, 4 species increased in abundance at intermediate levels of late-seral forest area (Table 2). Overall, given the compensatory relationship between late- and early-seral associates, species richness and diversity was greatest at intermediate levels of late-seral forest area, suggesting that intermediate levels of disturbance at the landscape level may sustain greater levels of diversity.

Contrary to the idea that habitat fragmentation is detrimental to species that specialize on a particular habitat, most species that exhibited significant relationships with late-seral forest configuration in our study were associated with the more fragmented distribution of late-seral forest (although see Scope and Limitations). Like late-seral forest area, the relationship between late-seral forest configuration and bird abundance varied dramatically among species. Not surprisingly, the strongest configuration affects were for species associated with the juxtaposition of late- and early-seral forest. But even for these species, the percent of variation in abundance among landscapes attributable to late-seral forest configuration was moderate (Fig. 3).

Although late-seral forest fragmentation affects the abundances of many species, conservationists are most concerned with negative impacts on species closely associated with late-seral forest. Eleven species demonstrated positive linear relationships with late-seral forest area at the landscape level (Table 2). Most of these species also demonstrated exclusive statistical selection for late-seral forest at the patch-level and thus can be considered late-seral forest associates (McGarigal and McComb 1995). These are the species most likely to be adversely impacted by late-seral forest loss. However, despite statistical selection for late-seral forest by most of these species, all of these species were detected frequently in younger seral patch types as well (McGarigal and McComb 1995). Moreover, much of the variation in these species' abundances at the landscape level cannot be explained by either late-seral forest area (R^2, 12-63%) or configuration (R^2, 0-27%). Given the broad realized niche of these species with respect to forest stand condition, and the relatively weak relationships between abundance and late-seral forest fragmentation, it is not clear how these populations

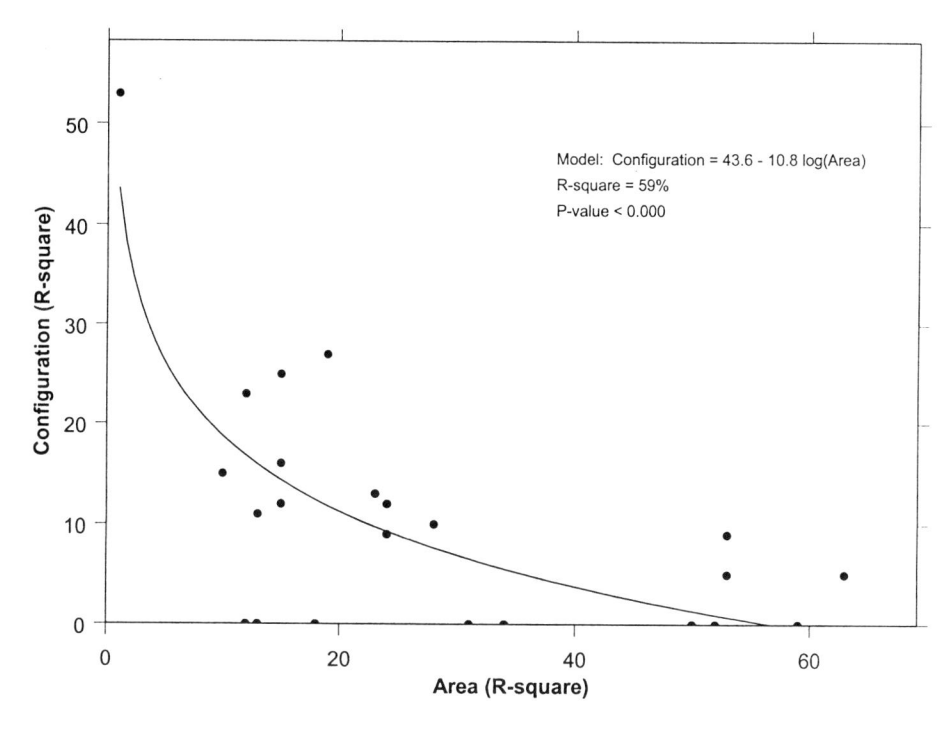

Figure 4. Relationship between the magnitude of late-seral forest area relationships and late-seral forest configuration relationships for breeding bird species with significant area relationships, where the area and configuration relationships represent the R^2 values associated with general linear or nonlinear models of species' abundances in 30 300-ha landscapes in Drift Creek, Lobster Creek, and Nestucca River basins in Benton, Lincoln, and Tillamook Counties, Oregon, 1990-92.

ultimately might respond to further habitat loss and fragmentation.

What is more important: late-seral forest area or configuration? The answer partly depends on the method used to analyze the data. Based on the ANOVA, late-seral forest area was clearly more important than configuration; 23 species had significant area effects, whereas only 5 species had significant configuration effects. Based on the regression procedures, the answer is more complex. Late-seral forest area alone explained between 10-63% of the variation in species' abundances among landscapes for 31 species. Late-seral forest configuration alone explained between 10-50% of the variation in species' abundances among landscapes for 45 species. However, many of these species were "edge" species associated with the juxtaposition of early- and late-seral forest patches. For species positively associated with late-seral forest area, with two exceptions, variation in abundance among landscapes was more strongly related to changes in area; configuration was of secondary importance. This finding is consistent with simulation results that suggest that habitat loss has a much larger effect than habitat fragmentation (configuration) on population extinction (Fahrig 1997), and we concur with Fahrig that current emphasis on habitat spatial pattern may be misplaced because the "details of how habitats are arranged" (Kareiva and Wennergren 1995) are unlikely to mitigate the risks of habitat loss.

Our findings are consistent with the few other studies from the PNW region (Rosenberg and Raphael 1986, Lehmkuhl et al. 1991) and from comparable avian studies conducted in other North American forest-dominated landscapes (e.g., Welsh and Healy 1993, Thompson et al. 1992, Keller and Anderson 1992, Derleth et al. 1989). Moreover, our findings are not surprising given the scale of our investigation and characteristics of the broader regional landscape context. Late-seral forest still comprises the matrix throughout much of our study area. Hence, large regional source populations of late-seral forest species may be able to offset any local habitat configuration-related effects, because local bird abundance patterns are produced not only by local processes or events, but also by the dynamics of regional populations or events elsewhere in the species' range (Wiens 1981, Vaisanen et al. 1986, Haila et al. 1987, Ricklefs 1987, Wiens 1989b). In our study area, if mechanisms are operating on species associated with late-seral forest to reduce population abundance in fragmented landscapes, immigration from large regional source populations may be sufficient to offset any tendency for local population declines. This would be particularly likely for vagile species such as birds that can disperse over large distances easily. Thus, the limited evidence gathered so far for these forest landscapes suggests that we should not blindly accept the fragmentation dogma. Habitat configuration and subdivision undoubtedly play a role in regulating population abundance, distribution, and dynamics, but the magnitude and nature of this role may vary geographically and over time in relation to changes in regional habitat conditions and other factors, and probably varies among species in relation to habitat selectivity, vagility, and scale.

6. SCOPE AND LIMITATIONS

It is critical that the scope and limitations of this study be made explicit before applying the results in a management context (see McGarigal and McComb 1995 for a more detailed discussion). First, the scale of our investigation placed upper and lower limits of resolution on our ability to detect habitat configurations and assess bird-habitat relationships (Wiens 1989a). Populations of the species we investigated undoubtedly extend over much larger areas and are subject to demographic influences operating over correspondingly larger areas than the small 300-ha landscapes we investigated. Thus, the bird species we analyzed may be more sensitive to variations in habitat area and configuration at a coarser (or finer) scale than we investigated.

Second, we focused this investigation on the fragmentation of late-seral forest. Although this broad class is believed to be important to a wide variety of wildlife in western Oregon and Washington (Bruce et al. 1985), other habitat features (e.g., snags, vertical foliage diversity, understory vegetation) not effectively captured in this simple classification might be more important in governing the abundance and spatial distribution patterns of any particular species. Thus, our findings do not preclude stronger fragmentation effects on species associated with habitats defined at a finer resolution.

Third, our analysis was limited to diurnal birds during the breeding season. It is unknown how late-seral forest area and configuration affect habitat use patterns for winter residents. Finally, for statistical reasons, our analysis excluded rare and uncommon species (e.g., spotted owl, Vaux's swift), those with patchy distributions in the study area (e.g., downy woodpecker), and those which have extremely low detection rates (e.g., Cooper's hawk, sharp-shinned hawk). Most of these species are not associated with late-seral forest; therefore, there is little reason to believe that late-seral forest fragmentation has any direct impacts on these species. However, a few of these species are clearly associated with late-seral forest (e.g., spotted owl) or some component of late-seral forest (e.g., Vaux's swift), and could very well be the ones most sensitive to landscape structure (Terborgh and Winter 1980, Pimm et al. 1988).

7. CONCLUSIONS

Given the scope and limitations of this field study, the many potential sources of error in measuring bird abundance, and the natural variation in wildlife populations, it is not too surprising that we did not detect stronger relationships. Indeed, given these considerations, the strength of our findings provides strong empirical evidence that landscape structure at the scale of 300-ha watersheds is probably important to several bird species in the central Oregon Coast Range. In addition, our study provides empirical evidence that late-seral forest area is probably more important than configuration. Thus, land managers should focus first on maintaining sufficient habitat area and secondarily consider the details of how habitat is arranged. Finally, our study demonstrates that landscape ecological relationships can and should be studied in the field to test theoretical concepts and challenge dogma.

Acknowledgments. Many individuals contributed toward the completion of this research and we are indebted to all of them. We are especially grateful to John Mullen for his dedicated efforts throughout the duration of the study and to Sue Schlosser, Jim Fairchild, Nobuya Suzuki, Melissa Platt, Ken Burton, Jim Kiser, Gody Spycher, Barbara Marks, and Scott Splean for assistance in various phases of this project. Lenore Fahrig and Steven Garman provided thoughtful reviews of an early draft of this chapter. Funding for this research was provided through the Coastal Oregon Productivity Enhancement (COPE) program; COPE is a cooperative research and technology transfer effort among Oregon State University, USDA Forest Service, USDI Bureau of Land Management, other state and federal agencies, forest industry, county governments, and resource protection organizations.

LITERATURE CITED

Brown, E. R. (tech. ed). 1985. Management of wildlife and fish habitats in forests of Western Oregon and Washington. Part 2--Appendices. U.S. Dep. Agric. Publ. No. R6-F&WL-192-1985. 302 pp.

Bruce, C., D. Edwards, K. Mellen, A. McMillan, T. Owens., and H. Sturgis. 1985. Wildlife relationships to plant communities and stand conditions. Pages 33-56 In E. R. Brown, tech. ed. Management of wildlife and fish habitats in forests of Western Oregon and Washington. U.S. Dep. Agric. Publ. No. R6-F&WL-192-1985.

Derleth, E. L., D. G. McAuley, and T. J. Dwyer. 1989. Avian community response to small-scale habitat disturbance in Maine. Canadian Journal of Zoology 67:385-390.

Fahrig, L. 1997. Relative effects of habitat loss and fragmentation on population extinction. J. Wildl. Manage. 61:603-610.

Fahrig, L., and G. Merriam. 1994. Conservation of fragmented populations. Conserv. Biol. 8:50-59.

FEMAT (Forest Ecosystem Management Assessment Team). 1993. Forest ecosystem management: An ecological, economic and social assessment. Washington DC. US Gov. Printing Office, no. 1993-793-071.

Franklin, J. F., and C. T. Dyrness. 1973. Natural vegetation of Oregon and Washington. U.S. For. Serv. Gen. Tech. Rep. PNW-118. 48 pp.

Franklin, J. F., and R. T. T. Forman. 1987. Creating landscape configuration by forest cutting: ecological consequences and principles. Landscape Ecology 1:5-18.

Haila, Y., I. K. Hanski, and S. Raivio. 1987. Breeding bird distribution in fragmented coniferous taiga in southern Finland. Ornis Fennica 64:90-106.

Hejl, S. J. 1992. The importance of landscape patterns to bird diversity: a perspective from the northern Rocky Mountains. Northwest Environmental Journal 8:119-137.

Hunter, J. E., R. J. Gutierrez, and A. B. Franklin. 1995. Habitat configuration around spotted owl sites in northern California. Condor 97:684-693.

Kareiva, P., and U. Wennergren. 1995. Connecting landscape patterns to ecosystem and population processes. Nature 373:299-302.

Keller, M. E., and S. H. Anderson. 1992. Avian use of habitat configurations created by forest cutting in southeastern Wyoming. The Condor 94:55-65.

Lehmkuhl, J. F., and L. F. Ruggiero. 1991. Forest fragmentation in the Pacific Northwest and its potential effects on wildlife. Pages 35-46 In L. F. Ruggiero, K.B. Aubry, A. B. Carey, and M. H. Huff, tech. coords. Wildlife and vegetation of unmanaged Douglas-fir forests. U.S. For. Serv. Gen. Tech. Rep. PNW-285.

Lehmkuhl, J. F., L. F. Ruggiero, and P. A. Hall. 1991. Landscape-scale configurations of forest fragmentation and wildlife richness and abundance in the Southern Washington Cascade Range. Pages 425-442 In L. F.

Ruggiero, K. B. Aubry, A. B. Carey, and M. H. Huff, tech. coords. Wildlife and vegetation of unmanaged Douglas-fir forests. U.S. For. Serv. Gen. Tech. Rep. PNW-285.

Li, H., and J.F. Reynolds. 1995. On definition of quantification of heterogeneity, Oikos 73: 280-4.

McGarigal, K., and B. J. Marks. 1995. FRAGSTATS: spatial pattern analysis program for quantifying landscape structure. U.S. For. Serv. Gen. Tech. Rep. PNW-351.

McGarigal, K., and W. C. McComb. 1995. Relationship between landscape structure and breeding birds in the Oregon Coast Range. Ecol. Monogr. 65(3):235-260.

Pimm, S. L., H. L. Jones, and J. M. Diamond. 1988. On the risk of extinction. American Naturalist 132:757-785.

Reynolds, R. T., J. M. Scott, and R. A. Nussbaum. 1980. A variable circular-plot method for estimating bird numbers. The Condor 82:309-313.

Ricklefs, R. E. 1987. Community diversity: relative roles of local and regional processes. Science 235:167-171.

Robbins, C. S., D. K. Dawson, and B. A. Dowell. 1989. Habitat area requirements of breeding forest birds of the middle Atlantic states. Wildlife Monographs 103. 34 pp.

Rosenberg, K. V., and M. G. Raphael. 1986. Effects of forest fragmentation on vertebrates in Douglas-fir forest. Pages 263-272 In J. Verner, M. l. Morrison, and C. J. Ralph, eds. Wildlife 2000: modeling habitat relationships of terrestrial vertebrates. University of Wisconsin Press, Madison.

Smith, D. M. 1986. Practice of Silviculture, 8th ed. John Wiley & Sons, New York. 527 pp.

Spies, T. A. and S. P. Cline. 1988. Coarse woody debris in forests and plantations of coastal Oregon. Pages 5-24 In C. Maser, R. F. Tarrant, J. M. Trappe, and J. F. Franklin, tech. eds. From the forest to the sea: A story of fallen trees. U.S. For. Serv. Gen. Tech. Rep. PNW-229.

Swanson, F. J., J. F. Franklin, and J. R. Sedell. 1990. Landscape patterns, disturbance, and management in the Pacific Northwest, USA. Pages 191-213 In T. S. Zohneveld and R. T. T. Forman, eds. Changing Landscapes: An Ecological Perspective. Springer-Verlag, New York.

Swanson, F. J., T. K. Kratz, N. Caine, and R. G. Woodmansee. 1988. Landform effects on ecosystem patterns and processes. BioScience 38:92-98.

Terborgh, J. W. 1989. Where Have all the Birds Gone? Princeton University Press, New Jersey. 207 pp.

Terborgh, J. W., and B. Winter. 1980. Some causes of extinction. Pages 119-133 In M. E. Soule, ed. Conservation biology: the science of scarcity and diversity. Sinauer Associates, Sunderland, MS.

Thomas, J. W., E. D. Forsman, J. B. Lint, E. C. Meslow, B. R. Noon, and J. R. Verner. 1990. A conservation strategy for the northern spotted owl. Interagency Scientific committee to address the conservation of the northern spotted owl. USDA FS, USDI BLM, USDI FWS, USDI NPS. Portland OR: US Govt. Printing Office. 427 pp.

Thompson, F. R., W. D. Dijak, T. G. Kulowiec, and D. A. Hamilton. 1992. Breeding bird populations in Missouri Ozark forests with and without clearcutting. Journal of Wildlife Management 56:23-29.

Vaisanen, R. A., O. Jarvinen, and P. Rauhala. 1986. How are extensive, human-caused habitat alterations expressed on the scale of local bird populations in boreal forest? Ornis Scandinavica 17:282-292.

Van Horne, B. 1983. Density as a misleading indicator of habitat quality. Journal of Wildlife Management 47:893-901.

Wallin, D.O., F.J. Swanson, B. Marks, J. H. Cissel, and J. Kertis. 1996. Comparison of managed and pre-settlement landscape dynamics in forests of the Pacific Northwest, USA. Forest Ecology and Management 85:291-309.

Welsh, C. J. E., and W. M. Healy. 1993. Effect of even-aged timber management on bird species diversity and composition in northern hardwoods of New Hampshire. Wildlife Society Bulletin 21:143-154.

Whitcomb, R. F., C. S. Robbins, J. F. Lynch, B. L. Whitcomb, M. K. Klimkiewicz, and D Bystrak. 1981. Effects of forest fragmentation on avifauna of the eastern deciduous forest. Pages 125-205 In R. L. Burgess and D. M. Sharpe, eds. Forest

Island Dynamics in Man-Dominated Landscapes. Springer-Verlag, New York.

Wiens, J. A. 1981. Scale problems in avian censusing. Studies in Avian Biology Number 6:513-521.

Wiens, J. A. 1989a. Spatial scaling in ecology. Functional Ecology 3:385-397.

Wiens, J. A. 1989b. The Ecology of Bird Communities: Volume 2, Processes and Variations. Cambridge University Press, Cambridge. 316 pp.

Wiens, J. A., J. T. Rotenberry, and B. Van Horne. 1987. Habitat occupancy configurations of North American shrubsteppe birds: the effects of spatial scale. Oikos 48:132-147.

Wiens, J. A., N. C. Stenseth, B. Van Horne, and R. A. Ims. 1993. Ecological mechanisms and landscape ecology. Oikos 66:369-380.

APPENDIX A

Number of detections and effective detection distance for breeding bird species detected in 30 300-ha subbasins in Drift Creek, Lobster Creek, and Nestucca River basins in Benton, Lincoln, and Tillamook Counties, Oregon, 1990-92. Species are ordered from most to least common based on total number of detections.

Species	Scientific name	N_t^a	N_{new}^b	N_{subs}^c	Edd[d]
Swainson's thrush	*Catharus ustulatus*	8758	8457	30	80
Wilson's warbler	*Wilsonia pusilla*	8588	8149	30	90
Winter wren	*Troglodytes troglodytes*	8291	7757	30	95
Pacific slope flycatcher	*Empidonax difficilis*	6031	5769	30	80
Varied thrush	*Ixoreus naevius*	5133	3960	30	100
Hermit warbler	*Dendroica occidentalis*	4096	3604	30	100
Evening grosbeak	*Coccothraustes vespertinus*	3470	3164	30	95
Steller's jay	*Cyanocitta stelleri*	3256	2416	30	130
Red crossbill	*Loxia curvirostra*	2779	2547	30	95
Chestnut-backed chickadee	*Parus rufescens*	2626	2596	30	50
Black-headed grosbeak	*Pheucticus melanocephalus*	2043	1631	30	100
Song sparrow	*Melospiza melodia*	1981	1844	30	90
Orange-crowned warbler	*Vermivora celata*	1653	1466	28	100
Golden-crowned kinglet	*Regulus satrapa*	1601	1593	30	40
American robin	*Turdus migratorius*	1573	1387	30	100
Band-tailed pigeon	*Columba fasciata*	1349	1131	30	105
Warbling vireo	*Vireo gilvus*	1311	1080	29	110
Brown creeper	*Certhia americana*	1164	1145	29	55
Macgillivray's warbler	*Oporornis tolmiei*	1057	930	27	100
Hammond's flycatcher	*Empidonax hammondii*	984	897	30	90
White-crowned sparrow	*Zonotrichia leucophrys*	954	803	24	115
Dark-eyed junco	*Junco hyemalis*	907	842	30	95
Hairy woodpecker	*Picoides villosus*	898	788	30	100
Rufous-sided towhee	*Pipilo erythrophthalmus*	828	743	28	100
Western tanager	*Piranga ludoviciana*	741	587	29	115
Rufous hummingbird	*Selasphorus rufus*	642	640	30	20
Wrentit	*Chamaea fasciata*	636	547	23	100
Pileated woodpecker	*Dryocopus pileatus*	611	439	30	160
Purple finch	*Carpodacus purpureus*	600	525	30	100
Olive-sided flycatcher	*Contopus borealis*	540	306	25	175
Hutton's vireo	*Vireo huttoni*	537	496	29	95
American goldfinch	*Carduelis tristis*	533	504	29	85
Red-breasted nuthatch	*Sitta canadensis*	481	430	30	100
Northern flicker	*Colaptes auratus*	455	338	28	150
Gray jay	*Perisoreus canadensis*	418	385	29	85
Common Raven	*Corvus corax*	402	288	30	170

Continuing Appendix A

Species	Scientific name	N_t^a	N_{new}^b	N_{subs}^c	Edd^d
House wren	*Troglodytes aedon*	366	294	15	100
Willow flycatcher	*Empidonax traillii*	356	311	25	100
Black-throated gray warbler	*Dendroica nigrescens*	316	281	27	100
Cedar waxwing	*Bombycilla cedrorum*	253	249	24	50
Mountain quail	*Oreortyx pictus*	209	146	26	195
Marbled murrelet	*Brachyramphus marmoratum*	190	184	18	---
Bewick's wren	*Thryomanes bewickii*	190	175	25	100
Western wood-pewee	*Contopus sordidulus*	154	107	23	150
Northern pygmy owl	*Glaucidium gnoma*	141	94	23	180
Red-tailed hawk	*Buteo jamaicensis*	131	103	26	200
Red-breasted sapsucker	*Saphyrapicus ruber*	129	118	21	130
Violet-green swallow	*Tachycineta thalassina*	105	86	12	85
Hermit thrush	*Hylocichla guttata*	101	89	8	95
Bushtit	*Psaltriparus minimus*	91	91	17	40
Pine siskin	*Carduelis pinus*	66	64	13	80
Townsend's solitaire	*Myadestes townsendi*	66	49	6	115
Ruffed grouse	*Bonasa umbellus*	64	52	17	90
American Crow	*Corvus brachyrhynchos*	64	48	8	---
Common nighthawk	*Chordeiles minor*	57	54	17	---
Yellow-rumped warbler	*Dendroica coronata*	56	50	11	70
Western bluebird	*Sialia mexicana*	52	48	12	95
Tree swallow	*Tachycineta bicolor*	49	42	8	90
Black-capped chickadee	*Parus atricapillus*	42	35	8	75
Turkey vulture	*Cathartes aura*	35	35	16	---
Blue Grouse	*Dendragapus obscurus*	35	18	3	---
Downy Woodpecker	*Dendrocopos pubescens*	30	29	10	85
Great-horned owl	*Bubo virginianus*	18	18	7	---
Belted kingfisher	*Ceryle alcyon*	18	15	9	---
Common yellowthroat	*Geothlypis trichas*	13	12	4	---
Mallard	*Anas platyrhynchos*	12	12	5	---
Screech owl	*Otus kennicottii*	11	11	10	---
Brown-headed cowbird	*Molothrus ater*	10	10	5	---
Great blue heron	*Ardea herodias*	9	9	6	---
Spotted owl	*Strix occidentalis*	9	5	2	---
Hooded merganser	*Lophodytes cucullatus*	8	8	1	---
European starling	*Sturnus vulgaris*	7	7	2	---
Wood duck	*Aix sponsa*	6	6	2	---
Sharp-shinned hawk	*Accipiter striatus*	6	6	6	---
Solitary vireo	*Vireo solitarius*	6	4	3	---
American dipper	*Cinclus mexicanus*	5	5	3	---
Barn swallow	*Hirundo rustica*	4	4	2	---
Common merganser	*Mergus merganser*	4	4	1	---

Continuing Appendix A

Species	Scientific name	N_t^a	N_{new}^b	N_{subs}^c	Edd^d
Vaux's swift	*Chaetura vauxi*	4	4	4	---
Bald eagle	*Haliaeetus leucocephalus*	2	2	1	---
Canada goose	*Branta canadensis*	2	2	1	---
Cooper's Hawk	*Accipiter cooperii*	2	2	2	---
Red-winged blackbird	*Agelaius phoeniceus*	2	2	2	---
Northern saw-whet owl	*Aegolius acadicus*	2	2	2	---
Scrub jay	*Aphelocoma coerulescens*	1	1	1	---

[a] N_t = Total number of detections at all distances, including new and repeat detections of individuals from separate sampling points within a subbasin.

[b] N_{new} = Total number of detections at all distances, including only detections of new individuals within a subbasin.

[c] N_{subs} = Total number of subbasins out of 30 that the species was detected in..

[d] EDD = 75% cumulative detection distance; 75% of detections with estimated distance were ≤ EDD.

Using Landscape Design Principles to Promote Biodiversity in a Managed Forest

David C. McAllister[1], Ross W. Holloway[2] and Michael W. Schnee[2]

The Oregon Department of Forestry is currently developing a plan for the management of 615,000 acres of forestland in northwest Oregon. These lands were logged or burned over and have been regenerated through an aggressive reforestation program. The department has developed an integrated management approach called structure based management (SBM) with a goal of restoring more diverse wildlife habitats to these forests, while producing sustainable levels of commodities and revenues. The approach focuses on the development of a variety of forest stand types in a planned landscape and on continuous management of down woody debris, snags and other important components. A key element of implementing SBM is a landscape planning process that can be operationally applied. The integration of SBM at the landscape level is described in this paper. This approach is based on the application of key landscape management concepts, including recognition of various scales of analysis, the notion of managing for interior habitat areas as a core element of habitat patches, the allocation of these patches across the forested landscape based on a log-normal size distribution, and the development of an adaptive management framework to assess attainment of biodiversity goals . Methods and guidelines are described to assist the forest manager in formulating appropriate landscape designs. A specific example of this guidance is provided for a portion of the Tillamook State Forest. Included is a discussion of the monitoring and adaptive management framework that will be used to assure that the goals for the plan are achieved.

Key words: Oregon Department of Forestry; forest management; landscape design; wildlife habitat; stand types; biodiversity; active management; habitat patch; interior habitat; structure based management; landscape management.

1. INTRODUCTION

The Oregon Department of Forestry (ODF) is in the process of preparing a long-range forest management plan for 615,000 acres of state-owned forest land in northwest Oregon. A key component of this plan is a set of landscape design strategies with goals of providing for a range of wildlife habitats through the application of silvicultural activities that create a more diverse forest landscape, more diverse stand structures, and also providing sustainable levels of timber harvest and revenue. The purpose of this paper is to describe how principles of landscape design were used to develop strategies that can meet these goals and

---

[1] Oregon Department of Fish and Wildlife, Lands Program Manager, Habitat Conservation Division
[2] Oregon Department of Forestry, Northwest Oregon State Forest Management Plan

the mandates provided for the management of these lands.

2. HISTORY AND LEGAL MANDATES FOR STATE FOREST LAND MANAGEMENT

Most of the land managed by the ODF lies in the north coast range of Oregon, with the Tillamook State Forest comprising the largest contiguous block of state forest land. These state forest lands were almost exclusively in private ownership earlier in this century. Through a series of events related to the economic depression in the 1930's, large blocks of private forest land reverted to county ownership through tax foreclosure. Many of these lands were subsequently deeded to the Oregon Board of Forestry by the counties, in exchange for a share of future revenues derived from the management of the lands. State laws governing the management of the lands and defining the counties interests in that management are contained in Oregon Revised Statutes, Chapter 530, and in administrative rules adopted by the Board of Forestry.

Oregon administrative rules on State Forest Management Policy and Planning (OAR 629-035-0000 through 629-035-0110) call for the lands to be actively managed in a sound environmental manner to provide sustainable timber harvest and revenues to the state, counties, and local taxing districts. This management must be pursued within a broader context that also provides for properly functioning aquatic habitats, native wildlife habitats, soil, air and water, and recreational opportunities. The rule defines active management to mean "applying practices, over time and across the landscape, to achieve site-specific forest resource goals using an integrated and science-based approach that promotes the compatibility of most forest uses and resources over time and across the landscape." The rule also requires that forest management plan strategies for these forests "Contribute to biological diversity of forest stand types and structures at the landscape level and over time". It is these principles and mandates that have governed the development of an integrated forest management plan for the state forests of northwest Oregon.

The forested landscape of the north coast range has been shaped by both natural disturbances and by a series of human influences, ranging from pre-European settlement (native American) to Euro-American influences in the last two centuries. Prior to European settlement fire was a source of both large - and small-scale disturbances and came from two sources: lightning, and fires set by native Americans to burn grasslands, which sometimes spread into forests. In the Coast Range, these forest fires were relatively infrequent, but could be very large. (USDA Forest Service et al., 1994) While fire was already part of the northwestern Oregon landscape, the evidence indicates that the frequency of large fires increased in the 1840s, with the increasing number of Euro-American settlers (Pyne, 1982).

European settlement and influences on the forest accelerated in the late nineteenth century, with homesteading in the river valleys and on the fringes of the forest. Early logging in the forest followed construction of the railroads in the early twentieth century, but was limited to the eastern and western fringes of the forest. Loggers burned the slash after harvest to reduce the fire hazard, but did not plant trees. Many acres of timberland were allowed to go tax-delinquent after timber harvest. This practice increased during the Great Depression, and was common in areas burned by forest fires, such as the Tillamook Burn (Fick and Martin, 1992). As a result, several Oregon counties acquired large tracts of forest land through foreclosure and lacked the staff or resources to properly reforest and manage these lands. This process resulted in the eventual transfer of these lands to state ownership and management under the statutory arrangement referred to earlier.

3. STRUCTURE BASED MANAGEMENT

The forest management plan describes an integrated and comprehensive approach to the management of a broad range of forest resources. A new approach for state forest lands, called structure based management (SBM), is a central theme in the development of the

Northwest Oregon State Forest Management Plan. At the core of the SBM concept is the idea that managing forests to produce a variety of forest stand types in a planned landscape context can produce a diverse and sustainable flow of benefits. It is based on the idea that timber management and management for wildlife habitats and biodiversity are not mutually exclusive. (Hunter, 1990; McComb et al, 1993; Oliver, 1992). SBM integrates management strategies for timber, wildlife, fish, plants, forest health, and biodiversity. The approach combines a set of landscape, aquatic and riparian, and forest health strategies to promote the development of landscape conditions that will produce habitats for the range of indigenous species through management activities that provide a substantial and sustainable level of forest products. This forest landscape then serves as the backdrop to meet goals for a multitude of other resources or uses including but not limited to recreation, cultural resources, and water quality.

It will take many decades to fully implement the strategies and produce the targeted array of stand types on the landscape. Over time, monitoring and research will indicate the extent to which the assumptions underlying the strategies are correct and if the strategies are accomplishing their intended purpose. As monitoring provides feedback, the plan will be fine-tuned and improved through application of an adaptive management process.

SBM includes a set of management approaches, techniques and practices that encourage the development of a desired array of forest stand types across the landscape. The array ranges from open areas where new trees are being established to older forest structure that would feature "old growth" characteristics including numerous large trees, multi-layered canopies, substantial number of down logs and large snags. The individual stands themselves continue to evolve and change, but the range of stand types and their relative abundance across the landscape would be reasonably sustained. Because the structures are in a dynamic balance across the landscape, the forest will provide a steady flow of timber volume, jobs, habitats, and recreational opportunities through time.

Using the SBM approach, stand density will be actively managed to accelerate stand development. This will be done through periodic thinning and partial cutting. Thinning and partial cutting applied in a variety of prescriptions to produce variable densities can be used to produce a variety of results. Some prescriptions will result in fast-growing, well-stocked stands with minimal understories. Other prescriptions will develop more complex stand structures, with rapid tree diameter growth, enough sunlight on the forest floor to maintain understory plants, and a complex forest canopy. Thinning and partial cutting can also be used to create or maintain other important structural components, such as snags, down wood, gaps in the canopy, and multiple canopy layers. (Hayes et al, 1997)

Regeneration harvests will also occur. Regeneration harvesting approaches will include clearcuts that retain snags, down wood and residual live trees; patch cuts; shelterwood cuts; group selection cuts and possibly seed tree cuts.

For the array of stand types to successfully provide habitats for the range of indigenous wildlife species, careful planning and development of a variety of habitat patch types, sizes and arrangement is necessary. This landscape design component of the strategies is discussed in the next section. Some background on the forest stand types that are described in the forest management plan is necessary to understand the landscape design approach.

There are five forest stand types that are described in the plan. These same types apply to conifer, hardwood, and mixed stand compositions. The five descriptions developed by ODF are "snapshots" along forest development continuums that are intended to represent the range of forest conditions that have existed in the planning area over the last several thousand years (Oliver, 1996). *All stand types* will be managed to retain or develop snags, down wood, and other important components. The following five stand types descriptions are short summaries of the descriptions contained in the forest management plan:

(1) *Regeneration* (REG). The site is occupied primarily by tree seedlings or saplings, and

herbs or shrubs. The tree composition can be either conifers or hardwoods. Competition among the trees and other vegetation is not yet resulting in widespread loss of herb or shrub layers. The herbs and/or shrubs are widespread and vigorous. This type includes first-year regenerated stands, and continues to the stage when the trees approach crown closure. At that point, increasing competition causes a significant loss of vigor or death of understory vegetation.

(2) *Closed Single Canopy* (CSC). Trees fully occupy the site and form a single, main canopy layer. There is little or no understory development. Where understory vegetation exists, there is low shrub and herb diversity. The shrub and herb layers may be completely absent or may be short and dominated by one or two shade-tolerant species, such as sword fern, Oregon grape, oxalis or salal.

(3) *Understory* (UDS). These stands have developed more diverse herb or shrub layers than CSC stands and have trees larger than sapling size. Tree canopies may range from a single species, single-layered, main canopy with associated dominant, co-dominant and suppressed trees, to multiple species canopies. However, significant layering of tree crowns has not yet developed.

(4) *Layered* (LYR). The vertical organization and structure of the living plant community are more complex than in the understory type. Vertical layering of herbs, shrubs and tree crowns is extensive. Plant communities are complex in terms of numbers of species and in vertical arrangement. Shrub or herb layers and tree canopies in two or more layers are present.

(5) *Older Forest Structure* (OFS). This stand type contains the variety of trees, shrubs, and other understory vegetation found in layered stands. In addition to the variety of trees typically found in a layered stand, Older Forest Structure includes each of the following four characteristics.

(1) At least 8 or more live trees per acre that are at least 32 inches in diameter at breast height. For site classes 3, 4, or 5 on the Santiam State Forest at elevations greater than 3,000 feet, the diameter standard is lowered to at least 8 or more live trees per acre that are at least 24 inches in diameter breast height.

(2) Two or more tree canopy layers. Frequently one of the layers will be a shade-tolerant species.

(3) Snags — at least 6 per acre, 2 of which must be at least 24 inches in diameter breast height; the remaining 4 must be at least 12 inches in diameter breast height.

(4) 600 to 900 cubic feet per acre of sound down logs (decay class 1 or 2), or 3,000 to 4,500 cubic feet of down logs in any or all decay classes 1-5.

In addition, multiple tree species are encouraged as are trees with deeply fissured bark, large limbs, broken tops, evidence of fungal decay or other characteristics commonly associated with older forests.

OFS is intended to provide some or all of the structural components commonly associated with old growth. OFS will not necessarily emulate all the processes and functions of very old forests. The definition was derived through consultation with foresters and biologists and represents their best professional judgment, based on experience and scientific literature review (Franklin and Spies 1991; Hansen et al. 1991; Spies and Franklin 1991; Spies and Cline 1988). Over time, research and monitoring will provide better understanding about the similarities and differences between OFS and older forests.

The stand type percentages in Table A describe the desired future condition target for each district in the planning area. These are long-range targets, described with upper and lower limits as well as a mid-range percentage that is used for technical analysis. There is no specific timeframe within which the targets must be met, as current stand conditions for each district are quite variable and will require different timeframes to achieve.

The ranges were selected as a basis to provide habitats for all indigenous wildlife species, promote biodiversity, and provide timber and revenue. The percentages are a reasonable estimate of what is desirable to achieve, based on what we know today. Foresters and biologists on the planning team agreed that the ranges represented a reasonable place to start. It is likely that these ranges will change as more is learned through research and monitoring.

Table A. Target Stand Type Percentages

Stand Types: Percent of State Forest Landscape	
Regeneration	5-15 percent (10% used for analysis)
Closed Single Canopy	10-20 percent (15% used for analysis)
Understory	15-35 percent (25% used for analysis)
Layered	20-30 percent (25% used for analysis)
Older Forest Structure	20-30 percent (25% used for analysis)

4. MANAGING FOREST LANDSCAPES FOR BIODIVERSITY

Consideration of multiple species and their functional relationships (biodiversity) is an important part of modern forest management. Historically, forest management focused on certain "indicator species" that were thought to represent broader species assemblages. Consideration of these species' habitat was generally restricted to the stand or unit scale. Little attention was given to the positive or negative effects from the different aggregation of habitats across the landscape. More recently, various researchers have expressed concern about the lack of consideration for larger spatial scales, particularly the role of habitat fragmentation in reducing habitat quality and quantity for wildlife species (Franklin and Forman, 1987; Soulé, 1991; Ruggiero et al, 1991). Landscape management planning as a means to address the conservation of biodiversity is a relatively new concept in forest management. Examples of landscape management planning on federal lands include the regional scale (USDA Forest Service et al., 1994) approach and the district level (Diaz and Apostol 1992) plan. Development of a forest management plan for lands managed by the ODF presents an opportunity to apply landscape considerations to its management practices in order to more effectively manage biodiversity.

4.1 Overview of Landscape Management Concepts

Landscapes contain a mosaic of different habitat patches (Dunning et al, 1992). There is no single landscape for all classes of wildlife since each organism's requirements are different. What constitutes a single habitat patch for a deer may be a landscape for a salamander. Planning for biodiversity at the landscape level needs to recognize a range of spatial scales.

Landscapes are not of a particular size or shape; rather they are defined by the amount and arrangement of various habitat patches. Habitat patches may be defined as habitat units differing in quality for one or several species (Wiens, 1976). While a stand may be a convenient management unit for silvicultural planning, it may or may not be synonymous with the habitat patch for a particular class or individual wildlife species. Habitat patches are dynamic, occurring on a variety of spatial and temporal scales. For forested landscapes, habitats typically change in response to anthropogenic or natural disturbances such as silvicultural management or fire.

At any given scale, finer scale habitat patches, each of which is capable of supporting some aspect of the forest biodiversity, can be recognized. The lower size limit of a habitat patch is that size at which the species in question no longer perceives it as suitable habitat. The upper limit of size is typically defined by the species' home range (Kotliar and Wiens, 1990). Patch size for populations or subsets of populations (metapopulations) will be larger. Patch boundaries separating suitable and unsuitable habitat are only meaningful when considered at a particular scale and for a particular species. An abrupt change between patches for one species may actually be a continuous gradient of suitable patches for a different species.

The term matrix represents the dominant landscape patch in which other habitat patches reside (Franklin and Forman, 1987). At a given scale, the matrix exerts the greatest influence on the landscape in question. The relationship of the matrix to other embedded patches is known as fragmentation (Franklin and Forman, 1987). As fragmentation increases, the matrix decreases in size; becomes geometrically more complex and more isolated. Maximum fragmentation occurs when no single habitat patch dominates the landscape.

In forests of the Pacific Northwest fragmentation of the older forest matrix is of great concern. While research in Northwest forests does not provide clear evidence of negative effects of fragmentation on wildlife populations, (McGarigal and McComb, 1995; Rosenburg and Raphael, 1984), studies from other areas indicating that such responses occur have been generalized to forest lands (Whitcomb et al, 1981; Robbins et al, 1998). Those classes of wildlife generally considered most sensitive to fragmentation in Northwest forests are habitat specialists that prefer late-seral forest interiors and wide ranging species with low reproductive rates (Thomas et al, 1990). Rather than representing a single trajectory, fragmentation in forested landscapes is both temporally and spatially dynamic. The mix of seral conditions across a forested landscape does not represent clear distinctions in habitat suitability but rather gradations in suitability. The degree to which any class of wildlife is affected depends on the amount of habitat fragmentation and the relative suitability and pattern of surrounding habitat patches (McGarigal and McComb, 1995).

Landscapes do not exist in isolation. There is always a larger scale context within which several landscapes exist. Context is most important when organisms can easily move between landscapes. Recognition of the relationship of a particular species to its landscape and surrounding landscapes is essential to provide the proper context for management.

To accomplish biodiversity goals, landscapes should be evaluated in terms of both pattern and composition. Landscape composition, which includes habitat patch size and total patch area, is important to many ecological processes and to many species. Composition alone may fulfill certain species' habitat requirements. However, other organisms require additional considerations including those of habitat shape and placement relative to other patch types within the landscape to fulfill their population requirements. These attributes refer to landscape pattern. Using computer modeling, McKelvey et al. (1992) has shown that both landscape pattern and composition are important in northern spotted owl use of northwest forests. Northern spotted owls require some total area of older forest patches within some maximum area for occupancy (Thomas et al, 1990). Other landscape metrics potentially important for evaluating landscape effects on certain wildlife populations include mean and variability of patch size, shape, core area, density, nearest neighbor distance and connectivity (McGarigal and Marks, 1995).

For wildlife populations that benefit from the juxtaposition of different habitat patches, it may be the combination rather than the type of individual patch that is most important. The response of wildlife to this type of landscape is referred to as landscape complementation and landscape supplementation. Landscape complementation occurs when the presence of one type of resource in one patch is complemented by the close proximity of a different resource in a second patch so that larger populations can be supported in a given area. Deer and elk are examples of species benefiting from different habitats in close proximity (Wisdom et al, 1986). These wildlife species require both older forests and early-growth for different life history requirements. Similarly, certain bird species such as olive-sided flycatcher in the Coast Range are most abundant when older forest patches and open-canopy patches are juxtaposed. Older forests provide suitable nesting habitat while foraging habitat is found in the open-canopy areas (McGarigal and McComb, 1995). Landscape supplementation occurs when the particular juxtaposition of patches (similar or different) provides sufficient amounts of a given resource to sustain a population level above that provided in an individual patch. Examples include brown creepers, which require some maximum amount of large saw timber over some area to successfully occupy and breed (McGarigal and McComb, 1995).

Certain landscapes can affect wildlife populations through source/sink relationships. In these landscapes, productive source

Table B. Similarity coefficients between designated pairs of structural stages for wildlife species using each stand type as primary habitat

	REG	CSC seedling/sapling	CSC Pole	UDS	LYR	OFS
REG	1.0	.91	.53	.52	.43	.42
CSC seedling/sapling		1.0	.60	.59	.47	.46
CSC pole			1.0	.96	.69	.67
UDS				1.0	.73	.70
LYR					1.0	.97
OFS						1.0

patches supply emigrants to less productive patches termed sinks. Sub-populations within the sink areas are considered unstable and subject to extinction without new immigration from the source areas. In this manner, the total landscape functions to increase overall populations from a relatively small amount of source habitat. Maintenance of local sink populations within the landscape is dependent on the continued presence and proximity to source areas. Both landscape composition and pattern of source and sink patches can have an influence on overall population size (Thomas et al, 1990).

Three factors have been postulated as defining the functional patch size for meeting biodiversity goals: 1) actual size; 2) distance from a similar patch, and; 3) degree of habitat difference of the intervening matrix (Harris, 1984). The presence of a species in a particular patch can be strongly affected by the composition of adjacent patches. The following table (Table B) taken from Harris (1984, pg.112) and adjusted to SBM stand types defined within the Northwest Forest Plan illustrates this relationship. Data from Table B indicates that SBM types with comparable structural characteristics contain greater species similarity than those types that are structurally more distinct.

These neighborhood effects or edge contrasts can be either positive or negative. In the case of habitat generalists such as deer and elk, the edge between different patches of habitat is generally considered important to the population. For other species, notably interior habitat specialists, high contrast edge can have negative effects. Rosenberg and Raphael (1984) found that for mature forest patch sizes less than 120 acres the frequency of interior habitat species observations was negatively correlated with the presence and amount of adjacent regeneration and young forest patches. The decrease in interior habitat specialists noted by these authors is not well understood. It could have resulted from several factors including predation, competition, and nest parasitism from species occupying adjacent patches. It could also be the result of changes in habitat quality due to microclimatic changes within older forest patches due to increased light intensities, wind, and other unbuffered climatic factors from surrounding open areas. Chen (1991) determined for Douglas-fir forests that high contrast edge affected certain biological and microclimatic factors from 20-240 meters into the forest depending on the variable examined.

Corridors are thought to act opposite to boundaries. Corridors can facilitate movement of individuals between habitat patches, serving to connect separate but similar habitat patches within the landscape mosaic. They may act to channel dispersing individuals into pathways between patches or provide "intermediate" habitat of sufficient quantity and quality for survival until the species can find suitable habitat in another patch. The presence and location of corridors may provide important contributions to the functionality of similar patches within a landscape.

In Oregon, the most important forested wildlife habitat to consider in landscape planning is mature forest habitat. It is considered important because of its limited supply and because of the reliance on this habitat by over 118 species for either some or all of their life history requirements (Harris, 1984). Emphasizing management for mature forest habitat also ensures that other habitats will be maintained through the course of expected forest development.

The quantity of effective mature forest habitat is often smaller than the total amount of this habitat within a given landscape because of edge impacts. Interior habitat area (IHA) is defined as that portion of the mature forest patch that remains functional for late successional species after the negative effects of high contrast edge are removed (Spies et al, 1994). Two factors influence the amount of IHA in relation to total patch size: the degree of edge contrast with adjacent patches and patch configuration which changes the amount of edge and hence the amount of IHA. For a given patch configuration, IHA is smallest when edge contrast is highest (Table B). IHA also decreases when patch shape increases the amount of edge relative to interior portions of the patch.

Not all mature forest-dependent wildlife need the same size IHA to assure population maintenance. To assure adequate IHA patch sizes are maintained to meet biodiversity goals across the landscape, a range of IHA patch sizes needs to be attained. Harris (1984) argues for a distribution of sizes based upon three factors: the form of distribution, a measure of the central tendency (mean), and a measure of dispersion (variance). Harris (1984) puts forth several arguments for using a log-normal distribution to define the size and number of habitat patches for maintenance of wildlife diversity. These arguments relate to the relationships within wildlife populations for home range size, abundance, and spatial movement which all tend to follow a log-normal distribution. A log-normal distribution also tends to represent the scale and frequency of certain landscape disturbance processes such as fire and windstorms, and it seems to conform to physical relationships at the landscape scale such as watershed area and stream length (Strahler, 1957; Shugart, 1984). A theoretical variance for many of these relationships has been calculated to be 0.2. Mean patch size of the distribution is dependent on the type of species to be emphasized. For those with larger home ranges, a larger mean patch size is necessary than for species with smaller home range sizes. A mean patch size somewhere in the middle of this range is best for conserving overall biodiversity.

4.2 Application of Landscape Concepts to Forest Planning

In the landscape plan that was developed, it was assumed that wildlife use discriminates between SBM stand types according to the general relationship presented in Table B. We further chose to emphasize IHAs in our landscape planning based upon the following:

- Biodiversity is optimized by providing a range of forest stand types. Because IHAs are only associated with mature forest patches, managing for IHAs ensures that other habitats will also be represented through the course of development in managed forest stands.
- Species associated with IHAs are usually the limiting factor in reaching biodiversity goals in forested landscapes.
- Mature forest acreage that makes up IHAs is limited within the planning area.

IHAs are allocated across the planning area using two principal criteria. The first criterion defines the parameters for IHA habitat. The following parameters are used:

- Based upon the relationships presented in Table B, IHA patches can be made up of SBM stand types -OFS, LYR, and UDS stands depending on patch size and placement;
- IHA patches should be centered on OFS stands;
- patch size must consider adjoining structural types. Microclimate changes are considered to reduce IHA patch size accordingly, with REG types having a greater influence than CSC;
- an average distance of 100 meters is a reasonable approximation of edge impacts when calculating IHA patch size when adjacent to REG and CSC.

Specifying patch characteristics to include structurally similar stand types allows the

Table C. *Summary of IHA Patch Sizes for a hypothetical 250,000 acre planning unit using a log-normal distribution of mean 250 acres, variance 0.2, and minimum size of 40 acres.*

Number of IHA Patches	IHA Patch Range (Acres)	Midpoint of IHA Patch Size (Acres)
63	0-80	(40)
128	80-120	(100)
85	120-200	(160)
68	200-320	(260)
41	320-520	(420)
19	520-840	(680)
7	840-1360	(1100)
2	1360-2180	(1780)
0.5	>2180	(2880)

opportunity to increase the amount and size of IHA patches across the landscape and reduce the average distance between units. This also allows the opportunity to maintain IHA patch size through the course of future management by recruiting adjacent managed stands to replace IHA components removed through management activities such as commercial thinning and regeneration harvesting.

The second criterion defined the number and range of IHA patch sizes for a given planning area. To determine this distribution we followed Harris (1984) using a log-normal distribution with the following parameters:
- average patch size 250 acres,
- variance 0.2,
- minimum patch size 40 acres.

An example of the application of the log-normal distribution can be found in Table C for a hypothetical 250,000 acre planning unit. The following assumptions were applied in creating the distribution: 90 percent of the area was managed habitat and 90,000 acres allocated to LYR and OFS stands. Table C indicates the central tendency of the distribution emphasizes smaller IHA patches relative to larger size patches. The largest size IHA patch calculated from the distribution was 2180 acres. This distribution of IHA patch sizes from the example landscape appears to conform to Harris (1984) theoretical patch size requirements for maintaining biodiversity over a given landscape.

4.3 Guidelines for Habitat Patch placement across the Landscape

Application of landscape planning to meet biodiversity goals depends on providing a full range of habitat conditions at a range of scales. We have chosen to call this coarse and fine scale planning. The coarse scale included landscape planning from the regional to the stand level. Fine scale planning generally emphasized stand level and smaller scales, but also recognized consideration of unique habitats. The number of different patches; their size, shape, location and relationship to other patches within a landscape represent coarse scale biodiversity planning. Fine scale planning includes down wood, snag retention, and structural targets, which are summarized in Table D. Also considered at the fine scale are unique habitats that comprise such landscape features as caves, seeps, wetlands and talus slopes. These areas provide important habitat conditions for a limited array of habitat specialists.

Coarse scale planning used individual stands of similar structure as the basic building blocks to form different sized patches of similar habitat value. These patches are then arranged across the landscape through time and place guided by the goals for biodiversity

Table D. Guidelines for Incorporating Structural Habitat Components at the Landscape Level

(1) Remnant old growth trees — Retain scattered remnant old growth trees or patches of old growth

(2) Snags — The target is at least 187 snags per 100 acres. These should include snag numbers by diameter class and decay class as listed below (modified from Brown et al. 1985). Larger snags may be substituted for smaller snags, but not the other way around. Snags should be retained in a variety of arrangements throughout the landscape. Uniform or random distributions as well as dispersed clumping should all be used to provide for a variety of habitat and predator/prey conditions. Excessive use of any one technique will not result in the greatest overall benefits for the broad range of wildlife.

Snag Targets by Diameter Class and Decay Class

Snag DBH Class[1]	Hard Snags (Class 2-3)	Soft Snags (Class 4-5)	Total Snags by DBH Class
11+ inches	—	10	10
15+ inches	27	115	142
17+ inches	—	29	29
25+ inches	6	—	6
Total snags	33	154	187

(3) Residual live trees — When regeneration harvesting, retain sufficient snags and residual live trees to maintain target snag levels throughout the next rotation. In addition, retain sufficient live trees to provide legacy structures to potentially allow targeted stands to develop into older forest structure. The following suggestions are offered as important considerations — not prescriptions.

Regardless of the targeted stand type for a particular stand, an average of 5 green trees/acre (in addition to existing snags) is recommended to enhance snag and down wood recruitment throughout the rotation. Where OFS is the targeted stand type consider leaving additional green trees to encourage even more recruitment of snags and down wood.

Retained trees should include a component of defective trees that contribute to wildlife habitat in the short term and the long term as they die to create future snags. Retained trees also should include a component of sound, healthy trees with good crowns, that will continue to grow and become "large residual trees", an important component of older forest structure. A component of hardwood leave trees also is desirable, especially bigleaf maple and/or Oregon white oak when available.

(4) Down woody debris — The retention target is 600 to 900 cubic feet of hard conifer logs per acre, including at least 2 logs per acre greater than 24 inches in diameter (at the largest end) and ultimately to achieve 3000 to 4500 cubic feet of down wood distributed throughout all decay classes.

(5) Multi-layered forest canopies — Complex layering of forest canopies generally creates diverse habitat niches and benefits biodiversity. Consistent with all management objectives, manage vegetative communities to create complex multi-canopied forests or at least to increase the amount of layering in most stands.

In order to meet the stand structure criteria for the layered and older forest structure stands, it is necessary to develop multiple canopies in many stands. Stands managed in the closed single canopy type will not have multi-layered canopies.

(6) Multiple native tree species (conifers and hardwoods) — Increased tree species diversity within and among stands generally creates more diverse habitat niches and benefits biodiversity. Consistent with sound silviculture and all management objectives, manage to include a variety of native species. Individual stands may be predominantly single species (conifers or hardwoods), and the forest overall may be predominantly conifer. However, maintaining or establishing components of other species (conifers and hardwoods) is desirable.

(7) Herb/shrub considerations — Diverse herb and shrub vegetation layers provide important forage for wildlife, provide diverse habitat niches, and benefit biodiversity. Consistent with all management objectives, manage vegetative communities to encourage diverse herb and shrub layers.

(8) Gaps — Gaps increase the horizontal diversity within stands, provide important forage for wildlife, provide diverse habitat niches, and benefit biodiversity. Consistent with all management objectives, manage stands for gaps to provide horizontal diversity. Natural openings due to windthrow, insects, and disease, etc. will suffice in many cases. However, where a deficiency exists, consider creating gaps through management activities.

and the strategies of the forest management plan. Direction to area foresters, coupled with adaptive management and monitoring plans that evaluate and modify that direction if necessary, provide the link between plan goals and landscape strategies, and implementation plans.

To effectively address biodiversity at both fine and coarse scales, a decision matrix was constructed to guide judgments at various landscape scales (Table E). At the regional scale landscape, decisions are designed to support regional conservation goals such as threatened species recovery strategies and are therefore generally broad. Decisions made at this level generally may not consider IHAs directly. Final determination is left to the implementation phase of the planning process. Consideration at regional scale provides a rational basis to assess the contribution of these forests to larger conservation issues and to determine the role of the forest plan within this larger context. Several threatened species or groups of species were addressed at this level. They are: 1) northern spotted owl, 2) marbled murrelet, 3) anadromous fish including coho salmon and steelhead.

The district level is where stand type targets for management basins are established and the frequency distribution of IHA patch sizes are defined. It is also at this level where decisions are made on how the overall frequency distribution of IHA patch sizes should be allocated across various basins based on current age structure, regional conservation objectives, and the relationship with other plan considerations including recreation, scenic quality, operational constraints, and others. Decisions at the district level can lead to allocation of certain basins to emphasize different parts of the frequency distribution. Thus one basin in a district could emphasize smaller IHA patches while another may emphasize larger IHA patches. This would depend on such considerations as current forest condition, regional contributions, operational constraints, and land ownership to name several.

It is at the management basin scale of landscape planning where most of the implementing decisions are made. Broad decisions have already been made at the district level that recognize relative contributions of each basin to district-wide distribution of patch sizes based upon certain constraints and management options. These decisions indicate, generally, how much fragmentation will be represented and the mix of large and small patches desired. Based upon this information, basin management planning will make refinements to define the desired range of stand types for the area. Specific guidance was developed to guide foresters in making and refining IHA patch size and placement decisions at the management basin level. Table F presents a summary of the implementation guidance developed to date.

5. AN APPLICATION EXAMPLE FOR THE TILLAMOOK STATE FOREST

5.1 Tillamook State Forest

The Tillamook State Forest stretches across 364,000 acres in the north coast range of Oregon, ranging from the coastal fog belt to the Willamette Valley foothills. Much of the area that is now the Tillamook State Forest was burned in a series of wildfires in the years 1933, 1939, 1945, and 1951. The Board of Forestry began to acquire land in the Tillamook Burn in 1940. Land acquisition accelerated after the Legislature authorized bonds to rehabilitate the Burn. Eventually, the Board of Forestry acquired roughly 255,000 acres of the Tillamook Burn, mostly from counties who had foreclosed on tax-delinquent lands. (Oregon Department of Forestry, 1993b)

The Department of Forestry carried out a massive reforestation and rehabilitation project in the Tillamook Burn between the years 1948 and 1973. Crews cut more than 220 miles of snag-free corridors as firebreaks, felling an estimated 1.5 million snags. Tree planting crews planted 72 million Douglas-fir seedlings, and 36 tons of Douglas-fir seed was spread on the burn through aerial seeding. In 1973, the former Tillamook Burn was dedicated as the Tillamook State Forest.

The forested landscape of the Tillamook State Forest today is dominated by young

Table E. Matrix Showing Resource Considerations Appropriate At Various Scales of Landscape Planning

Considerations	Region	District	Basin	Stand
Contribution to population goals for T&E and Sensitive Species	X	X		
Structural Targets		X		
Patch Size Distribution		X		
Recreational Sites		X		
Sites with Operational Constraints (Unstable/Steep slope)		X		
Unique Habitats such as wetlands, eagle sites etc.		X		
Scenic Corridors and Viewsheds		X		
Desired Basin Stand Structures		X	X	
Current Stand Condition			X	
Riparian Management Strategies			X	
Placement of Patch and Stand Structure Types			X	
Consideration of Isolated Stands			X	
Consideration of Adjacent landuses and Adjacent Basin Patch Location			X	
Edge Considerations			X	
Connectivity between Patches		X	X	
Patch Relationships between Aquatic and Upland Management Units			X	
Location of Replace Stands/Patches		X	X	
Big Game Management Considerations		X	X	
Timber Harvest Plans and Operation Specific Decsions			X	X
Structural Components (down wood, layed canopy, snag targets)			X	X
Within Stand Diversity (Gaps)				X
Species Composition				X

stands. Less than 3% of the forest is in stands greater than 85 years of age. 73% of the forest is in stands 26-55 years of age. Tree species composition is dominated by Douglas-fir. (Northwest Oregon State Forests Management Plan Draft, 1998). The Forest Grove District of the Department of Forestry manages the eastern one-third of the forest, comprising 117,339 acres of forest land. The following sections describe the process used by district personnel to translate the goals and strategies of the forest management plan into operational activities on the ground.

5.2 District Level Implementation Planning

Implementation planning is the process used by ODF to organize resource information, determine desired future conditions, identify and coordinate management activity, and assess progress toward meeting the goals identified in forest management plans, habitat

Table F. Guidelines for Implementing Interior Habitat Areas in Landscape Planning

Size, composition, and configuration	Corridors and Patch Placement
IHA patches consist of the suitable habitat after consideration of the negative influence of surrounding stand structures. The maximum effect can be 100 meters when adjacent to REG or CSC stands.	Patch placement will be a function of topography, relationship to corridors, and silvicultural considerations.
IHAs can include OFS/LYR/UDS when adjacent or in the immediate proximity to each other. OFS should be located near the center of the patch.	Consideration of IHA patches to serve complementary functions can best be achieved when considered jointly with riparian management area layouts, scenic, recreational, unique habitat, unstable slope and owl conservation areas (if applicable).
Minimum IHA patches (40 acres), must contain only OFS stand types. Small OFS patches <40 acres are not considered functional for meeting biodiversity goals and do not count toward the OFS targets.	IHA patch placement should include linkages across the landscape. Patches linking suitable habitats (corridors) may be smaller than the minimum patch size for IHAs but are still important to enhance the function of IHA patch network. Corridors can be as narrow as riparian management areas and as small as unique habitat areas. Riparian management areas can effectively link upland and riparian habitat patches. Corridors can also be dispersal habitat linking northern spotted owl conservation areas. For spotted owl emphasis areas, corridors can link a series of IHAs to form a patch of larger suitable habitat.
OFS/LYR/UDS in juxtaposition can be used to define IHA patch sizes of 100-420 acres when 50 % of the IHA patch is OFS. UDS cannot exceed 15 % and should be placed toward the outside of the patch.	Maintain IHA habitat until replacement patches become available. This can best be accomplished by planning for the entire patch and how the landscape will maintain similar habitats through time rather than on individual stands making up the patch.
Patch sizes greater than 420 acres require 40% OFS. The UDS patch contribution cannot exceed 35% of the total patch acreage.	Minimum patch distance between IHA patches should be a function of size and frequency within a management basin. Smaller patches should be placed closer together than larger patches to increase species movements between patch types.
IHA patches with greater edge to interior ratios need to reflect edge influences in determining target sizes. Particular attention needs to be given to riparian management areas because of their linear configuration. Functional riparian IHAs need to be extended up slope to decrease their edge to interior ratio.	For isolated patches, with greater than 50% of its boundary adjacent to REG/CSC or surrounded by forest land where future patch contributions are not anticipated e.g. non-state plantations, the minimum patch size should be increased to 120 acres. Retention of isolated patches below 120 acres should only be maintained when addressing short-term biodiversity goals. Long term biodiversity goals are best accomplished where corridors and similar habitats are in close proximity.
	As a general rule, the size of the IHAs should follow the size of other landform units within the basin. Smaller IHAs should be emphasized higher in the drainage associated with smaller stream and corridor networks. Locate IHAs near drainage divides to enhance species movements between watersheds

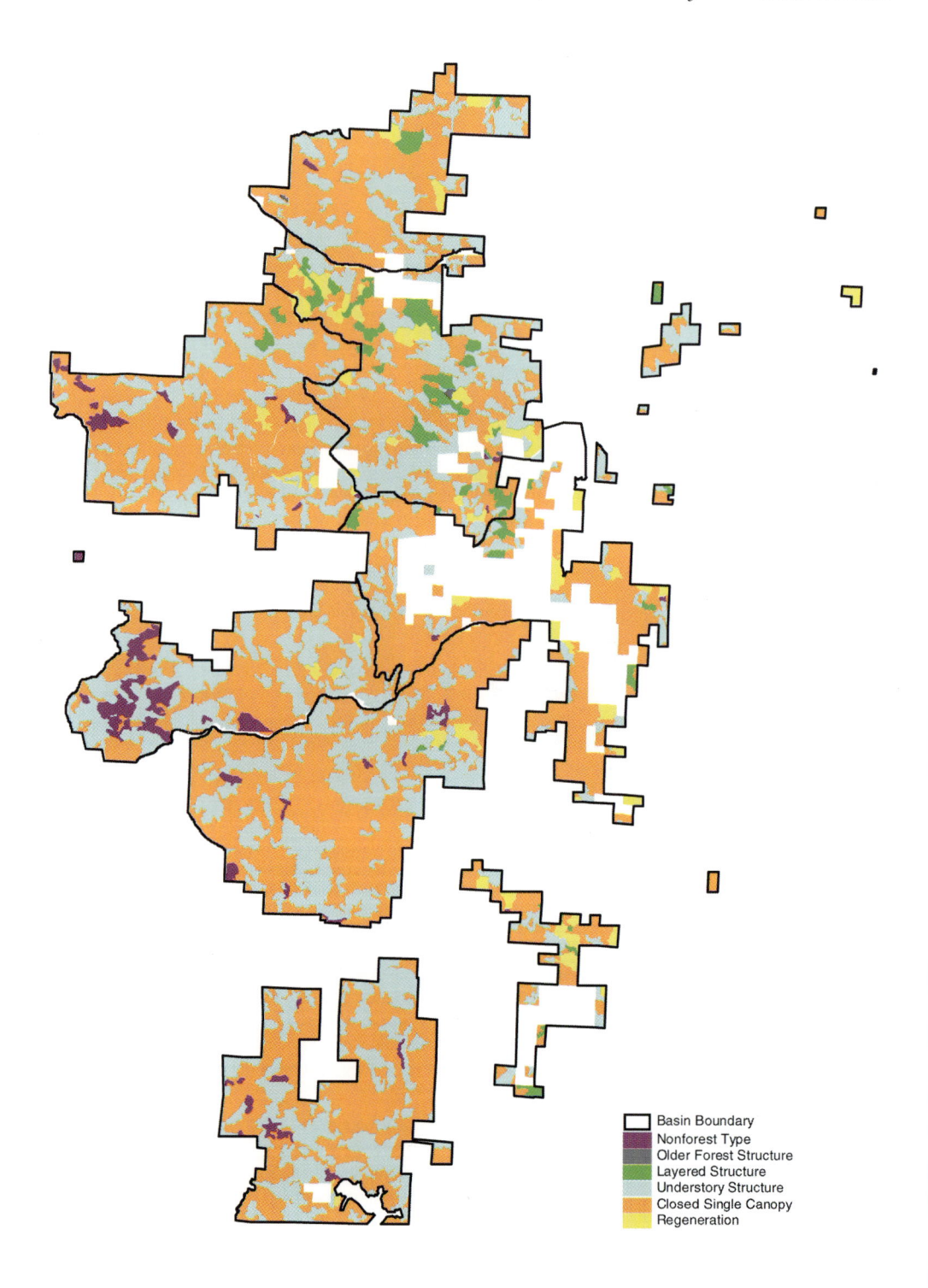

Figure 1. Current Stand Conditions for Forest Grove District

Using Landscape Design Principles 261

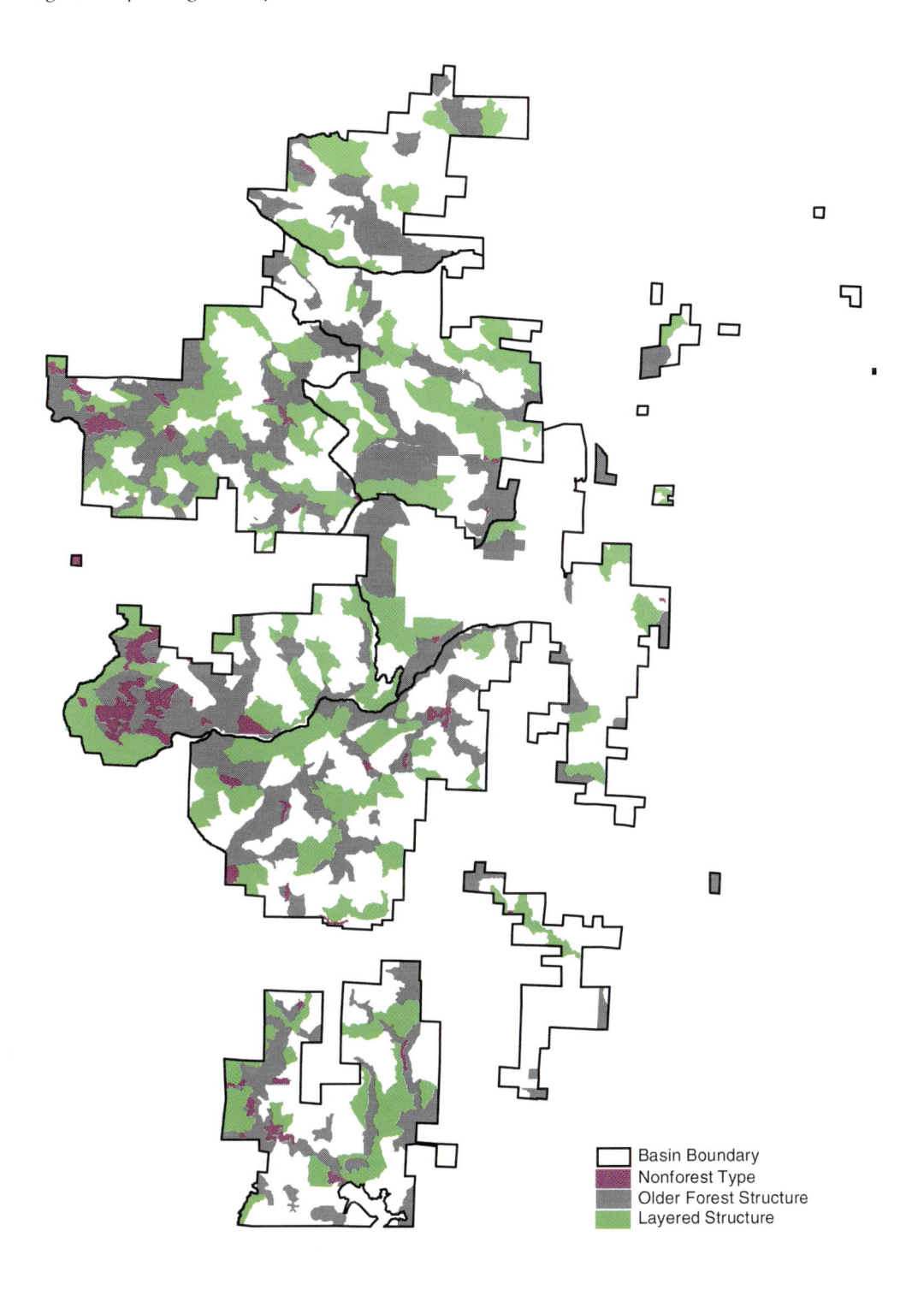

Figure 2. Desired Future Stand Conditions for Forest Grove District

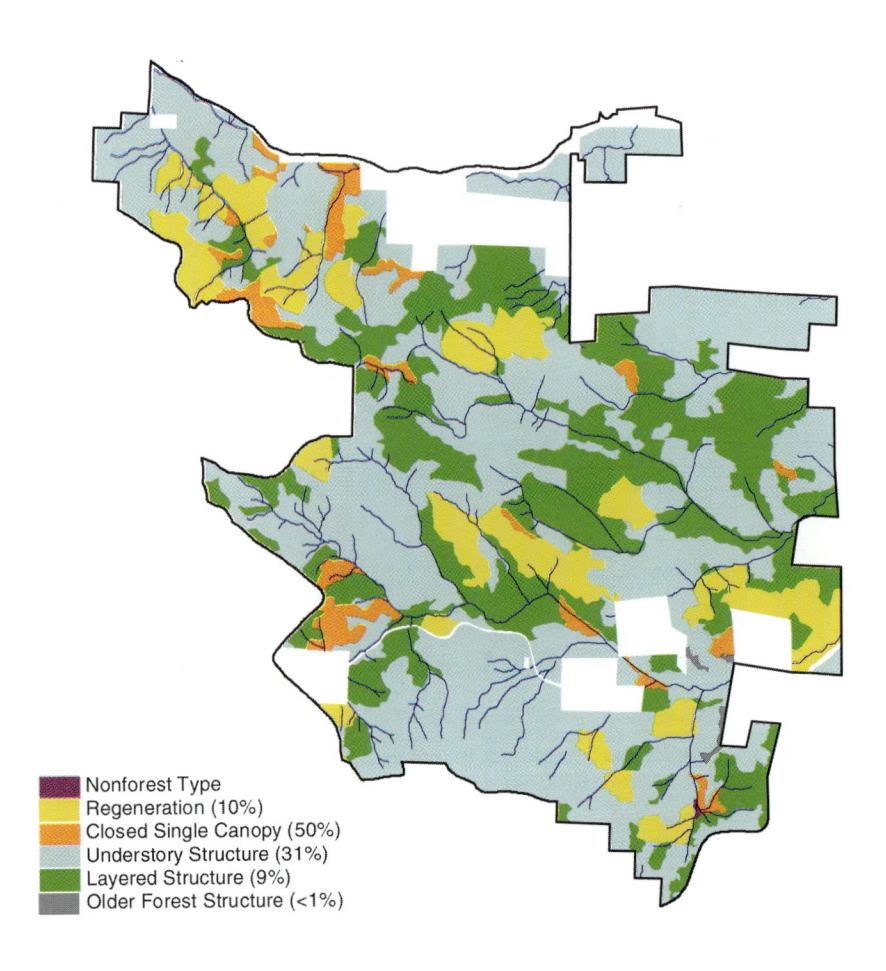

Figure 3. Current Stand Conditions for Wheeler Basin

Using Landscape Design Principles 263

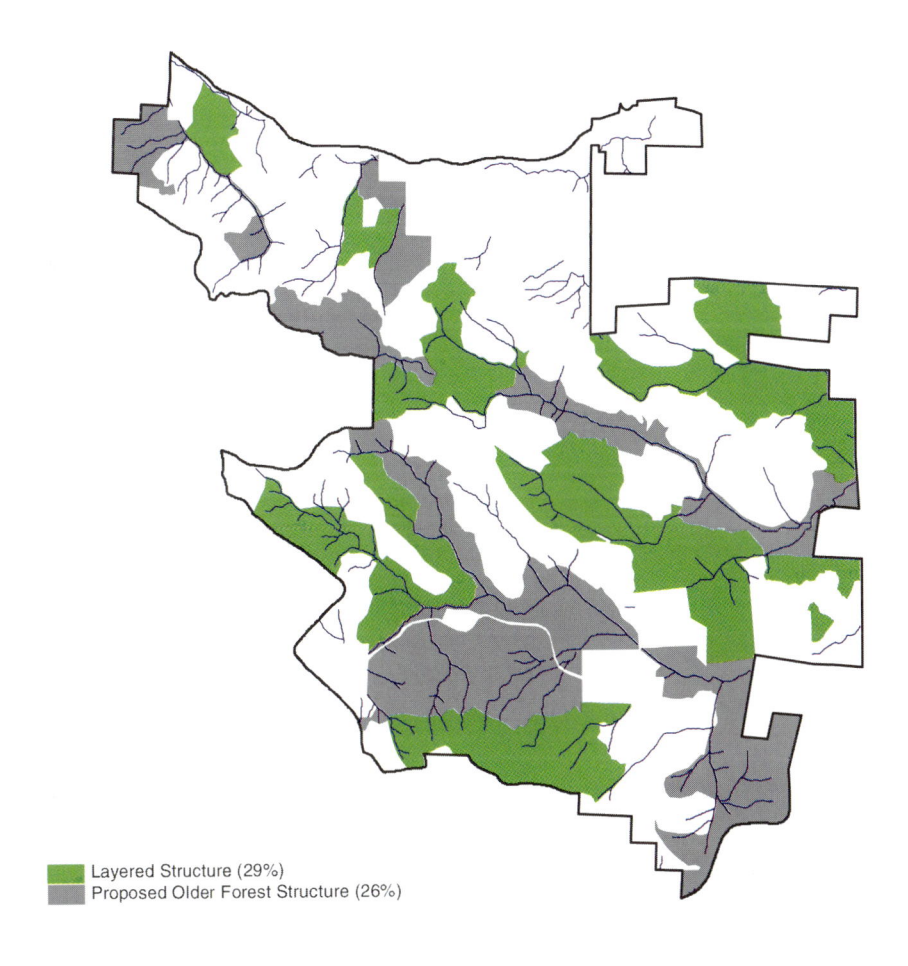

Figure 4. Desired Future Stand Conditions for Wheeler Basin

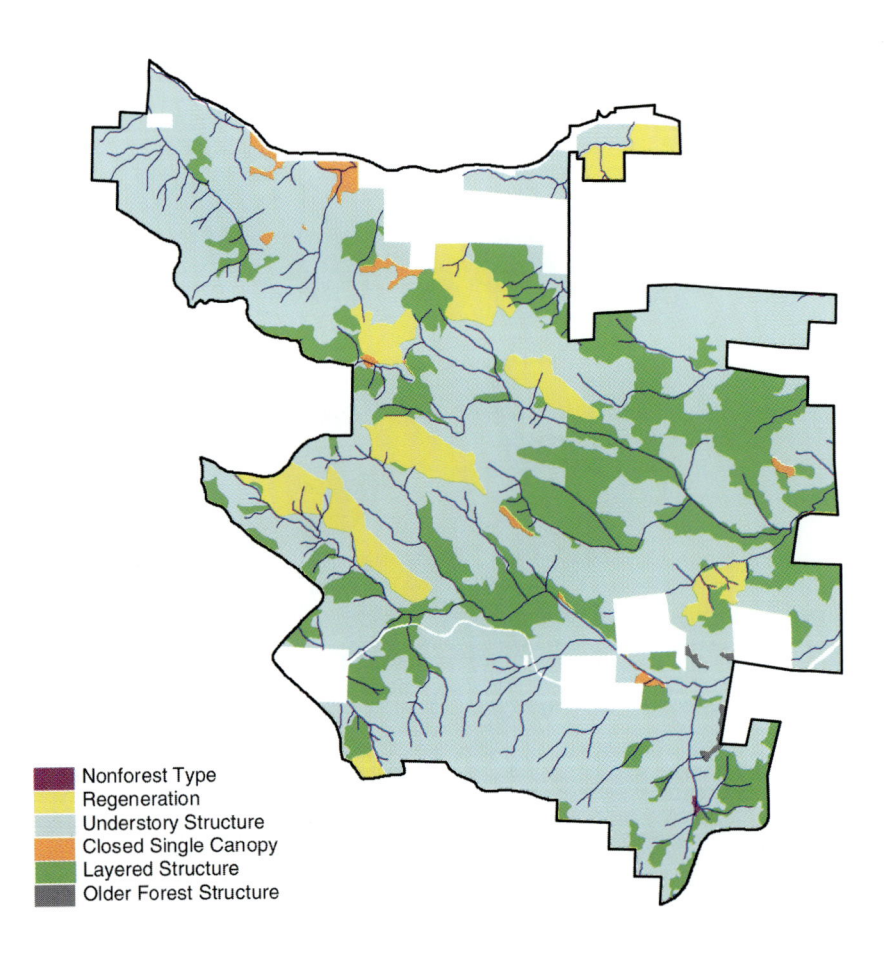

Figure 5. Projected Stand Conditions for Wheeler Basin End of First Decade

Using Landscape Design Principles 265

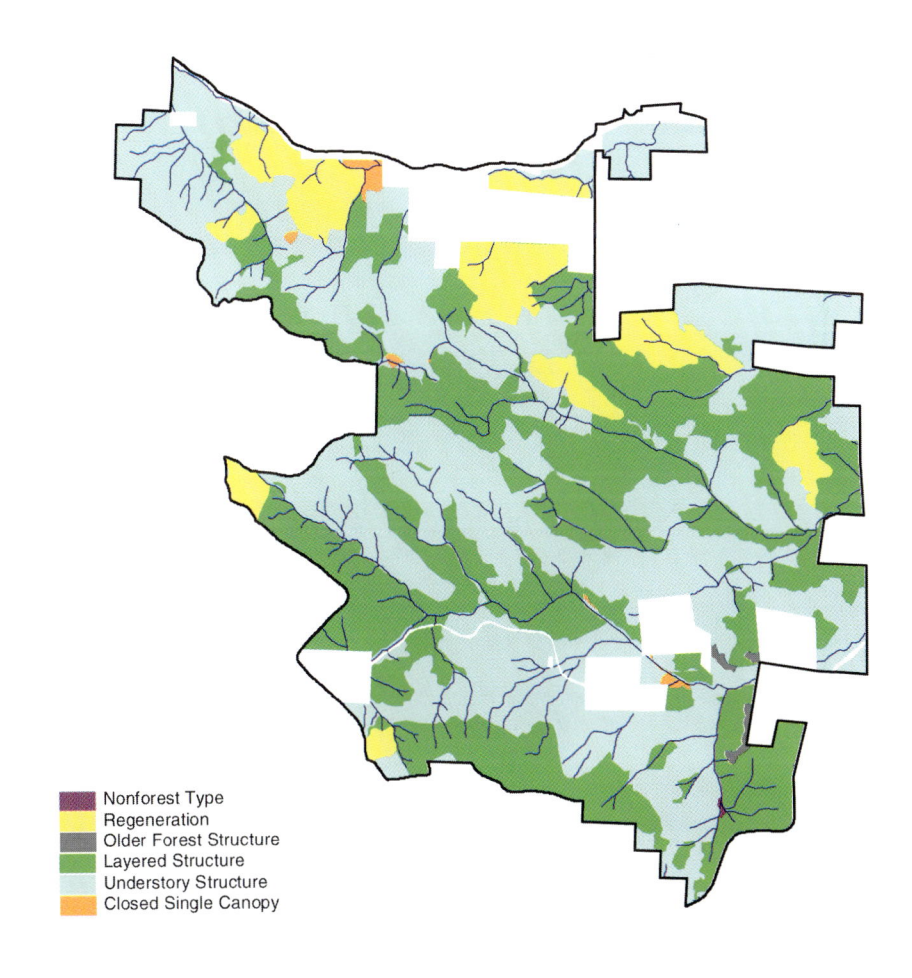

Figure 6. Projected Stand Conditions for Wheeler Basin End of Second Decade

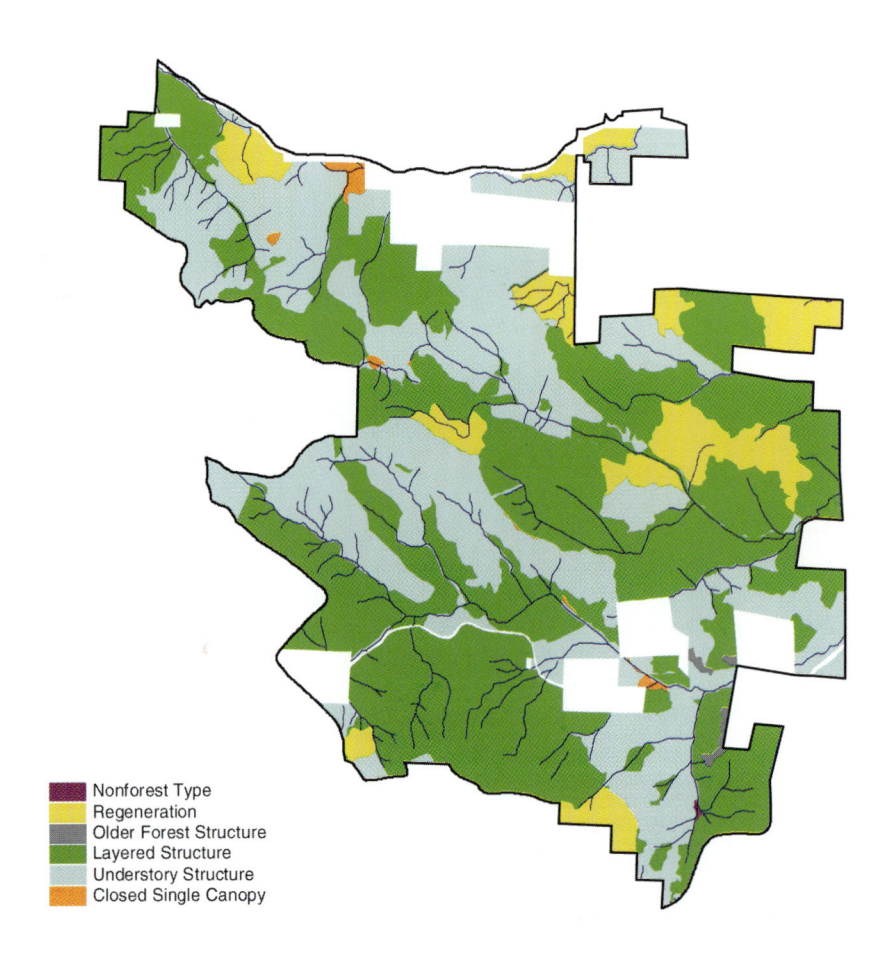

Figure 7. Projected Stand Conditions for Wheeler Basin End of Third Decade

conservation plans and other applicable plans. During implementation planning local field personnel and resource specialists apply the goals and strategies to real stand and forest conditions. Information is collated, processed and grouped within specific watersheds or groups of watersheds that comprise ODF Management Basins. The same information is aggregated and assessed at the district level (numerous management basins) and also examined within the context of all ownerships in regional context.

5.3 Implementation plans

- Identify the current status of resources.
- Identify desired future conditions of resources.
- Identify management opportunities to move the forest and stands from their current condition toward the desired future conditions.
- Serve as the primary basis for review of how the strategies will be applied in specific districts and management basins with program staff, other resource specialists, and interested parties.
- Describe how the proposed management activities will contribute to meeting the goals and strategies in the forest management plan.
- Describe the expected outputs and achievements for the next ten-year period and over the long term.

To implement the strategies for landscape design the local managers used a combination of inventory sorting techniques, management history and personal knowledge to assess and describe the current forest stand types that comprise the forest landscape in the district (Figure 1). They also assembled the best available information on other resources, including the physical landscape and key areas important for wildlife, recreation, or other uses.

They next applied the concepts and guidelines in the landscape strategies along with the current condition information to develop the desired future landscape condition for the district (Figure 2). In developing the desired future condition, the district chose to focus on the quantity and arrangement of the two most complex stand structures, LYR and OFS. This is because these structures are currently found in limited amounts and their future quantity and arrangement are key considerations in identifying stand management opportunities for developing existing stands toward those conditions. Areas not identified as LYR or OFS will contain a mix of the three other stand types (REG, CSC, and UDS). The district arranged LYR and OFS using the following key considerations:

- Place OFS and LYR stands on sites where structural components of mature forests could be rapidly achieved (30–40 years from present) and where resource risk is low.
- Place OFS and LYR stands on sites where resource protection can be maximized or where other resource values have a higher management priority. Examples of other key resources include: riparian areas, large interior habitat areas, northern spotted owl management area, amphibian emphasis areas, important salmonid streams, water quality, site quality, slope stability, scenic and recreational areas.
- Provide for connectivity or linkages between riparian/aquatic zones and upland areas, across basins, and across district boundaries.
- Seek a range of habitat development across elevational, north-south, and east-west gradients.

Information in the implementation plans is updated as new information is gathered or monitoring information becomes available. Watershed assessments and updated forest inventory are two sources of information that will be used to develop updated implementation plans. A new implementation plan is developed for review and approval every 5 to 10 years, or if some significant change in resource conditions or management approaches is identified

6. MONITORING

It is essential that the forest management plan strategies be implemented in an adaptive management context. Monitoring and research information will be planned for, obtained and used to assess:

- Historic, current and desired future conditions and likely trajectories.

- Ecological/cultural trends (How are resources changing outside of the management actions in the plan?)
- Management actions (How are the strategies in the plan being implemented?)
- Management effects (How is the resource changing in response to the strategies?)
- Assumptions and hypotheses inherent in the plan.

A rigorous and ambitious monitoring and adaptive management program is described in the forest management plan. It is clear that all the desired information would be operationally impossible to obtain in the short term and the financial resources to gather the information would be prohibitive. Thus ODF's approach has been to develop an exhaustive list of the monitoring and research needs, prioritize the items on the list, cooperate with research organizations and others working on forest management issues, and move forward in a very transparent and deliberate way to gather the desired information. As information becomes available it will be evaluated to determine additional information needs and necessary changes to the strategies.

An example of a series of monitoring questions related to landscape design that is included in the monitoring plan can be described for "spatial distribution of habitat patches". The example is not inclusive of all considerations for even this one aspect of landscape design, but is presented to demonstrate the thought process that is incorporated into the monitoring plan. The questions are the basis for developing specific monitoring projects or research efforts.

- Current condition – What is the spatial distribution of habitat patches and Interior Habitat Areas within and between management basins today?
- Trends – How does the size and spatial distribution of patches and Interior Habitat Areas change over time in managed stands? In unmanaged stands?
- Management actions - How are the objectives for patch types, sizes, and arrangement incorporated into Implementation Plans? Annual operations plans?
- Management effects – How do management actions affect the frequency distribution of patches and Interior Habitat Areas by district, over time?
- Assumptions - Are Interior Habitat Areas being used by wildlife species commonly associated with older forests, such northern spotted owls? How productive are the species using the habitat?

Some of the questions will be relatively easy to answer while others will demand long term or expensive monitoring or research efforts. For example, existing conditions and assessments of the extent to which the objectives are being incorporated into operational plans can be developed relatively quickly and inexpensively. However, an assessment of how effective management activities are in developing the desired array of patches and the determination of whether the patches are actually being productively used by the wildlife communities they are intended for may take several decades and be quite expensive. Cooperative efforts among research institutions, federal and state agencies, and other land managers will help make monitoring and research efforts more efficient and effective.

7. CONCLUSION

State forest land managers in Oregon are faced with the challenge of actively managing state forests to provide a variety of benefits to the citizens of the state. These benefits include sustainable levels of timber and revenue to benefit the state, counties, and local taxing districts. They also include providing for properly functioning aquatic habitats, habitats for native wildlife, productive soil, clean air and water, and recreational opportunities for a growing population. This multiple-resource mandate calls for an integrated system of management that can realize the value of these lands for the State of Oregon. The ubiquitous nature of many of these resources requires that forest management planning address resource issues at a variety of landscape scales, through a landscape design system.

The landscape design system must be one that results in broad benefits to biodiversity and indigenous wildlife habitats, can be applied in an efficient and cost-effective manner, and can be reasonably implemented by forest managers at the field level. The system must be flexible enough to provide field managers with the ability to address changing forest conditions through time and space. We believe that the system described in this paper can meet these tests and will result in a forest capable of achieving multiple objectives into the future.

Applying this landscape planning process will require careful monitoring through time and a willingness to adapt and change as we learn more. That has been the nature of state forest management over the past five decades, as the lands in northwestern Oregon have been reforested and nurtured into the tremendous resource that exists today. We believe this approach provides a pathway that continues this strong stewardship tradition.

LITERATURE LISTED

Carey, A.B., C.E. Elliott, B.R. Lippke, J. Sessions, C.J. Chambers, C.D. Oliver, J.F. Franklin, and M.G. Raphael. 1996. Washington Forest Landscape Management Project – A Pragmatic, Ecological Approach to Small Landscape Management. Report No. 2. Washington State Department of Natural Resources. Olympia, WA.

Chen, J. 1991. Microclimatic and biological pattern at edges of Douglas-fir stands. Dissertation. University of Washington, Seattle, Washington, USA.

Diaz, N. and D. Apostol. 1992. Forest landscape analysis and design, a process for developing and implementing land management objectives for landscape patterns. USDA Forest Service. PNW, R6 ECO-TP-043-92.

Dunning, J.B., B.J. Danielson, and H.R. Pulliam. 1992. Ecological processes that affect populations in complex landscapes. Oikos 65:169-175.

Fick, L., and G. Martin. 1992. The Tillamook Burn: Rehabilitation and Reforestation. Oregon Department of Forestry, Forest Grove District, Forest Grove, OR.

Franklin, J.F., and R.T.T. Forman. 1987. Creating landscape pattern by forest cutting: ecological consequences and principles. Landscape Ecology 1:5-18.

Franklin, J.F., and T.A. Spies. 1991. Composition, Function, and Structure of Old-Growth Douglas-Fir Forests. In: L.F. Ruggiero, K.B. Aubry, A.B. Carey, and M.H. Huff, technical coordinators, Wildlife and Vegetation of Unmanaged Douglas-Fir Forests. USDA Forest Service, Portland, OR. General Technical Report PNW-GTR-285.

Hansen, A.J., T.A. Spies, F.J. Swanson, and J.L. Ohmann. 1991. Conserving Biodiversity in Managed Forests — Lessons from Natural Forests. BioScience, 41(6):382-392.

Harris, L.D. 1984. The Fragmented Forest: Island Biogeogrpahic Theory and the Preservation of Biotic Diversity. University of Chicago Press, Chicago. 211 pp.

Hayes, J.P., S.S. Chan, W.H. Emmingham, J.C. Tappeiner, L.D. Kellogg, and J. D. Bailey. 1997. Wildlife Response to Thinning Young Forests in Western Oregon and Washington. Journal of Forestry. Vol. 95, No.8.

Hunter Jr., M.L. 1990. Wildlife, Forests, and Forestry – Principles of Managing Forests for Biological Diversity. Prentice Hall, Englewood Cliffs, NJ.

Kotliar, N.B. and J.A. Wiens. 1990. Multiple scales of patchiness and patch structure: a hierarchical framework for the study of heterogeneity. Oikos 59:253-260.

McComb, W.C., T.A. Spies, and W.H. Emmingham. 1993 Douglas-fir Forests: Managing for timber and mature-forest habitat. Journal of Forestry. 91(12):31-42.

McGarigal, K. and B. Marks. 1995. FRAGSTATS: spatial pattern analysis program for quantifying landscape structure. Gen. Tech. Rep. PNW-GTR-351. Portland OR. 122p.

McGarigal, K. and W.C. McComb. 1995. Relationship between landscape structure and breeding birds in the Oregon Coast Range. Ecol. Mon. 65(3):235-260.

McKelvey, K., B.R. Noon, and R. Lamberson. 1992. Conservation planning or species occupying fragmented landscapes: the case of the northern spotted owl. Pages 338-357 In J. Kingsolver, P. Kareiva, and R. Hyey, eds. Biotic interactions and global change. Sinauer Associates, Sunderland, MA.

Oliver, C.D. 1992. A Landscape Approach – Achieving and Maintaining Biodiversity and Economic Productivity. Journal of Forestry. Vol. 90, No. 9.

Oliver, C.D., and B.C. Larson. 1996. Forest Stand Dynamics. John Wiley and Sons, New York, NY.

Oregon Department of Forestry. 1993. Tillamook State Forest: Tillamook Burn to Tillamook State Forest. Oregon Department of Forestry, Salem, OR.

Oregon Department of Forestry. 1998. Northwest Oregon State Forests Management Plan – Draft, April 1998. Oregon Department of Forestry, Salem, OR.

Pyne, S.J. 1982. Fire in America: A Cultural History of Wildland and Rural Fire. Princeton University Press, Princeton, NJ.

Robbins, C.S., K.D. Dawson, and B.A. Dowell. 1989. Habitat area requirements of breeding forest birds of the middle Atlantic states. Wildl. Monogr. 103. 34 pp.

Rosenberg, K.V. and M.G. Raphael. 1984. Effects of forest fragmentation on vertebrates in Douglas-fir forests. Pages 263-272. In J. Verner, M.L. Morrison, and C.J. Ralph, eds. Wildlife 2000. Modeling habitat relationships of terrestrial vertebrates. Univ. Wisconsin Press.

Ruggiero, L.F., L.C. Jones, and K.B. Aubry. 1991. Plant and animal associations in Douglas-fir forests of the Pacific Northwest: an overview. Pages 447-462. In L.F. Ruggiero, K.B. Aubry, A.B. Carey, and M.H. Huff, technical editors. Wildlife and vegetation of unmanaged Douglas-fir forests. U.S. Forest Service General Tech. Rpt. PNW-285.

Shugart, H.H. Jr. 1984. A Theory of Forest Dynamics. Springer-Verlag. New York.

Spies, T.A., and J.F. Franklin. 1991. The Structure of Natural Young, Mature, and Old-Growth Douglas-Fir Forests in Oregon and Washington. In: L.F. Ruggiero, K.B. Aubry, A.B. Carey, and M.H. Huff, technical coordinators, Wildlife and Vegetation of Unmanaged Douglas-Fir Forests. USDA Forest Service, Portland, OR. General Technical Report PNW-GTR-285.

Spies, T.A., and S.P. Cline. 1988. Coarse Woody Debris in Forests and Plantations of Coastal Oregon. In: C. Maser, R.F. Tarrant, J.M. Trappe, and J.F. Franklin, editors, From the Forest to the Sea: A Story of Fallen Trees. USDA Forest Service, Portland, OR. General Technical Report PNW-GTR-229.

Spies, T.A, W.J. Ripple, and G.A. Bradshaw. 1994. Dynamics and pattern of a managed coniferous landscape in Oregon. Ecol. Appl. 4(3):555-568.

Soule, M.E. 1991. Conseration: tactics for a constant crisis. Science 253 744-750.

Strahler, A.N. 1957. Quantitative analysis of watershed geomorphology. Trans. Am Geophyssical Union 38:913-920.

Thomas, J.W., E.D. Forsman, J.B. Lint, E.C. Meslow, B.R. Noon, J. Verner. 1990. A conservation strategy for the Northern Spotted Owl. Interagency Scientific Committee to Address the Conservation of the Northern Spotted Owl. Portland OR. 427 pp.

USDA Forest Service, et al. 1994. Final Supplemental Environmental Impact Statement on Management of Habitat for Late-Successional and Old-Growth Forest Related Species Within the Range of the Northern Spotted Owl. Also known as the Clinton Forest Plan or the Final SEIS. USDA Forest Service, Pacific Northwest Region, Portland, OR. February 1994.

Whitcomb, R.F., C.S. Robbins, J.F. Lynch, B.L. Whitcomb, M.K. Klimkiewicz, and D. Bystrak. 1981. Effects of forest fragmentation on avifauna of the eastern deciduous forest. Pages 125-205 In R.L. Burgess and D.M. Sharpe, eds. Forest Island Dynamics in Man-Dominated Landscapes. Springer-Verlag, New York.

Wiens, J.A. 1976. Population response to patchy environments. Ann. Rev. Ecol. Syst. 7:81-129.

Wisdom, M.J., L.R. Bright, C.G. Carey [and others]. 1986. A model to evaluate elk habitat in western Oregon. R6-F&WL-216-1986. Portland, OR, U.S.D.A. Forest Service, Pacific Northwest Region. 36p.

Patch Sizes, Vertebrates, and Effects of Harvest Policy in Southeastern British Columbia

Fred L. Bunnell[1], Ralph W. Wells[1]
John D. Nelson[2], Laurie L. Kremsater[1]

We evaluate three harvesting scenarios over a 200-year period for a 32,000 ha landscape in southeastern British Columbia. The scenarios include a 'Default' option reflecting typical application of current guidelines, a 'Target' option applying a range of patch sizes, and a 'Zones' option specifying harvest and non-harvest areas. The Default option generated the most volume and the most active road use, but also resulted in the most fragmented landscape pattern, dominated by small patches. Old patches were substantially reduced in size and total area. While the Target option generated a range of patch sizes, it resulted in similar amounts and patterns of old patches as the Default option. The Target option also yielded about 8% less timber, with losses especially pronounced in the short term. The Zones option had the biggest impact on volume (almost a 10% reduction compared to the Default option), but it also had substantially more old forest (over 30% of forested area) and significantly less active road use (over 30% less than the Default option). Initial landscape patterns were highly persistent: to fully implement the Zone option, two full rotations (200 years) were required.

Evaluation of the effects of different patterns on terrestrial vertebrates was difficult because understanding of species' response to pattern is poor. Based on our review, we believe that the reduction in size and amount of old patches will affect some species negatively. There is, however, little evidence that old patches in western forests are acting as isolated patches for small terrestrial vertebrates and there is growing evidence that disturbed landscapes will support a richer community of vertebrates than do contiguous areas of old forest. Smaller organisms, such as mosses and lichens, may be more sensitive to patch size of older forest than are vertebrates.

We conclude that harvest patterns should be carefully conceived because harvest patterns are enduring and changes can incur substantial penalties to harvest volumes. Flexibility and variability are important because different patterns will favor different species, and improved understanding of species responses to landscape pattern will generate new landscape objectives. We further conclude that biological diversity cannot be sustained by doing the same thing everywhere. Zoning provides a compelling approach to developing different landscape patterns designed to favor different groups of species.

Key words: case study, patch size, simulation modeling, timber supply, vertebrates

---

[1]*Centre for Applied Conservation Biology;* [2]*Department of Forest Resources Management; both University of British Columbia, Vancouver, B.C.*

1. INTRODUCTION

We evaluate a case study of attempts to create a range of forest patch sizes through forest harvesting in an area of southeastern British Columbia, examining both economic and biological consequences. Indices of economic impacts are easily defined and include harvest volumes and total road length. Indicators of the biological impacts are less well defined, so we first provide a brief review of relevant literature on patches to aid interpretation. We continue by stating the problem, its context, and the proposed solution. We provide spatial analyses of alternative ways of using harvest policy to generate a range of patch sizes, and offer our conclusions on this case study.

2. WHAT WE KNOW ABOUT PATCHES

There is no consistent definition of patch. Applied to forest practices, "patch" typically refers to the opening or gap in the canopy created by logging (Bradshaw, 1992), and we have terms like "small-patch selection" referring to the removal of 3 to 10 trees in groups (e.g., Franklin et al., 1983). For conservation biologists, "patch" refers to standing trees, isolated or distant from the nearest similar patch of standing trees. Both are correct because an opening grows to become a stand of older trees. In a forest, however, the stand will not long be surrounded by untreed areas. Habitat patches can be defined as areas used for breeding or obtaining resources that are distinct from surrounding habitats (Fahrig and Merriam, 1994). An immediate complication is that habitat patches differ among species (e.g., review of Bunnell and Huggard, 1999). For conservation biologists, habitat "patches" often are equated with "islands" to allow exploration of the applicability of the theory of island biogeography and its derivatives (Harris, 1984; Bunnell this volume).

Three features of patches have been associated with species losses: small patch area (Galli et al., 1976; Freemark and Merriam, 1986; Robbins et al., 1989); isolation or distance from other suitable habitats (Lynch and Whitcomb, 1978; Urban et al., 1987); and edge-related phenomena that can reduce patch area or increase mortality within a patch, thus reducing population size (Gates and Gysel, 1978; Yahner, 1988). Edge-related phenomena are treated by Kremsater and Bunnell and by Marzluff and Restani (both this volume). We consider area and distance effects. Both concepts derive from the theory of island biogeography. Briefly, as the area of an isolated patch of suitable habitat is reduced, the numbers of individuals decline, especially among uncommon species and species with large area requirements (e.g., Diamond, 1984). Small populations are vulnerable to extinction due to chance environmental events, unexpected demographic changes, edge effects, human disturbance, and reduction in genetic variation (e.g., Shaffer, 1981, 1990; Janzen, 1986). As distance between patches increases, local extinctions are less likely to be rescued by recolonization from nearby patches. Bunnell (this volume) provides more detail. Most studies addressing patches have examined islands and parks or forest patches in agricultural areas. Few have looked at patches of older forests surrounded by younger forests. In forest management our primary concern is older stands surrounded by younger forests, but we review findings of other studies briefly.

There is abundant evidence that species disappear or go extinct from islands and parks faster than they do from mainlands or surrounding areas (e.g., MacArthur and Wilson, 1967; Diamond, 1984; Burkey, 1995; Newmark, 1987, 1995). Oceanic islands are surrounded by habitat distinctly different and usually much more hostile than the island itself. That is not consistently true for parks and may explain the lack of consistently strong effects in western parks (e.g., Glenn and Nudds, 1989). Researchers seeking to apply the theory of island biogeography to forest patches, have examined patches surrounded by very different habitat—commonly agricultural land. In North America, data are most abundant from eastern Canada and the eastern and midwestern United States (e.g., Freemark and Merriam, 1986; Askins et al., 1987: Opdam, 1991; Wenny et al., 1993). Despite the variability of early studies, they accumulated data suggesting that patch area was important.

From these data grew the concepts of "forest interior species" and "minimum critical areas" (e.g. Wenny et al., 1993). Forest interior species require some minimum area of contiguous forest to breed successfully. Individual species are not consistently area sensitive[1] (Table 1; Kremsater and Bunnell, this volume), but several workers have documented that patch area is related to population persistence and number of species (Lynch and Whigam, 1984; Paine, 1988; Verboom et al., 1991). Several studies in Table 1 report that diversity of vegetative structure within patches had a greater influence on species richness than did area itself; the effect was evident even among patches <1 ha (e.g., Yahner, 1983). Forested patches as small as 1 ha surrounded by fields sustained some forest interior species (e.g., Galli et al. 1976; Table 1). These findings imply value in retaining structurally diverse (commonly older), patches of forest. Generally, evidence from an agricultural matrix supports a positive relationship between patch area and population persistence or abundance, consistent with the theory of island biogeography (Table 1). Forests surrounded by vastly different vegetation often act as islands. Data also suggest that some species require larger patches of forest than do others (Table 1). Evidence from areas of where forestry is the dominant land use is not as compelling.

Several species associated with older seral stages appear less abundant than they were formerly. Forest practices have reduced the amount of favorable habitat. It is much less clear whether the decline in abundance is due to simple reduction in total "old growth" habitat, or to the arrangement of that habitat into small, isolated patches (see Bunnell this volume and Fahrig this volume). In western forests the term "forest fragmentation" does not refer to remnant patches of trees within an agricultural or urban matrix, but to scattered stands of old forest within areas of continuous forest cover of all ages.

Some studies in eastern forests report negative effects of decreasing patch area or increasing forest fragmentation (e.g., Hagan et al., 1996). In the Pacific Northwest there have been few observational studies and only one experimental study (Schmiegelow this volume). The seminal work for western forests is Harris (1984) who estimated consequences by reasoned guesses because studies were then lacking. Studies now available suggest that results of eastern studies in a non-forested matrix cannot be transplanted to western forests. Negative effects of small patch size could appear as elevated depredation rates in small patches or as reduced richness and abundance in small patches. Schieck et al. (1995) found no relationship between patch size and richness of corvid species (potential predators) at patch centres. McGarigal and McComb (1995) found that abundance of one corvid (gray jay) increased with total area of late seral forest. That finding suggests reduced depredation by gray jays in small patches. The two other western studies we reviewed (Rosenberg and Raphael, 1986; Lehmkuhl et al., 1991) found that abundance of corvids was not significantly related to patch size. It appears that avian depredation rates should not increase with decreasing patch size down to 4 ha. Squirrels also predate on nestling birds. Data of Rosenberg and Raphael (1986) suggest that squirrel abundance decreases with decreasing patch size.

Schieck et al. (1995) found no relationship between patch area of old growth and bird abundance or richness on Vancouver Island. In Oregon, McGarigal and McComb (1995) selected 15 bird species assumed to be associated with late seral forests and found only one, the winter wren, significantly associated with less fragmented landscapes. In mixed Douglas-fir hardwood stands, Rosenberg and Raphael (1986) evaluated the response of 74 vertebrate species to stand area of forests by two methods: correlation of abundance with stand area, and analysis of variance of abundance or frequency across 5 classes of stand area. Nine species showed a significant positive correlation with stand area ($p < 0.05$); seven species showed significant increases in abundance or frequency across the five area classes. Combined, available data suggest little influence of the area of either mature or old growth patches on bird richness or abundance

---

[1] "Forest interior" species are therefore defined empirically as those species absent from small patches but present in larger ones.

Table 1. Summary of literature on area and distance, or isolation, effects of forest patches.

Author	Matrix	Patch size	Patch age	Isolation	Comments
Martin et al. 1995	B.C.; forested islands in ocean	1 ha to 400 ha	Mature/old		Various vegetation structures correlated with island size. Average number of species per point count remained relatively constant. Total census > on larger islands. Impoverishment only on smallest islands. Therefore, structure (and area) both important.
Agriculture Matrix					
McIntyre 1995	Georgia; agriculture	10 ha–13.25 ha (large) <3.25 ha (small); "isolated" contiguous control	Deciduous forest	Not measured	Species associated with larger patches: black-throated green warbler, Kentucky warbler, northern parula
Usher et al. 1993	England; Fields	0.1 ha to 10.1 ha most between 0.1 to 4 ha	Mature 20 to 90 yrs	"Isolation" and "Shape" indices	Species richness of woodland ground beetles related to shape and area. Spiders related to isolation. Cites other authors who show little effect of patch size.
Ambuel & Temple 1983	Wisconsin; Fields	3 ha to 500 ha	Various deciduous		Species richness increased with area. Effect of area on more common species is weak, strong for uncommon long-distance migrants (ULDMs). ULDMs were: least Flycatcher; Acadian flycatcher; yellow-throated vireo; cerulean warbler; chestnut-sided warbler; hooded warbler; American redstart; mourning warbler; wood thrush; veery. Vegetation structure may be more important than patch area.
Galli et al. 1976	New Jersey; Fields	0 ha to 24 ha 30 forest islands	Oak; mature		Species richness increased with area. Did not find foliage diversity to affect species richness. Provide a table of area-dependent species. Forest interior birds began appearing in patches >0.8 ha.
Yahner 1983	Minnesota; Fields	Shelterbelts 3 to 9 rows wide 0.2 ha to 0.79 ha area	Conifer / deciduous mixes; 10 to 30 yrs old		Species increased with area only in winter. Other vegetation characteristics more important than area.
Butcher et al. 1981	Connecticut; Powerlines, homes, pond	29.6 ha	Mature forest	Time since isolation; 14 years.	7 mature forest species declined with time since isolation: red-eyed vireo, hooded warbler, ruffed grouse, eastern wood pewee, black-throated green warbler, Canada warbler, American redstart.
Martin 1980	S. Dakota;	69 shelterbelts		Distance to nearest	Species richness increased with area. Area more

Patch Sizes, Vertebrates, and Effects of Harvest Policy

Author	Matrix	Patch size	Patch age	Isolation	Comments
	Field			forest island	important than vegetation. Distance to forest island did not explain any more variation than did area.
Askins et al. 1987.	Connecticut; Fields	46 forest tracts	Composition changed in each tract	Isolation measured 2 ways - by doughnut - 2 km from outside of patch; and amount of forest within 2 km from centre of patch	For species richness of forest birds, size most important predictor in small tracts <72 ha. Over 72 ha, degree of isolation most important predictor. Species only in larger forests: worm-eating warbler; hermit thrush; brown creeper; blue-gray gnatcatcher; yellow-throated vireo; black-throated green warbler; cerulean warbler.
Wolff et al. 1997	Oregon; Experimental enclosure	70% decrease in habitat by mowing	Alfalfa		Populations of small mammals still increased in spite of reduced habitat and fragmentation into small patches.
Andren 1994	Sweden; boreal forest in farmland	3 study sites: 10% forest area 40% forest area 93% forest area	Areas included a variety of habitat: old and young forest, bog Agriculture		Total corvid density increased with increasing % farmland. Jay density increased with increasing forest. Predation on artificial nests increased with increasing farmland. Jays and ravens absent from smallest fragments.
Freemark & Merriam 1986	Ontario; forests in fields	21 patches 3 ha to 7620 ha	Forest age unknown, but measured several structural attributes		Increased area increased bird diversity (explained > than 50% of variation). Habitat heterogeneity also explained variation in bird numbers. See their Fig 2.
Bancroft et al. 1995	Florida; matrix type not stated	27 patches 0.2 ha to >100 ha	?		Selected species sensitive to area: - no white-eyed vireos < 2.3 ha - no northern flickers < 3.5 ha - no yellow-billed cuckoo < 7.5 ha - no mangrove cuckoo < 12.8 ha
Robinson et al. 1992	Kansas; field experiment	32 m² patch old field in clusters 288 m² patch old field 5000 m² patch old field	Old field		More individuals of larger small mammals in larger patches. Clonal plants did better in larger patches.
Bolger et al. 1991	California; chapparal in urban matrix	36 patches 0.25 ha to 68 ha unfragmented and fragmented of same size			- Small isolates had fewer species. - Faunas strongly nested (ie. larger fragments have same species as smaller ones and more).

Author	Matrix	Patch size	Patch age	Isolation	Comments
Forest Matrix					
Rosenberg & Raphael 1986	Clearcuts and forest	5 ha to 300 ha	55 to 370 yrs		Isolation: bird species richness increased with proximity to clearcuts, higher in stands with more edge. Amphibian species richness also increased with proximity to clearcuts. 10 species decreased with more insular stands (more edge). Area: correlations with area are weak but western toad, acorn woodpecker, pileated woodpecker, common raven, northern flying squirrel, western gray squirrel and fisher responded positively to increased area (see also Table 13).
Hagan et al. 1996	Maine; early forest regeneration and clearcuts	Boreal/hardwood 5 km x 5 km fragmented block; 5 km x 5 km unfragmented block			Blackburnian warbler was the only 1 of 6 species with greater abundance in less fragmented forest. No difference in ovenbird abundance, but males more likely to be paired in larger tracts. No measures of isolation or fragmentation–qualitative definitions only.
Schieck et al. 1995	Vancouver Island; clearcut	23 "old" remnants, 4 ha to 2500 ha	Old growth	Surveys in centre of patch; distance to nearest patch; amount of old growth within 2 km	Number of non-old growth birds decreased in patch centre as patch size increased. Old-growth species were no different in small vs. large patches. No cowbird parasitism. Corvid abundance at centre not related to patch size.
McGarigal & McComb 1995	Oregon forest and clearcuts	30 landscapes 250 ha–300 ha each	% in late seral		- Only winter wren associated with least fragmented landscapes. - 5 species strongly affected by changes in habitat area: gray jay, brown creeper, winter wren, varied thrush, chestnut-backed chickadee. - 5 moderately affected: evening grosbeak, Hammond's flycatcher, pileated woodpecker, western wood pewee, red-breasted nuthatch. - 5 unaffected: red crossbill, red-breasted sapsucker, western tanager, olive-sided flycatcher, red-tailed hawk.

Author	Matrix	Patch size	Patch age	Isolation	Comments
Keller & Anderson 1992	Wyoming; old growth clearcut	4 pairs fragmented & unfragmented stands: 1) 45 ha–100 ha old 2) 45 ha (50% cut) 40 ha old 3) 25 ha (20%cut), 20 ha old 4) 25 ha (20%cut), 20 ha old	Old	Note uncut 1) has many more plots; size and isolation confounded	- No brown creeper in fragmented stands. - More hermit thrush, mountain chickadee, yellow-rumped warbler, and red-breasted nuthatch in unfragmented stands. - Different results and different scale from Rosenburg and Raphael 1986.
Yahner 1992	Pennsylvania; forest/clearcut	Experimental; 4 ha blocks 0%cut 25%cut 50%cut			- Increased fragmentation resulted in increased abundance of deer mouse (*P. leucopus*) and red-backed voles (*P. gapperi*).

(over a range of patch sizes from 4 ha to 1000 ha). In short, it appears that depredation rates are not elevated within smaller patches of western forests, nor are there other negative effects of small patch sizes. Keller and Anderson (1992) in Wyoming suggested that the most important effect of early fragmentation was the absolute reduction of habitat rather than its distribution, but did not separate effects of isolation. The suggestion that total habitat area is more influential than distribution of habitat in current forested landscapes is consistent with observations of McGarigal and McComb (1995) and several studies of Table 1 (e.g., Rosenberg and Raphael, 1986; Wolff et al., 1997; see also Bunnell this volume). We found no data evaluating distance effects for older stands in western forests.

Fragmentation is less evident in western than in eastern forests for three potential reasons. First, different stand age classes may not have vertebrate resources unique to those age classes (Bunnell this volume). Second, even though distinctive and desired resources are provided in older stands, there is sufficient of these stands (enough area) and the intervening matrix does not provide an impediment to movement among them. Third, western forests have not been managed or fragmented as long as have eastern forests. Under intensive management younger age classes will come to differ more from old stands. Models of fragmentation suggest threshold effects (e.g., Kareiva and Wennergren, 1995); western forests may not have reached threshold conditions.

Current data offer no firm guidelines on how to assess effects of patch size for the entire range of species that inhabit forests. Several points are suggested: 1) total amount of habitat is likely more important than distribution, 2) species show different responses (suggesting the importance of a range of patch sizes), and 3) some species do better in larger patches of older forest. Despite lack of evidence for a strong effect of patch size, there are three good reasons to avoid a uniform distribution of small patches. First, is the fact, already noted, that western forests simply may not have been managed long enough for incipient effects to appear. Second, is the scarcity of data for larger species (e.g., caribou, wolverine) that may do better in larger tracts of unmanaged or gently managed forests. Third, is the lack of data for cryptogams (mosses, liverworts, lichens) that often disperse poorly (Rose, 1976; Soderstrom, 1989) and may be especially responsive to microclimatic effects, and thus do better in larger forest patches (e.g., Canters et al., 1991; Chen et al., 1993). Together, these observations imply little more than it is important to sustain or develop a range of patch sizes, including contiguous tracts of older forests.

3. THE PROBLEM, THE CONTEXT, AND PROPOSED SOLUTION

3.1 The problem

When the Chief Forester of British Columbia determines the annual allowable cut for large planning units (1,000,000 ha plus), projection of growth and yield is aggregated over the entire area (aspatial) rather than tracked by real locations of distinct stands. Conversely, regulations intended to sustain forest values, such as water quality or biological diversity, are enacted in a spatially-explicit fashion (BC Ministry of Forests, 1995a,b). For two large planning units in southeastern British Columbia[2] this combination proved unfortunate. Timber that was physically present, and included in estimations of annual allowable cut, was made inaccessible for harvest by spatially-explicit regulations intended to sustain biological diversity. Unfortunately, those same regulations were creating a pattern of harvest potentially inimical to biodiversity.

3.2 The context

We treat two broad elements of the context: the socio-economic and regulatory environment, and the physical and biological environment.

---

[2] West Kootenay-Boundary Land use Plan (CORE, 1995a) and East Kootenay-Boundary Land Use Plan (CORE, 1995b).

3.2.1 Socio-economic and regulatory environment

Over the past decade British Columbia has typically contributed about one-third of the entire world's exports of softwood lumber (COFI, 1997). That market position makes BC extremely vulnerable to international perceptions of how well sustainable forestry is practiced in the province. Recognizing that vulnerability, the government of British Columbia sought to establish a code of forest regulations that would be internationally recognized as protecting all forest values. The result was the Forest Practices Code (FPC; BC Ministry of Forests, 1994) that covers all aspects of forest practice. Compared to at least 14 other jurisdictions the FPC and associated Guidebooks are, indeed, more protective of combined forest values (Westland Resource Group, 1995; Bunnell, 1996). The FPC and Guidebooks are "rule-based"; that is, they specify actions to be taken under the belief that these actions will produce the desired outcome. A "performance-" or "results-based" approach would specify desired outcomes, without constraining ways of attaining outcomes. Existing regulations reflect historical practices prevalent at the time of their creation – clearcutting. Key elements of the regulations include the size of cutblocks, adjacency guidelines (governing the period in which adjacent cutblocks can be harvested), and limits on the proportions of area for three broad age classes, including old growth (BC Ministry of Forests, 1995a,b). Age class targets are intended to be met within units that have both biological meaning (biogeoclimatic units; MacKinnon et al., 1993) and physical meaning by reflecting land form (Landscape Units). Forests of the province have long been mapped into biogeoclimatic units (Meidinger and Pojar, 1991). To implement the FPC, Landscape Units were described for the entire province (median size of about 50,000 ha).

When the FPC was created there was neither the time nor the will to examine the landscape pattern that would develop through time. Moreover, the requisite techniques of spatial modeling were still relatively new. All forests have some disturbance history producing a spatial and temporal distribution of age structures unique to that forest. When rules that serve to allocate harvest spatially are imposed on the existing pattern, regulatory outcomes may not meet desired goals. Guidelines governing age targets, cutblock size, and adjacency interact with the existing landscape pattern over space and through time to produce unintended consequences (Bunnell, 1998). Theoretically, the undesirable outcomes could be avoided. As written, the FPC allows considerable flexibility around guidelines. Most of that flexibility has not been attained for two reasons. First, departures from default values encourage greater scrutiny, which delays plans. Second, in an environment of public distrust, statutory decision makers are risk adverse and deviate little from default values. Guidelines become rules.

Any set of rules, when applied repeatedly, enforces a specific pattern at some scale. The FPC and Guidebooks tended to create a pattern of small, scattered cutblocks, rather like a checkerboard, with little range in patch size. As enacted, regulations to sustain biodiversity not only reduced accessible timber, but were having unintended consequences on biodiversity, including a well dispersed distribution of openings of relatively uniform sizes and increased amounts of road construction and actively used road (Nelson and Finn, 1991; Sahajananthan et al., 1996; Bunnell, 1998).

3.2.2 Physical and biological environment

Combined, the West Kootenay-Boundary and East Kootenay-Boundary Land Use Plans cover about 8.2 million hectares in southeastern BC. Forests within the area are diverse and at least 7 conifer species are harvested commercially. The tree species harvested range dramatically in their autecology and their regeneration is favored by different opening sizes (e.g., small for Engelmann spruce, whereas lodgepole pine responds well to large openings). Given the diversity of forest types, the vertebrate richness also is high. More than 210 forest-dwelling vertebrate species are present. Habitat preferences of these species also vary greatly. Depending upon forest type 18-25% of the species prefer well-expressed openings and early seral stages; another 20% appear associated with late-successional stages. A total of 58 species respond positively to edges while 12 species avoid edges. In short, the sustenance of both productive forests and vertebrate richness would benefit from a range of patch sizes. Terrain is rugged and dominated by several mountain ranges (including the Rockies and the Selkirks), so costs of road building and maintenance are high. Whatever the solution to attaining a range of patch sizes, it should not increase the lengths of active road. Roads have both an economic cost and several harmful biological effects (Bunnell et al., 1998).

3.3 The proposed solution

There were two clear needs: find more wood, and increase the range of patch or opening sizes.

One potential solution appeared obvious. Incorporating the flexibility in the FPC would allow some larger openings. This solution appeared "win-win" because larger cut blocks would make more timber immediately available while creating new areas of open habitat desirable for species such as elk, blue grouse, MacGillivray's warbler, or long-tailed vole. In the future these cutblocks would become larger tracts of older timber favorable to other species. Given the social context described, any modification of regulations could not occur, or be perceived to occur, in an arbitrary fashion. Under the mandate of the Kootenay-Boundary Land Use Plan Implementation Strategy (KBLUP, 1996), a committee was formed, chaired by the Assistant Chief Forester and including members the Ministries of Forests, Environment, Lands and Parks, and local stakeholders. Relaxing guidelines governing block size and adjacency were expected to permit a range in block sizes (and more timber from the larger blocks). In reality, that goal is constrained by the limits on allowable proportions of particular seral stages and the disturbance history of the landscape. Both the exposition of the problem and attempts at solution required explicit spatial modeling. The Centre for Applied Conservation Biology at the University of British Columbia was requested to perform the analyses.

4. THE APPROACH

Atlas/Simfor, a spatially-explicit, decision support tool, was chosen. Atlas derives solutions to harvest scheduling problems within a specified array of regulatory and spatial constraints (Nelson and Finn, 1991; Nelson, 1998). Simfor addresses consequences of different spatial distributions of habitat (modified by forest practices) on forest-dwelling vertebrates (e.g., Daust and Bunnell, 1992; Daust and Sutherland 1997). Combined they provide an effective tool for addressing public concerns about economic opportunities and biological diversity, and had already demonstrated their value in the area (Nelson and Wells, 1996; Nelson and Price, 1997).

To facilitate timely but credible analysis, the committee restricted analyses to one Landscape Unit (LU 26). LU 26 has a disturbance history of fire and harvesting by both large and small clearcuts (Figure 1). The unit covers 31,729 ha, of which 16,166 ha (51%) is operable and 15,563 ha (49%) is inoperable or reserved from harvest. A total of 2,271 harvest units averaging 8 ha in size were manually designated for the area to reflect normal operating constraints (e.g., topography, unstable slopes, existing roads, and FPC guidelines). We analyzed only the forested portion of the LU (Figure 1) of which 16,166 ha (86%) are operable and 2,703 ha (14%) are inoperable or reserved from harvest. The harvest units (cutblocks) created to reflect operating constraints, could be combined into larger openings.

A total of 17 combinations of regulatory constraints governing accessibility of timber were projected using Atlas/Simfor (these are summarized by Nelson and Wells, 1998). Here we present outcomes of three broad policy options evaluated by the committee charged with finding a solution:

- default FPC guidelines (patches as unanticipated outcomes);
- target patch range (patches by decree); and
- zones (patches by design).

The first ('Default') option examined consequences of implementing default guidelines of the FPC as these were being used in most areas of the province. The maximum size of

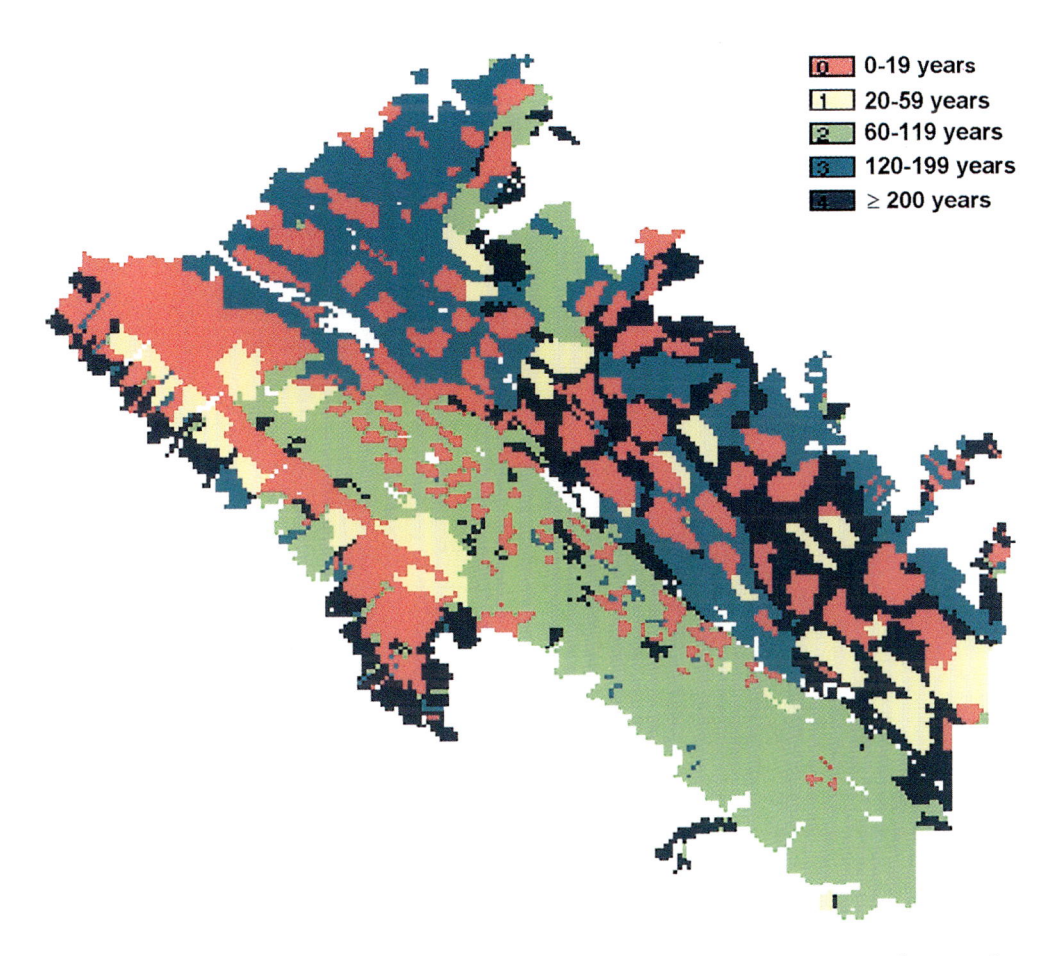

Figure 1. Forest age class structure of Landscape Unit 26 at year 1 (pattern existing at the time of the study).

cutblocks was 40 ha, adjacency rules specified 19 years before adjoining blocks could be harvested, and the proportion of area in old growth (>200 years) or mature + old forest (>100 years) could not be less than 10% and 30%, respectively.

The second ('Target') option examined consequences of imposing a specific target for distribution of patch sizes at the outset. The agreed-upon target range included small blocks (averaging 10 ha), and large blocks (averaging 102 ha). The range was constrained by past disturbance history. Planners aggregated small blocks of similar age to attain larger blocks. Adjacency rules were retained around the larger blocks.

The third ('Zones') option excluded some areas entirely from harvest and relaxed the default seral and adjacency constraints in other areas of production-oriented forestry. In the production-oriented zones no requirement for mature forest was applied. However, while zones were under development, we required a minimum of 13% old growth by area to be present within the Landscape Unit. An additional objective was to develop contiguous patches of old growth. For all scenarios a 100-

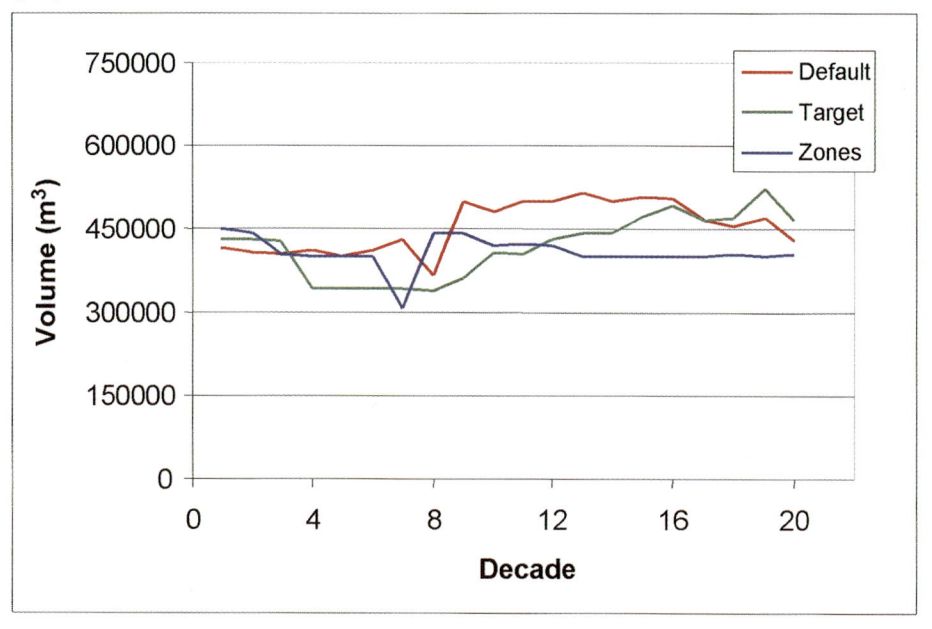

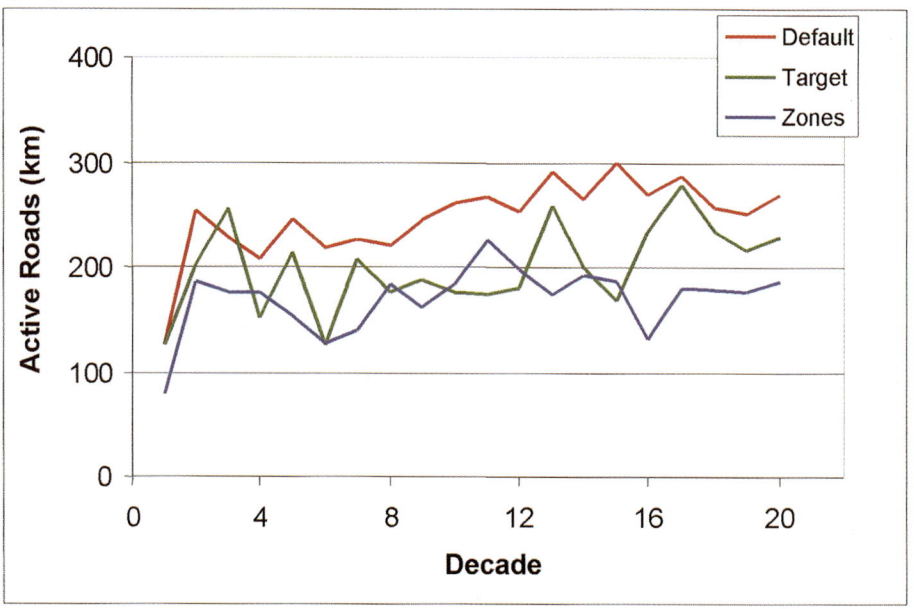

Figure 2. Timber volumes (a) and length of active road (b) under the three options. The values illustrated are decadal sums.

year rotation was used, projected for 200 years.

Three broad indicators were used to quantify the consequences of each option:
- timber available for harvest (m^3 per decade as projected over 200 years);
- roads (km of active roads); and
- landscape pattern (area in patch size classes).

5. RESULTS

Implications of analyses to economic opportunities and vertebrate richness are discussed separately.

5.1 Economic opportunities

Impact of the three options on the two major economic indicators was pronounced (Figure 2). Over the planning horizon, the Default option produced 9.1 million m^3 of timber volume, the Target option generated 8.4 million m^3, and Zones produced 8.2 million m^3. Because of the existing age structure and disturbance history, each option incurred a decline in volume available for harvest around the fourth to eighth decade (Figure 2a). The Target option, which imposed a range of patch sizes at the outset, generated the most enduring decline, lasting for about 40 years. This result was contrary to anticipated increases in yield. It reflects inefficiencies that can result from trying to impose a new patch pattern on a landscape that is already fragmented. It is less disruptive to work gradually towards a desired distribution of patch sizes, as in the Zones option. Despite this long-lasting decline in harvest volume, the Target option still extracted 92.4% of the volume over a 200-year period as was attained by the Default option. The Zones option stabilized below the Default option, but still extracted 90.3% of the volume that was attained under Default (a decline of about 4,500 m^3 annually). Under actual practice that difference likely would be less. With fibre production concentrated in zones, savings from the reduced costs of road building could be allocated to silviculture, producing greater yields (e.g., Binkley, 1997).

Differences in the amount of actively used road required to attain the harvest are more dramatic than differences in volume (Figure 2b). Total amounts of active road over the 200 year planning horizon were: Default (4939 km), Target (3995 km), and Zones (3390 km). Total active roads under the Zones option were only 69% of that under the default guidelines. The projected difference is underestimated. The simulation responded to the existing road network implemented during the previous harvest, results of which are illustrated in Figure 1. Had the road network been designed explicitly for the zoning option, road layout would have been more efficient and active road would have declined still further. In rugged terrain road construction is costly and a reduction in amounts of active road by >30% represents a substantial saving that could be allocated to more intensive forest tending in other areas. Further economic benefits can be attained by concentrating forest practices in a portion (zone) of the management area. As zones develop, initially stand tending is concentrated in the production-oriented zone; ultimately harvesting also is concentrated. Not only does this reduce costs of road construction and maintenance but it increases the likelihood of return from concentrated silvicultural investment.

Regardless of the option implemented, substantial time is required to modify existing patterns on the landscape. For example, to attain a large contiguous old-growth zone required about 200 years (Figure 3; see also Wallin et al., 1994).

5.2 Implications to vertebrate richness

We still know too little, to unequivocally specify the most desirable range of patch sizes. From basic natural history we know that some vertebrates do better in openings while some are favored by stands having late-successional attributes. Large contiguous tracts of tracts old growth do not appear necessary for the large majority of vertebrates (see also Kremsater and Bunnell this volume), but data for some widely ranging species are sparse. Moreover, there is growing evidence that some smaller organisms (mosses, lichens) do better in con-

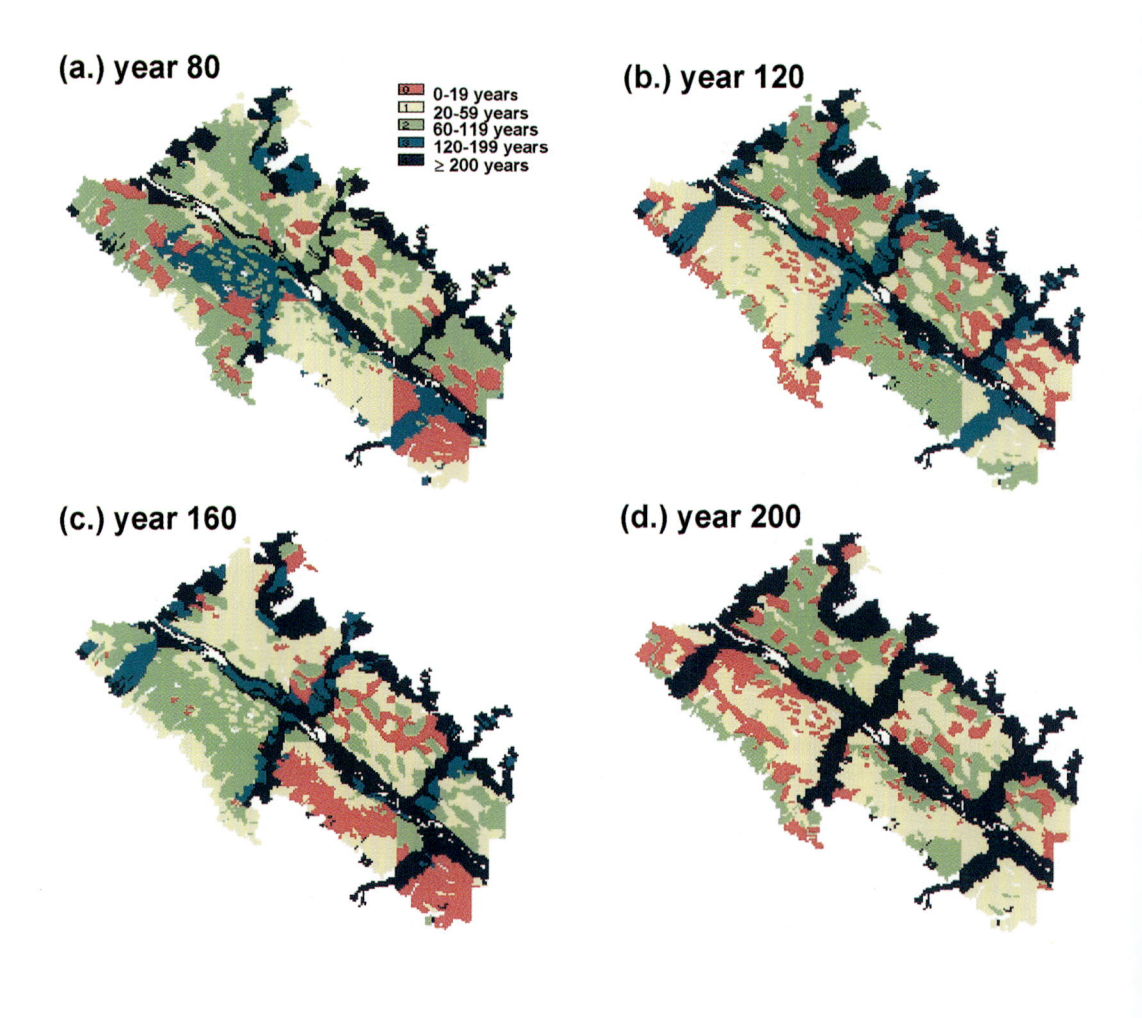

Figure 3. Development of the old-growth zone through time under the Zones option.

tiguous tracts of older forest. We appear to have only three criteria for evaluating the relative success of different distributions of patch size in sustaining vertebrate richness and biological diversity more generally. First, given the lack of compelling evidence for any particular distribution of patch sizes, management should not foreclose future options. That is, a particular patch distribution should not be entrained on the landscape; opportunities to create larger or smaller patches should be retained. Second, though still weak, there is accumulating evidence of some threshold effect where distribution of patches becomes more important than total area of patches (Fahrig this volume and With this volume). For example, as the total area of old growth declines it is increasingly important to aggregate that area; current estimates place the threshold at about 25 to 30% of total area (Andren, 1994; Kareiva and Wennergren, 1995; Fahrig, 1997). Third, a range of patch sizes is preferable to a uniform distribution of sizes.

We present outcomes of the three options visually in Figure 4 and as a summary of the patch size distribution after 200 years (Table 2). Any classification of patch sizes is necessarily arbitrary. We chose three class sizes for Table 2: 0-79 ha, 80-249 ha, and >249 ha.

At 200 years the three options yield very different patterns of patch size within the forest (Figure 4). Figure 4 illustrates several points. First, the Default option yielded a small range of patch sizes, a smaller range than was present under initial conditions (compare Figure 1 and 4; see also Table 2). Second, the Target option produced a greater range of patch size than did the Default option, but that range too is smaller than that under initial conditions. This occurred despite the explicit goal of a wider range of patch sizes. Third, both the Default and Target options acted to reduce the amount of old growth from initial conditions. Fourth, the Zones option sustained a range of patch sizes but increased the amount of old growth reserved from harvest about threefold.

One important lesson is that arbitrarily increasing the range of patch sizes (Target option) is ultimately and severely constrained by past disturbance history, including harvest practices. Adjacency rules simply eliminate harvest from larger blocks with a wider range of age classes, some of which would be accessible under smaller blocks of narrow age range. In short, spatially-explicit tools help greatly to reveal opportunities to attain particular objectives, such as expanding the range of patch sizes. Under the Target option it was difficult to move from the historical pattern, and road construction declined little while harvest declined as much as under the Zones option. The Default option also enforced existing patterns (Figure 4). While serving to sustain a higher volume harvested, the Default option also required more road and produced a distribution of patch sizes that appear undesirable and served to reduce total amounts of old growth (Table 2). Zoning served to increase connectivity and amount of old growth from about 10 to 30%, at a cost of less than 10% of the yield under the Default option. Moreover, the extent of roading and its negative effects on biological diversity were reduced. Planned zoning was able to create a range of patch sizes while generating a dramatic increase in area of contiguous old growth. Estimates of old growth retained are too high in each option simply because natural disturbance was not incorporated, but the relative ranking holds.

Ironically, the Default option, which was intended to sustain biological diversity, provides the least promising opportunities (Table 2, Figure 4). After 200 years the total amount of old growth under the Default option declined to about 12% of the area, and none was in patches >249 ha (Table 2). The patch distribution created in younger age classes provides little future option for modifying patch size as information on biological responses to patch patterns accrues (Table 2). Moreover, the increased amounts of active road (Figure 2b) have associated negative affects on both terrestrial and aquatic vertebrates. In part this failing results because of the manner in which past disturbance history interacts with guidelines as these are projected through time and space. However, even if the initial age distribution was uniform, the guidelines would tend to enforce a dispersed

(a.) Default Option

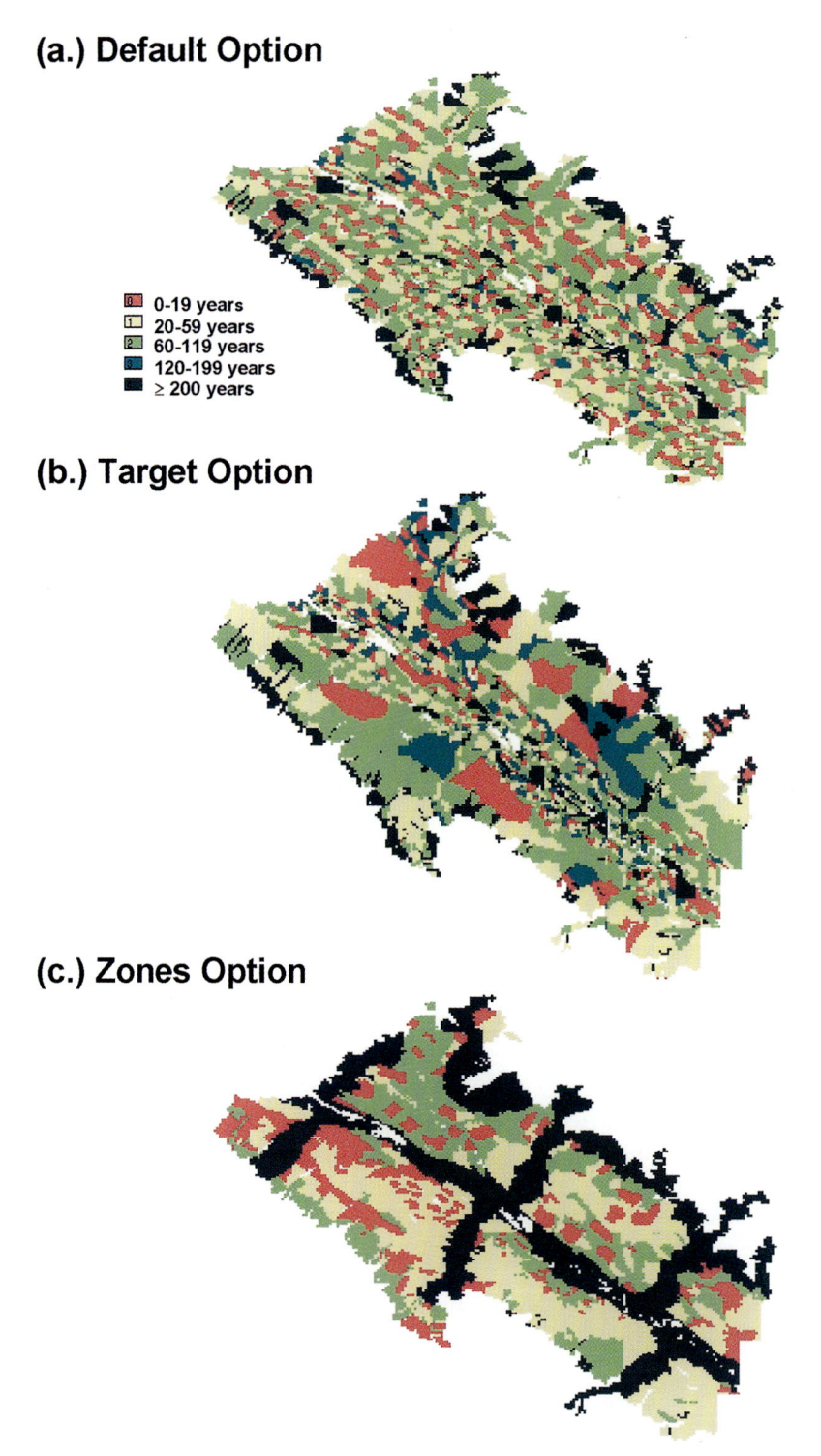

0-19 years
20-59 years
60-119 years
120-199 years
≥ 200 years

(b.) Target Option

(c.) Zones Option

Figure 4. Age class pattern at year 200 for Default option (a), Target option (b) and Zones option (c).

Table 2. Area in three patch-size classes (0-79, 80-249, and > 250 ha) after 200 years for the three harvest options as implemented on Landscape Unit 26

Option	Initial	Default	Target	Zones
Year	1	200	200	200
Young (0-19 yrs)				
0-79 ha	2053	2711	1598	1475
80-249 ha	988	0	512	558
>249 ha	1628	0	654	891
Mature (60-199 yrs)				
0-79 ha	184	2208	1898	304
80-249 ha	210	2536	1217	1449
>249 ha	4669	2586	2960	2249
Old (>200 yrs)				
0-79 ha	694	1282	1224	56
80-249 ha	541	774	1099	0
>249 ha	2340	260	0	6160

pattern of harvest, producing many smaller blocks and more active road.

The Target option serves primarily to illustrate the difficulties in imposing a different patch distribution on an existing disturbance history. Patches do show a greater range in size by the end of the 200 year planning horizon, which is likely beneficial (Table 2). However, despite sustaining a similar reduction in volume harvested as the Zones option, the amount of old growth reserved is only 12%. Because the guidelines do not explicitly require recruitment of old patches, almost all old growth requirements were met by reserves, thus old growth in the Target option is very similar to the Default option. The amount of active road decreased only 19% compared to Default. More future options exist for modifying patch size distribution than under the Default option, but they are not pronounced (Table 2). The Zones option has the biggest impact on volume (almost a 10% reduction), but would reduce road construction and maintenance costs substantially. It serves to sustain about 30% old growth, reduces roads by at least 30%, and creates more options for creating a range of patch sizes in the future. Not only is a large contiguous area of old growth patches created, but the young forest includes a range of patch sizes that permit flexibility among future options (Table 2).

It also is important to consider the nature of old growth reserved from harvest under each option. The Default and Target options both rely on the existing reserve system. While some of the reserved area included riparian and other identified environmentally sensitive areas, much was confined to subalpine forests. That is not unrealistic: without compelling evidence to the contrary, there is pressure to limit reserves designed to meet biodiversity objectives to inoperable or unproductive areas, which are biased towards subalpine forests. In the Zones option we specifically targeted productive forests as the areas to be reserved.

6. DISCUSSION

The mosaic of stand ages and structures always will be shaped by a combination of wind, fire, insects, fungi, and forest practices. This mosaic helps to determine habitat available to vertebrates and other components of biodiversity. The lack of directly applicable data and the variability of existing data permit few unequivocal statements about desirable patch-size distributions. We do know that a major effect of current practices is to reduce amounts of old growth and the size of contiguous, old-growth patches (Spies et al., 1994). For our review and analyses we can draw several implications relevant to efforts to guide the character of the forest mosaic.

6.1 Reducing patch area can have serious impacts on species.

Species disappear or go extinct from islands and parks faster than they do from mainland or surrounding areas (e.g., Diamond, 1984; Burkey, 1995; Newmark, 1995). Data from wooded patches surrounded by other land uses support a positive relationship between area and population persistence and abundance (Table 1; Opdam et al., 1984). Data also suggest that some species, forest interior species, require larger patches of forest than do others (e.g., Freemark and Merriam, 1986; McIntyre, 1995). Forests surrounded by vastly different vegetation do act as islands.

6.2 Natural disturbance regimes are of limited use in estimating desirable patch distributions.

Efforts to sustain biological diversity in managed forests have sought guidance from natural disturbance regimes (e.g., BC Ministry of Forests, 1995a). By emulating natural disturbance we attempt to sustain forest structures within the range to which organisms are adapted, but we need more clarity on the kinds of guidance we seek. There are several difficulties. One difficulty is the finding that for some forest types repeatable disturbance patterns apparently do not exist (e.g., Cumming et al., 1996; Bunnell and Huggard, 1999). Another is that some natural disturbances are simply too large and disruptive to willfully incorporate into planning (Bunnell, 1997). At the broadest level vertebrate faunas appear adapted to natural disturbance regimes (Bunnell, 1995), but our forest practices intervene in ways that are not expressed by these broad patterns. We will define better practices when we focus on total habitat (including structure) for the majority of species and patch size for those species that are demonstrably area sensitive.

6.3 There is no ideal arrangement of block sizes (openings or patches)

Managers need to invoke a variety of approaches. Emulating natural disturbance regimes makes sense only over periods of time long enough and areas large enough to accommodate the relatively rare but large-scale events that most impact a system. Any set of practices, when repeatedly applied, eventually homogenizes the landscape at some broader scale, to the detriment of some portion of biological diversity (e.g., Figure 3; Bunnell and Huggard, 1999). These observations imply that practices to maintain the entire range of biological diversity should be both aggregated or clustered, and sufficiently different from each other that markedly different forest structures result. Areas that retain a more diverse vegetative structure and reduce the density of barriers, such as roads, near older patches will increase effective patch area and reduce potential negative edge effects.

6.4 Some fragmentation is not bad

In western forests support for this observation is most apparent in data of MacGarigal and McComb (1995). They found most bird species associated with fragmented landscapes (see also Sallabanks et al. this volume). Evidence is implicit in the review of Bunnell (this volume) that western old-growth patches rarely function as isolated habitat islands. Given the diverse natural histories of vertebrates, these findings argue for a range of patch sizes.

6.5 Continuity of habitat (e.g., old growth) becomes increasingly important as total amounts of habitat decline.

This conclusion currently is derived primarily from studies using modeling (e.g., Fahrig this volume, With this volume), but has common sense appeal. The continuity of habitat appears to become important when total habitat declines to about 25-30% of the area. In Landscape Unit 26 we were able to attain 30% old growth at a cost of 10% of the harvest (Figure 2), by relaxing constraints in the area allocated to timber production. That cost is exaggerated because it ignores reduced costs of road construction, and maintenance and benefits of focussed silvicultural investment. Moreover, the bulk of available data suggest that harvest need not be completely excluded from areas managed to sustain late-successional features (review of Bunnell et al., 1998).

6.6 Once established the distribution of patch sizes is difficult to modify.

It can require more than one rotation to modify the distribution of patch sizes (Figure 3; see also Wallin et al., 1994). Even if partial cutting is widely adopted, incorporating harvested blocks into current plans tends to reinforce existing patterns. Because patch sizes created by current practices are enduring, the patterns should be carefully conceived. Flexibility and variability are important, as is spatially-explicit analysis. Implementing patch sizes by decree eliminates opportunities to work with the existing land base, and may enforce lower rates of harvest (Figure 2, Table 2).

6.7 There is a strong social element to desired patch distribution.

The social element extends beyond concerns about aesthetics or contributions to social infrastructure, to the relative numbers of species sustained. It is clear that extensive old growth was a less richly populated forest than are present forests. The most complete studies (e.g., Rosenberg and Raphael, 1986; McGarigal and McComb, 1995) document a greater richness of vertebrates in forests with wider ranges of age classes and more edge. Some species groups are particularly depressed within older forests; for example, the shrub nesters and understory foragers, many of which are neotropical migrants. The manager's dilemma is simple: closely emulating natural disturbance patterns is appropriate for some species, but if widely applied will markedly reduce species richness below levels to which the public has become accustomed. Biological diversity, particularly the current levels, cannot be sustained by doing the same thing everywhere (e.g., Bunnell, 1997; Bunnell et al., 1998). Together, these observations emphasize the importance of decision support systems that facilitate visualization of probable outcomes of alternative practices. That is particularly true for public lands.

6.8 Spatial analysis suggests utility in zoning intensity of forest practices

Current data suggest that species typical of open areas or younger forests have entered managed forests, and species more closely associated with older forests have persisted, but at declining numbers. That implies two broad alternatives: attempting to sustain all groups of species everywhere to the detriment of some, or deliberately zoning practices so that different groups are favored in different zones. Our findings suggest that when thoughtfully pursued, zoning the intensity of practices has potential to confer both economic benefits (Figure 2) and favourable biological outcomes (Table 2).

Acknowledgements. We thank Tim Shannon for technical assistance on ATLAS runs. The BC Forest Service, Canadian Forest Service, and Forest Renewal BC have supported our development of Atlas and Simfor. Collation of biological data was supported by MacMillan Bloedel Ltd and Lignum. This is Publication Number R-29 of the Centre for Applied Conservation Biology, University of British Columbia.

LITERATURE CITED

Ambuel, B. and S.A. Temple. 1983. Area-dependent changes in the bird communities and vegetation of southern Wisconsin forests. *Ecology* 64:1057–1068.

Andren, H. 1994. Effects of habitat fragmentation on birds and mammals in landscapes with different portions of suitable habitat: a review. *Oikos* 71:255-366.

Askins, R.A., M.J. Philbrick, and D.S. Sugeno. 1987. Relationship between the regional abundance of forest and the composition of forest bird communities. *Biological Conservation* 39:129–152.

Bancroft, G.T., A.M. Strong, and M. Carrington. 1995. Deforestation and its effects on forest-nesting birds in the Florida keys. *Conservation Biology* 9: 835–844.

BC Ministry of Forests. 1994. *Forest practices code of British Columbia.* Ministry of Forests, Victoria, BC.

BC Ministry of Forests. 1995a. *Forest practices code of British Columbia, Biodiversity Guidebook.* Ministry of Forests, Victoria, BC.

BC Ministry of Forests. 1995b. *Forest practices code of British Columbia, Green-up guidebook.* Ministry of Forests, Victoria, BC.

Binkley, C.S. 1997. Preserving nature through intensive plantation management: the case for forestland allocation with illustrations from British Columbia. *Forestry Chronicle* 73:553-559.

Bolger, D.T, A.C. Alberts, and M.E. Soulé. 1991. Occurrence patterns of bird species in habitat fragments: sampling, extinction, and nested species subsets. *American Naturalist* 137:155–166.

Bradshaw, F.J. 1992. Quantifying edge effect and patch size for multiple-use silviculture—a discussion paper. *Forest Ecology & Management* 48:249–264..

Bunnell, F.L. 1995. Forest-dwelling vertebrate faunas and natural fire regimes in British Columbia: patterns and implications for conservation. *Conservation Biology* 9:636–644.

Bunnell, F.L. 1996. Forest issues in BC.: the implications for zoning. *Truck Logger BC* 19(3): 49-55.

Bunnell, F.L. 1997. Operational criteria for sustainable forestry: focusing on the essence. *Forestry Chronicle* 73:679–684.

Bunnell, F.L. 1998. Next time try data: a plea for variety in forest practices. *In* Interior Forestry Conference: investing today in tomorrow's forests. University College of the Cariboo, Kamloops, BC. pp. 59-70.

Bunnell, F.L., and D.J. Huggard. 1999. Biodiversity across spatial and temporal scales: problems and opportunities. *Journal of Forest Ecology & Management* 115: 113-126.

Bunnell, F.L., L.L. Kremsater, and M. Boyland. 1998. An ecological rationale for changing forest practices on MacMillan Bloedel's forest tenure. Pub. No. R-22, Centre for Applied Conservation Biology, University of British Columbia, Vancouver, BC.

Burkey, T.V. 1995. Extinction rates in archipelagoes: implications for populations in fragmented habitats. *Conservation Biology* 9:527–541.

Butcher, G.S., W.A. Niering, W.J. Barry, and R.H. Goodwin. 1981. Equilibrium biogeography and the size of nature preserves: an avian case study. *Oecologia* 49:29–37.

Canters,K.J., H. Schöller, S. Ott, and H.M. Jahns. 1991. Microclimatic influences on lichen distribution and community development. *Lichenologist* 23:237-252.

Chen, J., J.F. Franklin, and T.A. Spies. 1993. Contrasting microclimates among clearcut, edge, and interior of old growth Douglas-fir forest. *Agricultural and Forest Meteorology.* 63:219-237.

Council of Forest Industries (COFI). 1997. British Columbia Forest Industry Fact Book - 1997. Council of Forest Industries, Vancouver, BC.

Commission on Resources and the Environment (CORE). 1995a. West Kootenay Land Use Plan. BC Land Use Coordination Office, Victoria, BC.

Commission on Resources and the Environment (CORE). 1995b. East Kootenay Land Use Plan. BC Land Use Coordination Office, Victoria, BC.

Cumming, S.G., P.J. Burton and B. Klinkenberg. 1996. Boreal mixedwood forests may have no "representative" areas: Some implications for reserve design. *Ecography* 19(2): 162-180.

Diamond, J.M. 1984. "Normal" extinctions of isolated populations. *In* M.H. Nitecki (Ed). *Extinctions.* Univ. Chicago Press, Chicago, IL pp. 191–246.

Daust, D.K., and F.L. Bunnell. 1992. Predicting biological diversity on forest land in British Columbia. *Northwest Environmental Journal* 8: 191-192.

Daust, D.K., and G.D. Sutherland. 1997. SIMFOR: software for simulating forest management and assessing biodiversity. *In* J.D. Thomson (ed.). *The status of forestry/wildlife decision support systems in Canada: proceedings of a symposium, 1994.* Natural Resources Canada, Canadian Forest Service, Great Lakes Forestry Centre, Sault Ste. Marie, ON. pp 15-29

Fahrig, L. 1997. Relative effects of habitat loss and fragmentation on population extinction. *Journal of Wildlife Management* 61:603-610.

Fahrig, L., and G. Merriam. 1994. Conservation of fragmented populations. *Conservation Biology* 8:50-59.

Franklin, J.F., W. Emmingham, and R. Jaszkowsi. 1983. True fir-hemlock. In R.M. Burns (tech. coord.). *Silvicultural systems for the major forest types of the United States*. USDA Forest Service, Agricultural Handbook No. 445, Washington, DC. pp. 13–15.

Freemark, K.E., and G. Merriam. 1986. Importance of area and habitat heterogeneity to bird assemblages in temperate forest fragments. *Biological Conservation* 36:115-41.

Galli, A.E., C.F. Leck, and R.T.T. Forman. 1976. Avian distribution patterns in forest islands of different sizes in central New Jersey. *Auk* 93:356-364.

Gates, J.E., and L.W. Gysel. 1978. Avian nest dispersion and fledging success in field-forest ecotones. *Ecology* 59:871-883.

Glenn, S.M., and T.D. Nudds. 1989. Insular biogeography of mammals in Canadian Parks. *Journal of Biogeography* 16:261-268.

Hagan, J.M., W.M. VanderHaegen, and P.S. McKinley. 1996. The early development of forest fragmentation effects on birds. *Conservation Biology* 10:188-202.

Harris, L.D. 1984. *The fragmented forest. Island biogeography theory and the preservation of biotic diversity*. University of Chicago Press, Chicago, IL.

Janzen, D.H. 1986. The eternal external threat. In M.E. Soulé (ed.). *Conservation biology: the science of scarcity and diversity*. Sinauer Associates Inc., Sunderland, MA. pp. 286–303.

Kareiva, P., and U. Wennergren. 1995. Connecting landscape patterns to ecosystem and population processes. *Nature* 373:299-302.

Keller, M.E., and S.H Anderson, 1992. Avian use of habitat configuration created by forest cutting in southeastern Wyoming. *Condor* 94:55-65.

Kootenay-Boundary Land Use Plan (KBLUP). 1996. Implementation Strategy. Land Use Coordination Office Victoria B.C. http://www.luco.gov.bc.ca/slupinbc/kootenay/kootnews/kblup.htm

Lehmkuhl, J.F., L.F. Ruggiero, and P.A. Hall. 1991. Landscape-scale configurations of forest fragments and wildlife richness and abundance in the southern Washington Cascade range. In L.F. Ruggiero, K.B. Aubry, A.B. Carey, and M.H. Huff (tech. coords.). *Wildlife and vegetation of unmanaged Douglas-fir forests*. U.S.D.A. For. Serv. Gen. Tech. Rep. PNW-285, Portland, OR. pp. 425–442.

Lynch, J.F.. and D.F. Whigam. 1984. Effects of forest fragmentation on breeding bird communities in Maryland, USA. *Biological Conservation* 28:287-324.

Lynch, J.F., and R.F. Whitcomb. 1978. Effects of insularization of the eastern deciduous forest on avifaunal diversity and turnover. In A. Marmelstein (ed.). *Classification and evaluation of fish and wildlife habitat*. USDI Fish & Wildlife Service Pub OBS-78176, Washington, DC. pp. 461.

MacArthur, R.H., and E.O. Wilson. 1967. *The theory of island biogeography*. Princeton University Press, Princeton, NJ.

Mackinnon, A., D. Meidinger, and K. Klinka. 1993. Use of the biogeoclimatic ecosystem classification system in British Columbia. *Forestry Chronicle* 69:100-120.

Martin, T.E. 1980. Diversity and abundance of spring migratory birds using habitat islands of the Great Plains. *Condor* 82:430-439.

Martin, J-L, A.J. Gaston, and S. Hitier. 1995. The effect of island size and isolation on old growth forest habitat and bird diversity in Gwaii Haana (Queen Charlotte Islands, Canada) *Oikos* 72:115-131.

McGarigal, K. and W.C. McComb. 1995. Relationships between landscape structure and breeding birds in the Oregon Coast Range. *Ecological Monographs* 65:235-260.

McIntyre, N.E. 1995. Effects of forest patch size on avian diversity. *Landscape Ecology* 10:85-99.

Meidinger, D. and J. Pojar (comp. & eds.). 1991. *Ecosystems of British Columbia*. B.C. Ministry of Forests, Victoria, B.C.

Nelson, J.D. 1998. Operations Manual - ATLAS/FPS. Faculty of Forestry, University of BC, Vancouver, BC.

Nelson, J.D. and Finn, S.T. 1991. The influence of cut block size and adjacency rules on harvest levels and road networks. *Canadian Journal of Forest Research* 21:595-600.

Nelson, J.D. and R.W. Wells. 1996. Forest planning using the ATLAS/SIMFOR models: A case study in the Revelstoke Timber Supply Area. Contract report prepared for the BC Ministry of Forests.

Nelson, J.D. and L. Price. 1997. Modelling regional land-use plans in British Columbia. Forest Scenario Modelling for Ecosystem Management at the Landscape Level. European Forest Institute, EFI Proceedings No. 19. Wageningen, The Netherlands. June 26-July 3, 1997. pp. 89-99.

Nelson, J.D. and R.W. Wells. 1998. The effect of patch size on timber supply and landscape structure in Landscape Unit 26 – Golden Timber Supply Area. Contract report prepared for the Kootenay-Boundary Land Use Plan Implementation Committee, Nelson, BC. 29p.

Newmark, W.D. 1987. Legal and biotic boundaries of western North American National parks: a problem of congruence. *Biological Conservation* 33:197-208.

Newmark, W.D. 1995. Extinction of mammal populations in western North American parks. *Conservation Biology* 9:412-426.

Opdam, P. 1991. Metapopulation theory and habitat fragmentation: a review of holarctic breeding bird studies. *Landscape Ecology* 5:93-106.

Opdam, P., D. Van Dorp, and C.J. F. Ter Braak. 1984. The effect of isolation on the number of woodland birds in small woods in the Netherlands. *Journal of Biogeography* 11:473-478.

Paine, R.T. 1988. Habitat suitability and local population persistence of the sea palm *Postelsia palmaeformis*. *Ecology* 69:1787-1794.

Robbins, C.S., D.K. Dawson, and B.A. Dowell. 1989. Habitat area requirements of breeding forest birds of the mid Atlantic states. *Wildlife Monographs* 103.

Robinson, G.R., R.D. Holt, M.S. Gaines, [and others]. 1992. Diverse and contrasting effects of habitat fragmentation. *Science* 257:524-257.

Rose, F. 1976. Lichenological indicators of age and environmental continuity in woodlands. *In* D.H. Brown, D.L. Hawksworth, and R.H. Bailey (eds.). *Lichenology: progress and problems*. Academic Press, London. pp. 279-307

Rosenberg, K.V., and M.G. Raphael. 1986. Effects of forest fragmentation on vertebrates in Douglas-fir forests. *In* J. Verner, M. Morrison, and C. Ralph, (eds.). *Wildlife 2000: modelling habitat relationships of terrestrial vertebrates*. Univ. of Wisconsin Press. Madison, WI. pp. 263–272.

Sahajananthan, S., Haley, D., and J.D. Nelson. 1996. Opportunities for forest land zoning. In Proceedings of the International Mountain Logging and Pacific Northwest Skyline Symposium, May 12-16, 1996, Campbell River, BC. FERIC Special Report SR-116, Vancouver, BC. pp. 118-123.

Schieck, J., K. Lertzman, B. Nyberg, and R. Page. 1995. Effects of patch size on birds in old-growth montane forests. *Conservation Biology* 9:1072-1084.

Shaffer, M.L. 1981. Minimum population sizes for species conservation. *BioScience* 31(2):131-134.

Shaffer, M.L. 1990. Population viability analysis. *Conservation Biology* 4:39-40.

Soderstrom, L. 1989. Regional distribution patterns of bryophyte species on spruce logs in northern Sweden. *Bryologist* 92:149-155.

Spies, T.A., W.J. Ripple, and G.A. Bradshaw. 1994. Dynamics and pattern of a managed coniferous forest landscape in Oregon. *Ecological Applications* 4:555-568.

Urban, D.L., R.V.O'Neill, and H.H. Shugart, Jr. 1987. Landscape ecology: a hierarchical perspective can help scientists understand spatial patterns. *BioScience* 37:119-127.

Usher, M.B., J.P. Field, and S.E. Bedford. 1993. Biogeography and diversity of ground-dwelling arthropods in farm woodlands. *Biodiversity Letters* 1:54-62.

Wallin, D.O., F.J. Swanson, and B.Marks. 1994. Landscape pattern response to changes in pattern generation rules: land-use legacies in forestry. *Ecological Applications* 4:569-580.

Wenny, D.G., R.L. Clawson, J. Foabory, and S.L. Sheriff. 1993. Population density, habitat selection, and minimum area requirements of three forest-interior warblers in central Missouri. *Condor* 95:968-979.

Westland Resource Group. 1995. A comparative review of the Forest Practices Code of British Columbia with fourteen other jurisdictions. Crown Publications, Victoria, BC.

Wolff, J.O., E.M. Schauber, and W.D. Edge. 1997. Effects of habitat loss and fragmentation on the behaviour and demography of gray-tailed voles. *Conservation Biology* 11: 945-956.

Verboom, J., A. Schotman, P. Opdam, and J.A.J. Metz. 1991. European nuthatch metapopulations in a fragmented agricultural landscape. *Oikos* 61:149-156.

Yahner, R.H. 1983. Seasonal dynamics, habitat relationships, and management of avifauna in farmstead shelterbelts. *Journal of Wildlife Management* 47:85-104.

Yahner, R.H. 1988 Changes in wildlife communities near edges. *Conservation Biology* 2:333-339.

Yahner, R.H. 1992 Dynamics of a small mammal community in a fragmented forest. *American Midland Naturalist* 127:381-391.

Forest Fragmentation: Wildlife and Management Implications Synthesis of the Conference

William C. McComb

This conference was convened as an opportunity to provide a synthesis of the current state of knowledge relating to forest fragmentation in managed forests. Such syntheses have value in providing land managers with a central source that can be used to initiate literature reviews and develop background materials for management plans. Further, such work sets the stage for development of further research so that decisions facing land managers 10 years from now will be based on an even more solid foundation of scientific inquiry. My task is to provide a summary of the conference and highlight important findings that may direct the readers to specific papers or literature citations. I will structure this paper by first presenting some of the basic issues facing biologists and managers regarding fragmentation as a process, discuss the role of connectivity in landscapes and then present some thoughts regarding the use of these concepts in management plan development and future research directions.

1. HABITAT LOSS OR HABITAT PATTERN?

Habitat fragmentation has been considered one of the greatest threats to biodiversity worldwide (Burgess and Sharpe 1981, Noss 1983, Harris 1984). But the term 'fragmentation' has been used in a variety of ways to describe a range of landscape conditions and processes. Habitat fragmentation is a landscape-level process in which a given habitat area is divided into smaller, geometrically more complex, and more isolated fragments as a result of both natural processes and human activities. Fragmentation usually co-occurs with habitat loss, but there is an important distinction between habitat loss and habitat fragmentation (Fahrig this book). For a given area of habitat, fragmentation refers to the pattern and isolation of patches embedded within a matrix. Fragmentation involves changes in landscape composition, structure, and function at many scales and occurs on the natural patch mosaic created by changing landforms and natural disturbance.

Habitat homogenization is a process of reducing habitat pattern complexity in naturally complex systems. Habitat may be gained or lost, but in systems such as those in the Interior west, natural disturbance processes such and fire and disease created a complex mosaic of patch types on the landscape (Agee this book). Fire suppression and insect control have led to coalescence of forest patches and reduction in the number of forest patches on the landscape. Landscape change beyond the natural range of variability, either through fragmentation or homogenization, can produce undesirable effects (Agee this book). Nonetheless, the dominant effect of landscape change on most vertebrate populations seems to be habitat loss primarily, and habitat pattern secondarily.

Much of the empirical work on which fragmentation theory is based comes from patch-level (not landscape-level) field studies on patch size, shape and isolation (Fahrig this book). Patch-level studies from the eastern deciduous forest of North America indicate

Department of Forestry and Wildlife Management, University of Massachusetts, Amherst, MA 01003

that the abundance of vertebrate species associated with forest interiors generally is low in areas dominated by agricultural development and urbanization (Whitcomb et al. 1981, Robbins et al. 1989, Terborgh 1989). Changes in vegetation, food resources, predation, brood parasitism, and competition have been noted as causes of these observed vertebrate community changes (Strelke and Dickson 1980, Kroodsma 1982, Brittingham and Temple 1983, Wilcove 1985, Noss 1988, Yahner and Scott 1988). In this form of forest loss, forest tracts are progressively reduced to smaller and more isolated patches embedded within a relatively permanent (barring reforestation) matrix of nonforest. The landscape structure becomes relatively static. Forest patches adjoin nonforest habitat and remain isolated from similar forest patches for long periods of time. From a forest-dwelling animal's perspective, forest fragments presumably become embedded in a matrix of largely unsuitable habitat that inhibits the movement of animals among isolated forest patches, depending on the species' habitat selectivity and vagility.

The few studies that have been conducted at the landscape scale (e.g., landscapes used as independent observations) suggest that many of the patterns observed at the patch level often are not clearly expressed at the landscape level (Fahrig this book). Although patch-specific studies often detect edge effects (Kremsater and Bunnell this book) and nest predation effects (Marzluff and Restani this book, Sallabanks et al. this book) in managed forests, these effects often are not seen at larger scales. Landscape-scale studies conducted in managed forests indicate that it is the sheer amount of habitat of a particular species present in the landscape that is the most significant associate of animal abundance and only secondarily does pattern influence abundance (e.g., McGarigal and McComb this book; Schmiegelow and Hannon this book, Welsh and Healy 1993).

The process of forest loss and fragmentation caused by forest management is somewhat different from the process in agricultural and suburban areas. Harvested forest patches regrow and eventually become mature forest again. Hence, as long as the forest is managed on a sustainable basis, the percent of the landscape in each stage of forest development remains relatively constant, but the distribution of forest patches shifts over time. Natural forest landscapes are spatially and temporally dynamic mosaics of forest patches caused by natural disturbances (e.g., fire and windthrow) and forest regrowth (Agee this book). Forest harvest disturbances can differ markedly from natural disturbances in their frequency, intensity and size, so landscape structure can be dramatically altered by timber management activities (Garman et al. this book). Vertebrate population dynamics in forests being fragmented by timber management activities seem to differ from population dynamics in forests that have been affected by urbanization and agricultural development. Sharp forest edges may produce microclimatic effects into remnant forests (Kremsater and Bunnell this book), but such edges are transient in managed forest landscapes because of forest regrowth. Regenerating plantations may function at least as dispersal habitat for many species, because they may represent stands of variable quality and permeability to animal movements. Mature forest patches may rarely be isolated or they may be only temporarily isolated from one another.

2. CONNECTIVITY AND THE LANDSCAPE MATRIX

It is clear that, depending on the abilities of organisms to cross gaps or otherwise low-quality habitat, some level of connectivity is necessary across landscapes over time (With this book). Connectivity becomes increasingly important as the loss of habitat proceeds to a level reaching 20-40% of a landscape that has a hostile matrix (With this book; Andren 1994; Fahrig 1997). It is important for managers to realize that habitats are not simply barriers or corridors, but in fact patches with varying levels of permeability to organisms moving through them. What is not clear, however, is that there are generalizations that can be made regarding species' needs for corridors. Indeed, it is the permeability and patch quality of the landscape matrix that may be more important to animal dispersal and subsequent persistence on dynamic landscapes than simply the presence or absence of corridors (Bunnell this book). Managers can play a key role in thoughtfully applying silvicultural approaches to patches across landscapes over time to ensure that permeability can be maintained. Simply not doing the same thing everywhere all the time can be an important step toward minimizing instances of habitat isolation.

3. USING SCIENTIFIC INFORMATION – LANDSCAPE PLANNING

Land use decisions often are driven primarily by economic concerns within regulations designed to protect soil, water, and air. The expression of these policies can have obvious and persistent imprints on the landscape (Garman et al. this book; Kremsater and Bunnell this book; Wells et al. this book). Managers charged with ensuring the persistence of organisms through space and time must move from being reactive to being proactive in design of land use plans that can lead to benefit for humans and other animals (McAllister et al. this book). Decisions made while considering areas of inappropriate spatial or temporal scale can lead to undesirable cumulative effects (Garman et al. this book; Wells et al. this book). Proactive management attempts to consider the effects over space and time and then allows the planners to consider alternatives that lead to desired results (McAllister et al. this book). If we do not develop information that can be used proactively now, the cumulative effects of past and current decisions may pre-empt the potential to achieve wildlife goals for a considerable period into the future.

As a first approximation of development of a planning process, at least the following points should be considered:

1. *Identify the extent of the plan and understand the context (current and future).* Many plans are applied specifically to one ownership without fully considering the influence of adjacent landowner activities on the likelihood of achieving habitat goals (e.g., McAllister et al. this book). At the very least, habitat change and function should be considered on all land holdings adjacent to those for which the plan is being developed. Ideally, the plan would be developed over an ecologically defined mixed-ownership landscape. That is not to say that all owners in the planning area will work together to achieve common goals, but it does mean that decisions regarding management on one ownership can be made with the knowledge of how neighboring patches are likely to develop.

2. *Identify the grain of the landscape and understand the underlying structure and composition as viewed by the organism.* Plans often are developed by identifying management activities among stands over time (McAllister et al. this book). The function of the stands to the organisms will be highly influenced by structure and composition that results from that management and recovery (McComb et al. 1993). Often, though, that fine-grained detail in structure and composition is either not specifically considered in the planning process, or it is classified and assumed to be constant among stands within each class. Classes often are developed through the human eye and may not be based on the key features selected by a species or group of species. In fact, site quality, stocking density, and tree and shrub composition can be considered for each stand on the landscape and the resulting landscape mosaic can become very complex if various silvicultural prescriptions are applied (McAllister et al. this book). This complexity could be classified in various ways to describe habitat quality for a number of species considered in a plan. Such silvicultural prescriptions influence aspects of habitat quality including matrix conditions, habitat area over time, edge contrast, and patch size.

3. *Identify the range of habitat area to be represented over the extent over time -- will this persist within the natural range of variability?* A coarse filter approach to managing for a variety of species over space and time may include management for a range of natural variability in forest conditions (McAllister et al. this book). Alternatively, the strategy may include management at a level that falls within the range but is relatively static and does not have the same amplitude of variability as might occur under natural disturbances (Garman et al. this book). Clearly the coarse filter approach is likely to be more

successful, especially if disturbance frequency and intensity are explicitly considered, but social pressures may prevent management over a full range conditions found naturally (Agee this book). In either case, management outside the range of natural variability may be desirable at times to achieve certain goals, but such management should raise questions regarding sustainability. Such management does not preclude the need for fine-filter assessment of species or processes that might be vulnerable at the extremes of the range.

4. *Pattern is persistent -- decisions made now may last for centuries.* Perhaps one of the most significant decisions made in land and water management over the past 2 centuries has been where lines were drawn on a map. Boundaries between landowners, between watersheds, between stands, or between counties can each reflect different policies and different land management approaches. Such differences are likely to be persistent for a rotation or more (Wells et al. this book). Although pattern effects may be secondary to amount of habitat within a species territory or home range, or as allocated over a landscape, pattern at some scales affects the potential for connectivity (With this book). Pattern at a scale larger than a species territory may also affect the potential for sufficient habitat to fall within a territory or home range of a species.

5. *Should edge effects be encouraged (Swiss cheese) or discouraged (Pac Man)? There will be winners and losers.* There are species that seem to respond to habitat complementation – that is the juxtaposition of two or more habitat conditions can lead to higher habitat quality than if these types are not adjacent. Elk (*Cervus elaphus*) are a good example. Olive-sided flycatchers (*Contopus borealis*) also seem to be edge associates in some forests. Consequently, management plans that develop landscape designed to encourage habitat for these species may reduce habitat availability for species associated with a particular patch type (e.g., Northern spotted owls, *Strix occidentalis caurina*) to the point where the habitat specialist is unlikely to persist in that landscape. Just as no single silvicultural system will meet the needs of all species, no single landscape design will meet the needs of all species. Variability in design over space and time will be key to ensuring that vertebrate species will persist throughout a region.

6. *What is the matrix? Can it be made more permeable?* Consideration of the within-stand features that are likely to increase stand permeability to dispersing organisms and which may accelerate development of acceptable habitat for some species through stand regrowth may be more important than the pattern of stands in a landscape at any one time (Bunnell this book, McComb et al. 1993). Dynamic systems such as forests do not function in the same way as islands in an ocean. Certainly some patches may become isolated for some species at certain times, but this isolation often is temporary. The effects of such isolation likely will be minimized if the duration of isolation is shortened relative to the generation length of the species. Consequently for species associated with old forests, stand management actions that allow second-growth stands to recover their function more rapidly (e.g., higher levels of legacy for the previous stand are retained) could play key roles in maintaining populations over space and time (McComb et al. 1993). And it is important to recognize that not all stands need to be managed to accelerate this recovery, but strategically located stands over space and time could accomplish this goal. This may be a key opportunity for private forest land managers to contribute to regional goals for maintaining wildlife species.

7. *How will the condition change through time given natural disturbances?* Clearly there is uncertainty regarding the effectiveness and persistence of stand conditions that might be arranged over space and time to achieve wildlife habitat goals. The continued effects of natural disturbances add to this uncertainty, and there is some likelihood that some large size, intense disturbance will reshape the landscape into a totally different structure. Consequently landscape plans must have some level of redundancy across the region to help minimize risk of species loss.

8. *Finally, there must be a sincere commitment to monitoring management action effects on habitat structure and function.* Although nearly every management plan includes a section on monitoring, this section is the one that is often least followed. Design of monitoring programs should be rigorous and ensure high quality data collection. It should allow managers and decision-makers to test hypotheses regarding management effects (McAllister et al. this book). Monitoring information should be coordinated among landowners within an ecologically meaningful unit of space and time.

4. NEW DIRECTIONS

There obviously is a nearly endless set of new questions and approaches that could and should be assessed to minimize land management risks and advance our knowledge in landscape ecology. I attempt to identify a few key areas of science, management and the interface between the two that might lead to improved management for wildlife in the region.

Because habitat area seems to be a more significant driving factor related to the abundance and distribution of organisms across landscapes than pattern, it becomes imperative that we begin to understand threshold effects of habitat area on population persistence. Landscapes with varying levels of habitat area between 0 and 50% will likely be the focus of such studies in order to detect such thresholds. Within that setting, not only should abundance be estimated, but processes that affect population dynamics should also be considered. Nest success, as affected by predation or brood parasitism, age-specific survival rates, reproduction rates, and mobility all should be investigated for species reflecting range of life history strategies. It is quite likely that the threshold effects detected will be highly dependent on the matrix condition. Consequently the studies outlined in the previous paragraph, in its simplest form, should have a two-way component:

	Habitat area (%)							
Matrix condition	0	5	10	15	20	30	40	50
Hostile	x	x	x	x	x	x	x	x
Permeable	x	x	x	x	x	x	x	x

It was clear from this conference that there is a significant need for improving communication between scientists and managers. Academia and management each have been prolific in use of jargon. Minimizing jargon and using examples to illustrate points and concepts seems a key first step to putting science into practice. However, given the information needs faced in this area, the most effective mechanism for increasing communication is for managers and scientists to work together to design landscape-scale studies that address management issues. Such a team approach also can lead to increased trust, a better understanding of the limitations of existing data, and an opportunity for scientists to make decisions with managers despite inadequate knowledge. One such study might be to estimate if landscapes managed within (or for) the range natural of variability of certain landscape conditions will likely maintain species over space and time.

Despite the uncertainty facing land managers, decisions must be made regarding land management actions. Decision support tools should be developed, improved, and used to aid managers in forest planning. Finding acceptable plans for multiple-species systems on multiple ownerships over time are so complex that planners will need tools to simulate change over space and time. Some tools are available now, but few if any include dynamic landscape simulations (including stochastic natural disturbances) and population viability assessments.

From a land management perspective large-scale plans will be most successful if multiple owners in an area agree to some common goals. Certainly not all landowners in an area will participate, but if even 2 parties agree to work together, there can be significant advantages to both groups. If they are successful, then other landowners may be likely to participate as well.

There are a number of questions that deal with spatial and temporal scaling among organisms and among levels of management actions that must be resolved to allow managers to achieve a desired landscape structure. Each organism scales its environment differently. Can we classify organisms reasonably into groups reflecting these scaling views (e.g., based on life history characteristics)? If so, what are the logical classes? How do they coincide with scales represented through land management activities? What information will be lost or gained by approaching management at one scale vs. another?

Last, but not least, nearly all of the information presented in this conference dealt with vertebrate animals. But it is clear from work conducted in the tropical systems with invertebrates that responses to patch characteristics are highly variable among taxa (With this book). What may be a permeable matrix for a salamander may be impermeable to a lichen. Or to view this another way, dispersal across complex landscapes may need to be viewed as a function of generations for many of the taxa representing our hidden diversity. Clearly we need a more complete understanding of the responses of hidden diversity to habitat loss and changes in landscape pattern.

5. SUMMARY

In summary, fragmentation or homogenization are processes that must be understood when developing landscape plans. Plans must look beyond ownership boundaries, however, and must reflect likely changes through time. Managers would do well to focus first on ensuring that adequate habitat area is provided for species of interest, but they also should consider potential pattern effects secondary to area effects. Pattern may have an effect on connectivity, especially in the matrix. But the matrix should be viewed as differentially permeable to dispersal in forested landscapes and the dynamic nature of the matrix and its permeability should be taken into account when planning.

One mechanism to achieve habitat area considerations and permeable matrix conditions is to apply silvicultural tools and reserves to the landscape in a manner that will achieve long-term objectives. These applications must be considered relative to their current and future value in achieving landscape objectives. Such projections of future conditions will likely come from decision support tools that can simulate change in conditions over space and time. Finally, the effectiveness of the chosen plan should be monitored. Those aspects of the landscape plan that are based on critical assumptions of animal response or habitat dynamics should be rigorously monitored to allow testing of the plan as a hypothesis. Done in a coordinated manner over space and time, such information could lead to a set of landscape plans that can be adaptable and effective. "Just don't do the same thing everywhere!" (Bunnell this book).

6. LITERATURE CITED

Agee, J. K. this book. Disturbance effects on landscape fragmentation in Interior West Forests. Chapter 3 in Forest Fragmentation: Wildlife and Management Implications.

Andren, H. 1994. Effects of habitat fragmentation on birds and mammals in landscapes with different proportions of suitable habitat: a review. Oikos 71:355-366.

Brittingham, M. C., and S. A. Temple. 1983. Have cowbirds caused forest songbirds to decline? BioScience 33:31-35.

Burgess, R.L., and D.M. Sharpe. 1981. Forest island dynamics in man-dominated landscapes. Springer-Verlag, New York.

Bunnell, F.L. this book. What habitat is an island. Chapter 1 in Forest Fragmentation: Wildlife and Management Implications.

Bunnell, F.L., R.W. Wells, and J. D. Nelson, and L.L. Kremsater. this book. Patch sizes, vertebrates, and effects of harvest policy in southeastern British Columbia. Chapter 14

in Forest Fragmentation: Wildlife and Management Implications.

Fahrig, L. 1997. Relative effects of habitat loss and fragmentation on species extinction. Journal of Wildlife Management 61: 603-610.

Fahrig, L. this book. Forest loss and fragmentation: which has the greater effect on persistence of forest-dwelling animals. Chapter 5 in Forest Fragmentation: Wildlife and Management Implications.

Garman, S. L., F. J. Swanson, and T. A Spies. this book. Past, present and future potential landscape patterns in the Douglas-fir region of the Pacific Northwest. Chapter 4 in Forest Fragmentation: Wildlife and Management Implications.

Harris, L. D. 1984. The fragmented forest: island biogeography theory and the preservation of biotic diversity. Univ. Chicago Press, Chicago. 211pp.

Kremsater, L. L. and F. L. Bunnell. this book. Edges: Theory, evidence and implications to management of westcoast forests. Chapter 7 in Forest Fragmentation: Wildlife and Management Implications.

Kroodsma, R. L. 1982. Edge effect on breeding forest birds along a power-line corridor. J. of Applied Ecology 19:361-370.

Marzluff, J. M. and M. Restani. this book. The effects of forest fragmentation on rates of avian nest predation and parasitism. Chapter 8 in Forest Fragmentation: Wildlife and Management Implications.

McAllister, D. C., R. W. Holloway, and M. W. Schnee. this book. Using landscape design principles to promote biodiversity in a managed forest. Chapter 13 in Forest Fragmentation: Wildlife and Management Implications.

McComb, W. C., T. A. Spies, and W. H. Emmingham. 1993. Stand management for timber and mature-forest wildlife in Douglas-fir forests. J. Forestry 91(12):31-42.

McGarigal, K. and W. C. McComb. this book. Forest fragmentation and breeding bird communities in the Oregon Coast Range. Chapter 12 in Forest Fragmentation: Wildlife and Management Implications.

Noss, R.F. 1983. A regional landscape approach to maintain diversity. BioScience 33:700-706.

Noss, R. F. 1988. Effects of edge and internal patchiness on habitat use by birds in a Florida hardwood forest. Ph.D. Thesis. Univ. of Florida, Gainesville.

Robbins, C. S., D. K. Dawson, and B. A. Dowell. 1989. Habitat area requirements of breeding birds of the Middle Atlantic States. Wildlife Monographs. 103. 34pp.

Sallabanks, R, P. J. Heglund, J. B. Haufler, W. Wall, B. A Gilbert, and W. Wall. this book. Forest fragmentation in the inland west: issues, definitions, and potential study approaches. Chapter 10 in Forest Fragmentation: Wildlife and Management Implications.

Strelke, W. K., and J. G. Dickson. 1980. Effect of forest clearcut edge on breeding birds in Texas. J. Wildl. Manage. 44:559-567.

Terborgh, J. 1989. Where have all the birds gone? Princeton University Press, New Jersey. 207pp.

Welsh, C. J. E., and W. M. Healy. 1993. Effect of even-aged timber management on bird species diversity and composition in northern hardwoods of New Hampshire. Wildlife Society Bulletin 21:143-154.

Whitcomb, R.F., J.F. Lynch, M.K. Klimkiewwicz, C.S. Robbins, B.L. Whitcomb, and D. Bystrak. 1981. Effects of forest fragmentation on avifauna of the eastern deciduous forest. Pages 125-205 in Burgess and Sharpe, eds., Forest island dynamics in man-dominated landscapes. Springer-Verlag, New York.

Wilcove, D.S. 1985. Nest predation in forest tracts and the decline of migratory songbirds. Ecology 66:1211-1214.

With, K. A. this book. Is landscape Connectivity necessary for wildlife management? Chapter 6 in Forest Fragmentation: Wildlife and Management Implications.

Yahner, R. H., and D. P. Scott. 1988. Effects of forest fragmentation on depredation of artificial nests. J. Wildl. Manage. 52:158-161.

Key Word Index

active management   247
agriculture   155, 187
artificial nests   155
avian productivity   187
biodiversity   247
biological diversity   117
birds   223
boreal mixedwood forest   201
case study   271
connectivity   1, 97, 201
conservation   33, 87
dispersal   97
ecosystem   33
ecosystem management   61
edge effects   117, 155, 187
experimental fragmentation   201
fire regimes   43
forest edges   117
forest fire   43
forest fragmentation   43, 61, 97, 187
forest management   61, 117, 247
forest-edge   87
forest-interior   87
forestry   1
fractal landscapes   97
fragmentation   1, 33, 155, 223
gaps   97
genetic variations   171
genetics   171
habitat   1
habitat configuration   87
habitat corridors   97
habitat destruction   87
habitat fragmentation   87
habitat loss   87
habitat patch   247
habitat restoration   87
hierarchy   97
historical conditions   187
inland west   187

interior habitat   247
Interior West forests   43
landscape composition   187
landscape design   247
landscape ecology   43
landscape management   247
landscape pattern   61, 223
landscape scale   87, 201
late-seral forest area   223
late-seral forest configuration   223
late-seral forest relationships   223
managed forest   155
management   247
microclimate   117
movement   97
natural disturbance   61
natural disturbance regimes   187
nest parasitism   117
nest predation   155, 201
neutral landscape models   97
Oregon Coast Range   223
Pacific Northwest   61
patch size   271
patch size effects   187
perceptual resolution   97
percolation theory   97
predation   117
reserve   33
scale   97
silviculture   187
simulation modeling   271
songbirds   201
stand types   247
structure based management   247
threshold   87
timber supply   271
urbanization   155
vertebrates   1, 271
western United States   43
wildlife habitat   247

Conference Summary Statement

FOREST FRAGMENTATION:
Wildlife and Management Implications

November 18-19, 1998, Portland, Oregon

Conference Summary Statement
Prepared by: James A. Rochelle

ACKNOWLEDGEMENTS

The conference, "Forest Fragmentation: Wildlife and Management Implications" was sponsored by seven organizations representing academia, government, forest landowners and conservation organizations. Approximately 300 individuals from various affiliations attended. The conference received financial support from the Oregon Forest Resources Institute and the National Council of the Paper Industry for Air and Stream Improvement. Co-sponsors included the Sustainable Ecosystems Institute, the Oregon State University College of Forestry, the Cooperative Forest Ecosystem Research program at Oregon State University, the University of British Columbia Centre for Applied Conservation Biology and Defenders of Wildlife.

James Rochelle served as chairman of the conference planning committee and conference coordinator. The members of the planning committee were Deborah Brosnan, Sustainable Ecosystems Institute; Fred Bunnell, Director of the Centre for Applied Conservation Biology at the University of British Columbia; Steven Courtney, National Council of the Paper Industry for Air and Stream Improvement; John Hayes, Associate Professor of Wildlife Ecology at Oregon State University; Leslie Lehmann, Executive Director, Oregon Forest Resources Institute and Sara Vickerman, Director, West Coast Office, Defenders of Wildlife.

Mike Cloughesy and the Conference Office of the College of Forestry at Oregon State University provided administrative support. Joe Wisniewski served as editorial coordinator. Jerry McIlwain prepared a preliminary draft of this summary statement.

ABOUT THE CONFERENCE AND SUMMARY STATEMENT

This report summarizes the findings of a scientific conference entitled "Forest Fragmentation: Wildlife and Management Implications" held in Portland, Oregon on 18-19 November, 1998. The report synthesizes key points from the authors' papers and conference presentations and the associated discussions, and does not necessarily represent the complete views of each of the individual contributors.

The conference was convened to provide a synthesis of the current state of knowledge related to fragmentation in managed forests of the Pacific Northwest. Conference sessions dealt with an overview of the fragmentation issue, the influence of

fragmentation on wildlife productivity and ecological processes, land use practices and forest fragmentation and case studies that illustrated some management approaches currently being used to address landscape-scale forestry effects. Participating authors and their scientific papers were as follows:
- What habitat is an island? Fred L. Bunnell, University of British Columbia.
- Vulnerability of Forested Ecosystems in the Pacific Northwest To Loss of Area. J. Michael Scott, University of Idaho, U.S. Geological Survey.
- Fire Effects on Landscape Fragmentation in Interior West Forests. James K. Agee, University of Washington
- Past, Present, and Future Landscape Patterns in the Douglas-fir Region of the Pacific Northwest. Steven L. Garman, Oregon State University, Frederick J. Swanson and Thomas A. Spies, USDA Forest Service, PNW Forest and Range Experiment Station.
- Forest Loss and Fragmentation: Which has the Greater Effect on Persistence of Forest-dwelling Animals? Lenore Fahrig, Carleton University.
- Is Landscape Connectivity Necessary and Sufficient for Wildlife Management? Kimberly A. With. Bowling Green State University.
- Edge effects: theory, evidence and implications to management of western North American forests. Laurie Kremsater and F.L. Bunnell, University of British Columbia.
- The Effects of Forest Fragmentation on Avian Nest Predation. John Marzluff and Marco Restani, University of Washington.
- The Role of Genetics in Understanding Forest Fragmentation. L. Scott Mills and David A. Tallmon, University of Montana.
- Forest Fragmentation of the Inland West: Issues, Definitions, and Potential Study Approaches for Forest Birds. Rex Sallabanks, Sustainable Ecosystems Institute, Patricia Heglund, Turnstone Ecological Research Associates, Jonathan B. Haufler, Boise Cascade, Brian A. Gilbert, Plum Creek Timber Company, and William Wall, Potlatch Corporation.
- Forest-level Effects of Management on Boreal Songbirds: the Calling Lake Fragmentation Studies. Fiona K.A. Schmiegelow and Susan J. Hannon, University of Alberta.
- Forest Fragmentation Effects on Breeding Bird Communities in the Oregon Coast Range. Kevin McGarigal and Willam C. McComb, University of Massachusetts.
- Using Landscape Design Principles to Promote Biodiversity in a Managed Forest. David C. McAllister, Oregon Department of Fish and Wildlife and Ross W. Holloway and Michael W. Schnee, Oregon Dept. of Forestry.
- Patch Sizes, Vertebrates, and Effects of Harvest Policy in Southeastern British Columbia. Fred L. Bunnell, Ralph W. Wells, John D. Nelson and Laurie L. Kremsater, University of British Columbia
- Forest Fragmentation: Wildlife and Management Implications. Synthesis of the Conference. William C. McComb, University of Massachusetts

ABOUT THE BOOK

A book containing 15 peer-reviewed chapters will be published by Brill Academic Publishers in the summer of 1999. For further information and/or to order a book, please contact: Dr. Janjaap Blom, Brill Academic Publishers, P.O. Box 9000, 2300 PA Leiden, in the Netherlands. (Phone 31 71 535 3540; Fax 31 71 531 7532; E-mail: blom@brill.nl).

INTRODUCTION

For years scientists have considered habitat fragmentation to be one of the greatest threats to wildlife survival and biodiversity worldwide. The fragmentation process generally is viewed as the breaking apart of a given area of habitat into smaller, geometrically more simple pieces as a result of natural processes as well as human activities. However, the term fragmentation has been used in a variety of ways to describe a range of landscape conditions and processes.

Concepts of fragmentation derive from the theory of island biogeography, in which patches of forest habitat are considered to be like islands, separated and isolated from each other by a sea of hostile land. With oceanic islands, consequences of this separation include reduced species richness, genetic isolation and species extinctions. However, the application of this theory to forests has been a source of confusion and controversy, with little in the way of clarifying research. Nevertheless, assumptions about impacts to biodiversity from forest fragmentation have affected forest policy, both in the private and public sectors.

The impetus for the conference, the key findings of which are summarized in this report, came from a scientific review and synthesis of current literature on wildlife relationships in western forests managed for timber production carried out in 1997 by Dr. Fred Bunnell of the University of British Columbia. This review found little evidence of fragmentation effects on vertebrates comparable to those observed in conjunction with forest land use changes in midwestern and eastern parts of North America. This conference was established to refine scientifically and clarify the issue of forest fragmentation. The major goal was to help researchers and managers both understand and address concerns related to forest management and wildlife at a landscape scale.

Although the subject of fragmentation has been addressed for some land uses (e.g. agriculture and urban development), this is the first time it has been comprehensively examined relative to forest management. Agreement on what constitutes fragmentation has

Forest landscape – Photo credit: J.A. Rochelle

been hindered by a lack of distinction in research design between habitat fragmentation, or change in landscape configuration, and habitat loss. Although the two processes can be separated in models, in field studies they are almost always confounded and the independent effects of habitat loss and separation of habitat are difficult to assess. This has resulted in conflicting views on the relative significance of these processes based on experimental evidence and the requirements of the species being studied.

Key findings of the conference are presented below. It's important to keep in mind that most fragmentation research to date has dealt with short-term effects on vertebrates, which appear to be minimal. As several authors suggest, there is uncertainty about the long-term effects, primarily because of the potential for time lags in habitat use patterns, demographic responses such as birth and death rates, and genetic changes. The inherent variation and frequently small sample sizes associated with wildlife population studies also present challenges to drawing absolute conclusions regarding effects of habitat change.

As an overview, the information presented provided little evidence of negative effects on vertebrate biodiversity from fragmentation in western forests. The increased predation and nest parasitism that are common when forest edge is surrounded by agricultural or suburban development are not evident where forest lands remain in forest use. However, both negative and positive influences have been observed, and are specific to particular vertebrate species.

There is also little evidence that lack of connectivity poses a threat to most vertebrate species in western forests, although use of corridors has been observed in some species. Leaving relatively small amounts of residual structure, (e.g. shrubs, snags and decaying wood), after harvest apparently makes the area between habitat patches (the "matrix") more hospitable, so that movement and dispersal can continue.

There is general agreement that the total amount of habitat is of greater significance to vertebrate survival and reproduction than its configuration, but that both are important. From a conservation perspective, this suggests prevention of permanent loss of forest habitat should be the first priority since opportunities remain to address habitat configuration issues over time in managed forests.

Finally, most scientists agree that widespread conversion of natural forests to forests managed for wood production modifies habitat and can cause adverse impacts to some species. What matters is that we take advantage of the latest available scientific information to shape a landscape in "working forests" that supports native species and provides other values such as wood products and recreation.

The negative impacts on wildlife observed with loss of forest habitat through conversion to agriculture or development raised serious concerns in the scientific community about similar effects in forests managed for wood production. However, this conference points to important management opportunities to maintain habitat for native wildlife in western managed forests by planning the size and distribution of habitat patches and by retaining key habitat structures (snags, shrubs, decaying wood, large live trees) in the surrounding landscape.

KEY FINDINGS

- Fragmentation has not been clearly defined by the research community, causing confusion in discussions of its ecological effects.
- Fragmentation (dividing a habitat into smaller, more isolated patches) usually co-occurs with habitat loss. The response of vertebrate populations differ, however, and for most species the effects of habitat loss are more significant than changes in habitat pattern.
- While the minimum area of habitat required is affected by needs of the individual species, the surrounding landscape and other factors, some research suggests survival will be affected if the area of suitable habitat falls below a threshold of 20-30 percent.
- Negative effects of fragmentation on vertebrates observed when forest land is converted to agriculture or suburban development are not apparent in western forests fragmented by timber production.
- Although research in western forests has not confirmed them, assumptions about negative effects of forest fragmentation have guided policy development.
- Northwest forests were naturally fragmented by disturbances such as fire and disease; small patches dominated east-side forests; larger patches characterized west-side forests. Habitat patterns changed following disturbance as young stands grew into mature forests. In drier eastside forest regions, fire suppression is, over time, "de-fragmenting" patterns of fuel distribution and increasing the potential for large wildfires.
- Discussions of forest fragmentation are only meaningful if made in the context of historical landscape conditions.
- Connectivity of the landscape is not uniformly important to forest vertebrates. Its importance is determined by the degree to which the land surrounding a forest patch (the "matrix") is hostile, and a species' ability to cross a matrix of low quality habitat. Forest vertebrates vary widely in their "gap-crossing" ability.
- The abundance of birds doesn't change significantly in a forest stand from 40 years old to maturity. A few vertebrate species are restricted to old growth.
- Leaving relatively small amounts of habitat structure, (e.g. shrubs, snags, decaying wood, live conifers and hardwoods), after harvest apparently makes the area between habitat patches (the "matrix") more hospitable, so that movement and dispersal of many species may be enhanced.

- Riparian areas often are richer in species and more densely inhabited than are upslope areas, however few forest-dwelling species are restricted to riparian areas.
- Both positive and negative effects of forest "edge" have been documented in recent research, although more species are positively- than negatively-associated with edge. Researchers in the Pacific Northwest have recorded a few species consistently associated with forest interior.
- The increased predation and nest parasitism seen when forest edge is surrounded by agricultural or suburban development are not evident where forest lands remain in forest use.
- There is no ideal habitat patch size. Management approaches that result in a variety of habitat conditions and patch sizes on the landscape will meet the needs of the greatest number of species.
- More research is necessary on determining the scale at which individual vertebrate species relate to habitat.
- Forest managers in the Pacific Northwest should not blindly apply the results of research studies from other regions such as the eastern and midwestern U.S.

WHAT IS FOREST FRAGMENTATION?

Concepts of fragmentation are based on the theory of island biogeography, where an island is surrounded by a sea hostile to the island wildlife. In order for the theory to apply to forests, forest stands must act as isolated islands in a sea of hostile land.[1]

Early research described both area and distance effects on oceanic islands.[2] Larger islands contain more species than smaller islands because small islands experience more extinctions and receive fewer immigrants. Equivalent-sized islands more remote from the mainland or source population will have fewer species because the extinction rate is the same but the immigration rate is lower. Larger areas also may have more diverse habitats and will be hospitable to more kinds of species.

Results of initial studies of forest fragmentation appeared to be consistent with island biogeographic theory. Forest study areas typically were surrounded by agriculture or urban land. The forest fragment behaved like an island in a sea of non-forest land use hostile to forest dwelling species. However, there is now evidence that island fragmentation theory does not hold true when extended to patches of older forest amidst stands of younger forests.[3]

Some of the confusion surrounding interpretation of fragmentation effects results from the way it has been defined in research. For example, definitions have ranged from "breaking the habitat into pieces" to "habitat loss combined with changes in habitat configuration." Perhaps because habitat loss and changes in habitat patterns frequently co-occur, many studies have failed to distinguish between them.

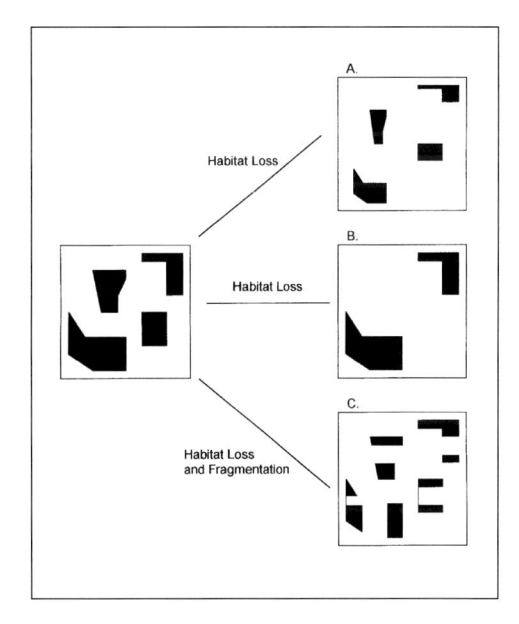

The concepts of habitat loss and fragmentation. From L. Fahrig (conference paper)

In one of the few studies which made this distinction, McGarigal and McComb concluded that "with the exception of a few edge species, variation in bird abundance among landscapes was more strongly related to changes in habitat area; habitat configuration was of secondary importance."[3] Fahrig also concluded that habitat loss has a far greater effect than the fragmentation or "breaking apart" of habitat. Some research suggests that if the overall amount of habitat falls below a critical threshold (20-30 percent), it will affect species survival. [4]

The influence of fragmentation also depends upon the individual species of interest. For species that use a range of forest habitats, the overall amount of habitat is the main issue. However, the more broken apart the forest, the less habitat there is available to forest interior species, and the more habitat there is available to species associated with the edge of a forest stand. [4] As With points out, "a landscape is not inherently connected or fragmented; the same landscape may be both from the perspective of two different species."[5]

In summary, we will pose the question again and try to answer it in a more concise fashion. What is forest fragmentation? It is the process of reducing size and connectivity of stands composing a forest. It may occur naturally or through human-caused activities. It may or may not be accompanied by habitat loss, depending upon the species of interest.

HISTORICAL DISTURBANCE REGIMES AND THEIR INFLUENCES ON FOREST FRAGMENTATION

Historically, Northwest forests were fragmented by natural disturbances such as fire and disease[9]. Wildfire has been the predominant natural force controlling landscape conditions over the past millennia.[6]

Fire effects differed considerably in terms of frequency and severity from one forest region to another. Historic fire patterns varied from frequent low-severity fires, which generally had little long-term effect on distribution of plants and animals across the burned area, to infrequent, high-severity burns, which resulted in replacement of the stands within the burned area.[7] This produced conditions in which some forests were

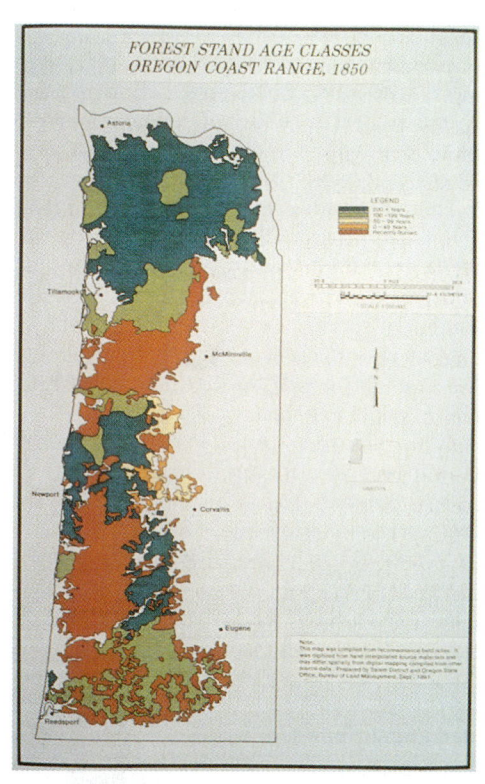

Oregon Coast Range – 1850: A mosaic of forest stand ages and habitat conditions. Photo credit: US Bureau of Land Management

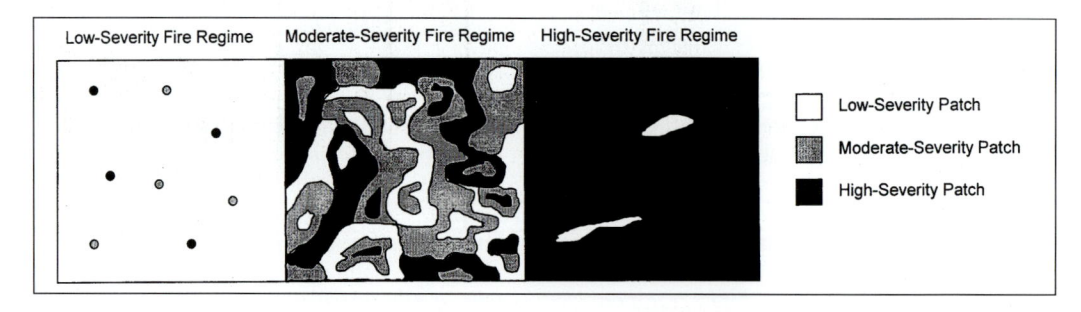

Landscape patterns associated with fire regimes of different severity. From J.Agee (conference paper)

dominated by small habitat patches while others were characterized by larger forest patches.

For example, assessments of forest cover conditions in the 1850's indicated "a mosaic" of forest habitat types. Over a third of the Oregon Coast Range was in a recently burned condition, for the most part distributed in very large patches (thousands of acres).[8] The intense fires left behind an abundance of snags and decaying logs on the ground that provided habitat in the areas cleared by burning.[6] In westside forests, old growth moved across the landscape over long periods of time.[7] Many forest animals apparently adapted successfully to these conditions.

HUMAN INFLUENCES ON PATTERNS OF FOREST FRAGMENTATION

Since the late 1800's, timber harvesting and fire suppression have replaced natural disturbance as the primary forces shaping forest landscapes. Perhaps the most important consequence of timber harvesting has been the significant reduction in amounts of old growth forest on private land and its high degree of fragmentation on federal land. Fire suppression over the last several decades is also altering natural disturbance patterns and is generally recognized to be "defragmenting" some northwest forests.[9]

Additionally, a significant proportion of low-elevation forest land in western Washington and Oregon has been converted to other uses, primarily agricultural and suburban development, resulting in both fragmentation and loss of forest habitat.

Interior western forests also have been radically altered by humans. Low-severity fire patterns have been greatly affected by grazing, fire exclusion and repeated timber harvests.[6] These factors have resulted in increased tree density and decreased herbaceous and low shrub understory, in effect, a conversion of these forests from low-severity fire regimes to high-severity fire regimes.

Recent harvest patterns in the Douglas-fir region are resulting in landscape patterns reflecting differences in harvest strategies among ownerships.[7] Spies demonstrated that, for comparable areas, the dispersed cutting pattern on public land resulted in about 10-30 percent less closed canopy interior and twice as much edge habitat as the aggregated cutting on private industrial land. At the landscape scale, however, private lands had much less interior and much more edge than public lands.[10]

New management approaches are projected to reduce the contrast between older and younger forests as federal land managers seek to establish and perpetuate "late succes-

Conversion of forest land to agriculture and residential uses is a major cause of forest habitat loss in the Pacific Northwest. Photo credit: J.A. Rochelle

Elk, considered to be generalists, utilize a variety of habitats in managed forests. Photo credit: J.A. Rochelle

The brown creeper, an old-growth associate, constructs its nest under loose tree bark.

sional" (old growth) reserves, state foresters manage in part for late successional habitat and private landowners apply enhanced habitat protection mandates. As a result, more spatially connected, older closed-canopy forests, with reduced contrast between older and young forest are expected in the future.[7]

CONTEXT: TIME AND SPACE

Because low elevation forests of the inland west were naturally fragmented by fires and other disturbances, the effects of forest management on fragmentation are meaningful only in the context of historical landscape conditions[9]. Interior forests have different fragmentation patterns than coastal and Cascade forests, because of their respective fire histories.

Human induced fragmentation has different effects in western forests than in midwestern and eastern forests where fragmentation is caused by the forest being converted to other uses and where forest habitat is permanently lost. In western forests, harvesting creates clearings of young forest surrounded by older forests, but the land remains in forest use.

The white-crowned sparrow is an early seral associate. Photo credit: Cornell Institute of Ornithology – M. Hopiak

Forest fragmentation affects various habitats differently and can best be understood from the perspective of individual species.[11] Harvesting temporarily returns the forest to the early seral stage (i.e. regeneration), and species associated with early successional habitats are favored. For species associated with later stages of forest development, the negative effects of fragmentation will be less if a population can move to a new habitat, survive in the surrounding matrix, or live in small patches of the original habitat until the surrounding habitat returns to more desirable conditions.

STAND AGE: WINNERS AND LOSERS

Forest vertebrates can be placed into one of three groups: generalists, which thrive in a broad variety of habitats; early-seral associates, which prefer younger forest habitats; and late-seral associates, which need features like large live trees, snags and decaying wood found in older forests. Forests managed for timber production typically provide abundant habitat for generalists and early-seral species. Of more limited supply, and perhaps greater concern, are the old seral stages or "old growth."

Several papers in the conference addressed the question: Do old growth patches in the Pacific Northwest follow island biogeographic theory and behave as islands in a sea of hostile habitat? Bunnell, in a comprehensive review of literature on species-habitat relationships, defined patches as areas of old forest surrounded by younger forest, and asked two broad questions about how wildlife responds to these patches: (1) Are there differences in the structure of the wildlife community across various ages of naturally regenerated stands? And (2) do individual species appear closely linked to late successional stages? His general conclusions were that "differences in abundance among forest age classes appear primarily among individual species rather than communities of species; there are no apparent " old growth" communities among amphibians, reptiles, birds or mammals." He also suggests that "there are species or small groups of species that are strongly linked to features found in late-successional forests."

For example, few significant differences in total abundance of birds among old growth, young and mature stands were noted by Bunnell. Amphibian species "richness" also differed little between old and young stands, with individual species relating more to downed wood, shrubs or other habitat features than to differences in stand age. The

Snags retained in young forests provide important habitat for some species associated with older forests. Photo credit: J.A. Rochelle

A patch of old-growth forest surrounded by younger stands. Photo credit: J.A. Rochelle

same is true for mammals, with the exception of bats. In eight of nine tests, bat activity was higher in old growth than in young stands. Available information on reptiles is limited, but most research suggests they benefit from openings large enough to create warmer, drier areas. Finally, neither furbearers nor forest ungulates (e.g. deer and elk), reported as associated with old growth, appear to require large, contiguous patches of old growth habitat.[1]

Another important observation was that small amounts of habitat structure maintain many species assumed to be late-successional associates at levels statistically inseparable from levels in old growth stands.[1] This finding is important from a management standpoint in that it suggests most late-successional species can be maintained in managed stands by retaining suitable levels of required habitat elements.

Thus, across a forested landscape, quality of the matrix (land surrounding a forest habitat patch) matters more to most vertebrate species than the patterns of stand distribution. Habitat patches will be isolated only to the extent that the matrix around them is inhospitable, and this varies greatly among individual vertebrate species. For many species, the quality of the matrix is enhanced by features (e.g. shrubs, snags and decaying wood, live trees), which can be provided through forest management. As Bunnell states, "Patches act as discrete entities or islands only when they provide favorable habitat and the surrounding habitat is hostile. Without a hostile matrix around the patch there is no island in the sea"[1].

EDGES, PREDATION AND PARASITISM

"Edge" is where different plant communities or forest stands of different ages come together, such as mature forest adjacent to a harvested area or an agricultural field. The creation of edge, and potential negative impacts on some species is one of the concerns expressed about forest fragmentation.

Observations, largely from eastern and midwestern North America, indicate that nest predation and parasitism may limit songbird reproductive success in forested landscapes fragmented by agriculture or urbanization. Creation of edge and reduction in size of forest fragments that favored nest predators and parasites were considered responsible for reduced survival and repro-

In managed forests, the most common type of edge occurs where forest stands of different ages come together. Photo credit: Rocky Mountain Elk Foundation

The presence of streams, wetlands and other special habitats adds to the variety of edges occurring in managed forests. Photo credit: J.A. Rochelle

duction.[12,13,14] However, in synthesizing information on these topics, Marzluff and Restani found that two thirds of earlier studies reviewed failed to find a significant decrease in nest predation or parasitism from the forest edge to the forest interior. They suggest "that, contrary to popular belief, edge effects may be an uncommon occurrence in forest fragments."[15]

Both positive and negative effects of edge have been documented in recent research. While most studies have focused on forest to field edges, fewer have examined forest to clearcut edges. Edge effects are similar in all studies: species richness generally increases with edge, and more species are positively, than are negatively associated with edge.[16] The ecological consequences of an individual species' response, positive or negative, to edge are less well understood. An example is the California red-backed vole[27], which responds negatively and has a potentially

Nest predators associated with edges in western forests include the Least chipmunk (left below) and Steller's jay (left). Photo credits: Rex Sallabanks (Least chipmunk) Steller's jay Cornell Institute of Ornithology – S. Bahart (left)

important role as a disperser of mycorrhizal fungi; its absence in cutover areas may have implications for tree nutrition and growth. Similarly, positive responses by some bird species to the creation of openings may result in patterns of seed dispersal which have important influences on patterns of vegetative development.

In studies conducted entirely within a forest landscape, Schmiegelow and Hannon found in three nest predation experiments that "fragmentation did not increase predation rates and there did not appear to be an increase in predation at the forest/clearcut edge.[17]

Kremsater and Bunnell also found little evidence of increased predation or nest parasitism in managed western coastal forests.[16] Juxtaposition of forest patches varying in age and structure was less likely to increase and focus predation at mature forest edges than was juxtaposition of forest patches with agricultural or urban parcels. In eastern and midwestern North America parasitism of nests by the brown-headed cowbird has been shown to negatively impact nesting success of songbirds in areas where wooded habitats are surrounded by agricultural fields. Cowbirds are rarely detected in western forests; as a result few instances of nest parasitism have been observed.[16] Marzluff and Restani predict that "forest fragments in western United States landscapes, dominated by regenerating managed forests will rarely show edge or fragmentation effects related to nest predation."[15]

CONNECTIVITY

Connectivity exists when organisms can move freely among separate patches of habitat.[1] Organisms differ in their movement ability. Some are extremely "gap sensitive," unable to move across a hostile habitat as narrow as 30 meters. Other, wide-ranging species are able to move freely across large

There is limited scientific evidence to show that animal movements are either enhanced or limited by the presence of corridors. Photo credit: J.A. Rochelle

expanses. When forest patches are disconnected and truly isolated for a given species, island biogeography theory predicts lower survival rates and possible extinction.

The use of corridors is commonly advocated to ensure connectivity.[18] Others have asserted the dangers of corridors.[19,20] Corridors are used by some species of wildlife. However, there is little evidence from the Pacific Northwest to indicate the degree to which animal movements are either limited by the absence or enhanced by the presence of corridors.

Species documented to use corridors are generally those that are unable or unwilling to cross gaps of unsuitable habitat.[5] However, most research on corridors has not examined the use of the adjoining matrix. Additional research, focused on individual species, is needed to better define the need for corridors.

Bunnell pointed out that management for connectivity within the matrix appears no more challenging than designing effective corridors, providing greater operational flexibility, and could prove much less costly.[1] Further, McComb noted that the quality of the forest "patches" and ease through which wildlife can travel through the landscape ("permeability") may be more important to animal movement and survival than simply the presence or absence of corridors. He also suggested that managers have the opportunity to maintain landscape "permeability" over time.[21]

Landscape modelling tools have been developed to help define connectivity from the perspective of different species and to calculate the amount of suitable habitat available to that species on the landscape based on their gap-crossing abilities and minimum area requirements. The models employ grid-based maps in which complex landscape patterns are created using theoretical spatial distributions. They are inspired by percolation theory, which concerns the flow of liquids through material aggregates.[25]

The models demonstrate how fine-scale fragmentation, such as selective logging, may enhance overall landscape connectivity at intermediate levels of habitat loss (40-60%) by providing "stepping stones" to dispersal within the matrix. Although connected, such landscapes have significantly less suitable habitat meeting minimum area requirements for some species, particularly those with large home ranges. Landscapes affected by fine-scale fragmentation may therefore support smaller populations that are more vulnerable to extinction.[26]

In summary, connectivity is important, but

The presence of habitat structures like snags can enhance the permeability of the matrix between forest stands. Photo credit: J.A. Rochelle

not uniformly so, and its importance depends on the movement abilities and hostility of the matrix to individual species. In theory, both the use of corridors and matrix management can be used to maintain or enhance connectivity. Understanding the need for special connectivity considerations in forest management requires additional on-the-ground research. However, Bunnell, in his review of research to date, found little evidence that lack of connectivity is a threat in forests of the Pacific Northwest.[1]

GENETICS

Genetic research has shown that when populations of organisms become isolated or too small, there can be a loss of genetic variation. This loss in genetic variation can cause inbreeding depression and decreased ability to adapt to long term changes.

To date, field studies have shown that genetic variation can be reduced as populations are isolated in forest remnants or fragments, although such changes may not occur for many years. Genetic tools have potential in assessing the effects of both decreased population size and decreased connectivity, and so are powerful supplements to ecological approaches to monitoring populations across fragmented landscapes.[22]

MANAGEMENT CONSIDERATIONS

Maintaining desired habitat characteristics for vertebrates along with sustained production of forest products and revenue requires landscape-level planning and an understanding of the processes of habitat loss and fragmentation. McComb, in his synthesis of conference findings, identified a number of specific relationships important to consider, as well as opportunities to address habitat needs, in the development of management plans as follows:[21]

- Focus first on maintaining sufficient habitat area and secondarily consider how habitat is arranged.
- Project expected changes in landscape structure through time; once established, the distribution of patch sizes is long lasting and difficult to modify.
- Plan to meet the habitat needs of most forest species by attention to size, placement and quality of habitat patches, but also the quality of the surrounding landscape (matrix).
- Retain existing habitat structures (snags, shrubs, decaying wood, large live trees), to the extent possible, and manage for adequate future amounts; these structures are crucial to survival of old growth-associated species in managed forests.
- Use modeling tools to explore conse-

quences of selected management options on patch size requirements and other habitat needs of gap-sensitive species.
• Develop management plans in the context of the total landscape; consider plan results in light of characteristics and anticipated future conditions on adjacent ownerships and seek opportunities for complementary planning.
• Consider the potential for natural disturbances in planning for desired habitat conditions; how might they affect habitat area and pattern?
• Monitor aspects of the plan where uncertainty exists to assess plan effectiveness and guide future plan modifications.
• Don't do the same thing everywhere and every time.

MANAGEMENT APPROACHES TO ADDRESS FRAGMENTATION CONCERNS: CASE STUDIES

Increased knowledge of wildlife habitat relationships at the landscape scale is leading to the development of management plans that address habitat availability and its distribution in space and time in managed forests. Several examples of landscape-scale management plans were presented at this conference that illustrate approaches in which incorporation of wildlife habitat needs is a primary determinant of plan structure.[23, 24] These approaches recognize that different habitat patterns favor different species and are directed at providing habitat distributions which address the needs of the spectrum of native wildlife, while continuing to produce sustainable levels of commodities and revenue.

Regardless of the forest treatment, there will be species that are winners and those that are losers. Formulation of wildlife objectives in forest management plans should take these variations in response into account.

UNDERSTANDING FRAGMENTATION: RESEARCH NEEDS

As is normally the case in attempting to understand complex ecological relationships such as forest fragmentation and how it affects vertebrates, we suffer from inadequate information. While the conference provided a synthesis of current knowledge, it also surfaced a number of information gaps that can only be met through additional research.

Some key research needs include:
• Determine how much habitat area is necessary for species and population survival (i.e., the "threshold effect").
• Assess the influence of matrix condition on landscape connectivity and vertebrate survival and dispersal, to identify management techniques for enhancing these processes.
• Determine the effects of habitat amounts and pattern on shrub- and canopy-nesting birds, since both groups are poorly studied.
• Improve understanding of ecology of nest predators and predator/prey relationships as influenced by habitat amount and configuration.
• Evaluate the implications of reduced levels of some old forest features such as deeply fissured bark, high epiphyte biomass, numerous cavity sites and deep organic matter layers in young stands.
• Assess the viability, over space and time, of species in landscapes managed within or for the natural range of variability.
• Develop support tools that include dynamic landscape simulations and population viability assessments to aid managers in forest planning.
• Assess genetic effects along with demographic and environmental responses to changes in amounts and configuration of habitat.
• Understand more fully the responses of hidden diversity, (i.e. non-vertebrates and plants) to habitat loss and pattern.
• Focus on landscape-level rather than patch-level research and monitoring, with emphasis on field studies in forests actively managed for timber production.

While many questions remain regarding the influence of forest fragmentation on vertebrate populations, the conference brought together information useful in clarifying a number of important habitat relationships. Understanding the relative importance of habitat loss and changed habitat configuration, and the need to distinguish between them in developing conservation measures should contribute to better resource management decisions. As was pointed out, not identifying and separating these processes may result in a failure to detect important responses to habitat change, such as threshold effects, with potentially significant adverse consequences for some vertebrate species.

An improved understanding of the nature of edge effects, predation, and other ecological processes, and the utility and limitations of corridors will also provide important direction to management plans. The conference pointed out opportunities for reducing the impact of habitat changes through retention of habitat structures in the forested matrix between habitat patches. Landscape-scale planning is increasingly being used to address multiple resource values; this conference brought together information that will help direct and improve that process.

LITERATURE CITED

1. Bunnell, F.L. (this conference) What habitat is an island?
2. McArthur, R.H. and E.O. Wilson. 1967. The theory of island biogeography. Princeton University Press, Princeton, N.J.
3. McGarigal, K. and W.C. McComb. 1995. Relationships between landscape structure and breeding birds in the Oregon coast range. Ecological Monographs 65:235-260.
4. Fahrig, L. (this conference) Forest loss or fragmentation: which has the greater effect on the persistence of forest-dwelling animals?
5. With, K.A. (this conference). Is landscape connectivity necessary and sufficient for wildlife management?
6. Agee, J.K. (this conference). Disturbance effects on landscape fragmentation in Interior West forests.
7. Garman, S.L., F. J. Swanson and T.A. Spies (this conference). Past, present and future potential landscape patterns in the Douglas-fir Region of the Pacific Northwest.
8. Teensma, P.D.A., J.T. Rienstra, and M.A. Yeiter. 1991. Preliminary reconstruction and analysis of change in forest stand age classes of the Oregon Coast Range from 1850 to 1940. USDI, Bureau of Land Management, Tech. Note T/N OR-9.
9. Sallabanks, R., P. Heglund, J.B. Haufler, B.A. Gilbert and W. Wall (this conference). Understanding relationships between landscape composition and nest success of forest songbirds: A study design and preliminary data from the Intermountain Northwest.
10. Spies, T.A., W.J. Ripple, and G.A. Bradshaw. 1994.

Dynamics and pattern of a managed coniferous forest landscape in Oregon. Ecological Applications. 4:555-568.
11. McGarigal, K., and W.C. McComb. (this conference). Forest fragmentation and breeding bird communities in the Oregon Coast range.
12. Laudenslayer, W.F. Jr. 1986. Summary: predicting effects of habitat patchiness and fragmentation. Pp. 331-333 in J. Verner, M. Morrison, and C. Ralph (eds.). Wildlife modeling habitat relationships of terrestrial vertebrates. Univ. Wisconsin Press, Madison, WI.
13. Reese, K.P., and J.T.Ratti. 1988. Edge effect: a concept under scrutiny. Trans. N. Am. Wildl. and Nat. Res. Conference 53: 127-136.
14. Harris, L.D. 1989. A rationale for wildlife habitat superiority of southern bottomland hardwood communities. In J. Gosselink, L. Lee, and T. Muir (eds.) 1990. Ecological processes and cumulative impacts illustrated by bottomland hardwood ecosystems. Lewis Publishers, Maryland.
15. Marzluff, J., and M. Restani. (this conference). The effects of forest fragmentation on rates of avian nest predation.
16. Kremsater, L., and F.L. Bunnell. (this conference). Edges: Theory, evidence and implications to management of west coast forests.
17. Schmiegelow, F.K.A., and S.J. Hannon. (this conference). Forest-level effects of management on boreal songbirds: the Calling Lake fragmentation studies.
18. Noss, R.F. 1991. Landscape connectivity: different functions at different scales. Pp. 27-39 in: W.E. Hudson, ed. Landscape linkages and biodiversity. Defenders of Wildlife and Island Press, Washington, D.C.
19. Simberloff, D.S. and J. Cox. 1987. Consequences and costs of conservation corridors. Cons. Biology 1:63-71.
20. Mann, C.C. and M.L. Plummer. 1995. Are wildlife corridors the right path? Science 1428-1430.
21. McComb, W.C. (this conference). Forest fragmention: Wildlife and Management Implications. A synthesis of the conference.
22. Mills, L.S. and D.A. Tallmon (this conference). The role of genetics in understanding forest fragmentation.
23. McAllister, D.C., R.W. Holloway and M.W. Schnee. (this conference). Using landscape design principles to promote biodiversity in a managed forest.
24. Bunnell, F.L., R.W. Wells , J.D. Nelson, and L.L. Kremsater. (this conference). Patch size, vertebrates, and effects of harvest policy in Southeastern British Columbia.
25. With, K. A. 1997. The application of neutral landscape models in conservation biology. Conservation Biology 11: 1069-1080.
26. With, K. A., and A. W. King. 1999. Extinction thresholds for species in fractal landscapes. Conservation Biology 13: in press.
27. Mills, L.S. 1995. Edge effects and isolation: red-backed voles on forest remnants. Conservation Biology 9:395-403.